Rewilding

Through a global and interdisciplinary lens, this book discusses, analyses and summarises the novel conservation approach of rewilding. The volume introduces key rewilding definitions and initiatives, highlighting their similarities and differences. It reviews matches and mismatches between the current state of ecological knowledge and the stated aims of rewilding projects, and discusses the role of human action in rewilding initiatives. Collating current scholarship, the book also considers the merits and dangers of rewilding approaches, as well as the economic and sociopolitical realities of using rewilding as a conservation tool. Its interdisciplinary nature will appeal to a broad range of readers, from primary ecologists and conservation biologists to land managers, policy-makers and conservation practitioners in NGOs and government departments. Written for a scientifically literate readership of academics, researchers, students, and managers, the book also acts as a key resource for advanced undergraduate and graduate courses.

Nathalie Pettorelli is a Senior Research Fellow at the Institute of Zoology, Zoological Society of London. She is a Senior Editor for *Journal of Applied Ecology* and the Conservation Specialist Interest Group Chair for the British Ecological Society.

Sarah M. Durant is a Senior Research Fellow at the Zoological Society of London and is affiliated with the Wildlife Conservation Society (WCS). She leads the Serengeti Cheetah Project, the longest ongoing study of wild cheetah, and the Range Wide Conservation Program for Cheetah and African Wild Dog.

Johan T. du Toit is a Professor in the Department of Wildland Resources at Utah State University. He studies the ecology of large mammals in terrestrial ecosystems and the integration of science and management in conservation.

Ecological Reviews

SERIES EDITOR Phillip Warren *University of Sheffield, UK*
SERIES EDITORIAL BOARD
Andrea Baier *British Ecological Society, UK*
Mark Bradford *Yale University, USA*
David Burslem *University of Aberdeen, UK*
Alan Gray *CEH Wallingford, UK*
Kate Harrison *British Ecological Society, UK*
Sue Hartley *University of York, UK*
Mark Hunter *University of Michigan, USA*
Ivette Perfecto *University of Michigan, USA*
Heikki Setala *University of Helsinki, Finland*

Ecological Reviews publishes books at the cutting edge of modern ecology, providing a forum for volumes that discuss topics that are focal points of current activity and likely long-term importance to the progress of the field. The series is an invaluable source of ideas and inspiration for ecologists at all levels from graduate students to more established researchers and professionals. The series has been developed jointly by the British Ecological Society and Cambridge University Press and encompasses the Society's Symposia as appropriate.

Biotic Interactions in the Tropics: Their Role in the Maintenance of Species Diversity
Edited by David F. R. P. Burslem, Michelle A. Pinard and Sue E. Hartley

Biological Diversity and Function in Soils
Edited by Richard Bardgett, Michael Usher and David Hopkins

Island Colonization: The Origin and Development of Island Communities
By Ian Thornton
Edited by Tim New

Scaling Biodiversity
Edited by David Storch, Pablo Margnet and James Brown

Body Size: The Structure and Function of Aquatic Ecosystems
Edited by Alan G. Hildrew, David G. Raffaelli and Ronni Edmonds-Brown

Speciation and Patterns of Diversity
Edited by Roger Butlin, Jon Bridle and Dolph Schluter

Ecology of Industrial Pollution
Edited by Lesley C. Batty and Kevin B. Hallberg

Ecosystem Ecology: A New Synthesis
Edited by David G. Raffaelli and Christopher L. J. Frid

Urban Ecology
Edited by Kevin J. Gaston

The Ecology of Plant Secondary Metabolites: From Genes to Global Processes
Edited by Glenn R. Iason, Marcel Dicke and Susan E. Hartley

Birds and Habitat: Relationships in Changing Landscapes
Edited by Robert J. Fuller

Trait-Mediated Indirect Interactions: Ecological and Evolutionary Perspectives
Edited by Takayuki Ohgushi, Oswald Schmitz and Robert D. Holt

Forests and Global Change
Edited by David A. Coomes, David F. R. P. Burslem and William D. Simonson

Trophic Ecology: Bottom-Up and Top-Down Interactions across Aquatic and Terrestrial Systems
Edited by Torrance C. Hanley and Kimberly J. La Pierre

Conflicts in Conservation: Navigating Towards Solutions
Edited by Stephen M. Redpath, R. J. Gutiérrez, Kevin A. Wood and Juliette C. Young

Peatland Restoration and Ecosystem Services
Edited by Aletta Bonn, Tim Allott, Martin Evans, Hans Joosten and Rob Stoneman

Rewilding

Edited by

NATHALIE PETTORELLI
Institute of Zoology, London

SARAH M. DURANT
Institute of Zoology, London

JOHAN T. DU TOIT
Utah State University

CAMBRIDGE UNIVERSITY PRESS

CAMBRIDGE
UNIVERSITY PRESS

University Printing House, Cambridge CB2 8BS, United Kingdom

One Liberty Plaza, 20th Floor, New York, NY 10006, USA

477 Williamstown Road, Port Melbourne, VIC 3207, Australia

314–321, 3rd Floor, Plot 3, Splendor Forum, Jasola District Centre,
New Delhi – 110025, India

103 Penang Road, #05-06/07, Visioncrest Commercial, Singapore 238467

Cambridge University Press is part of the University of Cambridge.

It furthers the University's mission by disseminating knowledge in the pursuit of education, learning, and research at the highest international levels of excellence.

www.cambridge.org
Information on this title: www.cambridge.org/9781108472678
DOI: 10.1017/9781108560962

© British Ecological Society 2019

This publication is in copyright. Subject to statutory exception and to the provisions of relevant collective licensing agreements, no reproduction of any part may take place without the written permission of Cambridge University Press.

First published 2019
3rd printing 2021

Printed in the United Kingdom by TJ Books Limited, Padstow Cornwall

A catalogue record for this publication is available from the British Library.

Library of Congress Cataloging-in-Publication Data
Names: Pettorelli, Nathalie, editor.
Title: Rewilding / edited by Nathalie Pettorelli, Institute of Zoology, London, Sarah M. Durant, Institute of Zoology, London, Johan T. du Toit, Utah State University.
Description: Cambridge, United Kingdom ; New York, NY : Cambridge University Press, 2019. | Series: Ecological reviews | Includes bibliographical references and index.
Identifiers: LCCN 2018033891| ISBN 9781108472678 (hardback) | ISBN 9781108460125 (paperback)
Subjects: LCSH: Wildlife reintroduction.
Classification: LCC QL83.4 .R49 2019 | DDC 639.9–dc23
LC record available at https://lccn.loc.gov/2018033891

ISBN 978-1-108-47267-8 Hardback
ISBN 978-1-108-46012-5 Paperback

Cambridge University Press has no responsibility for the persistence or accuracy of URLs for external or third-party internet websites referred to in this publication and does not guarantee that any content on such websites is, or will remain, accurate or appropriate.

Contents

List of Contributors	*page* ix
Acknowledgements	xiii

1. Rewilding: a captivating, controversial, twenty-first-century concept to address ecological degradation in a changing world   1
   Nathalie Pettorelli, Sarah M. Durant, Johan T. du Toit
2. History of rewilding: ideas and practice   12
   David Johns
3. For wilderness or wildness? Decolonising rewilding   34
   Kim Ward
4. Pleistocene rewilding: an enlightening thought experiment   55
   Johan T. du Toit
5. Trophic rewilding: ecological restoration of top-down trophic interactions to promote self-regulating biodiverse ecosystems   73
   Jens-Christian Svenning, Michael Munk, Andreas Schweiger
6. Rewilding through land abandonment   99
   Steve Carver
7. Rewilding and restoration   123
   James R. Miller, Richard J. Hobbs
8. Understanding the factors shaping the attitudes towards wilderness and rewilding   142
   Nicole Bauer, Aline von Atzigen
9. Health and social benefits of living with 'wild' nature   165
   Cecily Maller, Laura Mumaw, Benjamin Cooke
10. The psychology of rewilding   182
    Susan Clayton
11. The high art of rewilding: lessons from curating Earth art   201
    Marcus Hall
12. Rewilding a country: Britain as a study case   222
    Christopher J. Sandom, Sophie Wynne-Jones

13. Bringing back large carnivores to rewild landscapes 248
    John D.C. Linnell, Craig R. Jackson
14. Rewilding cities 280
    Marcus Owens, Jennifer Wolch
15. The role of translocation in rewilding 303
    Philip J. Seddon, Doug P. Armstrong
16. Top-down control of ecosystems and the case for rewilding: does it all add up? 325
    Matt W. Hayward, Sarah Edwards, Bronwyn A. Fancourt, John D.C. Linnell, Erlend B. Nilsen
17. Rewilding and the risk of creating new, unwanted ecological interactions 355
    Miguel Delibes-Mateos, Isabel C. Barrio, A. Márcia Barbosa, Íñigo Martínez-Solano, John E. Fa, Catarina C. Ferreira
18. Auditing the wild: how do we assess if rewilding objectives are achieved? 375
    Richart T. Corlett
19. Adaptive co-management and conflict resolution for rewilding across development contexts 386
    James R.A. Butler, Juliette C. Young, Mariella Marzano
20. The future of rewilding: fostering nature and people in a changing world 413
    Sarah M. Durant, Nathalie Pettorelli, Johan T. du Toit

Index 426

Colour plates can be found between pages 210 and 211

Contributors

DOUG P. ARMSTRONG
Wildlife Ecology Group
Massey University
Palmerston North
New Zealand
D.P.Armstrong@massey.ac.nz

ALINE VON ATZIGEN
University of Zurich; Ethnographic Museum; Swiss Federal Institute for Forest, Snow and Landscape Research WSL, Economics and Social Sciences; Social Sciences in Landscape Research Group
Birmensdorf
Switzerland
aline.vonatzigen@posteo.de

A. MÁRCIA BARBOSA
Centro de Investigação em Biodiversidade e Recursos Genéticos (CIBIO/InBIO)
Universidade de Évora
Évora
Portugal
barbosa@uevora.pt

ISABEL C. BARRIO
Institute of Life and Environmental Sciences
University of Iceland
Reykjavík
Iceland
icbarrio@gmail.com

NICOLE BAUER
Swiss Federal Institute for Forest, Snow and Landscape Research WSL, Economics and Social Sciences; Social Sciences in Landscape Research Group
Birmensdorf
Switzerland
nicole.bauer@wsl.ch

JAMES R.A. BUTLER
CSIRO Land and Water
Brisbane
Australia
James.Butler@csiro.au

STEVE CARVER
University of Leeds
Leeds
UK
S.J.Carver@leeds.ac.uk

SUSAN CLAYTON
Department of Psychology
College of Wooster
Wooster
USA
SCLAYTON@wooster.edu

LIST OF CONTRIBUTORS

BENJAMIN COOKE
Centre for Urban Research
RMIT University
Melbourne
Australia
ben.cooke@rmit.edu.au

RICHARD T. CORLETT
Center for Integrative Conservation,
Xishuangbanna Tropical Botanical
Gardens
Chinese Academy of Sciences
Yunnan
China
corlett@xtbg.org.cn

MIGUEL DELIBES-MATEOS
Instituto de Estudios Sociales
Avanzados Consejo Superior de
Investigaciones Científicas Córdoba
Spain
mdelibes@us.es

SARAH M. DURANT
Institute of Zoology
Zoological Society of London
London
UK
sdurant@wcs.org

SARAH EDWARDS
Cheetah Research Project, Leibniz
Institute of Zoo and Wildlife
Research & Centre for Wildlife
Management
University of Pretoria
Namibia
Sarah@cheetahnamibia.com

JOHN E. FA
Division of Biology and Conservation
Ecology, School of Science and the
Environment, Manchester
Metropolitan University
Manchester
UK
jfa949@gmail.com

BRONWYN A. FANCOURT
School of Biological Sciences
University of Tasmania
Hobart
Australia
Bronwyn.Fancourt@utas.edu.au

CATARINA C. FERREIRA
Department of Conservation Biology
UFZ – Helmholtz Centre for
Environmental Research
Leipzig
Germany
catferreira@gmail.com

MARCUS HALL
University of Zurich
Zurich
Switzerland
marc.hall@ieu.uzh.ch

MATT W. HAYWARD
School of Environmental and Life
Sciences
University of Newcastle
Callaghan
Australia
matthew.hayward@newcastle.edu.au

RICHARD J. HOBBS
School of Biological Sciences
University of Western Australia
Perth
Australia
Richard.Hobbs@uwa.edu.au

CRAIG R. JACKSON
NINA
Trondheim
Norway
craig.jackson@nina.no

DAVID JOHNS
School of Government
Portland State University
Portland, OR
USA
johnsd@pdx.edu
johnsd@embargmail.com

JOHN D.C. LINNELL
NINA
Trondheim
Norway
John.Linnell@nina.no

CECILY MALLER
Centre for Urban Research
RMIT University
Melbourne
Australia
cecily.maller@rmit.edu.au

ÍÑIGO MARTÍNEZ-SOLANO
Departamento de Biodiversidad
y Biología Evolutiva
Museo Nacional de Ciencias
Naturales (CSIC)
Madrid
Spain
inigomsolano@mncn.csic.es

MARIELLA MARZANO
Forest Research
Northern Research Station
Roslin
UK
Mariella.Marzano@forestry.gsi.gov.uk

JAMES R. MILLER
Department of Natural Resources
and Environmental Sciences
University of Illinois Urbana-Champaign
Urbana

USA
jrmillr@illinois.edu

LAURA MUMAW
Centre for Urban Research
RMIT University
Melbourne
Australia
laura.mumaw@rmit.edu.au

MICHAEL MUNK
Center for Biodiversity Dynamics in a
Changing World (BioCHANGE) and
Section for Ecoinformatics &
Biodiversity
Aarhus University
Aarhus
Denmark
munk@bios.au.dk

ERLEND B. NILSEN
NINA
Trondheim
Norway
Erlend.Nilsen@nina.no

MARCUS OWENS
College of Environmental Design
University of California
Berkeley
USA
mowens@berkeley.edu

NATHALIE PETTORELLI
Institute of Zoology
Zoological Society of London
London
UK
Nathalie.Pettorelli@ioz.ac.uk

CHRISTOPHER J. SANDOM
School of Life Sciences
University of Sussex & Wild Business
Ltd
Brighton

UK
chris@wildbusiness.org

ANDREAS SCHWEIGER
Center for Biodiversity Dynamics in a Changing World (BioCHANGE) and Section for Ecoinformatics & Biodiversity
Aarhus University
Aarhus
Denmark
Andres.Schweiger@uni-bayreuth.de

PHILIP J. SEDDON
Department of Zoology
University of Otago
Dunedin
New Zealand
philip.seddon@otago.ac.nz

JENS-CHRISTIAN SVENNING
Center for Biodiversity Dynamics in a Changing World (BioCHANGE) and Section for Ecoinformatics & Biodiversity
Aarhus University
Aarhus
Denmark
svenning@bios.au.dk

JOHAN T. DU TOIT
Utah State University
Logan
USA
johan.dutoit@usu.edu

KIM WARD
University of Plymouth
Plymouth
UK
kim.ward@plymouth.ac.uk

JENNIFER WOLCH
College of Environmental Design
University of California
Berkeley
USA
wolch@berkeley.edu

SOPHIE WYNNE-JONES
School of Natural Sciences
Bangor University
Bangor
UK
s.wynne-jones@bangor.ac.uk

JULIETTE C. YOUNG
NERC Centre for Ecology and Hydrology
Edinburgh
UK
jyo@ceh.ac.uk

Acknowledgements

The initial idea for this book grew out of a policy paper on rewilding published in the *Journal of Applied Ecology*, itself motivated by a special session organised by Pettorelli at the 2015 annual meeting of the British Ecological Society, and also a fun night out with the editorial team of the *Journal of Applied Ecology* at the 2015 International Conference on Conservation Biology in Montpellier, France. The reflections that underpin the structure of this book have thus taken years to mature, hopefully demonstrating that good books, like good wine, take time in the making.

We thank the 37 colleagues who joined us on the adventure of creating this edited volume. It has been an absolute pleasure to work with these deep thinkers distributed around the world and the overall experience has been particularly rewarding and enlightening for us. We especially value their willingness to explore ideas from multiple perspectives, and combine their experiences and expertise to produce an inclusive and representative overview of where current thinking on this topic sits.

We also thank the scientists who have given their time to review chapters, helping us strengthen the contributions and craft our book. Special thanks go to Toby Ackroyd, Alan Watson, Norman Owen-Smith, Kathy Hodder, Luigi Boitani, Bob Smith, Eric Higgs, Katherine Homewood, Tom Fry, Zoe Davies, Chris Sandbrook, Yasha Rower, Ian Convery, Luke Hunter, Arian Wallach, Jake Goheen, Alison Hester, and Georgina Cundill Kemp. Our thanks also go to Kate Harrison, Philip Warren, and Aleksandra Serocka for their help in the production of this book, as well as to our family and friends who have supported us during the ups and downs of putting together a contribution such as this.

CHAPTER ONE

Rewilding: a captivating, controversial, twenty-first-century concept to address ecological degradation in a changing world

NATHALIE PETTORELLI and SARAH M. DURANT
Institute of Zoology
JOHAN T. DU TOIT
Utah State University

Why a book on rewilding?

Rewilding is a novel and rapidly developing concept in ecosystem management, representing a transformative approach to conserving biodiversity. Originally defined as a conservation method based on 'cores, corridors, and carnivores' (Soulé and Noss, 1998), the term is now broadly understood as the repair or refurbishment of an ecosystem's functionality through the (re-)introduction of selected species. Although the term first occurred in print in 1990, its popularity only started to grow substantially over the past decade; during this time, rewilding has moved from a theoretical concept to a practical idea. It is currently being hailed by many as a potentially cost-effective solution to reinstate vegetation succession, reactivate top-down trophic interactions and predation processes, and improve ecosystem service delivery through the (re-)introduction of ecosystem engineers (Pettorelli et al., 2018). Several rewilding projects have now been implemented in multiple countries around the world (Figure 1.1), all being expected to hold potential for enhancing local biodiversity, ecological resilience, and ecosystem service delivery (see e.g. Lorimer et al., 2015; Pereira and Navarro, 2015; Svenning et al., 2016).

Rewilding has clearly attracted the attention of practitioners and the general public, as well as national and international bodies concerned with the management of our environment. Policy-makers are increasingly setting up inquiries, briefs, committees, and task forces to assess the potential opportunities associated with rewilding approaches. Similarly, the International Union for the Conservation of Nature (IUCN) Commission on Ecosystem Management has recently launched a task force on rewilding (IUCN, 2017). Yet the more sensational connotations of the early proposals for rewilding,

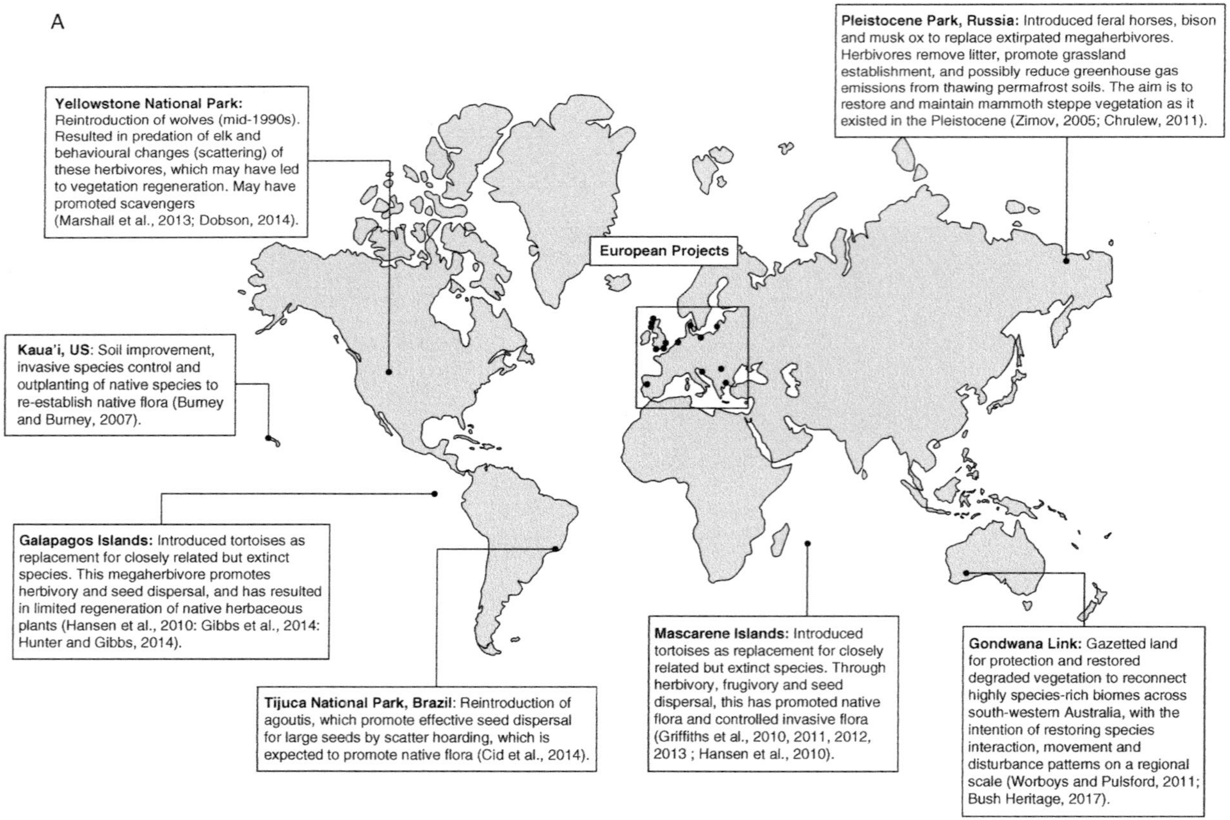

A

Yellowstone National Park: Reintroduction of wolves (mid-1990s). Resulted in predation of elk and behavioural changes (scattering) of these herbivores, which may have led to vegetation regeneration. May have promoted scavengers (Marshall et al., 2013; Dobson, 2014).

Kaua'i, US: Soil improvement, invasive species control and outplanting of native species to re-establish native flora (Burney and Burney, 2007).

Galapagos Islands: Introduced tortoises as replacement for closely related but extinct species. This megaherbivore promotes herbivory and seed dispersal, and has resulted in limited regeneration of native herbaceous plants (Hansen et al., 2010; Gibbs et al., 2014; Hunter and Gibbs, 2014).

Tijuca National Park, Brazil: Reintroduction of agoutis, which promote effective seed dispersal for large seeds by scatter hoarding, which is expected to promote native flora (Cid et al., 2014).

Pleistocene Park, Russia: Introduced feral horses, bison and musk ox to replace extirpated megaherbivores. Herbivores remove litter, promote grassland establishment, and possibly reduce greenhouse gas emissions from thawing permafrost soils. The aim is to restore and maintain mammoth steppe vegetation as it existed in the Pleistocene (Zimov, 2005; Chrulew, 2011).

Mascarene Islands: Introduced tortoises as replacement for closely related but extinct species. Through herbivory, frugivory and seed dispersal, this has promoted native flora and controlled invasive flora (Griffiths et al., 2010, 2011, 2012, 2013; Hansen et al., 2010).

Gondwana Link: Gazetted land for protection and restored degraded vegetation to reconnect highly species-rich biomes across south-western Australia, with the intention of restoring species interaction, movement and disturbance patterns on a regional scale (Worboys and Pulsford, 2011; Bush Heritage, 2017).

European Projects

Figure 1.1. Examples of currently ongoing projects overtly labelled as 'rewilding' (A) in the world and (B) in Europe. This figure was originally published by Pettorelli and colleagues (2018) in *Journal of Applied Ecology*, where the references mentioned in the figure are detailed.

B

Projects in Scotland:

Alladale Wilderness Reserve: Trees were planted, anti-deer fence built and boar were reintroduced to this site (to establish germination niches for seedlings by rooting). The aim is to restore a core area of native Caledonian pinewood forest (Sandom et al., 2013).

Glen Affric: Re-establishment of self-sustaining, native Caledonian pinewood forest. Current interventions include planting native trees and removing non-native trees, as well as excluding deer (Sandom et al., 2013; Trees for Life, 2015).

Knapdale Forest: The extirpated beaver was reintroduced in 2009 to create new wetland habitats and more diverse woodland structure (Jones et al., 2009; RZSS, 2014).

Projects in England:

Devon Beaver Project: Reintroduction of beavers, whose dams increased ponded water storage. This reduced peak discharge and pollutant load of downstream water, whilst increasing organic carbon load (Puttock et al., 2017).

Wicken Fen: Highland cattle and Konik ponies were introduced to this site to replace extirpated megaherbivores. Hydrological regime was restored to promote and maintain fen meadows and reduce scrub (Wicken Fen Project, 2017).

Knepp Castle: Introduced old breeds of pig, longhorn cattle, fallow deer and Exmoor ponies (Taylor, 2006; Hodder et al., 2014).

Wild Ennerdale: Galloway cattle were introduced, and sheep numbers were reduced, to restore browsing regime beneficial to regeneration of native trees. Restoration of waterways to allow fish migration and movement of sediment (Rewilding Britain, 2017).

Oostvardersplassen, NL: Extinct megaherbivores were functionally replaced by Heck cattle, Konik horses, and red deer, with the aim to install a Pleistocene community on reclaimed land. Their grazing maintains an open grassland, and important habitat for many other species (Vera, 2009; Cornelissen et al., 2014).

Vorup Enge, Denmark: European bison and Holstein Jutland dairy cows were reintroduced to this site to replace extirpated megaherbivores. The aim is to create a self-sustaining ecosystem which preserves Danish flora genetic variation (Randers Regnskoven, 2016).

Lake Pape, Lithuania: Introduced Konik horses as a replacement for extirpated wild horses in 1999 (Schwartz et al., 2005; Prieditis, 2012).

Oder Delta: New protected areas were established, with the aim to improve habitat quality so that regional wildlife can thrive (Rewilding Europe, 2017).

South Carpathians: Reintroduction of bison to promote herbivory; re-establishment of bark beetle disturbance (Rewilding Europe, 2017).

Rhodope Mountains: Introduced red and fallow deer, Konik and Karakachan horses to enhance herbivory, with the aim of controlling fire, creating a vegetation mosaic and sustaining scavengers and predators (wolves and several vulture species) (Rewilding Europe, 2017).

Velebit: Reintroduced Bosnian mountain horses, Konik horses and tauros (Helmer et al., 2015).

West Iberia: Introduced horses and a primitive cattle breed ('tauros') as a replacement for extinct megaherbivores to re-establish herbivore control of vegetation dynamics (Helmer et al., 2015).

Figure 1.1. (Cont.)

such as reintroducing native predators or introducing exotic megafauna (Donlan et al., 2005), fuel criticism on scientific, aesthetic, legal, political, economic, and cultural grounds (Lorimer and Driessen, 2014; Arts et al., 2016; Bulkens et al., 2016; Nogués-Bravo et al., 2016). Critics point to uncertainties and difficulties associated with the definition of rewilding and the practical implementation of rewilding projects. Of particular concern are issues related to the definition and consideration of appropriate ecological baselines and spatiotemporal scales when designing rewilding initiatives. There are also doubts about the extent to which ecological processes could resume significance in human-dominated landscapes. Other challenges include defining the role of humans in rewilded landscapes; aligning rewilding with legal, management and cultural categorisations and frameworks for species and lands; realistically evaluating costs and benefits of potential rewilding initiatives; as well as improving the monitoring and assessment of these projects (Pettorelli et al., 2018).

As applied scientists heavily involved with the management of natural resources in various countries and regularly confronted with the realities of planning for the delivery of ecological outcomes in human-dominated systems, we believe now is the time to synthesise available information on the benefits and risks, as well as the economic and sociopolitical realities, of rewilding as a conservation tool. Literature relevant to rewilding discussions has grown quickly over the past few years (Figure 1.2), yet until now there is no scientific book written by world leaders in the field that addresses rewilding with a global and inclusive perspective, or that examines rewilding in the context of social–ecological systems. To address that need, this book (1) introduces key rewilding definitions and initiatives and highlights their differences/similarities; (2) reviews matches and mismatches between the current state of ecological knowledge and the stated aims of rewilding projects; (3) discusses the role of humans in rewilding initiatives; and (4) highlights the merits and dangers of rewilding approaches. It does so by capitalising on the wealth of studies available in the fields of restoration ecology, reintroduction and conservation biology, social sciences, and conservation psychology to examine the concept of rewilding in a critical and objective light. This comes at a time when the field of conservation science is going through a difficult and controversial stage of redefinition, with pragmatism challenging purism (Kareiva and Marvier, 2012). The pace of global change throws the definition of restoration ecology into question (Rohwer and Marris, 2016) and novel ecosystems are gaining acceptance as inevitable and irreversible stages in some ecological transitions (Miller and Bestelmeyer, 2016). There is a need for new directions for environmental management to move in – going back is no longer an option – and rewilding stands as a candidate concept to be evaluated for certain systems under certain conditions. One could argue that

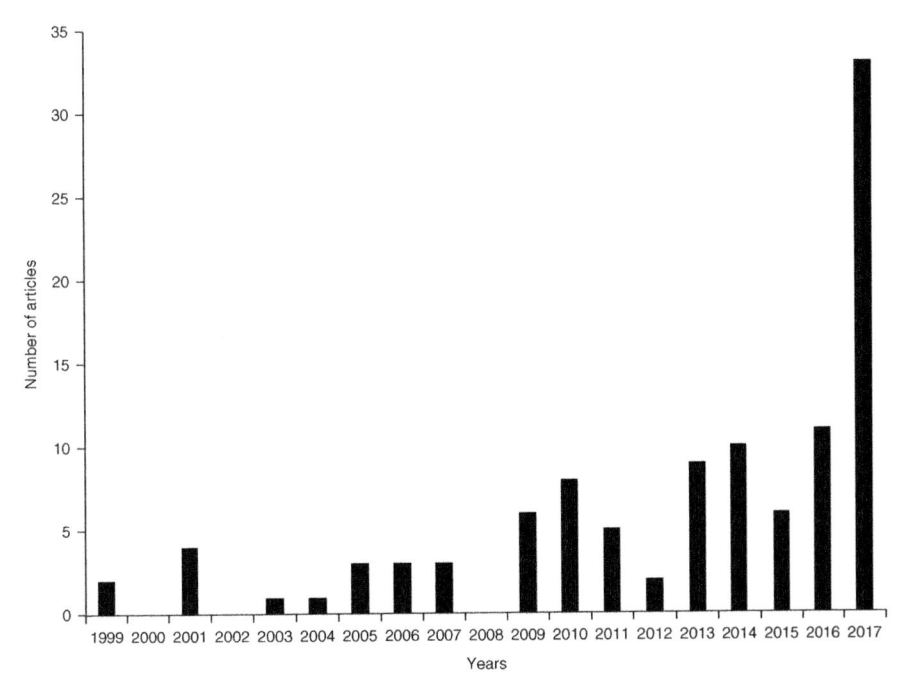

Figure 1.2. Number of ecological articles listed in Web of Science that mention 'rewilding' or 're-wilding' over the 1999–2017 period. The search led to 106 papers.

rewilding opens a fresh perspective on the practice of ecological conservation, challenging our relationship to the natural world, encouraging a more interdisciplinary approach to environmental management. However, deciding whether that argument holds merit requires a well-researched, comprehensive overview of the roots, meaning, applications, and challenges of the rewilding concept. Our goal here is to provide exactly that.

Where does rewilding originate from and what does it mean?
Rewilding is believed to have been first discussed by Dave Foreman in 1992, and its definition has been evolving ever since (Chapter 2). This evolution, to a certain extent, captures the changing trends that have shaped conservation biology over the past decades, providing a key outlook on how priorities and leading ideas have switched as our ecological understanding improved over time. Understanding current rewilding discussions is difficult without knowing about the history of the concept and without an appreciation of the link that connects rewilding to the concept of wilderness, an arguably subjective notion that tends to evoke landscapes where natural processes are permitted to operate without human interference. Articulating the link between wilderness and rewilding is indeed central to understanding the diversity of views

Table 1.1. *Main broad definitions of rewilding, as proposed over the past five years.*

Definition	Key points	Reference
'Rewilding has multiple meanings. These usually share a long-term aim of maintaining, or increasing, biodiversity, while reducing the impact of present and past human interventions through the restoration of species and ecological processes'	Focus on reducing impacts of management interventions Targets ecological processes and species restoration	Lorimer et al. (2015)
'Reintroduction of extirpated species or functional types of high ecological importance to restore self-managing functional, biodiverse ecosystems', 'emphasises species reintroductions to restore ecological function'	Focus on (re)introductions Targets ecological functions	Naundrup and Svenning (2015)
'Rewilding implies returning a non-wild area back to the wild ... This is the definition adopted in this review, except that I have followed normal usage in also including increases in relative wildness, i.e., from less wild to more wild'	Targets levels of wilderness	Corlett et al. (2016)
'A process of (re)introducing or restoring wild organisms and/or ecological processes to ecosystems where such organisms and processes are either missing or are "dysfunctional"'	Focus on (re)introductions Targets species composition and ecosystem processes	Prior and Brady (2017)
'The focus [of rewilding philosophy] is on benefits of renewed ecosystem function or processes (e.g. water storage, enhanced water quality, biodiversity support), rather than classic restoration thinking where a community converges towards a predefined target via a predictable trajectory'	Focus on non-predictable trajectory Targets ecosystem function/process	Law et al. (2017)
'The idea that unproductive and abandoned land can serve as new wilderness areas ("rewilding") i.e. self-sustaining ecosystems close to the "natural" state often supported by (re-)	Focus on (re)introductions and habitat protection Targets self-sustaining ecosystems	van der Zanden et al. (2017)

Table 1.1. (cont.)

Definition	Key points	Reference
introduction of large herbivores and habitat protection for carnivores and other species'	Supports low level of interaction between people and landscape	
'The reorganisation of biota and ecosystem processes to set an identified social–ecological system on a preferred trajectory, leading to the self-sustaining provision of ecosystem services with minimal ongoing management'	Acceptance of change, emphasis on reorganisation rather than restoration, focus on the social–ecological system and desired ecosystem services	Pettorelli et al. (2018)

on rewilding, and to exposing many of the values and politics that have been deep-rooted in modern conservation practice. What is 'wild' for some can be described as 'dominated' by others, and there is a vast diversity of perceptions of what the wild resembles and what natural means (Jørgensen, 2015). These perceptions vary geographically and culturally, can be linked to people's access to nature, but importantly are ultimately underpinned by clear social constructs that may influence how rewilding projects are being designed and implemented (Carver et al., 2002; Diemer et al., 2003; Bauer et al., 2009; Chapter 3).

The use of the term 'rewilding' is increasing in the peer-reviewed literature (Figure 1.2), but it has different meanings for different people, and also different framings, which we discuss later. There are three main themes in the current definitions (Table 1.1), the first being the resumption of wildness, by which degraded areas may regain biodiversity and develop into undefined future states without further interference, and not necessarily with any further utility to humans (Lorimer et al., 2015; Corlett et al., 2016). The second theme is about reintroducing extirpated species (or their substitutes) so that an ecosystem may resume a semblance of its former functionality, with potential benefits to humanity (Naundrup and Svenning, 2015; Prior and Brady, 2017; van der Zanden et al., 2017). Finally, an emerging theme recognises that biodiversity exists within constantly changing social–ecological systems in which perceived costs and benefits dictate which parts of wildness stay or go. The focus of this theme is the self-sustaining functionality of an ecosystem, which managers might not necessarily restore to a former state but could reorganise to provide ecosystem services with minimal intervention under prevailing environmental conditions (Law et al., 2017; Pettorelli et al., 2018). All three themes have applications in different places and circumstances, but they share a common departure that distinguishes rewilding from restoration: rewilding is about choosing new trajectories of change towards wildness in future undefined states; restoration is generally about reversing a trajectory of change to return to a defined previous state.

Introducing the different framings of rewilding

The concept of rewilding was originally framed as a call for large, connected wilderness areas to support wide-ranging keystone species such as apex predators (Soulé and Noss, 1998). Since then, the multiple definitions of rewilding (Table 1.1) relate to successive framings that have not necessarily replaced earlier ones. At present, there are four distinct framings that can be recognised in the literature: Pleistocene rewilding; trophic rewilding; passive rewilding; and ecological rewilding.

Pleistocene rewilding generally refers to restoring ecological processes lost because of the late-Pleistocene megafaunal extinctions. Josh Donlan and colleagues (2005) galvanised conservation biology with this bold and arguably overambitious framing of rewilding that invokes taxonomic substitution, using proxy species from other continents to serve the functions of extinct megafauna. Many describe Pleistocene rewilding as an absurd concept formulated by a small group of conservation biologists with little understanding of the practicalities and politics of animal translocations. Others, however, see this framing of rewilding as heuristically useful for developing the idea that extinct species leave vacant niches, and those vacancies have far-reaching ramifications through the ecosystem. Dealing with those ramifications requires an appreciation of the importance of conserving ecosystem processes and functions, and an acknowledgement that unorthodox management interventions may be required where all else fails.

Trophic rewilding specifically frames the reactivation of top-down trophic interactions. This framing is conceptually close to Pleistocene rewilding, but discards its historical benchmark and retains its main theoretical tenants: (1) megafaunal processes are important for ecosystem structure and functioning, promoting overall biodiversity in various ways, notably via top-down trophic effects fostering environmental heterogeneity; (2) rich megafaunas have been typical worldwide on evolutionary timescales and so modern species assemblages have evolved in, and are therefore adapted to, megafauna-rich ecosystems; (3) losses of megafauna from recent to distant times have led to ecosystem changes and biodiversity losses.

Passive rewilding refers to abandoned post-agricultural landscapes that are no longer actively managed, a framing that is current especially in Europe. It could be seen as an alternative to classic environmental management, substituting management *for* nature with management *by* nature. This framing of rewilding is conceptually close to *ecological rewilding*, which involves limited active management to facilitate natural processes and allow them to regain dominance.

Acknowledging the human dimension of rewilding

Rewilding does not happen in a vacuum. Social, cultural, psychological, economic, and political dimensions will all affect the ultimate success of any rewilding intervention. As such, it is impossible to discuss rewilding without considering its human dimensions, acknowledging that humans are key to the success, and the failure, of rewilding initiatives. Importantly, human responses to rewilding shed light on our responses and relationship with nature, providing us with important insights that can inform adaptive management and sustainable development.

Individual reactions to conservation actions are shaped by our perceptions of nature and our link to it, with people generally adopting one of four possible general attitudes towards nature: being a nature lover; a nature sympathiser; a nature-connected user; or a nature controller. These attitudes are not fixed in time and people may change their attitudes towards nature as their stage of life, place of residence, level of knowledge and experience change. Interestingly, rewilding is predominantly discussed in the context of developed countries, commonly in association with opportunities to increase nature's presence in urban settings. Yet living with nature in urban settings could have beneficial, but also harmful and unpredictable outcomes, which could ultimately affect people's support for rewilding initiatives. So far, little research has been done to deepen our understanding of the drivers shaping our relationship with wilderness, meaning that our current ability to predict and mitigate negative attitudes to rewilding projects is low.

Discussing the challenges associated with rewilding

Rewilding poses daunting ecological and societal challenges to practitioners who are left in charge of initiating and overseeing such projects. Any formulation of a rewilding project is underpinned by a number of ecological assumptions, which, if not met, could lead to damaging outcomes for the entire social–ecological system. For example, a badly designed rewilding project could increase the risk of new, unwanted ecological interactions (Nogués-Bravo et al., 2016). Ideally, the initiation of these projects should thus be preceded by a clear identification of the overarching goals, guiding principles, available management options, and key assumptions. Experience so far suggests that these foundational stages are rarely negotiated in full.

Carnivore (re-)introductions are often critical to rewilding discussions from the onset, because of their linkages to the restoration of ecological processes, yet these are known to be particularly challenging. Our general scientific understanding of the factors driving translocation success indeed remains relatively poor, which is a problem for rewilding initiatives placing translocations at the centre of their management approach. Recent experience from Europe has shown there is enormous scope for large carnivore recovery, even in shared and human-modified landscapes, but the extent to which we can expect large carnivores to resume their ecological functions in the rewilded landscapes of the Anthropocene is currently unknown. Additionally, evidence that rewilding approaches can restore top-down control of ecosystems remains equivocal.

To be successful, rewilding approaches need to demonstrate cost-effectiveness. Conservation funds are always limited and investments cannot be justified for projects that might fail or return low conservation benefits. As of present, rewilding is associated with fluid and unscripted targets as well as indeterminate outcomes. This lack of clarity extends to the monitoring and assessment of rewilding projects, begging critical questions such as 'how do we know that the rewilding project we paid for is successful?' or 'how do we know when success is met?'

Conclusions

This edited volume brings together, for the first time, leading authors in the rewilding literature who were each charged with synthesising the current thinking on their speciality within this field. The book was designed to provide a comprehensive, interdisciplinary overview of rewilding that outlines key concepts and details informative case studies. The need for an inclusive, scientifically rooted discussion on rewilding exists because of the unprecedented rates of environmental change in the Anthropocene, which call for a paradigm shift from focusing on the preservation of individual species to the enhancement of ecosystem health and processes, and for new and pragmatic options for mitigating the degradation of biodiversity and ecosystem services. Until now, however, rewilding has lacked the conceptual foundation needed for it to develop as a forward-looking, science-based, and policy-supported option. Our objective will be met if this book provides that foundation.

References

Arts, K., Fischer, A., and van der Wal, R. (2016). Boundaries of the wolf and the wild: a conceptual examination of the relationship between rewilding and animal reintroduction. *Restoration Ecology*, 24, 27-34.

Bauer, N., Wallner, A., and Hunziker, M. (2009). The change of European landscapes: human–nature relationships, public attitudes towards rewilding, and the implications for landscape management in Switzerland. *Journal of Environmental Management*, 90, 2910-2920.

Bulkens, M., Muzaini, H., and Minca, C. (2016). Dutch new nature: (re)landscaping the Millingerwaard. *Journal of Environmental Planning and Management*, 59, 808-825.

Carver, S., Evans, A., and Fritz, S. (2002). Wilderness attribute mapping in the United Kingdom. *International Journal of Wilderness*, **8**, 24-29.

Corlett, R.T. (2016). The role of rewilding in landscape design for conservation. *Current Landscape Ecology Reports*, **1**, 127-133.

Diemer, M., Held, M., and Hofmeister, S. (2003). Urban wilderness in Central Europe. *International Journal of Wilderness*, **9**, 7-11.

Donlan, J., Greene, H.W., Berger, J., et al. (2005). Re-wilding North America. *Nature*, **436**, 913-914.

IUCN. (2017). Rewilding. www.iucn.org/commissions/commission-ecosystem-management/our-work/cems-task-forces/rewilding (accessed 17 November 2017).

Jørgensen, D. (2015). Rethinking rewilding. *Geoforum*, **65**, 482-488.

Kareiva, P., and Marvier, M. (2012). What is conservation science? *BioScience*, **62**, 962-969.

Law, A., Gaywood, M.J., Jones, K.C., Ramsay, P., and Willby, N.J. (2017). Using ecosystem engineers as tools in habitat restoration and rewilding: beaver and wetlands. *Science of the Total Environment*, **605**, 1021-1030.

Lorimer, J., and Driessen, C. (2014). Wild experiments at the Oostvaardersplassen: rethinking environmentalism in the Anthropocene. *Transactions of the Institute of British Geographers*, **39**, 169-181.

Lorimer, J., Sandom, C., Jepson, P., Doughty, C.E., Barua, M., and Kirby, K.J. (2015). Rewilding: science, practice, and politics. *Annual Review of Environment and Resources*, **40**, 39-62.

Miller J.R., and Bestelmeyer, B.T. (2016). What's wrong with novel ecosystems, really? *Restoration Ecology*, **24**, 577-582.

Naundrup, P.J., and Svenning, J.-C. (2015). A geographic assessment of the global scope for rewilding with wild-living horses (*Equus ferus*). *PLoS ONE*, **10**(7), e0132359. doi:10.1371/journal.pone.0132359

Nogués-Bravo, D., Simberloff, D., Rahbek, C., and Sanders, N.J. (2016). Rewilding is the new Pandora's box in conservation. *Current Biology*, **26**, R87-R91.

Pereira, H.M., and Navarro, L.M. (2015). *Rewilding European landscapes*. Cham: Springer Open.

Pettorelli, N., Barlow, J., Stephens, P.A., et al. (2018). Making rewilding fit for policy. *Journal of Applied Ecology*, **55**, 1114-1125.

Prior, J., and Brady, E. (2017). Environmental aesthetics and rewilding. *Environmental Values*, **26**, 31-51.

Rohwer, Y., and Marris, E. (2016). Renaming restoration: conceptualizing and justifying the activity as a restoration of lost moral value rather than a return to a previous state. *Restoration Ecology*, **24**, 674-679.

Soulé, M., and Noss, R. (1998). Rewilding and biodiversity: complementary goals for continental conservation. *Wild Earth*, **8**, 19-28.

Svenning, J.-C., Pedersen, P.B.M., Donlan, C.J., et al. (2016). Science for a wilder Anthropocene: synthesis and future directions for trophic rewilding research. *Proceedings of the National Academy of Sciences of the United States of America*, **113**, 898-906.

van der Zanden, E.H., Verburg, P.H., Schulp, C.J., and Verkerk, P.J. (2017). Trade-offs of European agricultural abandonment. *Land Use Policy*, **62**, 290-301.

CHAPTER TWO

History of rewilding: ideas and practice[*]

DAVID JOHNS
Portland State University

> Wilderness without animals is dead – dead scenery. Animals without wilderness are a closed book.
>
> (Lois Crisler, 1958, p. 92)

Rewilding is a type of large-scale biological and ecological restoration, emphasising recovery of native wide-ranging species and top carnivores and other keystone animals in natural patterns of abundance, to regain functional and resilient ecosystems (Noss, 1992; Noss and Cooperrider, 1994). The term was coined by Dave Foreman, long-time path-breaking conservationist, founder of the Rewilding Institute and co-founder of the Wildlands Network (originally Wildlands Project), an effort to create a North American system of connected strictly protected areas able to support ecologically effective populations of all native species. Wildlands Network also sought to inspire continental rewilding globally. The term is now widely used, although its meaning has become varied with wider use. It retains its original meaning while also being used to describe restoration generally, restoration of Pleistocene species, and in some cases urban greenways (see Chapter 1). Reporter Jennifer Foote (1990), in a story on 'radical environmentalism', appears to be the first to use the term in print.

With this chapter, I aim to review how some advocates for wilderness and biodiversity orient themselves to the state of the Earth's lands and waters, the

---

[*] An earlier version of this chapter appeared as Johns, D., Rewilding, Reference Module in Earth Systems and Environmental Sciences, © Elsevier, 2016. 08-Feb-2016. doi: 10.1016/B978-0-12-409548-9.09202-2. The author gratefully acknowledges permission from Elsevier to reprint substantial portions of the earlier paper. The author also gratefully acknowledges those who made comments on the manuscript, including Toby Aykroyd, Sarah Durant, John Davis, Dominick DellaSala, Scott Elias, Alan Watson Featherstone, Dave Foreman, Wendy Francis, Cyril Kormos, Ellen Main, Vance Martin, Vivek Menon, Reed Noss, Bittu Sahgal, Michael Soule, and Virginia Young. They patiently answered questions and otherwise helped me avoid many errors and omissions. Any errors remaining are mine, not theirs.

circumstances of other species and human injuries to them, and how the wounds are to be healed, e.g. through rewilding. It is also worth noting that this chapter attempts to document a dual thread as Crisler (1958) comments above. However, not all those whose work meets the criteria of rewilding use the term. While trying to avoid putting words in the mouths of practitioners, it is important to identify their work. As for the term rewilding, it is often used differently outside of North America than in the continent of its origin.

Wildness and rewilding

Before discussing *re*wilding it is useful to explore the meaning of its root: wild, wildness, wild creatures and places and close derivatives such as wilderness.

The term

Jay Vest (1985) and Rod Nash (1982, 2014) trace the term 'wild' to the early German word for 'will', which was applied to places or creatures not under human control. Wild-doer-ness (or naess in Old English) was a place of self-willed creatures; it was undomesticated or not under human domination. Wildlands, then, are self-willed lands.

The distinction between wild and tame can be traced not to modern conservation and its arguments on behalf of protected areas, but to those who brought about the slow transition from gathering and hunting societies to agriculture and pastoralism (Nash, 1982, 2014; Johnson and Earle, 2000). These societies, as they sought to bring various landscapes under their control in order to extract more from them in response to population pressure (Harris, 1975, 1977), developed many binary distinctions concerning what is and is not domestic, safe, predictable, human (Nash, 1982, 2014; Shepard, 1982). These distinctions range across the many elements of culture, from understandings of how the world works (what is inside and outside of human control, what is useful or not, what is dangerous or not) to meanings, attitudes and what is considered appropriate emotional expression (which are good places or creatures, which are bad, which should be feared or embraced, and so on).

Human hierarchy is closely bound up with the notion of self-willed. Boehm (1999) and Harris (1977), for example, document how efforts to control nature eventually result in the creation of hierarchical human societies – the control of some people by others. The two aspects of control are unavoidably integral, which makes it very difficult to address problems associated with human hierarchy. Sanday (1981) traces the subjugation of women to this transition of humans from a world in which they experience themselves in nature – and without a wild–domestic distinction – to societies built on control. The wild and the controlled are opposite, just as love and control are opposites (Fromm, 1964).

US legal definition

A self-willed place – or wilderness – is not necessarily the same as Wilderness as designated by law. The latter has legal protection and must meet certain criteria discussed below.

After World War Two the USA experienced a boom in consumption, partly as a result in pent-up demand constrained during the war. The boom in consumption included housebuilding and suburbanisation which greatly increased demand for wood and the economics of logging favoured cutting big trees and lots of them. As a result, the destruction of forests occurred rapidly in many places; what in other parts of the world may have taken decades or even longer occurred in the space of a few years. It was noticeable. Although many had been critical since the 1930s of the US Forest Service's (USFS) administratively designated forest reserves as too susceptible to agency whims and industry pressure, massive logging after World War Two consolidated the drive to create legislatively protected wilderness areas. The first bill was introduced in 1956.

In 1964 the US Congress passed the Wilderness Act (PL 88-577; 16 USC 1131-6). The Act designated some USFS lands as wilderness and importantly defined what wilderness meant, the attributes of lands eligible for future designation, and the purposes of wilderness. Eventually other public lands were designated Wilderness, including parts of National Parks and Wildlife Refuges, and with the 1976 passage of the Federal Lands Policy Management Act (PL 94-579; 43 USC 1701-1785), land controlled by the Bureau of Land Management was eligible for Wilderness designation.

Section 2(c) of the Wilderness Act parallels notions of wildness discussed just above.

A wilderness, *in contrast with those areas where man and his works dominate* the landscape, is hereby recognized as *an area where the earth and its community of life are untrammeled by man*, where man himself is a visitor who does not remain. An area of wilderness is further defined to mean in this Act an area of undeveloped Federal *land retaining its primeval character and influence, without permanent improvements or human habitation*, which is protected and managed so as to preserve its natural conditions and which (1) generally appears to have been *affected primarily by the forces of nature, with the imprint of man's work substantially unnoticeable*; (2) has outstanding opportunities for solitude or a primitive and unconfined type of recreation; (3) has at least five thousand acres of land or is of sufficient size as to make practicable its preservation and use in an unimpaired condition; and (4) *may* also contain *ecological*, geological, or other features of scientific, educational, scenic, or historical value. (Italics added)

Wilderness areas, although designated and protected by human action, are places not dominated by humans; they are untrammelled or undomesticated, where the Earth's 'community of life' is subject to 'the forces of nature', not the forces of one species. Without demanding purity, the Act recognises the distinction between domesticated and undomesticated lands and the incompatibility between wildlands and industrial humans. Although designated wilderness areas must be a minimum size (5000 acres or 2023 hectares), there is no explicit requirement that such areas be self-sustaining or connected to other areas [§2(c)(3)]. (See discussions of island biogeography and connectivity below.)

It is also worth noting that the presence of all or most native species, or their preservation, is not a primary criterion or goal of the Act. Indeed, much US designated wilderness is high-elevation and comparatively less biologically productive; more productive lands tend to be dominated by humans. I will examine this issue more closely below, but note for now that many conservationists regard designated wilderness areas that lack certain species such as top predators as missing an important wilderness quality (Crisler, 1958; Foreman and Wolke, 1992).

Much ink has been spilled over what is truly wild or not, what wilderness means or ought to mean, and frequently the legal definition of wilderness is relied upon to try to rescue the debate from the ethereal philosophical labyrinth (e.g. Callicott and Nelson, 1998). Places actually designated wilderness certainly vary in their degree of 'untrammelled', with many the result of tough political bargaining: grazing, mining claims, water rights, inholdings and inholding access are often grandfathered in (see §§4(d) and 5(a) and (b) of the Act).

IUCN and related definitions

The latest version of the International Union for the Conservation of Nature's (IUCN, 2014) protected area (PA) category definitions and purposes recognises the duality that emerged from US and other efforts to protect wilderness and biodiversity. Neither was reducible to the other. PA Category 1a describes the attributes of those areas set aside for biodiversity protection, i.e. for maintaining functioning ecosystems and species communities: 'human visitation, use and impacts are strictly controlled'. Although lands needing restoration may be included in Category 1a reserves, the designation's focus is on areas that are largely intact, with populations of native species present and little evidence of human domination; non-human forces predominate. Such areas are mostly wild in the sense of self-willed, although the term is not used. Wilderness per se is addressed in Category 1b.

Although biodiversity protection is a 'distinguishing feature' of Category 1b (wilderness) reserves, their purpose and state at designation is their 'natural character', which is to be maintained and protected. Like the US Wilderness Act, absence of extensive human modification and infrastructure is a key feature. As with Category 1a, 'natural forces and processes predominate', but lands may also be restored to this state. Maintenance of 'wilderness qualities' is the priority for areas so designated. Categories 1a and 1b share many attributes, and are perhaps best thought of as complementary. Both categories of reserve rely on similar criteria of naturalness and absence of human modification and human-caused injury and loss, i.e. large, intact areas.

More recently, the Wild Europe Initiative (WEI, 2013) developed with dozens of other non-governmental organisations (NGOs) a definition of wilderness to be used throughout the continent to identify areas meeting criteria for wilderness and to set a standard for restoration or rewilding. The definition, which has since been adopted by the European Commission for its Guidelines on management of wilderness areas and its Wilderness Register, follows closely IUCN category 1b.

> A wilderness is an area governed by natural processes. It is composed of native habitats and species, and large enough for the effective ecological functioning of natural processes. It is unmodified or only slightly modified and without intrusive or extractive human activity, settlement, infrastructure or visual disturbance. (WEI, 2013)

Convergence of wilderness and biodiversity

Before the early 1990s in US conservation discussion, protection and restoration of wildlands and biodiversity protection were distinct in many respects and rarely conflated by those pursuing these objectives. Many early advocates did recognise strong linkages or overlap between wilderness and biodiversity: Aldo Leopold, Benton MacKaye, and Victor Shelford et al. at the Ecological Society of America (Foreman, 2004). Many wilderness advocates would not consider an area truly wild if certain iconic species are not present despite the absence of roads or other infrastructure, whatever the formal designation. And those seeking the conservation of wide-ranging species, top predators, and unencumbered disturbance regimes have usually recognised that large areas of self-willed land are necessary to maintain the foregoing. Indeed, areas designated as wilderness or which qualify as wilderness under most definitions are by virtue of those characteristics refuges for many species that humans would otherwise exploit, persecute, or rob of necessary habitat if they had the chance. Wild areas are refugia for biodiversity by default because of the relative lack of human presence and access – although humans do intrude and effectively persecute target species. Such areas have limits as refugia because in many cases, whether designated wilderness or not, they are areas of low biological productivity and otherwise areas humans have been content to leave alone for the time being because of their distance from urban centres, absence of roads, difficult terrain, or lack of 'resources'. Areas with the highest biological productivity are places humans covet and convert to their uses, with negative consequences for biodiversity.

At the beginning of the 1990s a group of conservation scientists and advocates (including the creator of the term rewilding), concerned that existing protected areas and other actions were failing to halt escalating extinctions or maintain wilderness qualities, called for the creation of a continental (North American) system of *connected* protected areas that would be adequate to maintain 'native biodiversity ... in perpetuity' by (1) representing 'all native ecosystem types and seral stages across their natural range of variation'; (2) 'maintain viable populations of all native species in natural patterns of abundance and distribution'; (3) 'maintain ecological and evolutionary processes' including disturbance regimes, tropic interactions and similar; and (4) be resilient in the face of anthropogenic and other change (Noss, 1992). Avowedly bio- or ecocentric, this proposal called for putting off limits to human exploitation those areas (terrestrial, aquatic, and marine) needed by the millions of non-human species; it was a positive vision describing as an overarching goal what life needed to thrive (Foreman et al., 1992).

Calling for the expansion of existing PAs, the designation of new ones, and their connection regionally and continentally, the Wildlands Network also redefined wilderness (cf. Wilderness) by incorporating both traditional US wilderness attributes and comprehensive biodiversity protection. Although reserves that did not qualify as wilderness would be important in this system, such as privately owned ranches managed for biodiversity, at the core of this system was existing and restored wilderness. In this case

(w)ilderness means: extensive areas of native vegetation in various stages off limits to human exploitation. Viable, self-reproducing, genetically diverse populations of all native plant and animal species, including large predators.

Vast landscapes without roads, dams, motorized vehicles, powerlines, overflights, or other artifacts of civilization, where evolutionary and ecological processes that represent four billion years of Earth wisdom can continue.

Such wilderness is absolutely essential to the comprehensive maintenance of biodiversity.
(Italics added; Foreman et al., 1992)

This vision of wilderness lies at the root of the original notion of rewilding. Based on the literature and experience, Noss (1992) states that about half of the landscape, depending on the region, would need to be strictly protected. Strict protection is a core attribute of wilderness. As early as 1994 E.O. Wilson lent his strong support to this vision (Wildlands Project, 1994). Safeguarding half of the North American landscape requires large-scale restoration or rewilding. A literature review by Jones (2011) found that about 25 per cent of the ice-free land remains wild and another 20 per cent semi-natural, so the gap is significant between half (the right half) and this.

A tension is apparent even at this early stage of development of the vision for a wild North America and before 'rewilding' came into wide use within conservation. It reflects issues arising from charting a strategy to implement the vision. Which of the several interrelated attributes to focus on? Historically defined wilderness attributes – large areas of untrammelled land and water – were beyond question. But which aspects of biodiversity? Although Noss (1992) advocates for the complete complement of biodiversity, he focuses on apex predators in a land conservation strategy because (1) they are sensitive to human presence and hostility and thus require wildlands, and (2) other groups shy away from the needs of controversial or inconvenient species such as predators.

Proceeding region by region towards a continental system of PAs, grassroots conservation groups delineated a common vision of protected cores and connectivity; deciding which species to take into account became a very practical concern. A variety of approaches were taken, some relying on 20–30 'focal species', representing a wide range of life-needs and processes and including keystones, umbrella, indicator, and iconic species; others focused on healing wounds to the land such as fragmentation, recovering regionally extirpated species, suppressed disturbance regimes, and similar (e.g. Wildlands Network, 2000, 2003, 2004; Wildlands Network et al., 2003).

Ongoing research on top-down regulation – the role that keystone species such as wolves, beavers, and bears play in shaping ecosystem function – generated stronger evidence of its essential role (Estes et al., 1978; Terborgh, 1988; McLaren and Peterson, 1994; Palomares et al., 1995; Noss et al., 1996; Miller et al., 1998; Terborgh et al., 1999; Terborgh and Estes, 2010; Estes et al., 2011). It came to be seen as important as connectivity (a response to the findings of island biogeography) in conservation planning. The findings also reinforced an emphasis on carnivores and other keystone or highly interactive species as essential to wildness/wilderness.

Rewilding

The first discussion of rewilding by Dave Foreman is in a Wild Earth essay (1992, inside front cover): 'It is time to rewild North America; it is past time to reweave the full fabric of life on our continent'. Two years later in an early outreach piece by Wildlands Project (1994), rewilding is defined in greater detail – as creating a system of cores, corridors, and buffers that further the four conservation objectives identified (see above) by Noss (1992) and Noss and Cooperrider (1994). To rewild is to ensure designated areas can support in perpetuity all native species – including top predators, other keystone species, and wide-ranging species – all ecosystem types and processes, disturbance regimes and resilience. Prior to this discussion the terms restoration or recovery are used (Foreman et al., 1992; Noss, 1992), but always modified to make clear that it is not limited to restoration of vegetation communities only, or restoration of small areas that would be overwhelmed by the surrounding human forces without significant ongoing human management. Whatever term is used – restore, restoration, recovery, or rewilding – the centrality of the very large-scale, the inclusion of all native species including highly interactive ones (keystone), and the importance of natural process, are part of a consistent thread. So is the long-term nature of the effort: decades and even centuries.

Ongoing strategic discussion resulted in refining the term to make it more precise, including the distinction between its value and scientific elements. The recovery of top predators and other keystone (highly interactive) species is not only good in itself, but essential to the protection and restoration of all biodiversity because of the central regulatory role the former play in ecological and community health – which is also considered good (Soulé and Noss, 1998; Soulé and Terborgh, 1999). Without keystone species, including top predators, ecosystems simplify and are degraded in other fundamental ways such as meso-predator and ungulate population increases and consequent deleterious effects such as degradation of riparian areas, increases in steam temperatures and sedimentation, loss of tree recruitment, and simplification of grasslands. Big, protected, and connected wilderness areas, another value held by the initial advocates of rewilding, is considered by many scientists to be required for the survival of healthy populations of wide-ranging species and top predators. Even large protected areas are inadequate to maintain ecologically effective populations of many species (MacArthur and Wilson, 1967; Noss, 1992; Noss and Cooperrider, 1994; Newmark, 1995; Berger, 2003) so regional and continental-scale connectivity is essential. Soulé and Terborgh (1999) sum it up by saying rewilding is '… maintaining viable populations of keystone species, particularly large carnivores'.

As an essential component of biodiversity, conservation rewilding deserves strategic emphasis because most NGOs shy away from the controversy associated with safeguarding or restoring predators and the large-scale (Soulé and Noss, 1998). They argue that humans also have a special moral obligation to heal the damage to non-human species and ecosystems they have caused. Notwithstanding the rewilding focus on the 3 Cs – cores, connectivity, carnivores – Soulé and Noss (1998) and others (Soulé and Terborgh, 1999; Foreman, 2004) say this is not an abandonment of the overarching goal of protecting and restoring all native species and functional ecosystem types. Instead, the focus of rewilding on top predators and other keystone species – those with a big influence on ecosystem structure and function despite their relatively small numbers – is extremely pressing given current persecution and habitat loss and fragmentation.

To distinguish rewilding from other forms of restoration, Soulé and Terborgh (1999) say that restoration's focus has been on ecological processes, not species, even though processes emerge from species interactions and needs; that restoration has mostly been bottom-up, ignoring top-down regulation; and it has been focused on addressing site-specific damage rather than large-scale injuries such as fire suppression, grazing, logging, monocultures, and other damage across large areas. Mainstream restoration has changed much since this was written, addressing many of these issues. Jordan and Lubick (2011) argue for a different distinction: that between ecological restoration which seeks to bring all the pieces together and recreate what was present prior to human degradation, and can be left to evolve if the area is large enough; and ameliorative restoration, which involves restoration of this or that aspect of the former system that humans favour – 'game' animals, clean water, or other ecological services.

In an interesting side note, Soulé and Noss (1998) think that the importance of big predators and some other keystones was less critical in North America prior to the large herbivore kill-off by the first North Americans 15–12,000 years ago. Prior to that, the enormous herds of large ungulates, giant camels, giant ground sloths, and species such as mammoths and mastodons played pivotal regulatory roles. It was only after these losses that top carnivores played a more central role.

Their comment raises perhaps the thorniest question of all in restoration, no matter what the approach: restore to what? Then what? Only the largest areas, in their view, can *reapproach* self-regulation. Smaller areas, once restored will be degraded almost immediately after restoration without some ongoing intervention even if the intent is not to hold a static state (which is the intent in some cases) but to allow eco-evolutionary processes to take over. Even the largest areas face human-caused degradation if left completely alone. For one, rapidly changing patterns of temperature, rainfall, snowfall, and intensity of precipitation are occurring for the first time on an Earth where habitat is fragmented. In addition to fragmentation, much of the landscape has been highly simplified or destroyed, heavily compromising ecological resilience.

Rewilding in Latin America

Rewilding is used in Latin America by Conservacion Patagonia to describe several successful efforts to create many new parks in Chile and Argentina, halt landscape fragmentation in the region by infrastructure development, and restore eight highly interactive species that have been in serious decline, including jaguars, giant river otters, peccaries, and others. The work also includes reconnecting local human populations with these species, who are often seen as lost relatives. Such reconnection makes conservation personal and links people to their sense of history, wild roots, and sense of beauty (Jimenez, 2017).

Thus far, five parks have been created with six more in the making, to total 13 million acres (5,261,000 ha). Latin America is also home to one of the very first large-scale projects in the world that included major restoration components: Paseo Pantera (Path of the Panther) – led by Mario Boza, Archie Carr III, and Jim

Barborak – extended from Panama to the Yucatan Peninsula of Mexico. Progress was uneven and unfortunately, a little more than 10 years after its founding, it was transformed into a development project – the Mesoamerican Biological Corridor – in part as a result of major UN funding. Paseo Pantera was an inspiration for the Wildlands Network – the goals were similar – and the two established an informal cooperative relationship. Despite goals similar to Wildlands Network, Paseo Pantera did not use the term rewilding, although commentators describe it using this term (e.g. Fraser, 2009; Sandom et al., 2013).

Other large-scale multijurisdiction projects have been and are being undertaken in Latin America with goals that are similar to rewilding but have not used the term, such as the Ecological Corridor of the Americas (Wildlife Conservation Society, 2000; Boza, 2001) and Cerrado–Pantanal corridors project in Brazil (Klink and Machado, 2005). Throughout Mexico, which is important as a mega-biodiverse country and also because it is both Latin American and North American, recovery of jaguars is underway. This will include the establishment of a system of connected reserves throughout the country, linking jaguar populations from Meso-America to the US borderlands. Jaguars and the reserve system created for them will not only recover the iconic and keystone jaguar but many other species as well (Moctezuma, personal communication, 18 May 2017).

Rewilding in Europe

Rewilding has become a central tenet of European conservation. It recognises that existing wilderness in the crowded continent is limited, and rewilding must play a very large role in addition to protecting existing wild areas and making them whole (WEI, 2013). So the European notion of rewilding combines what in North America started out as two threads that were only explicitly combined in the 1990s: wildlands and biodiversity.

The Wild Europe Initiative (WEI, 2013), however, conceives of wilderness differently than in North America. Their definition, which is widely shared in the EU, combines North American wilderness attributes (a large area only lightly if at all impacted by human infrastructure, occupation, or use) and biodiversity conservation attributes: functional ecological processes, presence of native species, or the possibility of restoring all native species.

Advocates for rewilding in Europe have received support from the European Parliament (2009) 'Resolution on Wilderness in Europe', which includes, inter alia, catalysing new wilderness areas by designating existing wildlands and rewilding and promoting the value of wilderness. At the same time, the European Parliament 'welcomed the establishment of the Wild Europe Initiative', a cooperative effort of the European Commission, Council of Europe, and many NGOs, including the former Pan Parks Foundation, the Europarc Federation, Bird Life International, WWF,

UNESCO, and others (WEI, 2013). In May of that same year 230 representatives of the EU, individual nation-state governments and NGOs met in Prague and made 24 recommendations to advance the cause of wilderness. WEI was tasked to coordinate follow-through, which included refining definitions of wild, wilderness, and rewilding so goals could be clearly stated and progress towards them assessed.

Rewilding, as used by WEI, is:

> ... the return of an area to its wild natural condition. As with restoration, re-wilding involves initiating, stimulating and allowing natural processes to occur (again), replacing human management and interference to shape new and wilder areas; it is applicable to any type of landscape and may not result in a predictable end-state, or restoration of an old state. A naturally functioning landscape that can sustain itself into the future without active human management is the ultimate goal of the rewilding approach. (at 7)

Rewilding is used interchangeably with restoration in many Wild Europe documents, but in practice and as there are many initiatives comprising the Wild Earth Initiative, it is not much different than North America usage where restoration encompasses a wide range of activities, including site-specific work on relatively small areas that require ongoing intervention. The point of this management in North America is sometimes to maintain a particular state, e.g. pre-European settlement grassland, but it is also undertaken to prevent the inevitable degradation of areas by ongoing human activities following restoration to a more natural state, such as fire suppression or absence of predators in areas too small to sustain them. The use of rewilding by Wild Europe is sometimes used as a promotional name for restoration of wild or wilderness areas that aims to reinstate natural processes and habitats together with reintroduction of native species formerly present. In many circumstances, however, other terms are used: 'ecological enrichment' or 'restoration of large, natural ecosystem areas' (Aykroyd, 2018, personal communication). Such terms accommodate a continuum of wildness – recognising that rewilding is a process that may take significant time – and they make less likely knee-jerk hostility from some communities.

The Tree for Life organisation, which is rewilding the Caledonian Forest of Scotland – and has been at it for almost three decades – uses a more detailed rewilding definition:

> In practical terms, rewilding consists of the restoration of healthy vegetation communities; reinstatement of key ecological processes, such as succession, predator prey relationships and natural disturbance; a reduction and/or cessation of human management of ecosystems; the removal of human infrastructure (e.g. roads, dams, fences) that have a serious impact on ecosystems and/or wildlife populations; and the reintroduction of key species (e.g. top predators) that have been removed. (Featherstone, 2014)

In both North America and Europe there is agreement that bigger PAs are better for a variety of biological reasons – they are able to accommodate wide-ranging species and apex predator, species dispersal, seasonal migration, allow ecological processes and evolutionary change, and rewilding generally – and also because the larger the protected area the more realistic the goal of ecological self-regulation and less reliance on human management with all its uncertainties. The experience of North America is that even the largest areas cannot be totally self-regulating given ongoing human pressures. Simply restoring all native species and allowing processes to recover in a specific area and 'letting it go' will not realise an area's evolutionary potential. Even the largest and most well-connected PAs are not immune from outside human activity, and few regions approach what Noss (1992) and colleagues (Noss et al., 2012) regard as realistic for regional ecological functionality: about half of the area under strict protection. Indeed, even large intact areas dominating the upper latitudes of North America are under assault by climate change, airborne toxic pollution, and other human forces.

WEI recognises minimum sizes for wilderness areas and speaks of their self-regulation as an ultimate goal, but discussion of how future deterioration of restored areas caused by human activities is to be prevented has proceeded somewhat slowly. Such prophylactic action might well be focused outside PAs, but probably not exclusively. Lobbying government for adequate laws and regulation is being discussed. These include the development of enforceable legal structures and incentives for privately held land, such as easements (WEI, 2017). Part of this strategy also includes the European Community's Wilderness Register, which as a requirement for listing requires individual protection plans for each area. The minimum size for wilderness areas under WEI and even much larger areas will not likely avoid deterioration after initial restoration of all native species and processes without additional human effort. Even if the vision of the now defunct Pan Parks Foundation, which had provided recognition of wilderness quality for a quarter of a million hectares and sought to do the same for another million hectares, and even if Rewilding Europe achieves its promoted goal of restoring 1 million hectares in 100,000-ha PAs, and even if all these areas were adequately linked as is the goal of WEI, it will only move European PAs *towards* self-regulation, not get them there. That being said, substantial movement along the continuum to greater wildness is positive. Rewilding is an ongoing task. The Yellowstone to Yukon (Y2Y) region, for example, has about 19.4 million hectares of PAs out of 137 million hectares; another 30.9 million hectares have some form of lesser protection (Francis et al., 2017) and significant parts of the rest of the region are largely intact (Y2Y, 2015). Nonetheless, constant human pressure remains on ecosystems and species within PAs requiring human response to maintain these areas (Noss, 1992; Terborgh, 1999; Pringle, 2017).

Size is not the only criterion, of course. Other factors such as productivity, degree of species richness and community diversity and other factors are important. WEI, but by no means all NGOs or commentators (such as George Monbiot, 2013), addresses the ongoing potential need to prevent future human-caused biological decline after so much hard-won rewilding work.

Rewilding in Asia

The term rewilding is often used in the literature (e.g. Fraser, 2009) to describe projects that the practitioners responsible for the projects do not use. The first use of the term in Asia by a practitioner appears to be by Dinerstein (2003). After several years of studying rhinoceroses and their habitat in the early 1990s, Dinerstein and his co-workers, building on earlier efforts by the government of Nepal to create effective parks to reduce wildlife decline, developed a plan to link 11 parks across the country and into India, called the Terai Arc Landscape Programme. The government backed the programme in 1993 and created a legal framework for community conservation projects, which enabled them to benefit from parks, buffers, and connectivity through tourism and similar projects. The focus, unlike many efforts (e.g. Oates, 1999), was directed first and foremost at conservation, not development. However, it ensured the benefits of conservation went to local people in a position to help or hurt conservation. In the end, Dinerstein wrote that 'We have accomplished what Michael Soulé and Reed Noss call "rewilding" – returning the land to creatures that once flourished here' (Dinerstein, 2003). He went on to express the hope that rewilding would become 'the mantra of the next generation of Asian conservationists, as it takes root in other regions of the world' (Dinerstein, 2003). Rewilding has taken root, but the term is infrequently employed elsewhere in Asia. Only in the last two or three years has the term been employed by conservation advocates to describe a variety of projects, most very small-scale, but nonetheless critical in a landscape dominated by humans.

In practice, rewilding has been underway in South Asia with many large-scale efforts to link tiger, elephant, and rhinoceros habitat and maintain forest and forest connectivity (Menon et al., 2005; Menon, 2015, personal communication). Among these efforts are the National Elephant Corridor project with at least 100 corridors; the Garo Green Spine project (90,000 ha), focused on elephant, gibbon, and fish; and the Greater Manas, which adds and links 100,000 ha to two existing parks for a total of 250,000 ha. However, the term has not been much employed by conservation advocates until the last several years. It is applied to a variety of projects, most not large-scale but nonetheless critical in a landscape so long dominated by humans.

One effort referred to as rewilding involves restoring native vegetation to a 70-ha area dominated by mesquite in the foothills of Mehrangarh

Fort which overlooks Jodhpur in central Rajasthan (Raina, 2014). Although such projects look much like meliorative restoration and it is tempting to discount such projects, in North America protection and recovery of fairly small and degraded areas has been important for securing connectivity for species such as wide-ranging grizzly bears. Other small-scale projects focus on increasing the size of protected areas to allow native species to have more habitat and larger populations, and make them safe from human injury. In eastern Maharashtra state, north of the Tadora-Andhari Tiger Reserve, a village council leader successfully petitioned to have his village relocated away from the reserve, to end human–wildlife conflicts and gain public services and better land (*Sanctuary Asia*, 2013).

In Kerala, the most south-westerly state of India and home to several tiger reserves and other wildlife sanctuaries, *The Times of India* proclaimed: 'Kerala's first rewilding project comes to fruition in Wayanad' (Rajeev, 2014). Two villages of several within the Wayanad Wildlife Sanctuary were relocated outside the sanctuary in response to government incentives. As with the Tadora villagers, moving away from conflicts with wildlife and to a village with electricity and water were big draws. Additional money has been appropriated to move others of the remaining 12,000 humans who live in the sanctuary. Those involved in the relocation have reported that large animals are already using the areas vacated by villagers, including deer, bison, and elephant. The area of vacated fields is about 40.5 ha, but if all villagers leave it would create significant contiguous habitat, reduce edge affects and conflict which results in behaviour hostile to wildlife.

Pioneered in Africa, Community Nature Conservancies are being utilised in India to achieve similar small-scale rewilding efforts, i.e. reconverting land from farm use to wildlife use. These efforts involve farmer cooperatives that stop farming and allow their land to be reinhabited by wildlife and in return are given authority and government support to set up tourism concessions. One such project involves 42.5 ha adjacent to the Umred-Karhandla Wildlife Sanctuary (William, 2015). There is little discussion in South Asia (or elsewhere in Asia) of precise definitions of rewilding or detailed criteria. The Nepalise Project Terai Arc Landscape Programme is the only larger-scale effort using the term; most other projects are quite small. Several large-scale projects addressing the needs of elephants, rhinos, and tigers and forest health are underway in South Asia, however (e.g. Dinerstein et al., 1997; Menon et al., 2005). In Kamchatka, along the Russian–Chinese border and the Amur River area, projects have been underway for some decades to secure threatened populations of tigers, leopards, brown bears, hooded cranes, and other species diminished by human hunting and resources extraction. None of these projects have been referred to as rewilding projects either by those undertaking them – e.g. the Khabarovsk-based Wildlife Foundation, the various government agencies in Russia, China, Mongolia, and North Korea, or regional, national, or international NGOs (Miquelle et al., 1999; Wildlife Foundation, 2000; Pikunov and Miquelle, 2001) or by outside observers. This is so even though these projects resemble various large-scale efforts that elsewhere are referred to as rewilding and that utilise the same tools: large protected areas and connectivity focused on apex predators and wide-ranging species.

Rewilding in Australia

Rewilding is not widely used by practitioners in Australia. The Wilderness Society of Australia (2002) called its 'Australia-wide program ... to protect our wild places and wildlife' and 'help define the path to restoration' WildCountry. 'WildCountry is a plan to re-wild Australia' (Wilderness Society Australia, 2002, p. 2). Much like rewilding in North America, WildCountry is science-based and '(s)eeks to maximize protection, rehabilitation and restoration of all native species and ecological communities across their range' and to 'ensure maximum survival and evolutionary potential of biodiversity' (Wilderness Society Australia, 2002, p. 2) by means of a system of connected cores surrounded by buffers. Noting that the island-continent has lost more mammal species to extinction in the last two centuries than other large countries, bringing back many species is not an option as is the case with North America, where remnant populations remain.

More recently, Rewilding Australia was created as a network of NGOs 'to support the reintroduction of apex species like devils and quolls to our ecologically impoverished landscapes' (Rewilding Australia, 2015). In 2016 the National Parks Association of New South Wales sponsored a National Rewilding Forum attended by 53 people from well over 30 NGOs (including Rewilding Australia), universities, and agencies. The meeting described what rewilding meant to the attendees and identified next steps. Among the central biological attributes of rewilding are re-establishing species across their former ranges, re-establishing ecosystem function, including predation, addressing exotic and feral species, minimising the need for human management (ceding control back to nature), working at a variety of scales, and re-establishing appropriate connectivity. These attributes are similar to North American notions of rewilding. Additionally, the National Rewilding Forum set out some social *goals* (not just as a means to biological rewilding), including engaging communities in decision making, consensus among stakeholders and public and inter alia accommodating the 'new nature'. Given the much larger number of human-oriented NGOs, many conservationists regard taking on social goals unrelated to achieving conservation goals as a dilution of their resources. It is unclear how the

National Rewilding Forum will address conflict between rewilding – reweaving all the pieces – with new species that are problematic for native species.

The Forum considered rewilding in Australia to be in its early stages and identified a range of next steps from experimental removal of fences and pilot restorations of Tasmanian devils and dingos to identification of priority areas for rewilding. Although other Australian NGOs do not seem to use the term, commentator Carolyn Fraser (2009) applies rewilding to the Gondwana Link, a major restoration effort underway in the south-east. It ties together '1,000 km of continuous habitat, from the dry woodlands of the interior to the tall wet forests of the far south-west corner', much of which is intact and is home to an extensive number of endemic species threated by habitat loss (Gondwana Link, 2015).

Rewilding and Africa

Last and far from least, African conservation NGOs do not much use the term rewilding to describe their many conservation efforts, including large-scale efforts aimed at restoring species, predators in particular or whole systems. As with North America, PAs were historically designated in most cases without regard for animal movement and some have become islands. Throughout Africa, however, there have long been efforts to increase protected areas' size by opening international boundaries, linking adjacent parks and coordinating conservation efforts across national boundaries, creating new parks, and recovering species such as elephants in areas such as large parts of Angola and Namibia, where their numbers were greatly reduced by conflicts (Fraser, 2012). One of the most ambitious projects is the Kavango Zambezi Transfrontier Conservation Area (or KAZA), which includes National Parks, communal lands, designated forest lands, private lands, and conservancies. KAZA includes areas of Angola, Namibia, Botswana, Zimbabwe, and Zambia. Not all lands within the conservation area (twice the size of the UK) are wild or slated for rewilding as usually defined. Enhancing tourism is a major goal, thereby enhancing income for local human communities through employment and subsistence opportunities (Kavango Zambezi, 2017). Insofar as a major focus of KAZA is maintenance and restoration of healthy wildlife populations and ecosystem function it is rewilding. Other large-scale efforts include the Greater Limpopo Park and Transfrontier Conservation Area, which includes Kruger and Limpopo National Parks among others. Not only will the area provide for larger populations of species by protecting secure habitat, it will connect parks that are otherwise subject to creeping isolation from rural development.

Growing protected areas via transboundary parks has contributed to a major rewilding goal of making more room for larger populations of keystone and iconic species. Uganda, Rwanda, and Congo, for example, have also combined efforts in their adjoining National Parks in the Virunga Mountains to protect mountain gorillas (Gray and Rutagarama, 2011). Gorongosa National Park in Mozambique, founded in 1960, was devastated by a long civil war (1977–1992). The park was a battlefield and source of food for both sides. The rebels deliberately targeted park buildings and infrastructure. After the war there was heavy poaching. Healing the park required restoring locally extirpated species and carefully nurturing small populations of remaining species until they could regrow ecological processes and re-establish themselves – especially trophic relations (Carroll, 2015).

Lion and other predator ranges – leopard, wild dog, and cheetah among others – are shrinking due to human population increase and the spread of agriculture and pastoralism. They are long gone from the Mediterranean, of course, but in sub-Saharan Africa lions were recently restored to Akagera National Park in Rwanda after they were extirpated by farmers to whom the government had given part of the park following the genocide (Republic of Rwanda, 2015). Akagera Park is one of 13 parks in nine countries managed by African Parks, which since its founding in 2000 works with national and local authorities to ensure protected area integrity and recovery of wildlife, in part by ensuring adjacent communities benefit from the parks. African Parks, like Wildlands Network, was founded in response to observed decline in biological diversity and wildlands (African Parks, 2017). In the case of many African parks, challenges included underfunding and poor management.

Connectivity is of increasing importance in Africa, especially in the east and south, as habitat is not only converted but fragmented by rising population, growing demand for land for agriculture, grazing, and massive infrastructure projects. Where predators have been exterminated it is generally locally or regionally, and their recovery often hinges on connectivity and natural recolonisation, as does maintaining ecologically effective populations of wide-ranging species. Connectivity is important for genetic diversity, seasonal migration, dispersal, and resilience in the face of human perturbation. Many connectivity efforts are focused on ameliorating human–wildlife conflicts that have been brought about by growing human demands on the landscape. Despite the obstacles, connectivity projects are making headway in many places (Worboys et al., 2010).

Africa is also the home of Community Nature Conservancies, which were pioneered in Namibia. The 1990 law, which allows communities to create these entities for conservation purposes and gain the benefit of tourism income, provides clear regulations and government support (Weaver and Skyer, 2003; Fraser, 2009). Not all countries have the legal or social infrastructure to make conservancies – or their variations – work. Corruption, anthropogenic pressure, and lack of funding and supportive laws militate against their formation or success in some countries. The conservancies, parks, and other wildlife areas depend heavily on eco-tourism – especially eco-tourism which benefits those living adjacent to parks.

Conclusions

As the human impact on the Earth has rapidly grown in the age of enormous energy subsidies, the rest of life and whole systems are being degraded and lost (Ewing et al., 2009; Brashares, 2010; Butchart et al., 2010; McKee, 2012). Many large mammal populations have suffered enormously, as the WWF Living Planet Reports (WWF, 2014, 2016) find: between 1970 and 2010 vertebrate populations declined an average of 52 per cent. Over the course of the twentieth century, tigers declined from an estimated 100,000 in 1900 to about 3200 today (National Geographic, 2014). Elephants in the same period declined from 10 million to 434,000 (*Daily Mirror*, 2014; Great Elephant Census, 2014). Many ecosystems are nearly gone: in North America alone, Long Leaf Pine have only 3 per cent remaining, Tall Grass Prairie only 4 per cent left and perhaps only 20 per cent of Ponderosa Pine (NatureServe/LandscapeAmerica, 2014). Only about 24 per cent of the Earth's original forests remain as primary forest, i.e. forests that are intact, unlogged, with native species and communities (Potapov et al., 2008).

Rewilding is a focused response to the loss of species that play an outsized role in making ecosystems work by restoring them to the places they once thrived and trying to ensure their persistence via large-scale networks of connected protected areas. The importance of top-down regulation does not ignore bottom–up regulation by photosynthesis or 'the little things that run the world' (Wilson, 1987). However, it recognises the particular vulnerability of wide-ranging species and keystone species, especially top predators, to human activities and emphasises their restoration. Rewilding recognises that many of these species, given their requirements for large wilderness areas and the limits of human management, are umbrellas for many other species and ecosystem function. Such areas also have intrinsic value.

Rewilding, where it has occurred, has been successful at restoring target species, other species, and system functioning (Sandom et al., 2013). The bigger question is whether rewilding alone, without confronting squarely the huge growth in human numbers and consumption, can prevail over enough of the Earth to halt or even slow the sixth great extinction underway.

References

African Parks. (2017). www.african-parks.org/about-us/our-story (accessed 23 December 2017).

Berger, J. (2003). Is it acceptable to let a species go extinct in a national park? *Conservation Biology*, **17**, 1451–1454.

Boehm, C. (1999). *Hierarchy in the forest*. Cambridge, MA: Harvard University Press.

Boza, M. (2001). *Activities of EcoAmericas*. San Jose, Costa Rica: Wildlife Conservation Society.

Brashares, J.S. (2010). Filtering wildlife. *Science*, **329**, 402–403.

Butchart, S.H.M., Walpole, M., Collen, B., et al. (2010). Global biodiversity: indicators of recent declines. *Science*, **328**, 1164–1168.

Callicott, J.B., and Nelson, M.P. (1998). *The great new wilderness debate*. Athens, GA: University of Georgia Press.

Carroll, S. (2015). *Serengeti rules*. Princeton, NJ: Princeton University Press.

Crisler, L. (1958). *Arctic wild*. New York, NY: Ballantine Books.

Daily Mirror. (2014). Poachers killing 35,000 elephants a year. www.mirror.co.uk/news/world-news/poachers-killing-35000-elephants-year-4595057 (accessed 1 September 2015).

Dinerstein, E. (2003). *The return of the unicorns*. Washington, DC: Island Press.

Dinerstein, E., Wikramanayake, E., Robinson, J., et al. (1997). *A framework for identifying high priority areas and actions for conservation of tigers in the wild*. Washington, DC: World Wildlife Fund/Wildlife Conservation Society.

Estes, J.A., Smith, N.S., and Palmisano, J.F. (1978). Sea otter predation and community organization in the western Aleutian Islands, Alaska. *Ecology*, **59**, 822–833.

Estes, J.A., Terborgh, J., Brashares, J.S., et al. (2011). Trophic downgrading of planet Earth. *Science*, **333**, 301–306.

European Parliament. (2009). Resolution on Wilderness in Europe. P6_TA(2009)0034 Adopted 3 February 2009; final version posted 13 October 2009. www.europarl.europa.eu/sides/getDoc.do?pubRef=-//EP//NONSGML+TA+P6-TA-2009-0034+0+DOC+PDF+V0//EN (accessed 3 September 2015).

Ewing, B., Goldfinger, S., Oursler, A., Reed, A., Moore, D., and Wackernagel, M. (2009). *The ecological footprint atlas 2009*. Oakland, CA: Global Footprint Network.

Featherstone, A.W. (2014). Rewilding in the Scottish Highlands. *Environmental Scientist*, **23**, 15–19.

Foote, J. (1990). Radical environmentalists are honing their militant tactics and gaining followers. *Newsweek*, **115**, 24.

Foreman, D. (1992). Around the campfire. *Wild Earth*, **2**, 3.

Foreman, D. (2004). *Rewilding North America*. Washington, DC: Island Press.

Foreman, D., and Wolke, H. (1992). *The big outside*, 2nd edition. New York, NY: Harmony Press.

Foreman, D., Davis J., Johns, D., Noss, R., and Soule, M. (1992). The Wildlands Project mission statement. *Wild Earth*, Special Issue, **1**, 3–4.

Fraser, C. (2009). *Rewilding the world*. New York, NY: Henry Holt/Metropolitan.

Fraser, C. (2012). Africa's Ambitious Experiment to Preserve Threatened Wildlife. http://e360.yale.edu/feature/kaza_ambitious_africa_experiment_to_preserve_threatened_wildlife/2527/ (accessed 26 September 2015).

Fromm, E. (1964). *The heart of Man*. New York, NY: Harper & Row.

Gondwana Link. (2015). www.gondwanalink.org/ (accessed 17 February 2018).

Gray, M., and Rutagarams, E. (2011). *20 years of IGCP: lessons learned in mountain gorilla conservation*. Kigali, Rwanda: International Gorilla Conservation Program.

Great Elephant Census. (2014). Elephant decline. http://news.nationalgeographic.com/news/2014/07/pictures/140729-tigers-conservation-cubs-hunting-environment-science/ (accessed 1 September 2015).

Harris, M. (1975). *Culture, people, nature*, 2nd edition. New York, NY: Crowell.

Harris, M. (1977). *Cannibals and kings*. New York, NY: Random House.

IUCN. (2014). www.iucn.org/about/work/programmes/gpap_home/gpap_quality/gpap_pacategories/ (accessed 30 June 2015).

Johnson, A.W., and Earle, T. (2000). *The evolution of human societies*, 2nd edition. Stanford, CA: Stanford University Press.

Jones, N. (2011). Human influence comes of age. *Nature*, **473**, 133.

Jordan III, W.R., and Lubick, G.M. (2011). *Making nature whole*. Washington, DC: Island Press.

Kavango Zambezi (KAZA) Transfrontier Conservation Area. (2017). Tourism without borders. www.kavangozambezi.org/index.php/en (accessed 14 December 2017).

Klink, C.A., and Machado, R.A. (2005). Conservation of the Brazilian Cerrado. *Conservation Biology*, **19**, 707–713.

MacArthur, R.H., and Wilson, E.O. (1967). *The theory of island biogeography*. Princeton, NJ: Princeton University Press.

McKee, J. (2012). The human population footprint on global biodiversity. In Cafaro, P. and Crist, E. (Eds.), *Life on the brink* (pp. 91–97). Athens, GA: University of Georgia Press.

McLaren, B.E., and Peterson, R.O. (1994). Wolves, moose and tree rings on Isle Royale. *Science*, **266**, 1555–1558.

Menon, V., Tiwari, S.K., Easa, P.S., and Sukumar, R. (2005). *Right of passage* (Conservation Reference Series No. 3). New Dehli: Wildlife Trust of India.

Miller, B., Reading. R., Strittholt, J., et al. (1998). Focal species in design of reserve networks. *Wild Earth*, **8**, 81–92.

Monbiot, G. (2013). *Feral*. London: Allen Lane.

Nash, R. (1982). *Wilderness and the American mind*, 3rd edition. New Haven, CT: Yale University Press.

Nash, R. (2014). *Wilderness and the American mind*, 5th edition. New Haven, CT: Yale University Press.

National Geographic. (2014). Tigers. http://news.nationalgeographic.com/news/2014/07/pictures/140729-tigers-conservation-cubs-hunting-environment-science/ (accessed 1 September 2015).

NatureServe/Landscope America. (2014). Disappearing landscapes. www.landscope.org/explore/ecosystems.disappearing_landscapes/ (accessed 3 September 2015).

Newmark, W.D. (1995). Extinction of mammal populations in western North America. *Conservation Biology*, **9**, 512–526.

Noss, R. (1992). The Wildlands Project land conservation strategy. *Wild Earth Special Issue*, **1**, 10–25.

Noss, R., and Cooperrider, A. (1994). *Saving nature's legacy*. Washington, DC: Island Press.

Noss, R.F., Quigley, H.B., Hornocker, M.G., Merrill, T., and Paquet, P.C. (1996). Conservation biology and carnivore conservation in the Rocky Mountains. *Conservation Biology*, **10**, 949–963.

Noss, R.F., Dobson, A.P., Baldwin, R., et al. (2012). Bolder thinking for conservation. *Conservation Biology*, **26**, 1–4.

Oates, J. (1999). *Myth and reality in the rainforest: how conservation strategies are failing in West Africa*. Berkeley, CA: University of California Press.

Palomares, F., Gaona, P., Ferreras, P., and Delibes, M. (1995). Positive effects on game species of top predators by controlling smaller predator populations: an example with lynx, mongooses, and rabbits. *Conservation Biology*, **9**, 295–305.

Pikunov, D.G., and Miquelle, D. (2001). Conservation of Amur tigers and Far Eastern leopards in the Tumen River Area, northeast Asia. Unpublished paper on file with author. Presented at Second Workshop on Environmental Peace in Northeast Asia, 28–31 August 2001. Vladivostok, Russia.

Potapov, P., Yaroshenko, A., Turubanova, S., et al. (2008). Mapping the world's intact forest landscapes by remote sensing. *Ecology and Society*, **13**, 51–67.

Pringle, R.M. (2017). Upgrading protected areas to conserve wild biodiversity. *Nature*, **546**, 91–99.

Raina, S. (2014). How an oasis was created in a rocky desert. *New Indian Express* (29 June). www.newindianexpress.com/nation/How-an-Oasis-was-Created-in-a-Rocky-Desert/2014/06/29/article2306271.ece (accessed 24 July 2015).

Rajeev, K.R. (2014). Kerala's first rewilding project comes to fruition in Wayanad. *Times of India* (7 July). http://timesofindia.indiatimes.com/city/kozhikode/Keralas-first-rewilding-project-comes-to-fruition-in-Wayanad/articleshow/37933932.cms (accessed 24 July 2015).

Rewilding Australia. (2015). Rewilding Australia – about. www.rewildingaustralia.com.au/about-rewilding-australia/ (accessed 24 August 2015).

Rwanda, Republic of. (2015). Seven lions released into Akagera National Park. www.gov.rw/news_detail/?tx_ttnews%5Btt_news%5D=1268&cHash=b87329041cc3166297005a5b34640463 (accessed 3 September 2015).

Sanctuary Asia. (2013). Special Tiger Award: Azhar Sheikh. *Sanctuary Asia*, **33**(6) (December). www.sanctuaryasia.com/people/earth-heroes/9657-azhar-sheikh.html (accessed 24 July 2015).

Sanday, P.R. (1981). *Female power and male dominance*. Cambridge: Cambridge University Press.

Sandom, C., Donlan, C.J., Svenning, J.C., and Hansen, D. (2013). Rewilding. In McDonald, D.W. and Willis, K. (Eds.), *Key Topics in Conservation Biology 2* (pp. 430–451). Hoboken, NJ: Wiley.

Shepard, P. (1982). *Nature and madness*. San Francisco, CA: Sierra Club Books.

Soulé, M., and Noss, R. (1998). Rewilding and biodiversity: complementary goals for conservation. *Wild Earth*, **8**, 19–28.

Soulé, M., and Terborgh, J. (1999). *Continental conservation*. Washington, DC: Island Press.

Terborgh, J. (1988). The big things that run the world – a sequel to E. O. Wilson. *Conservation Biology*, **2**, 402–403.

Terborgh, J. (1999). *Nature's requiem*. Washington, DC: Island Press.

Terborgh, J., and Estes, J.A. (2010). *Trophic cascades: predators, prey, and changing dynamics of nature*. Washington, DC: Island Press.

Terborgh, J., Estes, J.A., Paquet, P., et al. (1999). The role of top carnivores in regulating terrestrial ecosystems. *Wild Earth*, **9**, 42–56.

Vest, J.H.C. (1985). Will-of-the-land: wilderness among primal Indo-Europeans. *Environmental Review*, **9**, 323–329.

Weaver, L.C., and Skyer, P. (2003). Conservancies: integrating wildlife and land-use options into the livelihood, development and conservation strategies of Namibian communities. http://pdf.usaid.gov/pdf_docs/Pnacx280.pdf (accessed 3 September 2015).

Wild Europe Initiative (WEI). (2013). *A working definition of European wilderness and wild areas*. London: Wild Europe Initiative.

Wild Europe Initiative (WEI). (2017). Conference on old growth forest protection strategy. www.wildeurope.org/index.php/wild-areas/old-growth-forest-protection-strategy (accessed 18 February 2018).

Wilderness Act. (1964). Public Law 88-577 (16 U.S.C. 1131-1136) 88th Congress, Second Session 3 September 1964.

Wilderness Society Australia. (2002). *WildCountry* (brochure). Hobart, Tasmania: The Wilderness Society.

Wildlands Network. (2000). *Sky Islands Wildlands Network Conservation Plan*. Tucson, AZ: The Wildlands Network.

Wildlands Network. (2003). *New Mexico Highlands Wildlands Network Vision*. Richmond, VT: The Wildlands Network.

Wildlands Network. (2004). *Heart of the West Wildlands Network Vision*. Titusville, FL: The Wildlands Network.

Wildlands Network, Southern Rockies Ecosystem Project, and Denver Zoo. (2003). *Southern Rockies Wildlands Network Vision*. Golden, CO: Colorado Mountain Club Press.

Wildlands Project. (1994). *The Wildlands Project* (brochure). McMinnville, OR: The Wildlands Project.

Wildlife Conservation Society. (2000). *Ecological corridor of the Americas: linking landscapes for the new millennium*. New York, NY: Wildlife Conservation Society.

Wildlife Foundation. (2000). The Wildlife Foundation. Unpublished report on file at: Wildlife Foundation Office, PO Box 32/34, Khabarovsk, 680054, Russia.

William, S. (2015). Ecotourism project begins in Umred-Karhandla Wildlife Sanctuary. *Ecowanderlust*, 15 March. http://ecowanderlust.com/news/ecotourism-project-umred-karhandla-wildlife-sanctuary/3335 (accessed 24 August 2015).

Wilson, E.O. (1987). The little things that run the world (the importance and conservation of invertebrates). *Conservation Biology*, **1**, 344–346.

Worboys, G.E., Francis, W.L., and Lockwood, M. (2010). *Connectivity conservation management*. Abingdon, UK: Earthscan/Routledge.

World Wildlife Fund (WWF). (2014). *Living Planet Report 2014*. Gland, Switzerland: World Wildlife Fund.

World Wildlife Fund (WWF). (2016). *Living Planet Report 2016*. Gland, Switzerland: World Wildlife Fund.

CHAPTER THREE

For wilderness or wildness? Decolonising rewilding

KIM WARD
University of Plymouth

While rewilding has emerged as a popular and controversial conservation strategy in recent years, it has been lambasted by critics for advocating 'wilderness' preservation, and for its supposed preference of rewilding 'back' to a particular type of nature (see e.g. Jørgensen, 2015, who states Rewilders 'still want to re-create a wild without people and are oblivious to the problematic nature of the wilderness construct'). Both 'wilderness' and 'going back', critics argue, are bound up in a drive to create a 'pristine' nature, a nature before human inhabitation. While defenders of wilderness (see in particular Foreman, 1998) have historically hit back at such critiques as a 'war against nature', it is important to acknowledge that wilderness, while a seemingly innocent and objective material reality, is a concept indelible with symbolic meaning. In Western environmental narratives at least, the 'wilderness idea' was led by Euro-American men within the historical–cultural context of patriarchal colonialism, and wilderness preservation is, therefore, an artefact of colonialism that can (and has) act as a vehicle for the exclusion and erasure of people and their histories from the land (Merchant, 1980; Cronon, 1996; Plumwood, 1998; Adams and Mulligan, 2003). Critics of approaches to rewilding then have warned against the anti-humanist sentiment implicit in rewilding narratives that appear to advocate 'wilderness' rather than 'wildness', a linked but distinct concept that is qualified by non-human autonomy rather than the categorisation of humans and nature into the conceptually separate, distinct and pure sphere of wilderness (Prior and Ward, 2016; Prior and Brady, 2017).

This chapter begins by tracing the term 'wilderness' within Western environmentalism to reveal its conceptually problematic nature. It then proposes an alternative conceptualisation of 'wildness' as more useful in the practical and future-oriented implementation of rewilding initiatives. The final section explores how 'wilderness' and dualistic thinking are embedded within rewilding narratives in North America where rewilding has taken hold of public and conservation imaginations; however, in the European context, with its distinct cultural–historical expressions of the wild, a more hybrid, open-ended, and 'borderland' version of 'wildness' can be found, which appears to be more sensitive to existing cultural interpretations, social impacts, and indigenous livelihoods in designated 'wild' places. The chapter ends by addressing some of the tensions that arise in rewilding narratives due to the tendency for both some rewilding advocates and critics of rewilding to conflate wildness with wilderness.

Imagining 'wilderness'

Wilderness is a term that evokes a collective imaginary of 'wild' landscapes, areas vast and uninhabited by humans: spectacular National Parks, lush rainforests, desolate canyon valleys or mountainous forest regions. Most often we think of these spaces as being 'pristine' or untouched by humans. Indeed, these are imaginaries so deeply ingrained in human consciousness that they have been firmly embedded in the historical narratives driving forward Western environmental activism and global conservation policy over the last century. Yet beyond this immediate imaginary 'wilderness' is an elusive entity, something that assumes particular qualities or evokes particular feelings or moods (Nash, 1982). Wilderness is, critical social scientists would argue, our subjective and interpretive experience of the natural world, an interpretation that reflects the sociopolitical values and cultural hegemonies of our time (Merchant, 1980; Cronon, 1996; Plumwood, 1998; Adams and Mulligan, 2003). Yet despite this, to many scientists and environmentalists, wilderness is an unproblematic category of nature and to suggest that wilderness, as an idea, is a 'human creation', most famously by Cronon (1996), is not always well received by proponents of wilderness preservation.

In acknowledging that the idea of Western Wilderness (from herein this notion of Wilderness will be capitalised) is a 'human creation', this chapter explores the changing meanings and morality embedded within the concept. An examination of this social constructivist claim is vital if we are to bring to bare the types of human values and politics deep-rooted in modern conservation practice which are built upon wilderness narratives.

What is Nature? The Constructivist perspective

In order to understand how and why Wilderness has been so vital to modern environmentalism, we first need to understand how Nature itself has been conceptualised historically in the Western world. If the value of Nature is relative to the cultural context within which it is known (as social constructivists claim), then it is important to document how understandings of the concept have changed over time and how this has affected the way people recognise, know, and interact with 'wild' natures. The following sections briefly outline the historically situated cultural understandings of Nature and Wilderness.

Much modern conservation policy has been built upon an environmental ethics that holds normative assumptions about what Nature is. Studies of Nature informed by scientific philosophy often hold the position that Nature exists as a determinate entity, independent of human consciousness. This Rationalist position has instructed the protection and preservation-based approaches that we have seen historically in conservation policy and practice and upon which wilderness preservation is based (Callicott, 1984). In particular, Rationalism in environmental ethics has oft held up 'Wilderness' as the key site for protection and preservation of 'independent natures', where Wilderness is imagined as pristine, uninhabited and/or uninhabitable space, separate and autonomous from human thought (King, 1990; Cronon, 1996).

This human–nature distinction has a complex history in Western thought. While Classical scholars such as Aristotle sought to categorise and order forms of life into a hierarchical continuum, most importantly for the consolidation of the Rationalist dualism of humans and nature was the rise of modernity and the Enlightenment period, where duality was underpinned by Cartesian thinking (Whatmore, 1997; Wolch and Emel, 1998). This system of thought categorised the mind (as immaterial/thinking) and body (material/unthinking) as distinct and mutually exclusive entities and enabled the world to be constructed as subject–object relationships. The logical extension is a reductionist position of 'Self' as a rational, autonomous subject and 'Other' as the 'natural' radical negation of Self, whereby the Other must be passive and morally inferior (Gellner, 1992; Gerber, 1997; Plumwood, 1998). The importance of Cartesian dualisms in Western Culture, as 'a fault-line that runs through its entire conceptual system', cannot be understated in the development of contemporary human relationships with the natural environment in the Western context (Plumwood, 1993). In normative understandings of nature, nature is firmly positioned as the 'Other' in relation to 'Culture'. Such dualistic logics (re-)enforce cultural norms and moral justification that subjugate the Other in relation to the Self, e.g. Mind over Body, Self over Other, and Culture over Nature (Harraway, 1991; Rose, 1993).

From this discussion we can see that one issue with Rationalism in environmental ethics is that ideals of Nature are situated in opposition to Humans, as something distinctly material and separate from human consciousness. To consider Wilderness as socially constructed, in contrast, is based on the philosophical position that our perceptions and interpretations of the environment are relative to the social and cultural contexts within which they are made 'knowable'. This means that what we understand to be Nature is in itself is an interpretation based upon our societal, emotional, technological, and intellectual experiences and perceptions of the natural world (Cronon, 1996; Braun and Castree, 1998). This is to say that humans conceptualise, understand, and interpret meanings of Wilderness in reflection of the historically and morally situated frameworks within which they are thought and experienced. This is particularly important in understanding how wilderness has been conceptualised throughout the rise of modern environmentalism.

The rise of modern environmentalism: Romantic conceptions of Wilderness

The Wilderness concept to which this chapter refers was largely born out of America. Therefore, this chapter will focus on the evolution of the concept within Euro-American narratives. This is not to say that Wilderness is simply a Euro-American notion; indeed, metaphysical notions of wilderness are rooted in a variety of cultures across the world (Callicott and Ames, 1989; Braun, 2002). The American Wilderness concept is focused on here due to its influence on global environmental policy and the advent of rewilding narratives.

Romanticising Wilderness: from Europe to America

By the nineteenth century rural areas colonised by European settlers from the 1600s onwards were becoming emphatically transformed by urbanisation and industrialisation. North America had begun its 'urban revolution'. As large swaths of the population migrated to cities to live and work, people's perceptions of rural 'wild' lands began to change, from that of a desolate and God-less place (a powerful imaginary generated through European Judeo-Christian perceptions of Wilderness) (Merchant, 1980) to one that idealised and appreciated Wilderness as a sanctuary from modern urban life. Wilderness, as Merchant (1980) famously purports, became Eden on Earth. This heavenly conceptualisation of Wilderness was most visibly depicted through Euro-American Romanticism and formed the birth of the wilderness preservation movement.

The endeavour of romanticising wilderness drew heavily on the notion of sublimity, much influenced by the publication in 1757 of *A Philosophical Enquiry into the Origins of Our Ideas of the Sublime and the Beautiful* (Burke, 1757). Indeed, Burke's book was essential reading for European painters and travel writers of the early nineteenth century for whom the natural world was the focus of their art (Cronon, 1996; Fulford, 1999). For Burke, a sublime Wilderness then was one of terrifying rapture, awe, and reverence in the face of the infinity, power, and mystery of nature. The beauty and spiritual truths that could touch any Man who dared to linger in such environments was much celebrated in sublime Romantic writings and paintings of the period. For example, the sublimities of nature are expressed in the work of early Romantic poets such as Percy Bysshe Shelley, encapsulated in the first lines of his poem *Mont Blanc: Lines Written in the Vale of Chamouni*:

Mont Blanc yet gleams on high: –the power is there, The still and solemn power of many sights,
And many sounds, and much of life and death.
(Shelley, cited in Donovan and Duffy, 2017)

Shelley, like many Romantic poets of this period, stressed the wild and transcendental nature of Wilderness that inspired awe as well as spiritual melancholy. Shelley evokes a sense of divinity in nature and cannot avoid the terminology of religion so important to Classical conceptions of Wilderness. Yet while God is present, particularly in the early sublime Romantic writings such as this, the Wilderness itself is evoked as a place of awe-inspiring spirituality, a kind of substitute for the Christian religion, if you will. This substitution of Nature for God in part reflects the modern Enlightenment doctrine; by emphasising the belief that nature exists as a naturally ordered and hierarchical world, it eliminated the 'irrational' tendencies of religious doctrine in Classical conceptions of wilderness (Merchant, 1980; Cronon, 1996). These were the beginnings of the Euro-American cravings for Wilderness.

As Romanticism evolved in America, the terrifying aspects of sublimity gradually receded into a 'domesticated sublime' construction of Wilderness famously critiqued by Cronon (1996). American nature writing of the late Romantic period embodies this Romantic enthusiasm for the wilderness as a divine sanctuary from modern urban life; where the power of the sublime is softened and Wilderness is conceptualised as a beautiful, sentimental, and divine place. This 'domesticated sublime' is a sentiment best evoked by Scottish-American nature writer and conservationist John Muir. In his use of similes and metaphor like 'a lake of pure sunshine' and 'from the blue sky to the yellow valley smoothly blending as they do in a rainbow, making a wall of light ineffably fine' and comparing the Sierra to 'the wall of some celestial city' (1912, chapter 1, *The approach to the valley*), you could be forgiven for mistaking it as a love poem, although rather being written to a human companion, it is written to the wilderness of Yosemite. Olwig (1996) contends that Muir, like many nature writers of the Romantic period, equates love for Yosemite with the moral valorisation of nature. And like any star-crossed lover, Muir puts Yosemite, and the idea of Wilderness, on a moral pedestal. Merchant (1980) tells that gendered constructions of wilderness as pristine and virginal land to be tamed and domesticated can be understood in the colonial context, where the presence and agency of indigenous peoples is erased and the land is violently transformed through the 'Othering' of nature, legitimising land as 'empty' and therefore ripe for conquer and active domestication (Plumwood, 1998). These are critiques that shall be revisited when analysing wilderness preservation in practice.

Masculinity and Romanticism: the American Wilderness

Particular ideas of masculinity and femininity have also been a vital vehicle for driving forward particular conceptualisations of Wilderness. Nowhere is this more evident than in the American Frontier Myth. The American 'frontier' was officially closed from the late 1800s and subsequently numerous male writers sought to preserve and perpetuate a sense of the frontier imaginary in wilderness narratives (Brandt, 2017). Writings of the time show that much of the attraction to 'wild' areas by urban, white American men was nostalgia for an imagined frontier life, a life of freedom, rugged masculinity and individualism, physical endeavour and conquest (Plumwood, 1993; Rifkin, 2014). No man can better illustrate this perception of American wilderness than in the writings of Theodore Roosevelt, 26th President of the United States (Brinkley and Holland, 2009). Roosevelt wrote a number of books and essays that inscribe masculine endeavour into Wilderness narratives. As a politician and President he was a fundamental figure in the development of wilderness preservation legislation, heavily influenced by his close friend John Muir. The mythical 'wild pioneer' captured the American imagination with cowboys and pioneers upheld as national heroes and the rugged American hero continues to permeate American popular culture today (Wright, 1993). In these cultural artefacts masculine identity is linked with the conquest and subjugation, and/or protection of a domesticated wilderness, and was epitomised by masculine endeavours such as hunting. As such, hunting and military tactics codified wilderness as a colonial, masculine space, as Roosevelt himself strove to protect and preserve wilderness from human inhabitation while indulging in big game hunting. Indeed, the colonial tactic of preserving wilderness to secure property ownership and the right to hunt would become a key criticism of the global National Park movement (Neumann, 2002; Adams and Mulligan, 2003).

Romantic Wilderness as a basis for preservation policies

While the concept of Wilderness can be traced back to eighteenth-century European Romanticism, the concept came to the fore during the late nineteenth century, where preserving nature in its so-called wild and 'natural' state became the cornerstone of American environmental approaches. Leading advocates and architects of wilderness preservation and the Wilderness Movement in America were the writers and activists John Muir and Henry David Thoreau, as well as Emerson and Leopold. In particular, the power of Muir's emotive and sentimental prose, along with other writers of the time, cannot be underestimated (Callicott and Nelson, 1998; Phillipon, 2005). John Muir founded the Sierra Club in 1892, whose manifesto was built on the call to 'Save the Wilderness' from the increasing encroachment of infrastructure and industrial tourism (Woods, 2007). Paradoxically, tourism and the availability of 'wild nature' for the masses to enjoy was a driving factor in the political development of the preservationist movement in America.

The Sierra Club grew into a political force to be reckoned with. Romantic conceptions of Wilderness, for example, were pivotal in establishing the protection of Yosemite Valley via the Yosemite Grant in 1864, and The Wilderness Movement, led by Muir, was a driving force behind its successful designation as a National Park in 1890. In particular, John Muir along with other writers belonging to The Wilderness Society and Sierra Club drove forward the American (and global) appetite for the protection and preservation of what critical social scientists would describe as domesticated sublime and romantic natures which formed the basis of the American and global National Park movement. Indeed, once Yellowstone became America's first National Park on 1 March 1872, it stood as a model for the global expansion of America's 'big idea' (Stevens, 1997; Wuerthner, 2015). After the designation of Yellowstone, National Parks were developed across the world, particularly in colonial locations across Australia, Africa, Canada, and New Zealand in the late 1800s.

Critical reflections

So far, we could be forgiven for understanding Romantics' conceptions of Wilderness as innocent musings to give space to autonomous and beautiful nature. Yet these narratives are far from innocent, and embed within them a particular way of understanding *what* and *who* belongs in Nature. The cultural attitudes and values embedded in Romantic writing and painting informed and mobilised a particular environmental ethics evoking Wilderness as 'true nature' set apart from humans. This in turn informed conservation efforts of the late nineteenth and early twentieth centuries on the preservation of large tracts of land or, more specifically, wilderness preservation. The solution to environmental crisis under this environmental philosophy is to prioritise a human-less wilderness over a nature-less civilisation. Wilderness in this sense is dangerous, if it is pursued with a sense of absolutism. The philosophical separation of humans in these preservation narratives hinges on the idea of nature as a binary opposite of society, therefore symbolically and materially placing humans strictly, and often violently, outside of preservation areas.

Wilderness legislation was inscribed with philosophical foundations that articulate and defend particular values: embodying notions of a domesticated sublime, rugged masculinity, and individualism as well as capitalist and exclusionary colonial tactics (Sutter, 1999). These latter two will now be discussed further in relation to the American National Park project.

It is important not to conflate National Parks with wilderness. Wilderness areas can exist within National Parks, but National Park status isn't a prerequisite for wilderness. However, it should be noted that the Wilderness imaginary was vitally important to the American National Park concept and preservationist policies to follow in America and globally. Yellowstone, and what became known as the 'Yellowstone Model', was characterised by an exclusionary nature emanating from Euro-American ideas on property rights, colonialism, and Nature (Mackenzie, 1988; Adams, 2004). Rather than a move purely to protect an undisturbed wilderness, scholars note that Yellowstone was established with a political concern for establishing federally owned lands in a protectivist move from the private exploitation of land seen under America's Gilded Age (Germic, 2001). This move was underlined by a profit motive, to create a 'wilderness experience' to be enjoyed by American society through tourism and the beginnings of private–federal partnerships that fostered nature-based tourism based on romantic Wilderness imaginaries (Farrell, 2015).

One of the central features of Yellowstone National Park at the time of inception was erasure of the indigenous cultural landscapes, as well as now infamous instances of removal of indigenous Americans themselves (Meyer, 1996). Spence (1999) details the political backdrop of the Park Act, situated among heightened concerns that land be protected against potentially violent indigenous claims of 'ownership', which, in Yellowstone's case, would impact upon and frighten tourists. Consequently, in a move to prevent indigenous Americans entering the Park, a military post was installed at Yellowstone's western boundary in 1879. In this sense the Wilderness sold to tourists through the imagery of Yellowstone as a pristine and uninhabited space of nature is a politicised imaginary that erases the historical presence of indigenous people on such land for centuries prior to European colonisation. Consequently, wilderness preservation narratives within the National Park Model have the potential to negate the association of (indigenous) people from the landscape (Stevens, 1997; Dove, 2006). Indeed, beyond Yellowstone, attempts to own, protect, and preserve wilderness have often been accompanied by historical exclusion and dispossession from the land of indigenous people and accompanied by the profit motivation for nature-based tourism, recreation, and resource management (Stevens, 1997; Watson et al., 2003; Binnema and Niemi, 2006; West et al., 2006; Igoe et al., 2010).

The Romantic notions of Wilderness Muir inspired, as pristine and uninhabited spaces, assumes a virgin land before European conquest as well as unacknowledging the thousands of years of impact that pre-Columbian cultures had on the American landscape. This has the effect of valorising a mythical 'pristine' Wilderness and removes ownership and agency over the land from indigenous peoples. The Yellowstone model – inscribed with particular cultural politics and moralities that have the potential to violently erase the history of indigenous people from the land, both materially and culturally – was subsequently rolled out on the colonial map. Kruger National Park, the oldest National Park in Africa, was designated in 1898 as part of the broader appropriation of land and natural resources and named after Paul Kruger, an important military figure and statesman in Afrikaner history (Neumann, 1998). Indigenous people living within the boundaries of Kruger National Park were dispossessed and dislocated from their homeland under colonial conservation laws, such as the criminalisation of traditional hunting, wood collection, and cattle grazing on National Park land as part of a wider move to secure the land for tourist and recreational activities (Neumann, 1998). Not only did the expansion of the National Park movement into South Africa dramatically restructure property relations, but it also alienated indigenous people from the historically available resources to support their livelihoods as well as the gradual erosion of traditional knowledge (regarding hunting and grazing). More recently within and beyond academia, indigenous efforts to reclaim territories and rights have been both highly visible and controversial, especially when they come into conflict with state/corporate interests (see Standing Rock). As Neumann (1998) notes, '[r]epresentations of a harmonious, untouched space of nature, [which] mask the colonial dislocations and obliterate the history of these dislocations, along with the history of those spaces that existed previously'.

For wildness not Wilderness: decolonising rewilding

The preceding account has unpacked how deeply held social, cultural, and political values have projected and inscribed onto 'wilderness' idealised forms of Nature that consequently have been (re)made through environmental policy narratives. In doing so it has shown that the value we place on 'wilderness' is not construed from Nature/wilderness in itself as an independent entity, but from the sociocultural and political matrixes within which we have mediated our understanding of the term. In this mediation, Wilderness has been positioned as the 'Other' in the Nature–Culture dualism. This oppositional conceptualisation works to construct the wilderness into a subject–object relationship that is embedded with hegemonic ideologies. Critically analysing Wilderness narratives asks us to reflect on a series of questions when thinking about rewilding conservation practice. The first, of course, is whether rewilding also collapses Nature and Society into binary dualisms and, if so, to understand the consequences of this for how we morally 'view' and rewild nature in this guise. What values and assumptions are embedded within rewilding conservation? How useful is the term Wilderness to rewilding narratives?

Defining Wilderness as solely a social construct does not give space to non-human autonomy within its narrative (Whatmore, 2002; Booth, 2011). While Romanticism, frontierism, and religiosity are fundamental to the way in which wilderness is – and has been – constructed and classified by humans, to allow the semantic dominance of the term Wilderness, and its associated symbolic valence, within rewilding narratives is to deny agency to more-than-human forms of life within these narratives (Whatmore, 2002). While Plumwood (1998) argues that concepts of nature and wilderness should not be abandoned, but need to be situated 'within the context of a renewed, radical ecology committed to healing the nature/culture split and ending the war on the Other', others argue that a conceptual apparatus that acknowledges cultural expressions of Nature while allowing more room to more-than-human agency and intersubjectivity in the co-production of 'wild' spaces is necessary (Whatmore, 2002; Castree, 2003; Lorimer, 2015).

Both 'wildness' and wilderness are to some extent important to the ethos of rewilding; however, I argue that understanding and valorising 'wildness' as the key element of rewilding, rather than wilderness, would be useful in moving the rewilding agenda towards practices that do not fall back on an imaginary space of purity, but instead open up the possibility for co-producing spaces of 'wild' nature in a radical ecology committed to healing, as Plumwood argues, the philosophical split between Nature and Culture in Western environmental narratives.

'Wildness' has its own historical set of sociocultural and political associations. Henry David Thoreau famously asserted: 'In Wildness is the preservation of the world' (Thoreau, 1974 [1862]). It is true that Thoreau valorised the idea of wilderness in a way that cemented him as an icon of American Environmentalism (Bennett, 2000). Yet Thoreau does not use the term wilderness here, as is often stated; instead, he uses the term 'wildness'. While initially these terms might appear interchangeable, it should be argued that wilderness and wildness should not be understood as the same thing (Chapman, 2006; Prior and Brady, 2017). If wilderness is underpinned by a dualistic separation of Culture from Nature, then conflating wildness with wilderness also implies this dichotomy. This subsequently implies that 'wild places' or 'wildness' can only be realised when humans are excluded in time and space from Nature. This is a troublesome narrative that is marked by many of the same problems/critiques I've alluded to. Consquently, in the endeavour of developing more positive and socially just conservation practice in rewilding we can valorise Wildness rather than Wilderness to renegotiate our understanding and relationships with non-human nature in ways that are not dualistic, exclusionary, or indeed, loaded with cultural baggage. In order to do this we must conceptually understand Wildness as embodying three qualities: (1) wildness as a relational concept, (2) wildness as a borderland, and (3) wildness as autonomy.

Wildness as relational

By better distinguishing between Wilderness and Wildness as different terms in this way, we are able to understand 'wildness' in a relational, rather than binary, sense. As Chapman (2006) purports:

> Unhappily, environmental restoration turns out to be paradoxical under the current identification of wilderness with wildness where wildness is, at least, a necessary condition for the possession of natural value. The solution to the paradox is to separate wilderness from wildness both conceptually and ontologically by enlarging the domain of wildness to include certain human activities.

It is true that the term 'wildness' contains implicit historical cultural assumptions that categorise the 'wild' as Other, a process which has encompassed the 'placing' of 'wild things' in human-ordered spaces of belonging or not belonging (Urbanik, 2012). Part of this process of distancing of 'wild things' from civilisation has been the association of wildness, or wild animals, as belonging to wilderness, as spaces of a pristine nature (Woods, 2005; Chapman, 2006; Prior and Brady, 2017). Yet as many contemporary social scientists have argued, contemporary understandings of 'wildness' or 'wild-life' need to be more nuanced (Whatmore and Thorne, 1998; Latour, 2004). 'Wild-life', according to Whatmore and Thorne (1998), should be reconceived as

> a relational achievement spun between people, animals plants and soils, documents and devices, in heterogeneous social networks that are performed in and through multiple places and fluid ecologies.

In this account, wildness is abiotic, biotic, and a social relational achievement within human and more-than-human worlds. This understanding acknowledges the autonomy of more-than-humans, the social networks, multiple places and ecologies within which 'wild' life is brought into being. By reconfiguring wildness as quality exercised in relational exchanges across and within fluid ecologies rather than Other within an imaginary space of purity held at a distance from humans, we are able to open up possibilities that rewilding offers for co-producing experimental natures with non-humans outside of wilderness narratives.

Wildness as a 'borderland'

By defining Wildness as a border*land* concept, rather than the totalising border*line* concept of Wilderness–Culture, we are able to challenge the ontologies that conflate 'wildness' as something that is pure, separate, and Other from the human realm; and instead wildness is something that can be realised through topological borderlands – spaces. By engaging with understandings of wildness in a relational, topological sense, the notions of power and exclusion over wild-life configured in geometrical space is 'loosened' and wildness becomes a quality unconfined by territorial borderlines seen in wilderness management (Whatmore, 2002; Hinchliffe et al., 2013). Therefore, while wilderness is a 'border concept', a cultural concept that separates wildness from the sphere of human society, 'wildness' is not. Rewilding based on the premise of 'wilding' rather than Wilderness assumes the (re)creation of borderland spaces, through relational configurations of the human and more-than-human world. It is, then, these borderland spaces which offer the most promise for Rewilders.

Wildness as autonomy

While ecological restoration is enacted through practices of intervention and stewardship, rewilding is grounded in an ethos that relinquishing direct human management of wild organisms or ecological processes will generate better functioning ecosystems. Rewilding – unlike other ecological restoration practices – is premised on non-human autonomy; the self-willed and self-sustaining qualities of non-human Nature (Prior and Ward, 2016). In characterising wildness as autonomy, 'the more-than-human world where events, such as animals moving about, plants growing, and rocks falling occur largely because of their own internal self-expression' (Woods, 2005). Under this definition, then, wildness (of animals, plants, landscapes, or ecosystems) is premised on non-human autonomy within plural and hybrid spaces rather than the material realisation of vast pristine environments. Defining wildness in these terms again allows Rewilders to create 'wild spaces' rather than wilderness.

Wilderness or wildness in rewilding conservation?

Rewilding itself has been described by some of its fiercest proponents as a paradigmatic shift in conservation, and broadly speaking it aims to restore and/or regenerate ecosystems through reintroduction programmes (Monbiot, 2013). The ecological specifics of rewilding will not be discussed here (as they will be explained in detail elsewhere in the book); however, it is important to note that behind the fervent environmental discourse from pro-wilding scientists and activists the concept of rewilding is deeply imbued with cultural, political, and ethical values. Many rewilding practices, generally speaking, do not outwardly reproduce the aims of wilderness management, or indeed seek to construct pure spaces of Nature (see Prior and Ward, 2016). However, critics of rewilding have claimed that rewilding conservation seeks to reconstruct a mythical and fundamentally flawed Nature–Culture binary. According to Jørgensen (2015), Rewilders:

want to re-create a wild without people and are oblivious to the problematic nature of the wilderness construct. Rewilding as activist practice attempts to erase human history and involvement with the land and flora and fauna, yet nature and culture cannot be easily separated into distinct units.

Perhaps Jørgensen has a point. Emanating from North America, the first call of Rewilders was to 'restore' and 'rewild' large tracts of land in order to generate the space and connections necessary to reintroduce large carnivores. Such calls were put forward by Michael Soulé and Dave Foreman, the latter of whom co-founded the Wildlands Project in 1991, now known as the Wildlands Network, and who went on to become co-founder of the Rewilding Institute. Both organisations have been pivotal in developing the scientific ideas underlining continental-scale rewilding networks (Carver, 2013). Foreman (1981) historically co-founded the environmental group 'Earth First!', a radical movement with roots in deep ecology that had wilderness protection at the heart of its vision. Michael Soulé, a conservation biologist, while not a radical environmental activist in the same sense as Foreman, can still be recognised as a wilderness proponent, seeking to provide the scientific basis for the realisation of large core areas of protected wilderness and landscape connectivity.

The scientific ideas behind early calls to rewild America were first published in a landmark paper by Soulé and Noss (1998), where the reintroduction of megafauna, and the creation of what they term the 'three C's argument: cores, carnivores and corridors', were fundamental strategies for creating 'self regulating land communities' (Soulé and Noss, 1998):

> Our principal premise is that rewilding is a critical step in restoring self-regulating land communities ... Once large predators are restored, many if not most of the other keystone and 'habitat-creating' species (e.g., beavers, prairie dogs) ... and natural regimes of disturbance and other processes will recover on their own ... 'wide-ranging predators usually require cores of protected landscape for secure foraging, seasonal movement, and other needs; they justify bigness.
>
> (Soulé and Noss, 1998)

For Soulé and Noss, Foreman, and other proponents of the three C's approach, the protection, restoration, and connection of core areas of American 'big wilderness' would be vital to the success of rewilding projects on the North American continent, as is the introduction of wild-life. Rewilding in this context 'places' wild animals into large wilderness spaces, distancing wildlife from human-life and in doing so catagorising life forms into human-ordered spaces of belonging and not belonging.

Following Soulé and Noss' call to action, Donlan and colleagues (2006) produced a Manifesto for Pleistocene rewilding, claiming rewilding as a conservation priority for North America. In a similar vein to its predecessors, it calls for a multicontinent system of wilderness reserves and the introduction of charismatic pre-Columbian wild-life as an ecological framework. This approach has argued against the fragmentation of land seen in much sustainable management initiatives of contemporary conservation (Donlan et al., 2006).

The emphasis on designating and protecting large areas of land of both 'types' of rewilding pits wilderness conservation as the saviour of the global extinction crisis. Both the Pleistocene manifesto and the three C's approach of Soulé and Noss seek to rewild back to 'pre-human' state, a preference that radically negates humans from nature. The anti-humanism of these approaches is clear. Reflecting on earlier analysis of the way wilderness has been used to inscribe particular values on nature, there are several reasons to be cautious of such an approach to rewilding. Vast areas of protected wilderness landscape are necessary for this type of rewilding, creating enclosed and exclusionary places without regard for the people dwelling and seeking livelihoods in such places. Questions of land ownership, access rights, and elite power are raised in these narratives as vast areas of private land are marked for continental rewilding (see Donlan et al., 2006, p. 674: 'Private lands probably hold the most immediate potential' for rewilding).

While the idea of rewilding that emerged from North America has wilderness at its heart, this has not always been conceptually translated in the same way beyond the American context. It is true that most rewilding narratives take a critical stance towards the (over)management and regulation of non-human nature. However, there are significant philosophical differences between the types of rewilding advocated in America and those in Europe. Europe has a much longer history of human land use than America, and much less wilderness to 'preserve'. Indeed, according to Frank and colleagues (2007), in Europe as a whole (excluding the Russian Federation), only 1.4 per cent of forested areas are identified as 'untouched'. A long history and appreciation of cultural landscapes is also apparent within European conservation contexts (Drenthen and Keulartz, 2014). Recent academic literature has sought to put a lens on the ideology and practices of rewilding in Europe. For Europe, rewilding has emphasised the importance of the naturalistic grazing of large herbivores in the ecological restoration of European landscapes, inspired by Frans Vera's (2010) mosaic forest hypothesis. Vera suggested that Europe's ecology was characterised by a mosaic of woodland–pastures sustained by the grazing of large herbivores. In order to test this hypothesis, Vera and colleagues were able to conduct what has been termed a 'wild experiment' at Oostvaardersplassen (Lorimer and Driessen, 2014).

Oostvaardersplassen (OVP) is an 'experimental' nature reserve contained within an area of reclaimed polder, just a few miles north-east of Amsterdam. The polder had been initially marked out as an industrial development site but was gradually colonised by graylag geese as development plans fell through. The colonisation and intensive grazing of the polder by graylag geese created an 'accidental ecology' which secured its demarcation as an official nature reserve

(Lorimer and Driessen, 2014). Subsequent to this, the land was further diversified and de-domesticated as part of Vera's 'wild experiment' and 35 Heck cattle were introduced to the reserve in 1983, followed by a number of red deer in 1992. While the restoration of natural processes may be the aim of rewilding at OVP, the means to this end is through specific reintroduction of Heck cattle. The 'Heck' cattle aren't by any means natural; instead, they are the culmination of 35 years of 'back-breeding' and genetic manipulation by two German zoologists, Lutz and Heinz Heck, in a programme to recreate the aurochs[1] (Lorimer and Driessen, 2013). Such animals unsettle the supposed boundaries between Wildness and Culture, creating 'borderland beings' that challenge the ontologies that characterise 'wildness' as something that is distinct and separate from the human realm. The move to 'rewild' OVP had little to do with the concept of wilderness creation and management then, or the recreation of past 'natural' landscapes or animals in a utopic form. Instead, OVP can be placed within the 'nature development narrative' in the Netherlands, and rather than rewilding back to a pristine state aims to 'restore' or 'rewild' natural processes while creating 'new natures'. The latter point is important here as it underlines the open-ended and experimental approach instigated at OVP. This space, then, is certainly not one that equates to anti-human wilderness management or explicitly disavows humans from nature. Instead, it's a space of potential ecological surprises born out of the desire for creating 'wildness' in new natures.

Rewilding Europe (RE) is one of the leading rewilding organisations in Europe and is influenced by many of Vera's ideas. The reintroduction of keystone species (usually large herbivores or carnivores) and the availability of space and connectiveness are important to their narrative, as is the idea of rewilding through land abandonment, a particularly European inflection of rewilding (Keenleyside et al., 2010).[2]

According to Rewilding Europe's Annual Review (2016), wilderness protection is key to European approaches. They state:

Initial approaches in rewilding have shown that European ecosystems have a high potential for regeneration, while existing wilderness benefits from strict protection. Europe now has the chance to catch up with the global approach, where conservation is intrinsically linked to wilderness protection and wild nature.   (p. 7)

The implication here is that Europe has a chance to 'catch up' with Northern American approaches to wilderness protection through rewilding. However,

---

[1] An ancient bovine species that ranged across Eurasia during the glacial/interglacial intervals of the Pleistocene (Lorimer and Driessen, 2014).

[2] Indeed, debates about what to 'do' with the significant levels of farmland abandonment across Europe, and criticisms that EU subsidies paid to farmers for the upkeep of traditional pastoral practices were outdated and costly, has led rewilding proponents to advocate for rewilding through land abandment as an ecological practice (Merckx and Pereira, 2015; Navarro and Pereira, 2015).

the European conservation context has historically valorised and conserved 'semi-natural landscapes' prized for their long-standing cultural histories; histories that 'write' cultural meaning into landscapes and show an appreciation for the historical co-production of environments by both humans and nature. Indeed, Drenthen (2005) notes that European rewilding projects may not have the explicit *goal* of erasing human history and human involvement in the land, but have the potential to destroy historically important and meaningful places through the removal of particular cultural artefacts from the land. This may be a particular case in point when foregrounding rewilding strategies in ideas of Wilderness.

Tanasescu's study (2017) of Rewilding Europe's Romanian Danube Delta (RDD) project explores Rewilding Europe's vision for the RDD and the tensions and difficulties encountered on the ground in the conceptual and practical meaning of the term rewilding. According to Tanasescu, Rewilding Europe is an organisation curating the creation of a 'wilderness *spectacular*'; a representation of rewilded natures that romanticise particular versions of nature built upon cultural hegemonies that valorise 'wild' and people-less places. This spectacular uses the emotional appeal of 'wild spaces', the aesthetic-ethical appreciation of so-called 'wild landscapes' not only to drive forward a populist vision for rewilding, but more importantly as a driving force in the nature-based tourism aspect of Rewilding Europe. Indeed, the development of nature-based economies, largely through tourism, is key to the Rewilding Europe Vision (Rewilding Europe, 2016). In rewilded narratives, then, it is important to be aware of the past lessons learned from nature-based tourism and scholars who argue that nature-based tourism driven by marketisation of nature also comes with concerns over land ownership, uneven power relations between local people and conservation businesses, and the potential for an obsession with economic growth that can lead to socioecological damage driven by imaginary wilderness narratives (see Brockington and Igoe, 2006; Neves and Igoe, 2012).

As noted by Vasile (2018), the characterisation of Rewilding Europe as a 'wilderness spectacular' may actually be at odds with the reality of rewilding in Europe on the ground, which *is* founded on notions of 'wildness' and ecological surprises, rather than wilderness creation. In particular, the introduction of autonomous grazing herbivores, pivotal to European landscapes, may also come with 'wild consequences' not palatable to the tourists seeking the emotional wonders of wilderness as identified in promotional devices of Rewilding Europe (Tanasescu, 2017). One 'wild consequence' of rewilding can be seen during the severe winter of 2010 at OVP, where images of starving animals provoked national outrage at the moral implications of allowing animals to live autonomous lives (International Commission on the Management of the Oostvaardersplassen, 2010). In the context of Rewilding Europe, this tension is encapsulated by Vasile (2018), who explores the ethical tensions for local people in negotiating their behaviour towards introduced bison in the Romanian Carpathians, as well as the Rewilders themselves. In 2016, two years after reintroduction, four of the 30 reintroduced bison were found dead and an evaluation by the team identified the cause to 'a mixture of weakness, natural selection and predation by feral dogs' (Vasile, 2018). While, according to Vasile, the project team were 'devastated' by the loss of animals, locals viewed the loss as regrettable, but perhaps a consequence of their hybrid nature; the bison were not fully 'wild' or 'domestic' in the eyes of the locals, but what I term a 'borderland being'. An animal afforded autonomy through rewilding, but an autonomy that also leads to uncertain 'ecological surprises' that might present aesthetically challenging situations for locals, Rewilders, and tourists, ranging from untidy woodlands to the visible death and decay in what Prior and Brady (2017) term 'unscenic and terrible [wild] beauty of rewilding'.

There is reason to be hopeful in these narratives also. Tanasescu's research indicates that the rewilding project in the Romanian Danube Delta has been premised on meaningful engagement with the local community, a strategy that has helped to mitigate for highly tense and emotive conflicts based on the erasure of cultural histories (Tanasescu, 2017). Vasile's research also indicates that the introduction of grazing herbivores in the form of Tauros cattle on communal lands is well liked by locals, mainly because of the animals' aesthetic charisma (Vasile, 2018). However, Vasile (2018) also documents that there was resistance within some sections of the community to the Tauros, for fear of the 'wild unknown'. This fear of 'wildness' and uncertainty has been documented in relation to other reintroduction initiatives, particularly in relation to biosecurity concerns of agricultural communities (Buller, 2008, 2013), and it will be important for Rewilders to meaningfully engage with and understand local fear narratives in relation to human–wildlife conflict rising from the autonomous character of rewilded animals.

Yet another reason to be hopeful is the recognition of wildness and wild places as relational. For example, Rewilding Europe states:

> Rewilding is not geared to reach any certain human ended 'optimal situation' or end state, nor to only create 'wilderness' – but it is instead meant to support more natural dynamics that will result in habitats and landscapes characteristic of specific area(s), with abiotic, biotic and social features that together create the particular 'Sense of the Place'.

Acknowledging that abiotic, biotic, and social features are relational actors in creating a 'Sense of Place' rejects Nature–Culture binaries and allows for a rewilding in which collaboration and co-production of a multiplicity of human and more-than-human intersubjectivities is key to the (re)creation of 'wild' nature. This is a way of thinking that acknowledges the agency and autonomy of the Other.

For wildness: going forward critically

While both ideas of Wilderness and Wildness are important to the rewilding debate on both continents, this chapter argues that it is the notion of wildness, not Wilderness, that offers Rewilders the most potential in moving towards an inclusive, future-orientated conservation approach that doesn't seek the ever-elusive goal of creating pure spaces of Nature. Understanding wildness as natural autonomy is useful in acknowledging and allowing for the independence and self-governance of non-human nature while not restricting 'wild nature' to wilderness spaces. Understanding wildness as natural autonomy, then, means that rewilding can (and should) take place in a myriad of places beyond wilderness, a move that is particularly future-orientated in a rapidly urbanising world. However, allowing for the autonomy of nature means Rewilders will also have to grapple with ways of *living with* the 'unscenic and terrible beauty of rewilding' and potential human–wildlife conflicts in new relational exchanges and 'wild borderlands'.

References

Adams, W. (2004). *Against extinction: the past and future of conservation*. London: Earthscan.

Adams, W.M., and Mulligan, M. (2003). *Decolonizing nature: strategies for conservation in a post-colonial era*. London: Earthscan.

Bennett, J. (2000). *Thoreau's nature: ethics, politics, and the wild* (Modernity and Modern Thought, Vol. 7). London: Sage.

Binnema, T.T., and Niemi, M. (2006). 'Let the line be drawn now': wilderness, conservation, and the exclusion of aboriginal people from Banff National Park in Canada. *Environmental History*, **11**, 724-750.

Booth, K. (2011). In wilderness and wildness: recognizing and responding within the agency of relational memory. *Environmental Ethics*, **33**, 283-293.

Brandt, S. (2017). The wild wild world. In Armengol, J.M., Vilarrubias, M.B., Carabí, À., and Requena, T. (Eds.), *Masculinities and literary studies: intersections and new directions* (pp. 133-153). New York, NY: Taylor and Francis.

Braun, B. (2002). *The intemperate rainforest: nature, culture, and power on Canada's west coast*. Minneapolis, MN: University of Minnesota Press.

Braun, B., and Castree, N. (1998). *Remaking reality: nature at the millennium*. London: Routledge.

Brinkley, D., and Holland, D. (2009). *The wilderness warrior: Theodore Roosevelt and the crusade for America* (p. vii). New York, NY: HarperCollins.

Brockington, D., and Igoe, J. (2006). Eviction for conservation: a global overview. *Conservation and Society*, **4**, 424.

Buller, H. (2008). Safe from the wolf: biosecurity, biodiversity, and competing philosophies of nature. *Environment and Planning A*, **40**, 1583-1597.

Buller, H. (2013). Introducing aliens, reintroducing natives: a conflict of interest for biosecurity? In Dobson, A.N.H., Barker, K., and Taylor, S.L. (Eds.), *Biosecurity: the sociopolitics of invasive species and infectious diseases* (pp. 183-198). Abingdon: Earthscan/Routledge.

Burke, E. (1757). *A philosophical inquiry into the origin of our ideas of the sublime and beautiful.* Dublin.

Callicott, J.B. (1984). Non-anthropocentric value theory and environmental ethics. *American Philosophical Quarterly*, **21**, 299–309.

Callicott, J.B., and Ames, R.T. (1989). *Nature in Asian traditions of thought: essays in environmental philosophy.* Albany, NY: SUNY Press.

Callicott, J.B., and Nelson, M.P. (1998). *The great new wilderness debate.* Athens, GA: University of Georgia Press.

Carver, S. (2013). Rewilding and habitat restoration. In Howard, P., Thompson, I., and Waterton, E. (Eds.), *The Routledge companion to landscape studies* (p. 383). Abingdon: Routledge.

Castree, N. (2003). Environmental issues: relational ontologies and hybrid politics. *Progress in Human Geography*, **27**, 203–211.

Chapman, R.L. (2006). Ecological restoration restored. *Environmental Values*, **15**, 463–478.

Cronon, W. (1996). The trouble with wilderness; or, getting back to the wrong nature. *Environmental History*, **1**, 7–28.

Donlan, C., Berger, J., Bock, C.E., et al. (2006). Pleistocene rewilding: an optimistic agenda for twenty-first century conservation. *The American Naturalist*, **168**, 660–681.

Dove, M.R. (2006). Indigenous people and environmental politics. *Annual Review of Anthropology*, **35**, 191–208.

Drenthen, M. (2005). Wildness as a critical border concept: Nietzsche and the debate on wilderness restoration. *Environmental Values*, **14**, 317–337.

Drenthen, M., and Keulartz, J. (2014). Introduction. In Drenthen, M. and Keulartz, J. (Eds.), *Old World and New World perspectives in environmental philosophy* (pp. 1–14). Cham: Springer.

Farrell, J. (2015). *The battle for Yellowstone: morality and the sacred roots of environmental conflict.* Princeton, NJ: Princeton University Press.

Foreman, D. (1998). Wilderness areas for real. In Callicott, J.B. and Nelson, M.P. (Eds.), *The great new wilderness debate* (p. 395). Athens, GA: University of Georgia Press.

Foreman, D. (1981). Earth first! In Keller, D. (Ed.), *Environmental ethics: the big questions.* Oxford: Blackwell. Reprinted by permission of *The Progressive*, **45**, 39–42.

Frank, G., Parviainen, J., Vandekerkhove, K., Latham, J., Schuck, A., and Little, D. (2007). COST Action E 27. Protected forested areas in Europe. Analysis and harmonisation. Results, conclusions and recommendations. (PROFOR) conference. Vienna.

Fulford, T. (1999). *Romanticism and masculinity: gender, politics and poetics in the writing of Burke, Coleridge, Cobbett, Wordsworth, De Quincey and Hazlitt.* Basingstoke: Macmillan.

Gellner, E. (1992). *Reason and culture.* London: Blackwell.

Gerber, J. (1997). Beyond dualism – the social construction of nature and the natural and social construction of human beings. *Progress in Human Geography*, **21**, 1–17.

Germic, S. (2001). *American green: class, crisis, and the deployment of nature in Central Park, Yosemite, and Yellowstone.* Lexington, KY: Lexington Books.

Haraway, D. (1991). *Simians, cyborgs and women: the reinvention of nature.* New York, NY: Routledge.

Hinchliffe, S., Allen, J., Lavau, S., Bingham, N., and Carter, S. (2013). Biosecurity and the topologies of infected life: from borderlines to borderlands. *Transactions of the Institute of British Geographers*, **38**, 531–543.

Holgate, M. (1999). *The green web: a union for world conservation.* London: Earthscan.

Igoe, J., Neves, K., and Brockington, D. (2010). A spectacular eco-tour around the historic bloc: theorising the convergence of biodiversity conservation and capitalist expansion. *Antipode*, **42**, 486–512.

International Commission on Management of the Oostvaardersplassen (ICMO2). (2010). Natural processes, animal welfare, moral aspects and management of the Oostvaardersplassen.

Jørgensen, D. (2015). Rethinking rewilding. *Geoforum*, **65**, 482-488.

Keenleyside, C., Tucker, G., and McConville, A. (2010). *Farmland abandonment in the EU: an assessment of trends and prospects*. London: Institute for European Environmental Policy.

King, R.J.H. (1990). How to construe nature: environmental ethics and the interpretation of nature. *Between the Species*, **6**(3), 3.

Latour, B. (2004). *Politics of nature*. Cambridge, MA: Harvard University Press.

Lorimer, J. (2006). What about the nematodes? Taxonomic partialities in the scope of UK biodiversity conservation. *Social and Cultural Geography*, **7**, 539-558. https://doi.org/10.1080/14649360600825687

Lorimer, J. (2015). *Wildlife in the Anthropocene: conservation after nature*. Minneapolis, MN: University of Minnesota Press.

Lorimer, J., and Driessen, C. (2014). Wild experiments at the Oostvaardersplassen. Rethinking environmentalism in the Anthropocene. *Transactions of the Institute of British Geographers*, **39**, 169-181.

Lorimer, J., and Driessen, C. (2016). From 'Nazi cows' to cosmopolitan 'ecological engineers': specifying rewilding through a history of Heck cattle. *Annals of the American Association of Geographers*, **106**, 631-652.

Mackenzie, J. (1988). *The empire of nature: hunting, conservation, and British imperialism*. Manchester: Manchester University Press

Merchant, C. (1980). *The death of nature: women and ecology in the scientific revolution*. New York, NY: HarperCollins.

Merckx, T., and Pereira, H. (2015). Reshaping agri-environmental subsidies: from marginal farming to large-scale rewilding. *Basic and Applied Ecology*, **16**, 95-103.

Meyer, J. (1996). *The spirit of Yellowstone: the cultural evolution of a national park*. London: Rowman & Littlefield.

Monbiot, G. (2013). *Feral: searching for enchantment on the frontiers of rewilding*. London: Penguin UK.

Muir, J. (1912). *The Yosemite*. New York, NY: The Century Company.

Nash, W. (1982). *Wilderness and the American mind*. New Haven, CT: Yale University Press.

Navarro, L.M., and Pereira, H.M. (2015). Rewilding abandoned landscapes in Europe. In *Rewilding European landscapes* (pp. 3-23). Cham: Springer.

Neumann, R.P. (1998). Imposing wilderness: struggles over livelihood and nature preservation in Africa. *California Studies in Critical Human Geography*, **4**(xii), 256.

Neumann, R.P. (2002). *Imposing wilderness: struggles over livelihood and nature preservation in Africa* (Vol. 4). Berkeley, CA: University of California Press.

Neves, K., and Igoe, J. (2012). Uneven development and accumulation by dispossession in nature conservation: comparing recent trends in the Azores and Tanzania. *Tijdschrift voor economische en sociale geografie*, **103**, 164-179.

Olwig, K. (1996). Reinventing common nature: Yosemite and Mt. Rushmore - a meandering tale of a double nature. In Cronin, W. (Ed.), *Uncommon ground: toward reinventing nature* (pp. 379-408). New York, NY: W.W. Norton & Co.

Philippon, D.J. (2005). *Conserving words: how American nature writers shaped the environmental movement*. Athens, GA: University of Georgia Press.

Plumwood, V. (1993). The politics of reason: towards a feminist logic. *Australasian Journal of Philosophy*, **71**, 436-462.

Plumwood, V. (1998). Wilderness skepticism and wilderness dualism. In Callicott J.B. and Nelson, M.P. (Eds.), *The great new wilderness debate* (pp. 652-690). Athens, GA: University of Georgia Press

Prior, J., and Brady, E. (2017). Environmental aesthetics and rewilding. *Environmental Values*, **26**, 31-51.

Prior, J., and Ward, K.J. (2016). Rethinking rewilding: a response to Jørgensen. *Geoforum*, **69**, 132-135.

Rewilding Europe. (2016). *Annual Review 2016*. Netherlands.

Rifkin, M. (2014). The frontier as (movable) space of exception. *Settler Colonial Studies*, **4**, 176-180.

Rose, G. (1993). *Feminism and geography: the limits to geographical knowledge*. Minneapolis, MN: University of Minnesota Press.

Shelley, P., Donovan, J., and Duffy, C. (2017). *Selected poems and prose*. Oxford: Penguin Classics.

Soulé, M., and Noss, R. (1998). Rewilding and biodiversity: complementary goals for continental conservation. *Wild Earth*, **8**, 18-28.

Spence, M.D. (1999). *Dispossessing the wilderness: Indian removal and the making of the national parks*. Oxford: Oxford University Press.

Stevens, S. (1997). *Conservation through cultural survival: Indigenous peoples and protected areas*. Washington, DC: Island Press.

Sutter, P. (1999). The great new wilderness debate: an expansive collection of writings defining wilderness from John Muir to Gary Snyder. *Environmental History*, **4**, 280-282.

Tanasescu, M. (2017). Field notes on the meaning of rewilding. *Ethics, Policy & Environment*, **20**, 333-349.

Thoreau, H.D. (1974). *The illustrated Walden*. Princeton, NJ: Princeton University Press.

Urbanik, J. (2012). *Placing animals: an introduction to the geography of human-animal relations*. Lanham, MD: Rowman & Littlefield.

US Congress. (1964). Wilderness Act. *Public Law*, **577**, 1131-1136.

Vasile, M. (2018). The vulnerable bison: practices and meanings of rewilding in the Romanian Carpathians. *Conservation and Society*; **16**(3), 217-231.

Vera, F. (2010). The shifting baseline syndrome in restoration ecology. In Hall, M. (Ed.), *Restoration and history* (pp. 98-110). New York, NY: Routledge.

Watson, A., Alessa, L., and Glaspell, B. (2003). The relationship between traditional ecological knowledge, evolving cultures, and wilderness protection in the circumpolar north. *Conservation Ecology*, **8**(1).

West, P., Igoe, J., and Brockington, D. (2006). Parks and peoples: the social impact of protected areas. *Annual Review of Anthropology*, **35**, 251-277.

Whatmore, S. (1997). Dissecting the autonomous self: hybrid cartographies for a relational ethics. *Environment and Planning D: Society and Space*, **15**, 37-53.

Whatmore, S. (2002). *Hybrid geographies: natures cultures spaces*. London: Sage.

Whatmore, S., and Thorne, L. (1998). Wild (er) ness: reconfiguring the geographies of wildlife. *Transactions of the Institute of British Geographers*, **23**, 435-454.

Wolch, J.R., and Emel, J. (1998). *Animal geographies: place, politics, and identity in the nature-culture borderlands*. London: Verso.

Woods, M. (2005). Ecological restoration and the renewal of wildness and freedom. In Heyd, T. (Ed.), *Recognizing the autonomy of nature: theory and practice* (pp. 170-188). New York, NY: Columbia University Press.

Woods, M. (2007). Wilderness. In Jamieson, D. (Ed.), *A companion to environmental philosophy* (pp. 349-361). Chichester: Wiley-Blackwell.

Wright, W. (1993). Gunfighter nation: the myth of the frontier in twentieth-century America. *Contemporary Sociology*, **22**, 854-855.

Wuerthner, G. (2015). Yellowstone as model for the world. In *Protecting the wild* (pp. 131-143). Washington, DC: Island Press.

CHAPTER FOUR

Pleistocene rewilding: an enlightening thought experiment

JOHAN T. DU TOIT
Utah State University

With a focus on North America, a group of boldly optimistic ecologists (Donlan et al., 2005, 2006) launched the idea of introducing modern proxies for at least some of the terrestrial megafauna that became extinct around the time the Pleistocene epoch ended. The rationale was that the activities of megafauna have ecologically consequential effects, which were presumably important for the functioning of Pleistocene ecosystems. Therefore, extant ecological analogues of extinct megafaunal species should be substituted into defaunated Anthropocene ecosystems to serve those functions in the future. The idea sparked debate in the popular and scientific press, serving as a thought experiment that ran its course in the literature (Svenning et al., 2016a). It is now generally accepted that Pleistocene rewilding has too many logical flaws and practical problems to be implemented at spatial scales larger than a well-financed private ranch or small nature reserve. Nevertheless, the thought experiment has proved enlightening. It has driven adaptations of the rewilding concept and added momentum to a new phase of pragmatism in conservation ecology that emphasises function over taxonomic correctness.

Origins of Pleistocene rewilding

The rewilding concept was first defined by Soulé and Noss (1998) as 'the scientific argument for restoring big wilderness based on the regulatory roles of large predators'. The argument was that large predators are often keystone species (but see Mills et al., 1993; Caro, 2010) exerting top-down controls on ecosystems, and so they should be restored to large, well-protected core reserves surrounded by buffer zones and connected in a network of habitat corridors. Thus, rewilding was originally defined as a conservation method organised around 'the three C's': cores, corridors, and carnivores. The operational context was 'big wilderness', calling for a regional conservation strategy to place land under protection at a spatial scale larger than a landscape and perhaps as large as a continent (Soulé and Terborgh, 1999).

Developing this argument for a continent such as North America requires a decision on which historical state should be used as the restoration benchmark. Humans had occupied and modified the ecosystems of the continent from coast to coast long before the end of the fifteenth century – when the Spanish explored the Caribbean shores under the leadership of Christopher Columbus – and so the commonly used pre- and post-Columbian designations are arbitrary in an ecological context. Consequently, Donlan et al. (2005, 2006) proposed pushing the benchmark right back to 13,000 years BP (before present), into the late Pleistocene, to restore as much as possible of the North American 'ecological and evolutionary potential' destroyed by humans. And so the thought experiment of Pleistocene rewilding began, in a lively debate that still attracts wide interest due to the dramatic images it conjures in the imagination.

The thought experiment is based on two propositions: first, that extinct ecological processes could be restored by introducing proxy megafauna from other continents; second, that this type of rewilding could globalise the conservation of extant megafauna currently threatened in their native ranges in Africa and parts of Asia (Donlan et al., 2005, 2006). The suggested proxies for extinct North American megafauna included elephants (*Elephas maximus* from Asia and/or *Loxodonta africana* from Africa), camels (*Camelus bactrianus* from Asia), large cats (lions *Panthera leo* and cheetahs *Acinonyx jubatus* from Africa), horses and asses (*Equus caballus* and *E. asinus*, which are already feral in North America, and *E. hemionus* and *E. ferus* from Asia), and the largest remaining North American tortoise (*Gopherus flavomarginatus* from Mexico).

Ecological foundations

An ecosystem is defined not only by its components and its environment, but also the interactions among its coexisting organisms. Those interactions were pondered by Hairston et al. (1960), who concluded that coexisting populations of producers, carnivores, and decomposers are regulated by density-dependent resource limitation and interspecific competition, whereas herbivore populations are usually limited by predation alone. Despite the many exceptions to that now-classic generalisation, it serves to illustrate the difference between bottom-up control (the rate at which the abiotic environment enables organisms to sequester resources) and top-down control (the rate at which organisms are consumed by others higher up the food chain), with the state of an ecosystem determined by the relative strengths of those controls (Figure 4.1). It follows that if an ecosystem loses major components of its top-down control system then it will reorganise in an alternative state, and if humans were responsible then the new state might be considered unnatural. That was what happened, to a greater or lesser extent, across all continents other than Africa (and of course Antarctica) through the latter stage of the Pleistocene epoch, which ended 11,700 years BP.

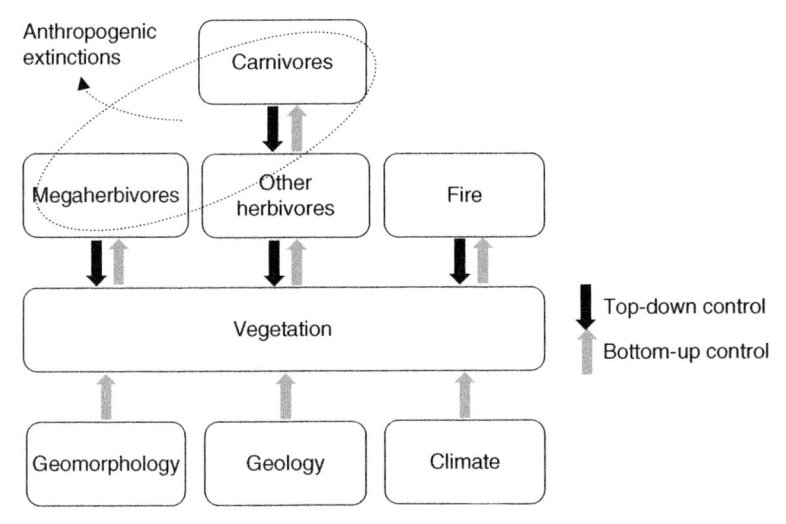

Figure 4.1. A simplified diagram of the main components of a terrestrial ecosystem showing where top-down and bottom-up controls operate, and where anthropogenic extinctions in the late Pleistocene (and ongoing) have removed a major component of the control system. Note that Pleistocene megacarnivores might have regulated megaherbivore populations through predation on juveniles (Van Valkenburgh et al., 2016), but humans are the only predators known to influence the dynamics of megaherbivore populations. (A black and white version of this figure will appear in some formats. For the colour version, please refer to the plate section.)

Paleontological records show that terrestrial faunal assemblages across all non-glaciated regions during the Pleistocene were comprised of species distributed across a wider body-size range than occurs today. What followed was a disproportionately high rate of extinction of the megafauna: megaherbivores ($\geq$ 1000 kg), megacarnivores ($\geq$ 100 kg), large herbivores (45–999 kg), large carnivores (21.5–99 kg), as categorised by Malhi et al. (2016). The wave of extinctions on each continent must have resulted in a major release of top-down control, with the main effects being increased woody cover (Owen-Smith, 1987, 1988, 1989; Bakker et al., 2016; Barnosky et al., 2016), reduced nutrient transport (Doughty et al., 2016), reduced soil compaction in summer and albedo in winter at high latitudes (Zimov et al., 2012), and multiple disruptions of predator–prey–scavenger interactions (Malhi et al., 2016). Those effects would have been additive to – and interactive with – climatic effects, causing ecosystems to ultimately reorganise in altered states.

Mounting paleontological evidence is shifting opinions on the main cause of the megafaunal extinctions away from climate change and firmly towards the agency of *Homo sapiens*. Worldwide, the arrival of humans on new continents and islands was coincident with major faunal extinction events (Martin, 1973;

Surovell et al., 2016). Species with slow reproductive rates are most vulnerable to over-hunting and, among mammals, such species are generally large, so in the face of sustained non-selective hunting it is to be expected that the megaherbivores and large herbivores would have been driven to extinction first (Johnson, 2002). The megacarnivores would have lost their main prey and cascading ecological effects of shrinking megaherbivore populations would have led to regime shifts across biomes, transforming habitats to which multiple species were adapted (Owen-Smith, 1987, 1988; Bakker et al., 2016). An obvious question, which remains conjectural, is why Africa's megaherbivores were not all over-hunted by humans (some were, locally) long before the end of the Pleistocene, considering they had at least a million years to do it in. Perhaps naiveté of prey to a new predator, and relatively advanced weaponry and cooperative hunting techniques, were enough to tip the balance when humans invaded new continents. Nevertheless, humans in Africa and elsewhere are now making up for lost time, to the extent that rapidly declining populations of large mammals are a feature of the Anthropocene (Ripple et al., 2015).

Extinction is a natural process, as is ecosystem change, but by convention it is generally considered unnatural when the causes of a major ecological disturbance are the actions of humans. For such cases, a subdiscipline of ecology – restoration ecology – emerged in the early 2000s to assist the recovery of ecosystems that have been degraded, damaged, or destroyed (Martin, 2017). That was the context in which the idea of Pleistocene rewilding emerged, focusing on those ecosystems most affected by megafaunal extinctions.

Existing projects associated with Pleistocene rewilding

In addition to managed and feral livestock, there is presently an abundance of exotic large animals maintained in various ways in North America, and on other continents, in situations ranging from domestic pets to free-ranging wildlife on private ranches (Lundgren et al., 2018). There is thus no question that representatives of each species identified as a proxy for Pleistocene rewilding could be sourced, introduced, and maintained in some way or another in a project area. Bigger and as yet unanswered questions are whether they could become free-living in self-sustaining populations and serve the ecological and evolutionary functions desired of them without causing intolerable conflicts and ethical concerns to human societies. Meanwhile, there are several projects underway that have become associated with Pleistocene rewilding and which maintain some level of popular support for the concept.

The Oostvaardersplassen is a 60-km^2 fenced nature reserve in the Netherlands that originated when a polder was created in 1968 by pumping out the sea and cultivating terrestrial vegetation on the former seabed. Greylag geese (*Anser anser*) soon flocked in and then managers introduced various large mammalian herbivores including cattle in 1983, horses in 1984, and red deer (*Cervus elaphus*) in 1992. The objective was to promote vegetation heterogeneity through differential herbivory. Also, the opportunity was used to assemble an approximation of a Pleistocene grazing guild with the types of cattle (Heck breed) and horse (konik breed) chosen as the closest analogues of extinct aurochs (*Bos taurus primigenius*) and tarpan (*Equus ferus ferus*), respectively. The populations were allowed to grow until becoming regulated by density dependence and interspecific competition, with management limited to that required to meet basic animal welfare standards (ICMO2, 2010). In just a few decades the Oostvaardersplassen has self-organised into a novel ecosystem of importance for biodiversity conservation in western Europe (DFC, 2010). Because of the particular breeds of cattle and horse involved, and the mainly 'hands-off' management approach, it has become labelled as an example of Pleistocene rewilding (Marris, 2009). The validity of the Pleistocene part of that label is debatable, but rewilding certainly occurred, not through *restoration* – the place is below sea level, after all – but through *reorganisation* of the biotic and abiotic components of the system for a desired outcome (*sensu* Pettorelli et al., 2018).

Pleistocene Park, located in the Republic of Yakutia in northern Siberia, was initiated in 1989 as a pilot project to restore the mammoth steppe ecosystem that existed there during the Pleistocene. The guiding hypothesis is that the Pleistocene megafauna maintained a productive grassland ecosystem, which became degraded and covered by mosses and shrubs after humans over-hunted the large herbivores (Zimov, 2005). A fenced enclosure now confines musk oxen (*Ovibos moschatus*), moose (*Alces alces*), reindeer (*Rangifer tarandus*), bison, and horses to a 16-km^2 treatment area. Their concentrated feeding, trampling, and nutrient cycling are predicted to change vegetation, snow-pack, and soil properties in ways such that permafrost thawing (caused by global warming) will be reduced (Zimov et al., 2012). However, limitations on the research effort, which is not replicated and obviously lacks the dominant member of the Pleistocene grazing guild, the woolly mammoth (*Mammuthus primigenius*), will make it difficult to draw ecological inferences from the outcomes.

Another conservation effort that has become associated (although not by design in this case) with Pleistocene rewilding is concerned with restoring Bolson tortoises (*Gopherus flavomarginatus*) to their former range in the Chihuahuan Desert. Locations of remains from the late Pleistocene indicate that this species once occurred across an area that now includes at least Arizona, New Mexico, and western Texas in the USA, and northern Mexico. Human predation seems to have caused its virtual extinction, with a few remnant populations having survived in the wild in a small distribution that only received protection in 1977 with the formation of the Mapimí Biosphere Reserve in northern Durango, Mexico (Truett and Phillips, 2009). Through the efforts of concerned individuals using private land and funding, several reintroduced Bolson tortoise subpopulations now occur in southern Arizona and New Mexico. The objective, however, has always been to save the species from extinction in its native range rather than to restore ecological functions through rewilding, as was suggested by Donlan et al. (2006). Nevertheless, there is no evidence to diminish the possibility that direct and indirect effects of Bolson tortoises digging burrows, feeding on plants, dispersing seeds, etc. are of functional significance to the desert ecosystem.

Potential effects on ecological processes

Thought experiments allow assumptions to be set for specific scenarios, even if unrealistic, and so we might consider a continuous expanse of land – at least 5000 km^2 – in a suitable climate envelope with sufficient productivity and space to support viable populations of megaherbivores and megacarnivores. A megafaunal assemblage should have occurred there up until the end of the Pleistocene. We assume the main functional types within that now-extinct assemblage could be represented by proxy species from the extant megafauna of Africa and/or parts of Asia, and we further assume they could be introduced together in a quick and effective rewilding operation. Then, assuming sociopolitics, ethics, economics, and evolution could be ignored to allow this 'Pleistocene park' to exist without further interference, how would it come to differ from the state it was in prior to rewilding?

A generalised answer can be sketched out – essentially by reverse-engineering – from combinations of modern and ancient evidence. The biomass density (kg/km^2) of large herbivores would increase substantially because, based on data from African protected areas, megaherbivores constitute 40–70 per cent of the total biomass of large herbivores in their community (Owen-Smith, 1988). That portion of biomass cannot be fully substituted by smaller herbivores, including livestock (Fritz and Duncan, 1994), because megaherbivores have such wide dietary tolerances that they can utilise fibrous and woody vegetation that smaller species cannot. They are thus able to feed in a wider range of habitats than smaller species and thereby reach disproportionately higher biomass densities in spatially heterogeneous ecosystems (du Toit and Owen-Smith, 1989). A substantial increase in herbivory and structural damage to trees would cause woody cover to fragment, herbaceous cover to expand, and fuel loads to decline so that fire intensity and frequency would also decline (Bakker et al., 2016). Long-distance dispersal of seeds, especially those of large-fruited trees, would be restored (Janzen and Martin, 1982; Pires et al., 2018), with elephants, for example, carrying seeds in their guts over distances up to 65 km (Bunney et al., 2017).

Pathways and rates of nutrient flux would change as plant biomass that would otherwise burn or decompose in the litter layer gets processed into dung, urine, and animal biomass. Evidence from an African woodland–savanna indicates that elephants maintain a fast–shallow nutrient cycle where they feed in zones of heavily utilised and rapidly regenerating shrubland, whereas the chemically defended foliage in the matrix of closed-canopy woodland regulates nutrient flux in a slow–deep cycle (du Toit et al., 2014). Patches of soil fertility promote fast-growing, herbivore-tolerant plants such as stoloniferous grasses and thorny shrubs and trees (Bakker et al., 2016). In turn, those attract grazers and browsers, which pass nutrients through the 'green cycle' at a faster rate than the 'brown cycle' that involves the slow decomposition of recalcitrant litter (du Toit and Olff, 2014). Over time, the feeding actions of megaherbivores (and other large herbivores) would directly and indirectly lead to plant-available nutrients being conserved near the soil surface, with reduced leaching because of faster cycling.

A recent focus of interest in ecosystem ecology is the functional significance of large-bodied animals as agents of nutrient transport between aquatic and terrestrial environments and then diffusion through ecosystems. One hypothesis is that, prior to the Pleistocene extinctions, whales, seabirds, and anadromous fish conveyed nutrients from ocean depths to continental shores and river headwaters, from where various large terrestrial animals dispersed the nutrients widely inland (Doughty et al., 2016). However, anadromy is mainly restricted to the higher latitudes and so this 'nutrient distribution pump' could not operate in the lower latitudes where freshwater ecosystems are more productive than the oceans (Gross et al., 1988). Also, some large herbivores are agents of large-scale nutrient transport in the opposite direction, from terrestrial to aquatic ecosystems (McCauley et al., 2015; Subalusky et al., 2017). It remains unclear, therefore, whether there is any general pattern by which large animals might drive the lateral diffusion of nutrients across a landscape. More obvious is that megaherbivores assimilate nutrients over decades as their massive bodies grow until, when they die, their carcasses become point sources of nutrient release with potentially long-term effects on plant community patterns, which require further study.

There are no living analogues of the largest Pleistocene felids, such as the sabretooth cats and massive lions of North America and Europe. Nevertheless, Donlan et al. (2005, 2006) proposed introducing African lions and cheetahs to augment cougars (*Puma concolor*), wolves (*Canis lupus*), and bears (*Ursus* spp.) for an approximation of a Pleistocene large-predator guild in North America. The ecological rationale was that missing components of top-down control should be restored to stabilise food webs, and predator–prey interactions are important drivers of evolution through natural selection. Also, public interest in conservation would be stimulated by the charismatic nature of large predators, with the added benefit of attracting eco-tourism revenues to rural economies.

Reduced natural predation, and in many cases its virtual absence, has contributed to unprecedentedly high densities of deer in parts of the temperate forest biomes of North America and Europe where over-browsing is a serious management concern (Côté et al., 2004). A return of top-down control – ignoring considerations of human–wildlife conflict for now – would likely have a remedial effect in those ecosystems. However, native canid and felid predators could probably serve that function if allowed to, whereas the cheetahs and lions proposed for Pleistocene rewilding would require more open habitats and different prey. Cheetahs could perhaps survive in the sagebrush–steppe ecosystem of western North America where they would prey mainly on mule deer (*Odocoileus hemionus*) and pronghorn (*Antilocapra americana*). It is questionable, though, whether cheetahs could reach sufficient densities to regulate populations of either ungulate, neither of which is currently over-abundant anyway. For larger ungulate prey (> 150 kg) the evidence from African savannas is that their populations are regulated more through bottom-up control by food limitation and less through top-down control by predation (Sinclair et al., 2003). Similar evidence is emerging from Yellowstone, where wolves reintroduced in 1995–1997 have not actually had as much impact on the elk (*Cervus canadensis*) population, and are thus less responsible for the ecological consequences attributed to a trophic cascade, than previously hypothesised (Ripple and Beschta, 2012). Drought and concurrent population increases of bears and cougars are now being understood as major factors driving the elk decline (MacNulty et al., 2016), while the bison (*Bison bison*) population is steadily increasing without much effect from wolves (Tallian et al., 2017). And, assuming African lions could survive the winters of the North American interior, it is unlikely that they could regulate populations of bison and feral horses. For a 'rewilded' region of North America we might imagine a predator–prey interaction between lion and bison comparable to that between lion and buffalo (*Syncerus caffer*) in an African savanna. There, however, such as in central Kruger where the lion density is comparatively high at ~15 animals/100 km^2 (Mills and Funston, 2003), long-term monitoring has shown the buffalo numbers to be controlled from the bottom up, by rainfall (du Toit, 2010). Population modelling indicates that predation by lions only reduces their prey numbers in closed systems (Tambling and du Toit, 2005). In open systems in which large ungulates move extensively across the landscape in herds, population regulation is by bottom-up control through food limitation in response to rainfall, as shown for multiple ungulate species in Kruger (Owen-Smith and Ogutu, 2003). In turn, lion populations are also regulated by bottom-up control, with lion recruitment in Serengeti, for example, being correlated with the recruitment of wildebeest (*Connochaetes taurinus*), which is regulated by rainfall operating through the food supply (Fryxell et al., 2015).

It is thus unlikely that the ecosystem-level effects of a 'rewilded' large-predator guild would emerge primarily through top-down control of herbivore population sizes. Rather, the effects would be complex, diffuse, and far-reaching through the food web (Bowyer et al., 2005). For example, anti-predator behaviour of prey would result in landscapes of fear (Laundré et al., 2001) with altered patterns of habitat use. Niches for scavengers would be extended as regular access to large carcasses becomes facilitated over longer periods, with less variability, than the late-winter (or dry-season) pulse that occurs in the absence of large predators (Wilmers et al., 2003). Disease transmission within and between prey and non-prey populations would be reduced by predators removing infected individuals from herds (Packer et al., 2003) and scavengers removing infective tissues from the landscape (Sekercioglu, 2006; Maichak et al., 2009). Apex predators would control mesopredators, potentially allowing growth in some prey populations, as has been found with wolves controlling coyotes (*Canis latrans*) and thereby reducing juvenile mortality in a pronghorn population (Berger et al., 2008).

Potential effects on ecosystem dynamics and evolutionary potential
From all of the above interactions it is reasonable to expect that if an approximation of a Pleistocene megafaunal assemblage could be achieved through rewilding and maintained over several decades, then measurable changes would occur in the ecosystem. It is questionable, however, whether those changes would culminate in a re-emergence of the functional characteristics of a Pleistocene ecosystem (McCauley et al., 2017). Ecosystem dynamics involve non-linear trajectories, commonly with hysteresis (Chapin et al., 2009), so it is unlikely that any substantially altered ecosystem would restabilise back in any former state after any period following rewilding. Ecosystems are complex-adaptive systems and no interaction within a food web operates in isolation from others, or from abiotic drivers. That is evident from research on the establishment and growth of riparian willows (*Salix* spp.) in Yellowstone, where hypothesised effects of a wolf-driven trophic cascade do not actually stand out from the effects of variations in climate, topography, and stream flow (Marshall et al., 2014).

A key element of the Pleistocene rewilding proposal was the restoration of not only ecological function but also evolutionary potential of lost North American megafauna (Donlan et al., 2006). However, attempting to use extant tropical species as proxies for extinct temperate and boreal species – such as elephants for mammoths – would run into the problem of climatic adaptation

(Richmond et al., 2010). Natural or even artificial selection for cold tolerance would require multiple generations and the largest species are the slowest breeders. Advances in genetic engineering raise the technical possibility of de-extinction (Shapiro, 2016), but even if successfully breeding mammoths or elephant/mammoth chimeras could be created, the next challenge would be to develop and maintain populations with sufficient genetic diversity for evolutionary processes to operate on. The requisite space, time, and intensive management nullify de-extinction as a viable option for Pleistocene megafauna (McCauley et al., 2017). A monolithic obstacle is the human population, which occupies so much land area that there is insufficient remaining for even extant megafauna to occupy species geographic ranges big enough to maintain their evolutionary potential.

Under pre-Anthropocene conditions, any continental assemblage of mammals would be expected to have most small species occurring in small geographic ranges with some in large ranges, and all large species occurring only in large ranges (Figure 4.2). That pattern is evident from the historical ranges of land mammals in North America (Brown and Maurer, 1989) and in African savannas and deserts (du Toit and Cumming, 1999). Large species generally require large geographic ranges as population density scales negatively with body mass, so large species confined to small ranges generally have small populations, which are vulnerable to stochastic disturbances that precipitate extinction. In Anthropocene assemblages the geographic ranges of the remaining large species are drastically reduced and so their diminishing populations are increasingly vulnerable to extinction. This 'flipped' pattern (Figure 4.2) is a transient artefact in evolutionary time because large species can neither persist in small ranges nor recolonise their formerly large ranges,

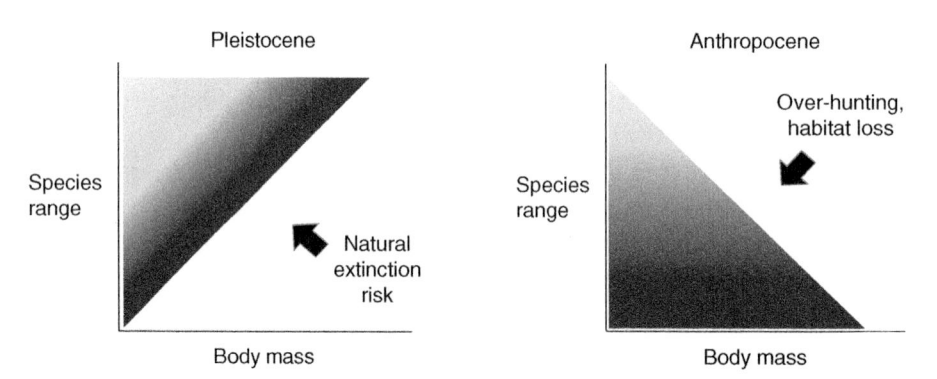

Figure 4.2. Conceptual representations of the relationship between species geographic range and body mass (logarithmic) for a continental assemblage of terrestrial mammals in Pleistocene and Anthropocene epochs. (A black and white version of this figure will appear in some formats. For the colour version, please refer to the plate section.)

which humans have occupied and transformed. It is thus impossible for those large species to recover their evolutionary potential as long as the global populations of humans and megafauna continue to change with opposite trajectories.

Globalising megafaunal conservation: practical, ethical, and political considerations

An ancillary component of the Pleistocene rewilding proposal was to globalise the conservation of extant megafauna currently threatened in their native ranges in Africa and parts of Asia (Donlan et al., 2005, 2006). That, once again, raises the crucial problem of space for sufficient and suitable habitat. An elephant breeding herd in southern Africa uses, on average, a home range area of 10,738 km^2 (van Aarde et al., 2008), which places the problem in perspective for megaherbivores, at least. Even where there are large expanses of potentially suitable habitat, the policies governing public lands in most industrialised countries, such as the USA, would restrict introductions of non-native wildlife species to private lands. Large exotic animals can be, and often are, maintained in relatively small private reserves in many parts of the world, but the spatial scale of such endeavours is restricted by the substantial running costs involved (fence erection and maintenance, compensation for depredation, disease control, etc.). For large animals, small areas of habitat mean small subpopulations requiring intensive management to maintain genetic diversity within metapopulations, and when megaherbivores such as elephants are confined to limited areas their impacts on local biodiversity can be devastating (Kerley and Landman, 2006). Then, managing population density introduces a suite of ethical concerns surrounding the treatment of large, empathic animals (Bates et al., 2008; Lötter et al., 2008).

Practical and ethical considerations aside, the suggestion of globalising biodiversity conservation, by moving groups of selected charismatic and endangered species from poor countries to rich ones, raises the objection of eco-imperialism (Driessen, 2003). The megafauna of Africa and parts of Asia contribute fundamentally to the identities of the countries in those regions, even if corruption and other aspects of weak governance hinder in-country conservation efforts. With the illicit trade in ivory, rhino horn, and other animal parts being driven by globalised economic forces too powerful for poor countries to withstand, it can be argued that global conservation would be better served if those demand-side problems were addressed first. Finally, the suggestion that Pleistocene rewilding could globalise megafaunal conservation assumes a responsibility to actually deliver on the promise of effective conservation. The track records of most rich countries are demonstrably poor in that regard, so conservation practitioners in the tropics may be excused for dismissing Pleistocene rewilding as an arrogant idea. Bridging the credibility gap with new attempts to do better with borrowed species would fail at the first signs of even the most common pitfalls in conservation practice, such as genetic problems caused by founder effects, artificial selection, inbreeding, or introgression.

Outcomes from the thought experiment with functional applications

The main outcome is a general acceptance that linking the rewilding concept to any historical benchmark is a mistake – because the world is continually changing – and so Pleistocene rewilding has evolved into 'trophic' rewilding (Rubenstein and Rubenstein, 2016; Svenning et al., 2016a, 2016b). Nevertheless, the very suggestion of Pleistocene rewilding, as a thought experiment with ambitious and dramatic connotations, energised discussion regarding the broader implications of the rewilding concept (Seddon et al., 2014). Quite obviously megafaunal extinctions did not end in the late Pleistocene: defaunation is actually accelerating in the Anthropocene (Young et al., 2016), and conservation biology has evolved over the past 30 years into a science with a more pragmatic and forward-looking perspective (Kareiva and Marvier, 2012). In particular, Pleistocene rewilding has promoted the notion that the functional properties of large animals are ecologically more important than their taxonomic identities, and that the functional properties of some extinct species could be replaced through taxonomic substitutions.

Taxonomic substitution is a bold, last-resort approach to repairing the 'machinery' of an ecosystem that is missing a 'part', analogous to substituting a part that fits and does the job but is of a different make and is possibly even salvaged from a different machine. It is to be considered only under specific conditions, for specific objectives, guided by adaptive management. The fundamental requirement is an understanding of how an ecosystem would function if one or more of its missing parts were replaced. For large vertebrates – the particular set of parts we are considering here – this requires a combination of paleoecology to identify what is missing, and modern ecology to catalogue the functional types and their biomass contributions in co-evolved assemblages (e.g. Hempson et al., 2015).

As previously discussed, however, taxonomic substitution is challenged by so many practical, ethical, and political considerations that other technological solutions might be more immediately effective. Options exist and are currently implemented (even if unwittingly) in various parts of the world to artificially simulate megafaunal effects on vegetation structure, as well as on nutrient transport, concentration, and redistribution. In North America, for example, the problem of woody encroachment in rangelands is commonly tackled using machines such as 'bull hogs' (Figure 4.3) and tractors pulling chains and harrows to perform treatments that were probably effected, to at

Figure 4.3. It is standard practice to reduce woody cover on rangelands in the western USA using mechanical tools, such as the 'bull-hog mulcher' (left), as an artificial substitution for the ecosystem function once performed by Pleistocene megaherbivores in the same way as elephants in Africa today (right). Photo credits: Fecon Inc. (left) and Stein Moe (right). (A black and white version of this figure will appear in some formats. For the colour version, please refer to the plate section.)

least some extent, by Pleistocene megaherbivores. Yet, there is little empirical evidence that these mechanical treatments actually improve ecosystem function. That could be attributed to research limitations caused by weak monitoring and lack of experimental design (Bombaci and Pejchar, 2016; Fulbright et al., 2018), but success might be improved if the mechanical treatments were applied in ways that (even vaguely) simulate megaherbivore foraging patterns. In the same way, hunting by modern humans could be used to simulate some of the effects of now-extinct predators. In most countries, sport hunting occurs in defined seasons and thus has a temporally discrete (pulse) effect on prime-age adults, whereas natural predation has a continuous (press) effect mainly on juveniles and senescent individuals. Hunting traditions are hard to change, but some time- and place-specific adjustments to licence-issuing and quota-setting could accommodate ecosystem-level objectives. For example, sport hunting presents opportunities for using fear as a tool to move wild ungulates around the landscape (Cromsigt et al., 2013), forcing them to use some habitats more and others less, in ways that Pleistocene predators might have done.

Conclusion

The concept of Pleistocene rewilding catalysed an enlightening thought experiment that attracted an extraordinary level of interest and debate among scientists, journalists, and environmentally concerned commentators around the world. Disparaging responses were provoked by the exclusive focus on charismatic megafauna and the obvious flaw that Pleistocene benchmarking has little ecological relevance in the Anthropocene, but those are actually just details. The important outcome is wider recognition of the fact that co-evolved species, after radiating into multiple trophic guilds and body-size classes, have functional properties that, if lost, have far-reaching ramifications through the ecosystem. A key lesson of the thought experiment is that ecosystem functions are worth conserving, and if they are lost, creative and possibly unorthodox solutions should be considered to reorganise the extant biota in ways that allow those functions to resume. Pleistocene rewilding drew attention to the rewilding concept, which has now shed the distraction of historical benchmarking and, as demonstrated in several chapters in this book, is moving on with more pragmatic applications for the Anthropocene.

References

Bakker, E.S., Gill, J.L., Johnson, C.N., et al. (2016). Combining paleo-data and modern exclosure experiments to assess the impact of megafauna extinctions on woody vegetation. *Proceedings of the National Academy of Sciences of the United States of America*, **113**, 847–855.

Barnosky, A.D., Lindsey, E.L., Villavicencio, N.A., et al. (2016). Variable impact of late-Quaternary megafaunal extinction in causing ecological state shifts in North and South America. *Proceedings of the National Academy of Sciences of the United States of America*, **113**, 856–861.

Bates, L.A., Lee, P.C., Njiraini, N., et al. (2008). Do elephants show empathy? *Journal of Consciousness Studies*, **15**, 204–225.

Berger, K.M., Gese, E.M., and Berger, J. (2008). Indirect effects and traditional trophic cascades: a test involving wolves, coyotes, and pronghorn. *Ecology*, **89**, 818–828.

Bombaci, S., and Pejchar, L. (2016). Consequences of pinyon and juniper woodland reduction for wildlife in North America. *Forest Ecology and Management*, **365**, 34–50.

Bowyer, R.T., Person, D.K., and Pierce, B.M. (2005). Detecting top-down versus bottom-up regulation of ungulates by large carnivores: implications for conservation and biodiversity. In Ray, J.C., Redford, K.H., Steneck, R.S., and Berger, J. (Eds.), *Large carnivores and the conservation of biodiversity* (pp. 342–361). Washington, DC: Island Press.

Brown, J.H., and Maurer, B.A. (1989). Macroecology: the division of food and space among species on continents. *Science*, **243**, 1145–1150.

Bunney, K., Bond, W.J., and Henley, M. (2017). Seed dispersal kernel of the largest surviving megaherbivore – the African savanna elephant. *Biotropica*, **49**, 395–401.

Caro, T. (2010). *Conservation by proxy: indicator, umbrella, keystone, flagship, and other surrogate species*. Washington, DC: Island Press.

Chapin, F.S. III, Kofinas, G.P., and Folke, C. (Eds.) (2009). *Principles of ecosystem stewardship: resilience-based natural resource management in a changing world*. New York, NY: Springer.

Côté, S.D., Rooney, T.P., Tremblay, J-P., Dussault, C., and Waller, D.M. (2004). Ecological impacts of deer overabundance. *Annual Review of Ecology, Evolution, and Systematics*, **35**, 113–147.

Cromsigt, J.P.G.M., Kuijper, D.P.J., Adam, M., et al. (2013). Hunting for fear: innovating management of human–wildlife conflicts. *Journal of Applied Ecology*, **50**, 544–549.

DFC. (2010). *The Oostvaardersplassen: beyond the horizon of the familiar*. Driebergen, the Netherlands: Dutch Forestry Commission.

Donlan, J., Greene, H.W., Berger, J., et al. (2005). Re-wilding North America. *Nature*, **436**, 913-914.

Donlan, C.J., Berger, J., Bock, C.E., et al. (2006). Pleistocene rewilding: an optimistic agenda for twenty-first century conservation. *The American Naturalist*, **168**, 660-681.

Doughty, C.E., Roman, J., Faurby, S., et al. (2016). Global nutrient transport in a world of giants. *Proceedings of the National Academy of Sciences of the United States of America*, **113**, 868-873.

Driessen, P.K. (2003). *Eco-imperialism: green power, black death*. Bellevue, WA: Free Enterprise Press.

du Toit, J.T. (2010). Considerations of scale in biodiversity conservation. *Animal Conservation*, **13**, 229-236.

du Toit, J.T., and Cumming, D.H.M. (1999). Functional significance of ungulate diversity in African savannas and the ecological implications of the spread of pastoralism. *Biodiversity and Conservation*, **8**, 1643-1661.

du Toit, J.T., and Olff, H. (2014). Generalities in grazing and browsing ecology: using across-guild comparisons to control contingencies. *Oecologia*, **174**, 1075-1083.

du Toit, J.T., and Owen-Smith, N. (1989). Body size, population metabolism, and habitat specialization among large African herbivores. *The American Naturalist*, **133**, 736-740.

du Toit, J.T., Moe, S.R., and Skarpe, C. (2014). Elephant-mediated ecosystem processes in Kalahari-sand woodlands. In Skarpe, C., du Toit, J.T., and Moe, S.R. (Eds.), *Elephants and savanna woodland ecosystems: a study from Chobe National Park, Botswana* (pp. 30-39). Chichester: Wiley-Blackwell and the Zoological Society of London.

Fritz, H., and Duncan, P. (1994). On the carrying capacity for ungulates of African savanna ecosystems. *Proceedings of the Royal Society of London, B*, **256**, 77-82.

Fryxell, J.M., Metzger, K.L., Packer, C., Sinclair, A.R.E., and Mduma, S.A.R. (2015). Climate-induced effects on the Serengeti mammalian food web. In Sinclair, A.R.E., Metzger, K.L., Mduma, S.A.R., and Fryxell, J.M. (Eds.), *Serengeti IV: sustaining biodiversity in a coupled human–natural system* (pp. 175-191). Chicago, IL: University of Chicago Press.

Fulbright, T.E., Davies, K.W., and Archer, S.R. (2018). Wildlife responses to brush management: a contemporary evaluation. *Rangeland Ecology and Management*, **71**, 35-44.

Gross, M.R., Coleman, R.M., and McDowall, R.M. (1988). Aquatic productivity and the evolution of diadromous fish migration. *Science*, **239**, 1291-1293.

Hairston, N.G., Smith, F.E., and Slobodkin, L.B. (1960). Community structure, population control, and competition. *The American Naturalist*, **94**, 421-425.

Hempson, G.H., Archibald, S., and Bond, W.J. (2015). A continent-wide assessment of the form and intensity of large mammal herbivory in Africa. *Science*, **350**, 1056-1061.

ICMO2. (2010). *Natural processes, animal welfare, moral aspects and management of the Oostvaardersplassen*. Report of the second International Commission on Management of the Oostvaardersplassen (ICMO2). The Hague/Wageningen: ICMO2.

Janzen, D.H., and Martin, P.S. (1982). Neotropical anachronisms: the fruits the gomphotheres ate. *Science*, **215**, 19-27.

Johnson, C.N. (2002). Determinants of loss of large mammal species during the Late Quarternary 'megafauna' extinctions: life history and ecology, but not body size. *Proceedings of the Royal Society of London B: Biological Sciences*, **269**, 2221-2227.

Kareiva, P., and Marvier, M. (2012). What is conservation science? *BioScience*, **62**, 962-969.

Kerley, G.I.H., and Landman, M. (2006). The impacts of elephants on biodiversity in the Eastern Cape subtropical thickets. *South African Journal of Science*, **102**, 395-402.

Laundré, J.W., Hernández, L., and Altendorf, K.B. (2001). Wolves, elk, and bison: reestablishing the 'landscape of fear' in Yellowstone National Park, USA. *Canadian Journal of Zoology*, **79**, 1401-1409.

Lötter, H.P.P., Henley, M., Fakir, S., Pickover, M., and Ramose, M. (2008). Ethical considerations in elephant management. In Scholes, R.J. and Mennell, K.G. (Eds.), *Elephant management: a scientific assessment for South Africa* (pp. 406-445). Johannesburg, South Africa: Wits University Press.

Lundgren, E.J., Ramp, D., Ripple, W.J., and Wallach, A.D. (2018). Introduced megafauna are rewilding the Anthropocene. *Ecography*, **41**, 857-866.

MacNulty, D.R., Stahler, D.R., Wyman, C.T., Ruprecht, J., and Smith, D.W. (2016). The challenge of understanding northern Yellowstone elk dynamics after wolf reintroduction. In Smith, D.W., Stahler, D.R., MacNulty, D.R., and Haas, S. (Eds.), *Yellowstone science* (pp. 25-33). Mammoth, WY: Yellowstone National Park.

Maichak, E.J., Scurlock, B.M., Rogerson, J.D., et al. (2009). Effects of management, behavior, and scavenging on risk of brucellosis transmission in elk of western Wyoming. *Journal of Wildlife Diseases*, **45**, 398-410.

Malhi, Y., Doughty, C.E., Galetti, M., Smith, F.A., Svenning, J.C., and Terborgh, J.W. (2016). Megafauna and ecosystem function from the Pleistocene to the Anthropocene. *Proceedings of the National Academy of Sciences of the United States of America*, **113**, 838-846.

Marris, E. (2009). Reflecting the past. *Nature*, **462**, 30-32.

Marshall, K.N., Cooper, D.J., and Hobbs, N.T. (2014). Interactions among herbivory, climate, topography and plant age shape riparian willow dynamics in northern Yellowstone National Park, USA. *Journal of Ecology*, **102**, 667-677.

Martin, D.M. (2017). Ecological restoration should be redefined for the twenty-first century. *Restoration Ecology*, **25**, 668-673.

Martin, P.S. (1973). The discovery of America. *Science*, **179**, 969-974.

McCauley, D.J., Dawson, T.E., Power, M.E., et al. (2015). Carbon stable isotopes suggest that hippopotamus-vectored nutrients subsidize aquatic consumers in an East African river. *Ecosphere*, **6**(4), 52. http://dx.doi.org/10.1890/ES14-00514.1

McCauley, D.J., Hardesty-Moore, M., Halpern, B.S., and Young, H.S. (2017). A mammoth undertaking: harnessing insight from functional ecology to shape de-extinction priority setting. *Functional Ecology*, **31**, 1003-1011.

Mills, L.S., Soulé, M.E., and Doak, D.F. (1993). The keystone-species concept in ecology and conservation. *BioScience*, **43**, 219-224.

Mills, M.G.L., and Funston, P.J. (2003). Large carnivores and savanna heterogeneity. In du Toit, J.T., Rogers, K.H., and Biggs, H.C. (Eds.), *The Kruger experience: ecology and management of savanna heterogeneity* (pp. 370-388). Washington, DC: Island Press.

Owen-Smith, N. (1987). Pleistocene extinctions: the pivotal role of megaherbivores. *Paleobiology*, **13**, 351-362.

Owen-Smith, R.N. (1988). *Megaherbivores: the influence of very large body size on ecology*. Cambridge: Cambridge University Press.

Owen-Smith, N. (1989). Megafaunal extinctions: the conservation message from 11,000 years B.P. *Conservation Biology*, **3**, 405-412.

Owen-Smith, N., and Ogutu, J. (2003). Rainfall influences on ungulate population dynamics. In du Toit, J.T., Rogers, K.H., and Biggs, H.C. (Eds.), *The Kruger experience: ecology and management of savanna heterogeneity* (pp. 310-331). Washington, DC: Island Press.

Packer, C., Holt, R.D., Hudson, P.J., Lafferty, K.D., and Dobson, A.P. (2003). Keeping the herds healthy and alert: implications of predator control for infectious disease. *Ecology Letters*, **6**, 797–802.

Pettorelli, N., Barlow, J., Stephens, P.A., et al. (2018). Making rewilding fit for policy. *Journal of Applied Ecology*, **55**, 1114–1125.

Pires, M.M., Guimarães, P.R. Jr, Galetti, M., and Jordano, P. (2018). Pleistocene megafaunal extinctions and the functional loss of long-distance seed-dispersal services. *Ecography*, **41**, 153–163.

Richmond, O.M.W., McEntee, J.P., Hijmans, R.J., and Brashares, J.S. (2010). Is the climate right for Pleistocene rewilding? Using species distribution models to extrapolate climatic suitability for mammals across continents. *PLoS ONE*, **5**(9), e12899. doi:10.1371/journal.pone.0012899

Ripple, W.J., and Beschta, R.L. (2012). Trophic cascades in Yellowstone: the first 15 years after wolf reintroduction. *Biological Conservation*, **145**, 205–213.

Ripple, W.J., Newsome, T.M., Wolf, C., et al. (2015). Collapse of the world's largest herbivores. *Science Advances*, **1**, e1400103. doi: 10.1126/sciadv.1400103

Rubenstein, D.R., and Rubenstein, D.I. (2016). From Pleistocene to trophic rewilding: a wolf in sheep's clothing. *Proceedings of the National Academy of Sciences of the United States of America*, **113**, E1.

Seddon, P.J., Griffiths, C.J., Soorae, P.S., and Armstrong, D.P. (2014). Reversing defaunation: restoring species in a changing world. *Science*, **345**, 406–412.

Sekercioglu, C.H. (2006). Increasing awareness of avian ecological function. *Trends in Ecology & Evolution*, **21**, 464–471.

Shapiro, B. (2016). *How to clone a mammoth: the science of de-extinction*. Princeton, NJ: Princeton University Press.

Sinclair, A.R.E., Mduma, S., and Brashares, J.S. (2003). Patterns of predation in a diverse predator–prey system. *Nature*, **425**, 288–290.

Soulé, M., and Noss, R. (1998). Rewilding and biodiversity: complimentary goals for continental conservation. *Wild Earth*, **8**, 18–28.

Soulé, M.E., and Terborgh, J. (Eds.) (1999). *Continental conservation: scientific foundations of regional reserve networks*. Washington, DC: Island Press.

Subalusky, A.L., Dutton, C.L., Rosi, E.J., and Post, D.M. (2017). Annual mass drownings of the Serengeti wildebeest migration influence nutrient cycling and storage in the Mara River. *Proceedings of the National Academy of Sciences of the United States of America* (early view, online).

Surovell, T.A., Pelton, S.R., Anderson-Precher, R., and Myers, A.D. (2016). Test of Martin's overkill hypothesis using radiocarbon dates on extinct megafauna. *Proceedings of the National Academy of Sciences of the United States of America*, **113**, 886–891.

Svenning, J.C., Pedersen, P.B.M., Donlan, C.J., et al. (2016a). Time to move on from ideological debates on rewilding. *Proceedings of the National Academy of Sciences of the United States of America*, **113**, E3.

Svenning, J.C., Pedersen, P.B.M., Donlan, C.J., et al. (2016b). Science for a wilder Anthropocene: synthesis and future directions for trophic rewilding research. *Proceedings of the National Academy of Sciences of the United States of America*, **113**, 898–906.

Tallian, A., Smith, D.W., Stahler, D.R., et al. (2017). Predator foraging response to a resurgent dangerous prey. *Functional Ecology*, **31**, 1418–1429.

Tambling, C.J., and du Toit, J.T. (2005). Modeling wildebeest population dynamics: implications of predation and offtake in a closed system. *Journal of Applied Ecology*, **42**, 431–444.

Truett, J., and Phillips, M. (2009). Beyond historic baselines: restoring Bolson tortoises to Pleistocene range. *Ecological Restoration*, **27**, 144–151.

van Aarde, R., Ferreira, S., Jackson, T., et al. (2008). Elephant population biology and ecology. In Scholes, R.J. and Mennell, K.G.

(Eds.), *Elephant management: a scientific assessment for South Africa* (pp. 84–145). Johannesburg, South Africa: Wits University Press.

Van Valkenburgh, B., Hayward, M.W., Ripple, W.J., Meloro, C., and Roth, V.L. (2016). The impact of large terrestrial carnivores on Pleistocene ecosystems. *Proceedings of the National Academy of Sciences of the United States of America*, **113**, 862–867.

Wilmers, C.C., Crabtree, R.L., Smith, D.W., Murphy, K.M., and Getz, W.M. (2003). Trophic facilitation by introduced top predators: grey wolf subsidies to scavengers in Yellowstone National Park. *Journal of Animal Ecology*, **72**, 909–916.

Young, H.S., McCauley, D.J., Galetti, M., and Dirzo, R. (2016). Patterns, causes and consequences of Anthropocene defaunation. *Annual Review of Ecology, Evolution and Systematics*, **47**, 333–358.

Zimov, S.A. (2005). Pleistocene Park: return of the mammoth's ecosystem. *Science*, **308**, 796–798.

Zimov, S.A., Zimov, N.S., Tikhonov, A.N., and Chapin, F.S. III. (2012). Mammoth steppe: a high productivity phenomenon. *Quaternary Science Reviews*, **57**, 26–45.

CHAPTER FIVE

Trophic rewilding: ecological restoration of top-down trophic interactions to promote self-regulating biodiverse ecosystems

JENS-CHRISTIAN SVENNING, MICHAEL MUNK,
and ANDREAS SCHWEIGER
Aarhus University

There is rapidly increasing interest in rewilding as an alternative to conventional, more human-controlled approaches to biological conservation and nature management (Donlan et al., 2005; Navarro and Pereira, 2012; Jepson, 2016; Svenning et al., 2016). Reflecting the wide use of the term and its intuitive meaning, its usage has not adhered to strict criteria. Nevertheless, there are consistent characteristics for what has been referred to as rewilding, namely a focus on restoring natural processes and reducing human management with the aim of achieving more self-managing ecosystems. Most rewilding projects have a focus on re-establishing missing large-bodied animals and fit under the definition of trophic rewilding, defined as an ecological restoration strategy (*sensu* Hobbs and Cramer, 2008) that uses species introductions to restore top-down trophic interactions and associated trophic cascades to promote self-regulating biodiverse ecosystems (Svenning et al., 2016; Figure 5.1). 'Self-regulating biodiverse ecosystems' refer to situations where ecosystems via natural (non-human-controlled) processes are able to generate and maintain relatively high biodiversity without human management. Trophic cascades can be broadly defined as the propagation of consumer impacts across multiple trophic levels downwards through food webs (Paine, 1980; Estes et al., 2011). The emergence of trophic cascades is linked to top-down control of communities, with the general importance of either subject to debate (Allen et al., 2017). However, trophic rewilding does not assume dominance of top-down control, but just that top-down effects – by herbivores as well as by carnivores – have importance (Svenning et al., 2016).

Megafauna is often defined as $\geq$ 45 kg body mass, but actual rewilding projects often include smaller but still relatively large species (Griffiths et al., 2011; Ripple and Beschta, 2012; Law et al., 2017). Here, we use megafauna in

REAL-WORLD TROPHIC REWILDING PROJECTS

Figure 5.1. Real-world trophic rewilding projects. (A,B) Yellowstone National Park (USA) – large-scale (~9000 km^2) project with wolf (*Canis lupus*) reintroduction, with large herbivore populations and strong vegetation impacts continuing, but perhaps modulated by restored predation (2011). (C) Experimental beaver (*Castor fiber*) reintroduction in Scotland, with strong vegetation impacts (Knapdale, ~44 km^2, 2013). (D) Free-ranging beaver (*C. fiber*) reintroduction in western Jutland (Denmark), also with strong vegetation impacts (2010). Small to moderately sized fenced projects in densely occupied European regions, with semi-wild horses (*Equus ferus*) and cattle (*Bos primigenius*), and sometimes also other species, in (E,F) Mols Bjerge (120 ha, Denmark, 2017), (G) Knepp Estate (1400 ha,

the latter, broader sense. The focus on relatively large-bodied species reflects two phenomena, discussed in detail below: (1) human-driven defaunation has been strongly size-selective, especially affecting large-bodied species, leading to down-sizing of animal communities (Estes et al., 2011; Dirzo et al., 2014); (2) large-bodied species overall have large ecological and often unique effects on ecosystem structure (Estes et al., 2011; Jorge et al., 2013; Dirzo et al., 2014; Ripple et al., 2014b). An additional factor that also plays a role is the socio-cultural importance of megafauna (Donlan et al., 2006).

The aim of this chapter on trophic rewilding is to present the scientific basis for this approach, to provide guidelines for implementation, and to outline where research should be focused. To this end, we first introduce the ecological and historical background for trophic rewilding and, subsequently, its goals. We then review the empirical evidence from rewilding projects and outline some general expectations for the outcomes of trophic rewilding. After this, we discuss ecological design considerations for rewilding projects, followed by a section dedicated to societal perspectives and risks, as these will be key to the long-term success of any rewilding project. Finally, we provide a research agenda in relation to trophic rewilding to improve its scientific basis, highlighting where we see the biggest research needs and opportunities.

Ecological and historical background

A key background for trophic rewilding is the ecological importance of megafauna (Figure 5.2). It is debated whether ecosystems are controlled bottom-up by abiotic conditions or top-down by consumers, and the answer is clearly complex, with interacting effects and geographic and environmental context dependence (Hopcraft et al., 2010). Nevertheless, it is clear that top-down forcing by large herbivores and carnivores is widely important for ecosystem structure and dynamics (Estes et al., 2011; Dirzo et al., 2014; Bello et al., 2015), with many large-bodied species long identified as keystone species (Mills et al.,

Caption for Figure 5.1. (cont.)

England, 2017), (H) Kasted Mose (62 ha, Denmark, 2017), (I,J) Sydlangeland (120+ 25 ha, Denmark, 2017), and (K,L) Gelderse Poort (700 ha, Netherlands, 2016). (M) Large-scale project with introduction of tapirs (*Tapirus terrestris*) and other historically extirpated species in a subtropical savanna setting (Iberá, ~1500 km², Argentina), with emerging restoration of seed dispersal and herbivory (2017). (N) Fenced project with Eurasian elk (moose, *Alces alces*), extirpated since mid-Holocene, to restore browsing in raised bog ecosystem (Lille Vildmose, 2100 ha, Denmark, 2017). (O) Use of functional analogue (*Bubalus bubalis*) of Pleistocene extinct water buffalo (*B. murrensis*) in small fenced project (Kasted Mose, 62 ha, Denmark, 2017). Photos: J.-C. Svenning. (A black and white version of this figure will appear in some formats. For the colour version, please refer to the plate section.)

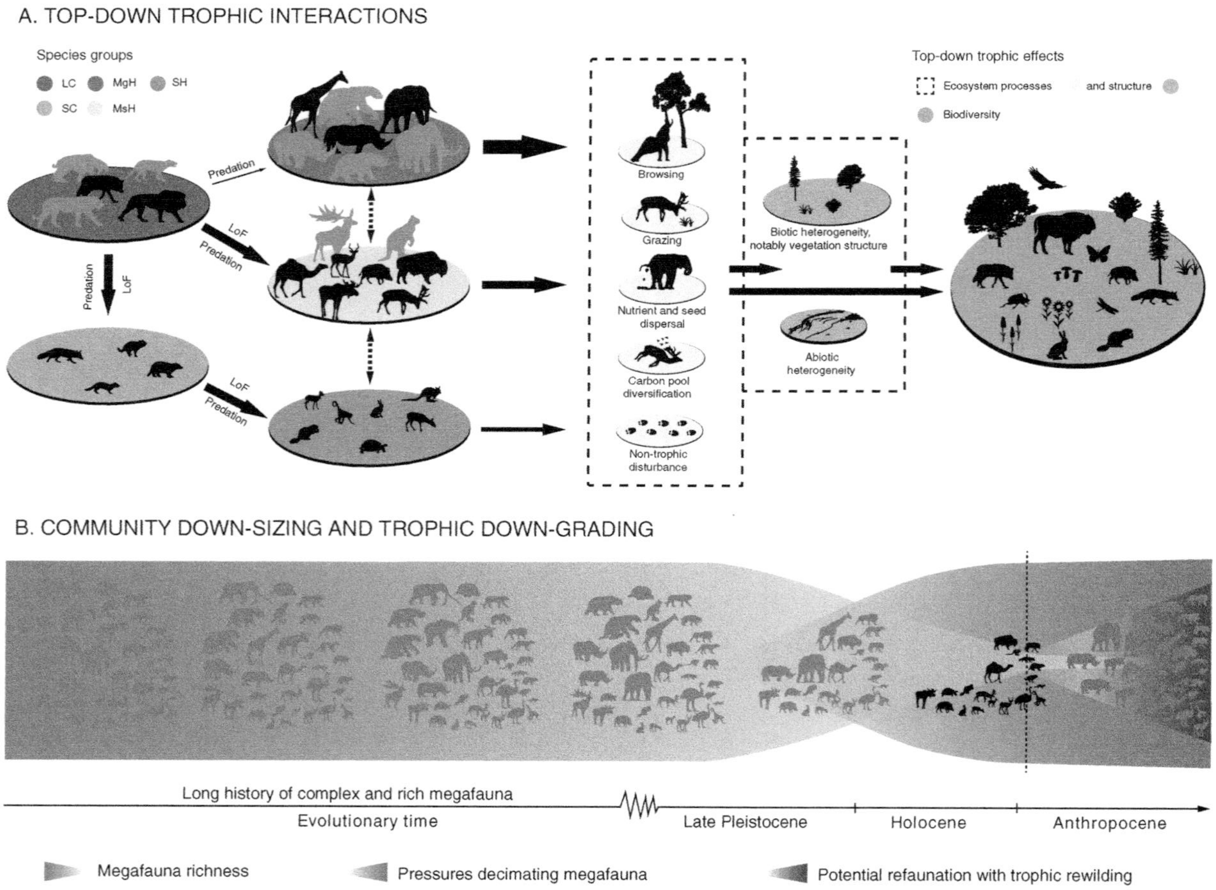

Figure 5.2. (A) Outline of how trophic rewilding may promote biodiversity via effects on ecosystem processes and structure. Species groups: large carnivores (LC), small carnivores (SC), megaherbivores (MgH), mesoherbivores (MsH), and small herbivores (SH), with representatives of extinct (grey) and extant (black) species are shown with their top-down trophic interactions. LoF, landscape of fear. (B) Species-rich and complex megafaunas have been the norm of the evolutionary timescales (transparent figurines), where current (dotted line and black figurines) species diversity evolved, but have undergone strong decimation during the last 50,000 years, with potential recovery with trophic rewilding as one possible Anthropocene scenario and continuing pressures and losses as another. (A black and white version of this figure will appear in some formats. For the colour version, please refer to the plate section.)

1993) or even ecosystem engineers (Jones et al., 1994). Overall, the main ecosystem effects of large herbivores are their direct effects on vegetation structure via herbivory, disturbance, and seed dispersal, while the main effects of large carnivores are via their effects on herbivores, modulating the latter's ecosystem effects, and mesocarnivores, shaping their effects on smaller species, including small herbivores (Estes et al., 2011; Dirzo et al., 2014; Ripple et al., 2014b; Bello et al., 2015; Bakker et al., 2016; Galetti et al., 2018). There is much variation in the details of these effects, and there are also more complex ecosystem impacts, e.g. beaver effects on landscape structure via hydrological change (Jones et al., 1994). In any case, it is clear that large herbivores can have strong effects on vegetation (Bakker et al., 2016), and these effects are expected to be especially strong under relatively warm, moderately moist climates where vegetation tends towards bistability between tall-growing woody and low-growing herbaceous dominance (Bond, 2005). However, even where climate determines the vegetation type, for example, in tropical rainforests, herbivory can still affect vegetation structure (Bello et al., 2015; Terborgh et al., 2016). It is sometimes assumed that herbivore effects will be strongly reduced when top carnivores are restored. This is an over-simplification. First, herbivores may escape regulation by predation via large body size ($\geq$ 1000 kg (megaherbivores), or $\geq$ 100 kg body mass in some settings) or certain behaviours (e.g. forming large migratory herds) (Owen-Smith, 1988; Sinclair, 2003; Hopcraft et al., 2010). Second, top-down regulation does not necessarily mean that a herbivore population will have little effect on vegetation, but just that it will be smaller than without predation; for example, small- to medium-sized ungulates also have vegetation effects on African savannas despite strong predation (Pringle et al., 2014; van der Plas et al., 2016).

Top-down trophic effects by megafauna promote biodiversity in various ways, but most importantly by fostering environmental heterogeneity (Figure 5.2). Increased environmental heterogeneity strongly correlate with species richness (Stein et al., 2014), with the mechanisms at local to landscape scales mainly being greater capacity for species to co-occur via matching of specific requirements and coexistence mechanisms with increased niche space (Stein et al., 2014; Brunbjerg et al., 2017). A large body of literature shows that large herbivores promote environmental heterogeneity, notably via their effects on vegetation (Cromsigt and te Beest, 2014; Bakker et al., 2016). Spatiotemporal variation in herbivory allows plants with different adaptations to coexist and generate increased variation in vegetation structure, providing a diversified habitat template for numerous organisms (Brunbjerg et al., 2017), for example, birds (Hovick et al., 2014). Diversity in large herbivore communities is important for these vegetation effects (Bakker et al., 2016), but needs to be better understood. Where single species become highly abundant, they may homogenise the environment and, thus, diminish species diversity, as often reported for highly abundant deer (Côté et al., 2004) and elephants (Asner and Levick, 2012). The generality of these negative impacts is nevertheless open to debate (Guldemond et al., 2017). Predation may increase vegetation variability by promoting herbivore diversity and subsequently herbivory patterns as well as by increasing spatiotemporal variation in herbivory pressure via herbivore population dynamics and behavioural responses (Ripple and Beschta, 2004; Ford et al., 2014).

Trophic rewilding also considers effects – of the species to be re-established – that are not or at least not purely top-down trophic (Svenning et al., 2016; Figure 5.2). Megafauna increase niche space via direct diversification of carbon pools (Brunbjerg et al., 2017), via their living and dead bodies and their dung, i.e. for parasites and decomposers and associated food webs (Galetti et al., 2018), i.e. bottom-up trophic effects. They also generate environmental heterogeneity via non-trophic disturbance, such as generating wallows, compacted soils, and wells (Knapp et al., 1999; Haynes, 2012; Howison et al., 2017; Lundgren et al., 2018). Further, given the positive link between body size and mobility, megafauna increase propagule dispersal distances (e.g. Fragoso, 1997; Bunney et al., 2017) and local immigration rates for many organisms, which should lead to increased local biodiversity as well as facilitating movements to adjust to climate change. Their feeding and movement behaviours may also allow for greater numbers of plant species to coexist given more diversified propagule arrival patterns and modes, e.g. allowing for so-called megafauna fruits to be a viable dispersal adaptation (Beaune et al., 2013; Doughty et al., 2016b; Galetti et al., 2018). The high mobility of large-bodied animals will also affect nutrient dispersal patterns (Doughty et al., 2016a), but effects on biodiversity are poorly understood.

A second argument for trophic rewilding is that rich megafaunas have been typical worldwide on evolutionary timescales (Stegner and Holmes, 2013; Nenzén et al., 2014; Figure 5.2). In many areas, the largest wild-living terrestrial animal today is a medium-sized deer or similar, a highly atypical situation relative to the biotic settings in which most of current species diversity has evolved. Most of our current species and genera have evolved during the Pleistocene (2.6 million to 11.7 thousand years ago), Neogene (23–2.6 million years ago), or even further back (Rull, 2008). Most of this time has been characterised by diverse megafaunas (Stegner and Holmes, 2013; Nenzén et al., 2014), similar to protected areas in Africa today (Faurby and Svenning, 2015). Rich faunas of large-bodied vertebrates built up on all continents within 10–20 million years after the Cretaceous–Palaeogene mass extinction 66 million years ago (Smith et al., 2010), and characterised all major land masses and terrestrial ecosystem types ever since until the late Quaternary. Strong attrition of megafaunas started in prehistoric time, starting 50,000–40,000 years ago in some areas (Australia), later elsewhere, and continuing into the early Holocene, notably, with massive losses in South America 14,000–7000 years ago (Barnosky et al., 2004; Sandom et al., 2014a; Bartlett et al., 2016; Johnson et al., 2016a; Ubilla et al., 2018; Figure 5.2). The reason has been controversial, with the only two credible hypotheses in our opinion being the climate and human overkill hypotheses, with the former suggesting that extinctions happened due to the climate changes of the last glacial cycle, and the latter suggesting that it was modern humans (*Homo sapiens*) that caused this extinction during their global spread (Martin, 1967). Recent broad-scale studies provide strong support for human overkill and find little or no link to climate (Sandom et al., 2014a; Bartlett et al., 2016; Johnson et al., 2016a). From a functional perspective, however, the key point is not whether losses were human-driven, but rather that current ecosystems are unusually poor in megafauna relative to the conditions under which current species diversity has evolved (Nenzén et al., 2014; Faurby and Svenning, 2015) and, hence, would be adapted to.

As expected from their functional importance, losses of megafauna have led to ecosystem changes and biodiversity losses. Notably, recent defaunation is often associated with strong ecological impacts and negative effects on a range of organisms (Estes et al., 2011; Dirzo et al., 2014). Defaunation has been linked to a broad range of ecological effects, from effects on processes such as seed predation, seed dispersal, tree regeneration, and herbivory (Beaune et al., 2013; Daskin et al., 2016; Pérez-Méndez et al., 2016) over effects on plant communities, small mammals, and parasites (Young et al., 2013; Weinstein et al., 2017) to effects on vegetation structure and carbon storage (Bello et al., 2015). We are still data-limited with regards to impacts of the prehistoric megafauna losses, but there is increasing evidence for strong effects in some cases (Gill, 2014). In Europe, palaeoecological evidence shows high vegetation diversity, with a mix of closed forest and more open vegetation under temperate climate prior to the extinctions, and higher tree dominance afterwards (Sandom et al., 2014b). An Australian study has shown strong increases in fires in response to megafauna extinctions and associated loss of fire-sensitive trees at one locality, but not at another (Johnson et al., 2016b). There is also evidence for co-extinctions among species directly dependent on megafauna, such as dung beetles and scavenging birds, and plants depending on megafauna for seed dispersal (Galetti et al., 2018).

In summary, key components to the ecological and historical background for trophic rewilding are: (1) megafauna is important for ecosystem structure and functioning, promoting biodiversity in various ways, notably via top-down trophic effects fostering environmental heterogeneity, but also via other effects; (2) rich megafaunas have been typical worldwide on evolutionary timescales; hence, current species diversity has evolved in and, hence, would likely be adapted to megafauna-rich ecosystems; and (3) losses of megafauna on a range from recent to distant timescales have led to ecosystem changes and biodiversity losses.

Goals for trophic rewilding

Given its focus on self-regulating ecosystems, trophic rewilding constitutes an inherently open-ended restoration approach (*sensu* Hughes et al., 2011), recognising that long-term ecosystem behaviour involves continual change, and that goals therefore need to be defined in terms of trajectories rather than specific equilibrium end-points. Hence, generally speaking, goals would not be focused around maintaining a certain static ecosystem structure or species composition, e.g. in terms of spatial distribution or population abundances.

Given that trophic rewilding is a biodiversity conservation strategy, successful projects should contribute to generating and maintaining biodiversity. Hence, outcomes that remove species from the global species pool would be clear failures. At the same time, trophic rewilding should also maintain or increase biodiversity locally. This is important for maintaining or restoring the ecological effects of species, for the long-term survival and evolution of biodiversity, and, of course, for people's experience of biodiversity and the delivery of ecosystem services. In their outline of open-ended restoration, Hughes and colleagues (2011) suggest that goals should be centred on the restoration of natural processes, the development of spatiotemporal habitat mosaics, as well as benefits to society. The first two align well with ideas in trophic rewilding (Svenning et al., 2016), representing measures of the functional restoration success, in terms of improving an area's capacity for biodiversity, i.e. via increasing environmental heterogeneity and restoring natural processes identified to be especially important for promoting biodiversity in a given setting (Figure 5.2). Hence, we suggest that the core goals for trophic rewilding can be formulated in terms of maintaining or increasing an area's realised biodiversity as well as its capacity for biodiversity. In line with Hughes and colleagues' (2011) third goal, we agree that in real-world projects it will often be advantageous to also consider additional socioecological aims (Jepson and Schepers, 2016). The exact goals would vary depending on local and broader-scale societal circumstances, and in all cases it would be important to ensure compatibility with biodiversity goals. Given ongoing, and likely accelerating, future global change as well as political ephemerality, the effects of trophic rewilding – as well as any other conservation and restoration approaches – should progress towards positive effects on biodiversity, or at least capacity for biodiversity, already on short timescales, e.g. 5–10 years, to continuously place biodiversity in a better position to withstand or recover from future stresses. Trophic rewilding lends itself easily to a short-term perspective, given its use of initial interventions (introductions) is precisely done to overcome long-term historical constraints on restoration, i.e. long immigration delays.

Despite its inherent open-endedness, trophic rewilding is often seen as an attempt to reconstruct a certain past state. This partly reflects the 're-' in the term 'rewilding', partly the often-made reference to past baselines. However, we see this as a misunderstanding. The use of past baselines is not relevant for arguing for restoration of ecosystem conditions to the state at a certain prior time period, but rather to understand how ecosystems functioned on the evolutionary timescales where current biodiversity evolved, as this would better allow us to understand the potential for biodiversity (e.g. are species or functional groups missing that could find suitable living conditions), the adaptations of current species (e.g. direct or indirect dependencies on megafauna), and what factors are important for promoting biodiversity (e.g. feeding, disturbance, and movement activities of large herbivores) (Jepson and Schepers, 2016). Hence, past baselines are mainly useful for assessing what are the key biodiversity-promoting factors to restore (Higgs et al., 2014). In contrast, the open-ended function-defined character of trophic rewilding makes it a well-suited strategy for restoration under global change, where it will become increasingly unfeasible to maintain a fixed ecosystem state in many cases, while biodiversity-promoting conditions can be favoured despite changes in climate and other factors. Trophic rewilding is thereby also applicable to novel ecosystems (Hobbs et al., 2006), as it can be implemented to promote biodiversity in these systems (Bowman, 2012; Hansen, 2015; see also Hobbs and Cramer, 2008). The latter objective cannot be simply assumed to be fulfilled though in such cases, as top-down trophic effects may sometimes favour invasive species (e.g. herbivory by ungulates favouring exotic plants; Vavra et al., 2007), which may or may not be associated with biodiversity losses. Hence, this would need careful assessment.

Empirical evidence and expected impacts

A recent systematic review showed that there are only limited amounts of published peer-reviewed empirical studies dealing explicitly with rewilding, with geographic bias towards Europe, North America, and oceanic islands (Svenning et al., 2016). Among the studies detailing a position on the desirability of trophic rewilding as a conservation strategy, the majority were positive, whether for reintroductions of species extirpated < 5000 years ago, > 5000 years ago, or functional analogues of extinct species, albeit more were negative of the latter. Empirical evidence from real-world rewilding projects is too sparse and heterogeneous to allow easy generalisation at this point (Svenning et al., 2016). Still, for oceanic islands large tortoises can be concluded to be low-risk, high-impact rewilding candidates that provide key functions, notably seed dispersal, herbivory, and disturbance (Svenning et al., 2016).

The review points to two underexploited sources of information on effects on trophic rewilding, namely unintentional rewilding (introduction of a species or functional type of relevancy for restoring trophic cascades done without functional restoration as the purpose) and wildlife comebacks (where megafauna restores itself by immigration). An important source of information for unintentional rewilding comes from studies of exotic megafauna populations established outside their historical native range (Lundgren et al., 2018). Cases of unintentional rewilding and wildlife comebacks provide opportunities to study the dynamics and effects of megafauna recoveries similar to what is aimed for in trophic rewilding, e.g. effects of megafauna re-establishment on resident biodiversity and ecosystem structure and functioning, interactions with society, modulating factors such as landscape structure and climate change, as well as population dynamics of the restored species themselves.

Looking in more detail at individual rewilding projects, the reported effects are generally consistent with the overall goals of increasing an area's realised biodiversity as well as its capacity for biodiversity, notably via increased environmental heterogeneity (Figure 5.2). Trophic rewilding by reintroduction of wolves (*Canis lupus*) to Yellowstone has been reported to be successfully restoring top-down trophic effects on American elk (*Cervus canadensis*) and an associated broad range of direct and indirect effects on other animals and ecosystem structure also reported (Ripple and Beschta, 2012; Ripple et al., 2014a). Importantly, elk browsing is reported to have become lowered and more heterogeneous in space through wolf-induced reduction of the elk population and changed elk foraging behaviour, notably avoidance of areas with high predation risk, resulting in browse species growing taller and canopy cover increasing in some areas (Ripple and Beschta, 2012; Figure 5.1), hence generating greater vegetation heterogeneity and promoting overall biodiversity in the area in line with general trophic rewilding goals (as discussed earlier). Positive diversity effects on certain groups such as songbirds are also directly reported (Ripple and Beschta, 2012). However, some of the effects have been questioned, and there are clearly complex interactions with other drivers, notably American bison (*Bison bison*), climate, topography, and human activities (Mech, 2012; Ripple and Beschta, 2012; Marshall et al., 2014). Introduction of large herbivores such as de-domesticated breeds of cattle and horse as well as red deer (*Cervus elaphus*) strongly reduced woody vegetation cover in a 56-km^2 rewilding site in the Netherlands (Oostvaardersplassen; Cornelissen et al., 2014). Similar transitions from woody- to grass-dominated systems are reported for the introduction or population enhancement of large herbivores such as bison (*Bison bonasus*), musk ox (*Ovibos moschatus*), elk/moose (*Alces alces*), horses (*Equus ferus*), and reindeer (*Rangifer tarandus*) in an experimental setting in Siberia

(Zimov, 2005). The extent to which these rewilding implementations have led increases in biodiversity or the capacity for biodiversity (see earlier discussion on goals) is not fully clear. However, it can be noted that Oostvaardersplassen, established on land reclaimed from an inland bay in 1968, harbours a heterogeneous ecosystem with high bird diversity (Vera, 2009). Re-establishment of beaver (*Castor fiber*) at a degraded wetland site in Scotland led to increased habitat heterogeneity and species diversity of plants and invertebrates, in line with rewilding goals, but also increased nutrient retention (Law et al., 2016, 2017).

Based on the above and the general ecological and historical background (outlined earlier), we provide here some general expectations for impacts of trophic rewilding. First and foremost, the expected impacts cover the core goals for trophic rewilding, i.e. maintaining or increasing an area's realised biodiversity as well as its capacity for biodiversity, reflecting the direct and indirect evidence that trophic rewilding can deliver them. Here, we restate them in more detail, plus providing additional predictions, e.g. concerning modulating factors (Figure 5.2A, Figure 5.3B).

- Functional ecosystem impacts:

  - Increased local- and landscape-scale environmental heterogeneity and vegetation diversity. These effects are especially driven by the direct effects of large herbivores, but may be modulated by large carnivores.
  - Positive effects on heterogeneity will increase with increasing herbivore species and functional diversity, with the strongest effects with megaherbivores present.
  - Heterogeneity increases will be stronger when initial vegetation structure and topographic–edaphic conditions are heterogeneous, due to non-random megafauna habitat use.
  - Effects will be strongest in relatively warm, moderately humid climates where vegetation tends towards bistability between woody and herbaceous dominance, but also important elsewhere.
  - Greater rates of long-distance dispersal of propagules and nutrients, but also increased heterogeneity in deposition and hence dispersion patterns.

- Biodiversity outcomes:

  - Increased species diversity at local and landscape scales.
  - Increases in abundance and diversity of organisms with strong links to megafauna, via top-down trophic effects as well as disturbance and bottom-up trophic effects, such as plants with strong anti-herbivore defence, grazing-tolerant, disturbance-adapted, and megafauna-fruit plants, scavenging vertebrates, and dung- and carcass-dependent invertebrates and fungi.

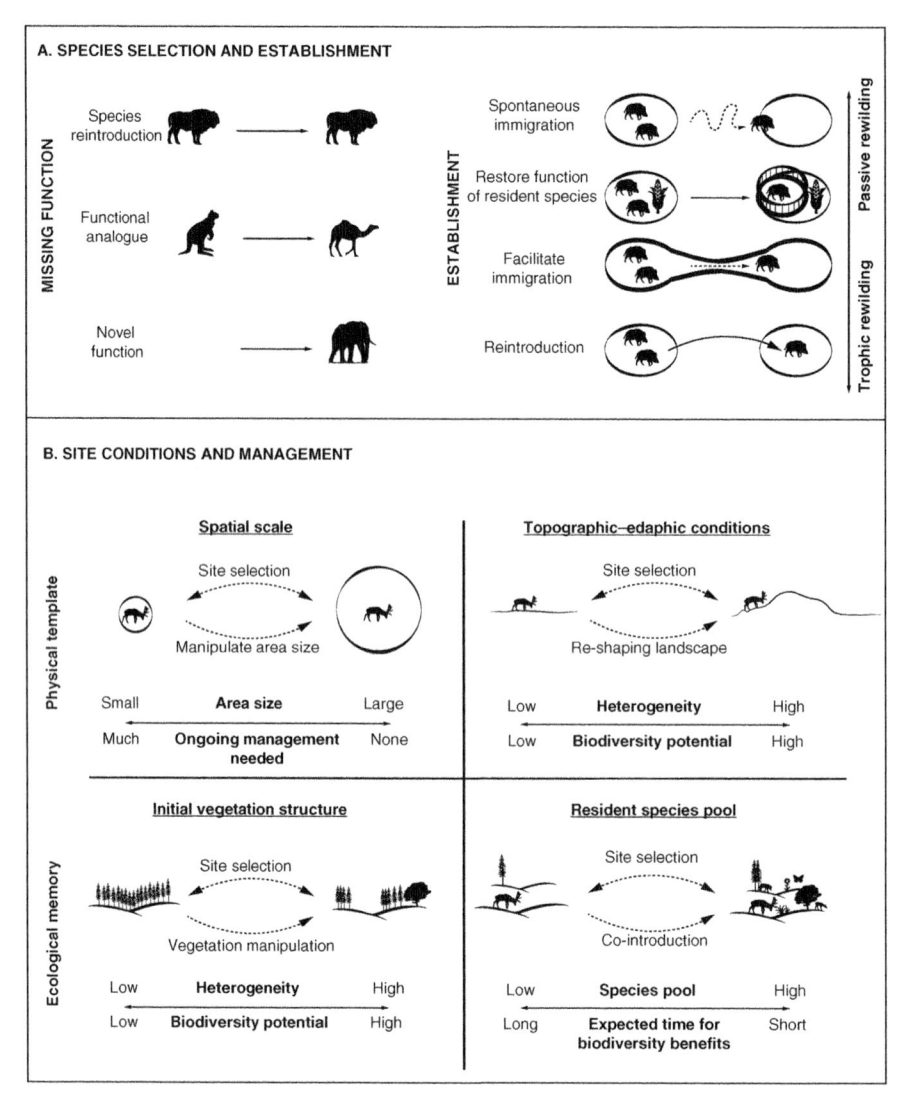

Figure 5.3. Key considerations for implementing trophic rewilding: (A) species selection and establishment, and (B) site conditions and the need for ongoing management.

Ecological design considerations

Active species introduction is the key difference between trophic rewilding and passive rewilding, which is simply the reduction or cessation of human management (Navarro and Pereira, 2012). However, facilitation of species immigration or local population expansion may also be involved to restore top-down trophic interactions, for example, establishment of habitat corridors or legislative changes, causing the two to intergrade (Figure 5.3A). Further, restoration may also focus on altering the behaviour of resident species to restore their trophic role, such as fencing to shift feeding from agricultural fields to adjacent natural areas (Fløjgaard et al., 2017b). From a functional perspective, the method whereby species or functional types are restored is not crucial, just that the re-establishment is successful. Selection of species for introductions should consider how the functions of the candidate species match the processes to be restored (Svenning et al., 2016). Further, phylogenetic relatedness to lost taxa should also be considered, notably to capture subtle functional aspects that may not be well described by the available trait information, but arguably also to restore the evolutionary potential of a lineage that belonged to a region or ecosystem prior to human impacts (Donlan et al., 2005; Svenning et al., 2016). Other factors to consider include suitability of the present and forecasted future climate, ecosystem, and societal circumstances with respect to accommodating a given candidate species (Svenning et al., 2016). The primary focus would generally be species with a recent to deeper-time history in the region, i.e. biogeographically native species (Figure 5.3A). However, non-native functional analogues may also be considered, notably for extinct species, but also for species that are otherwise non-available or ineligible, for example, for societal reasons (Figure 5.3A). Species used for or proposed as functional analogues have so far mainly included non-native relatives of extinct species and more or less de-domesticated forms of species with no or few surviving originally wild populations (Bunzel-Drüke, 2001; Donlan et al., 2005; Hansen et al., 2010). Analogues could also be species only distantly related to extinct former natives, but able to replace these functionally or more generally provide positive biodiversity effects via their functions (Bowman, 2012; Hansen, 2015; Figure 5.3A). In the future, de-extinct species or extant species genetically modified to have characteristics of relevant extinct species could also become an option (Seddon, 2017).

Oceanic islands are a special case for trophic rewilding. Isolated islands naturally did not harbour similarly large terrestrial animals as mainland, with smaller-sized effective megafaunas often including dwarfed members of mainland megafauna groups or giant members of non-megafauna groups (Faurby and Svenning, 2016), as well as other deviant forms, for example, flightless birds (Wright et al., 2016). Most island faunas have undergone severe human-driven species losses, increasing the need for functional analogues for rewilding (Hansen et al., 2010; Hansen, 2015). This requires special care on isolated islands as they are especially susceptible to invasions. Importantly, herbivory by flightless birds and giant tortoises is not equivalent to that by ungulates, and plant species from islands with no native large mammalian herbivores may be very sensitive to ungulate herbivory (Lee et al., 2010). A number of island projects exist where functional analogues are being tested, notably extant tortoises as replacements for extinct ones (Hansen et al., 2010; Hansen, 2015).

Spatial scale is a key consideration for implementing trophic rewilding (Figure 5.3B). If the scale of the restoration area is large, initial introduction may suffice as management, and ongoing management may be largely avoided. An example is the Yellowstone National Park, which is 9000 km^2, where the restoration of wolves and their ecological functions only required their reintroduction (Ripple and Beschta, 2012; see earlier discussion of the debate around this case). Areas down to 10 km^2 may in many cases offer similar opportunities, especially for herbivores (Fløjgaard et al., 2017a). Smaller areas will often require some level of ongoing management of the megafauna, such as some level of genetic management to avoid inbreeding, intermittent management to overcome limited possibilities for large-scale movements in response to weather or other environmental changes as well as limited possibilities to maintain viable populations (Miller et al., 2015). However, these constraints will be modulated by the extent the megafauna species are limited to a given nature area, or can also use or migrate across the surrounding landscape. Overall, as the scale becomes smaller, the ongoing need for management will increase, even if it will be smaller than for much conventional nature management, e.g. seasonal livestock grazing or mowing-based management (J.-C. Svenning, personal observation). Hence, trophic rewilding is actually a continuum that grades into controlled human management (Jepson and Schepers, 2016).

Another facet to the success of trophic rewilding initiatives is environmental variability provided by the physical template in the focal area (Figure 5.3B). A central factor for biodiversity outcomes of restored trophic effects lies in the generation of expanded niche space, notably via varied habitat conditions. Although such variation may be generated endogenously, it arises more easily if there is existing topographic or other habitat variability, as this should generate spatial variability in megafauna space use and their impacts, for example, steep topography may be avoided by herbivores (Terborgh et al., 2016).

Ecological memory, legacies of prior ecosystem states (Johnstone et al., 2016), is also central for ecosystem responses to trophic rewilding (Figure 5.3B), especially given the importance of short-term gains. Ecological memory may both facilitate and obstruct successful outcomes; for example, initial restoration activities may be needed to escape non-favoured alternative stable states. Initial vegetation structure may shape the activities of megafauna species (Ford et al., 2014; Franklin and Harper, 2016); for example, *Chelonoidis hoodensis* giant tortoises have been re-established on Española Island in the Galapagos Archipelago, but with limited ecological impact due to mobility constraints from dense woody vegetation, likely established as a consequence of now-eradicated feral goats and prolonged absence of tortoises (Gibbs et al., 2014), and clearing some of this vegetation would likely facilitate the functional restoration of the tortoises. Trophic rewilding might also be hampered by impoverished resident species pools, and co-introductions of relevant species might enhance biodiversity effects. For example, many plants are strongly dispersal limited in modern agricultural landscapes (Poschlod and Bonn, 1998; Ozinga et al., 2009), and may have low short-term spontaneous arrival possibilities even if suitable habitat conditions are restored. Here, initial introduction could be considered. Further, in the context of climate change, rewilding outcomes may benefit from assisted colonisation on larger scales (Hoegh-Guldberg et al., 2008), i.e. to help ecosystem species pools track climate and hence to realise potential biodiversity increases in response to rewilding effects on ecosystem structure and processes.

Societal perspectives and risks

Trophic rewilding needs to be developed in an Anthropocene context, i.e. to carefully consider its interactions with society as it will have to be implemented on a densely human-occupied and -used planet (Ellis et al., 2013). In regions where defaunation is ongoing, trophic rewilding is less relevant, but could still be implemented in subregions with better governance or protection (Galetti, 2004; Galetti et al., 2017). Similarly, it may be argued that trophic rewilding is not feasible in densely human-occupied landscapes due to limited habitat availability and human–wildlife conflicts. However, as the widespread comeback of large-bodied mammal and bird species in Europe shows (Deinet et al., 2013), reflecting the ecological generalism of most megafauna species, these can thrive in such landscapes if accepted by society. Furthermore, as discussed earlier, trophic rewilding can be implemented flexibly and hence also to small nature areas within intensely used landscapes (Figures 5.1 and 5.3).

Trophic rewilding will need to be integrated into policy and to become culturally accepted to be broadly implemented. As trophic rewilding represents a novel restoration approach (*sensu* Hobbs and Cramer, 2008), there are challenges for policy and societal acceptance (Jepson and Schepers, 2016). Its inherent open-endedness will often be one of the most challenging aspects, as it conflicts with conservation policies with compositional targets – for example, the European Union's Habitat Directive's orientation to preserve specific habitat types in favourable conservation status (Jepson and Schepers, 2016). However, in countries where nature conservation is more wilderness-oriented, open-endedness will not be as controversial (while trophic rewilding is often still relevant due to past defaunation; Galetti et al., 2017). Another major policy challenge concerns the use of species that are not considered native because they have not occurred in the area in recent centuries, i.e. because they were extirpated in prehistoric times or are truly exotic species used as functional analogues. This clashes with conservation's classic focus on native species (Lundgren et al., 2018), and may conflict with laws and regulations aimed at invasive species. A further challenge concerns laws protecting animal welfare (Jepson, 2016), especially when it comes to projects with some level of ongoing human management, such as fenced nature areas (Fløjgaard et al., 2017b), as well as projects involving de-domesticated animals or formerly captive individuals of wild species (Klaver et al., 2002). Here, conflicts may arise in relation to variation in animal health, for example in relation to starvation and death. Finally, there may be conflicts in relation to agricultural production, such as real or perceived risks of disease spread between wild and domestic populations, biohazard regulations that demand destruction of carcasses (Jepson, 2016), crop damage, and predation on farm animals.

It is important to consider the ecosystem services that society may derive from trophic rewilding. Expected societal benefits have inspired many rewilding projects (Jepson, 2016; Jepson and Schepers, 2016), notably direct benefits to people's quality of life as well as enhanced possibilities for eco-tourism and other means of income. However, under which societal circumstances is trophic rewilding then most relevant? In terms of socioecological factors, Hughes and colleagues (2012) highlight nature areas that are large and remote, where management costs are high, and where the public demands wildness as particularly suitable for open-ended restoration. The potential for conflicts with society is likely smaller in remote, large nature areas. However, management costs and public demands for wildness are often high in intensely human-used landscapes (Jepson, 2016). In promoting self-regulating ecosystems, rewilding is expected to facilitate ecosystem services at reduced economic costs (Navarro and Pereira, 2012). This is a major inspiration for much practitioner interest where management costs are high (J.-C. Svenning, personal observation). Further, it is where human population density is high, such as near or even in large urban centres that most people will be able to benefit from the ecosystem services provided (Jepson, 2016). We also note that there may also be scope for developing hybridised production-rewilding approaches (Root-Bernstein et al., 2017), e.g. integrating some harvesting of 'wild' meat into rewilding projects, perhaps mimicking predator effects on large herbivores. Reflecting such considerations, trophic rewilding is being implemented even in many small nature areas in densely inhabited landscapes (Figure 5.1, Figure 5.3B), and we suggest that trophic rewilding can be applied in landscapes ranging from remote areas with low human population densities and little intense land use to highly urbanised landscapes.

Trophic rewilding involves perceived risks, notably risks to native biodiversity and human–wildlife conflicts. These are primarily related to species introductions, notably the potential for negative effects on resident species and risks associated with co-introduction of pathogens and parasites (Nogués-Bravo et al.,

2016; Rubenstein et al., 2006). Overall, at broader scales, we believe risks are small and can be benchmarked against numerous realised conservation translocations and hunting introductions, and frequent cross-border movement of domestic, pet, and zoo animals. In many cases, risks can be assessed by looking at nearby regions, as most introductions involve species resident or used there. For more novel species or settings, we suggest an experimental approach to assess risks. Here, one would do testing under controlled circumstances, typically in small, fenced areas, before starting broader applications. This said, we note that megafauna represent some of the least risky species to introduce as they are mostly relatively easy, or at least possible, to control due to generally high visibility and relatively low population sizes and reproductive rates (i.e. relative to, e.g., invertebrates or small vertebrates). Still, all wild-living megafauna can create human–wildlife conflicts. These should be carefully assessed and addressed before implementing rewilding projects, and monitored and adaptively managed afterwards, building on the large literature on human–wildlife conflicts to reduce conflicts (Pooley et al., 2017). Overall, a key issue is to develop projects via participatory processes to facilitate acceptance. The main biodiversity risk may be in small biodiversity-rich localities depending on a certain ecosystem structure and often ongoing management, e.g. a small species-rich meadow. Here, open-ended restoration of any kind could be risky and would need careful monitoring (Hughes et al., 2011, 2012).

Rewilding projects should generally have an associated monitoring programme. How intense this should be depends on how novel and potentially risky the project is, the presence of species of special conservation concern, as well as a range of other factors, including societal interests and legislation. As mentioned earlier, monitoring should be designed in the context of open-ended projects and hence focused on trajectories rather than equilibrium end-points (Hughes et al., 2011), and should focus on functional restoration, biodiversity outcomes, as well as socio-ecological aims.

Outlook

Overall, we see trophic rewilding as having large potential as an ecological restoration approach for the Anthropocene. This is especially due to its focus on restoring the often megafauna-linked top-down trophic and associated processes that have characterised ecosystems on the evolutionary timescales during which Earth's rich current species diversity has evolved, with the aim of promoting biodiversity in a function-oriented, pragmatic manner also in human-used landscapes, under climate change, and in novel ecosystems. However, empirical literature on trophic rewilding is limited, and there is a need for a systematic research programme to develop its scientific basis (Svenning et al., 2016). A key opportunity lies in the rising number of practical trophic rewilding projects (Figure 5.1), if monitoring and research programmes become associated with them. There is also much more to learn from a range of related fields, from palaeoecological studies to research on wildlife ecology and managements, notably as related to megafauna and their ecological impacts. As trophic rewilding has focused on new establishment of species in ecosystems, research on settings where megafauna species are relatively new in a given system will be particularly informative. These can range from wildlife comebacks (Deinet et al., 2013); over classic conservation reintroductions; to sports game introductions; and cases of exotic and invasive species, with some of these qualifying as unintentional trophic rewilding, i.e. introduction of a species or functional type of relevancy for restoring trophic cascades done without knowledge that it is the case (Wilder et al., 2014; Svenning et al., 2016). The following topics have previously been highlighted as needed research foci for trophic rewilding research (Svenning et al., 2016; Figure 5.4A).

(1) Global scope, as research in the area is geographically biased.
(2) Role of trophic complexity, to assess the importance of having functional diversity of herbivores and carnivores for rewilding outcomes (Bakker et al., 2016; Figure 5.2).
(3) Landscape setting and interplay with society, highlighting that trophic rewilding is being implemented in a range of settings from small fenced reserves to wildlands integrated into intensely used landscapes to large wilderness areas (Figures 5.1 and 5.3B). Some of the key aspects to consider here are the importance of area size, topographic–edaphic heterogeneity, and ecological memory (initial vegetation structure, resident species pool) for rewilding outcomes (Figure 5.3B). Related to this, it is important to study how trophic rewilding interplays with abiotic rewilding, for example, restoration of hydrological dynamics.
(4) Management tools and targets, as while trophic rewilding is inherently an open-ended restoration approach (Hughes et al., 2011), goals for functional restoration success, biodiversity outcomes, and societal impacts should be set and monitored, and strategies for intervention to increase rewilding benefits and handling disservices such as human–wildlife conflicts (Pooley et al., 2017) need to be assessed.
(5) Climate change will strongly affect all ecosystems in the coming decades and centuries, and trophic rewilding will need to integrate this in its design. A related issue is that more and more natural areas will harbour novel ecosystems (Hobbs et al., 2006) due to human-caused species introductions and environmental change (Ordonez et al., 2016), and we need to understand how trophic rewilding can help promote biodiversity in such situations (Bowman, 2012).

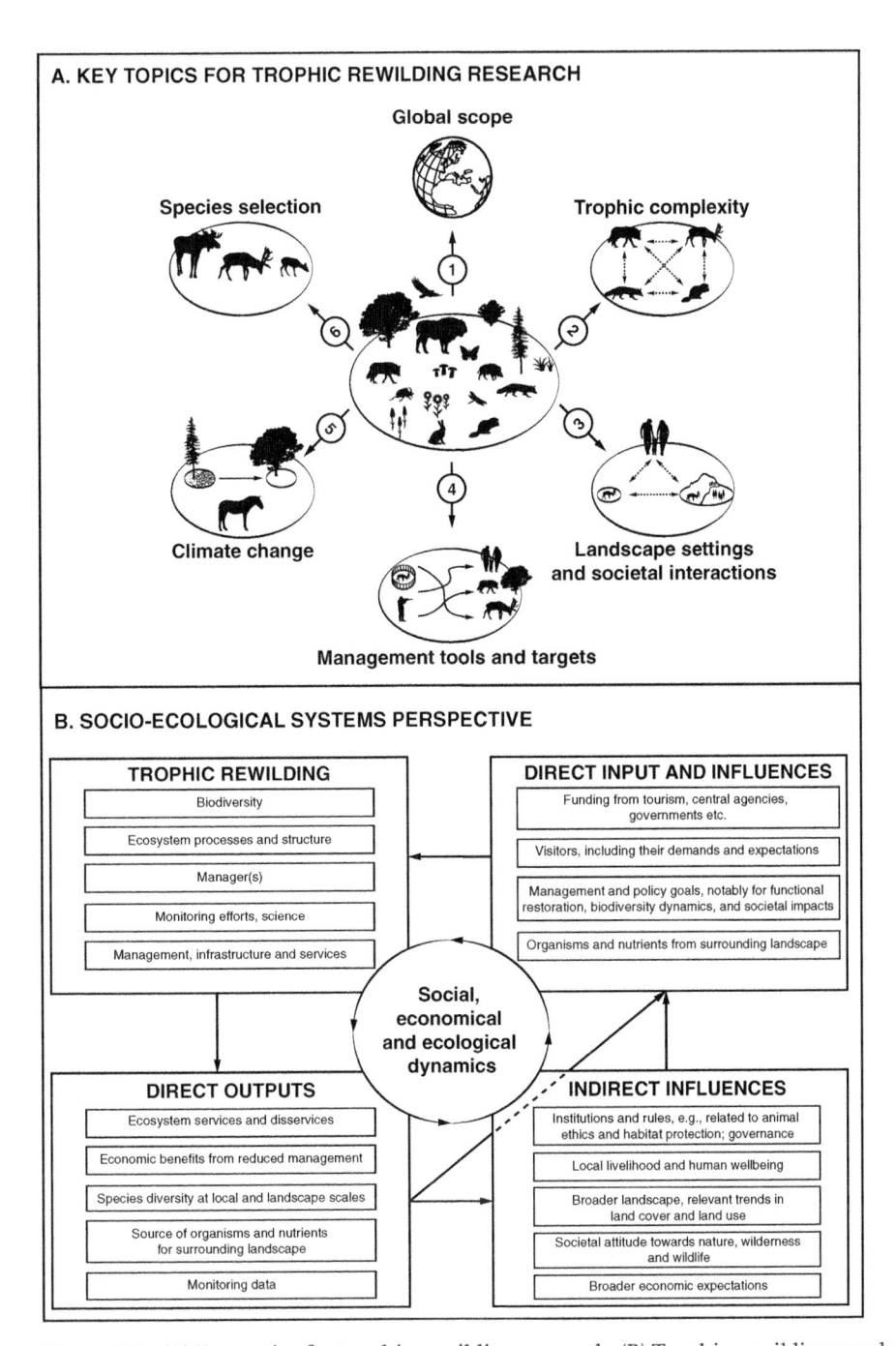

Figure 5.4. (A) Key topics for trophic rewilding research. (B) Trophic rewilding needs to be implemented in an Anthropocene context, and the socioecological systems perspective will be a useful framework to develop (figure adapted from Cumming and Allen, 2017).

(6) Species selection, pointing to the benefits of developing species-selection guidelines for trophic rewilding, although we also see it as important to ensure societal space for rewilding as a bottom-up-driven, experimental approach (Jepson, 2016; Jepson and Schepers, 2016).

These topics are all still key research foci. However, we would highlight the need to expand on the societal perspectives in (3) and (4), as most of the world is human-used and -occupied (Ellis et al., 2013). Hence, the socioecological systems perspective (Cumming and Allen, 2017) offers a highly useful framework to develop for trophic rewilding, to understand its societal facilitators and constraints as well as ecosystem services and disservices in the context of interacting social, economic, and ecological dynamics (Figure 5.4B).

Conclusion

Trophic rewilding is an ecological restoration strategy that uses species introductions to restore top-down trophic interactions and associated trophic cascades to promote self-regulating biodiverse ecosystems. Most real-world rewilding projects fit this definition, with a focus on re-establishing missing large-bodied animals. Key background for trophic rewilding is (1) the ecological importance of megafauna, with top-down forcing by large-bodied herbivores and carnivores a widespread driver of ecosystem structure and dynamics, promoting biodiversity in various ways, notably by fostering environmental heterogeneity; and (2) that rich megafaunas have been typical worldwide on the evolutionary timescales at which current species diversity has evolved. The main goal for trophic rewilding is to contribute to the maintenance of biodiversity at global and local scales, but given its focus on self-regulating ecosystems, it is inherently an open-ended approach, recognising that ecosystem behaviour involves continual change, and aims need to be defined in terms of trajectories rather than end-points. Linked to this, the main use of past baselines is for understanding the potential for biodiversity, the adaptations of current species, and what factors are needed to promote biodiversity, rather than for restoring to a certain prior state. This said, the megafauna itself of course is also part of biodiversity and hence part of the restoration goal. There are only limited amounts of empirical studies explicitly on rewilding, but these generally suggest beneficial effects. Important underexploited information sources are cases of unintentional rewilding, wildlife comebacks, exotic megafauna, and the rich literature on the ecology of present and past megafauna. Based hereon, some general expectations for impacts of trophic rewilding can be made, notably increased environmental heterogeneity, with effects increasing with increasing species and functional diversity of herbivores, and when initial vegetation structure and topographic–edaphic conditions are heterogeneous; increased species diversity at local and landscape scales; and increases in abundance and diversity of organisms with links to megafauna.

Important factors to consider for implementing trophic rewilding include species selection and establishment, as well as site conditions and management, with the latter linked to area, environmental heterogeneity, and ecological memory. Furthermore, interactions with society also need careful attention, e.g. ecosystem services and human–wildlife conflicts. We see trophic rewilding as having large potential as an ecological restoration and land management approach for the Anthropocene, focusing on restoring top-down trophic and related processes, but in a function-oriented, pragmatic manner. However, there is a need for a systematic research programme to develop the scientific basis, including the socioecological systems perspective.

Acknowledgements

This work is a contribution to the Carlsberg Foundation Semper Ardens project MegaPast2Future (CF16-0005) and to the VILLUM Investigator project 'Biodiversity Dynamics in a Changing World' funded by VILLUM FONDEN (grant 16549).

References

Allen, B.L., Allen, L.R., Andrén, H., et al. (2017). Can we save large carnivores without losing large carnivore science? *Food Webs*, **12**, 64–75.

Asner, G.P., and Levick, S.R. (2012). Landscape-scale effects of herbivores on treefall in African savannas. *Ecology Letters*, **15**, 1211–1217.

Bakker, E.S., Gill, J.L., Johnson, C.N., et al. (2016). Combining paleo-data and modern exclosure experiments to assess the impact of megafauna extinctions on woody vegetation. *Proceedings of the National Academy of Sciences of the United States of America*, **113**, 847–855.

Barnosky, A.D., Koch, P.L., Feranec, R.S., Wing, S.L., and Shabel, A.B. (2004). Assessing the causes of late Pleistocene extinctions on the continents. *Science*, **306**, 70–75.

Bartlett, L.J., Williams, D.R., Prescott, G.W., et al. (2016). Robustness despite uncertainty: regional climate data reveal the dominant role of humans in explaining global extinctions of Late Quaternary megafauna. *Ecography*, **39**, 152–161.

Beaune, D., Fruth, B., Bollache, L., Hohmann, G., and Bretagnolle, F. (2013). Doom of the elephant-dependent trees in a Congo tropical forest. *Forest Ecology and Management*, **295**, 109–117.

Bello, C., Galetti, M., Pizo, M.A., et al. (2015). Defaunation affects carbon storage in tropical forests. *Science Advances*, **1**, e1501105.

Bond, W.J. (2005). Large parts of the world are brown or black: a different view on the 'Green World' hypothesis. *Journal of Vegetation Science*, **16**, 261–266.

Bowman, D. (2012). Conservation: bring elephants to Australia? *Nature*, **482**, 30–30.

Brunbjerg, A.K., Bruun, H.H., Moeslund, J.E., Sadler, J.P., Svenning, J.-C., and Ejrnæs, R. (2017). Ecospace: a unified framework for understanding variation in terrestrial biodiversity. *Basic and Applied Ecology*, **18**, 86–94.

Bunney, K., Bond, W.J., and Henley, M. (2017). Seed dispersal kernel of the largest surviving megaherbivore – the African savanna elephant. *Biotropica*, **49**, 395–401.

Bunzel-Drüke, M. (2001). Ecological substitutes for wild horse (Equus ferus Boddaert, 1785 = E. prezewalski Poljakov, 1881) and aurochs (Bos primigenius Bojanus, 1827). *Natur- und Kulturlandschaft*, **4**, 240–252.

Cornelissen, P., Bokdam, J., Sykora, K., and Berendse, F. (2014). Effects of large herbivores on wood pasture dynamics in a European wetland system. *Basic and Applied Ecology*, **15**, 396-406.

Côté, S.D., Rooney, T.P., Tremblay, J.-P., Dussault, C., and Waller, D.M. (2004). Ecological impacts of deer overabundance. *Annual Review of Ecology, Evolution, and Systematics*, **35**, 113-147.

Cromsigt, J.P.G.M., and te Beest, M. (2014). Restoration of a megaherbivore: landscape-level impacts of white rhinoceros in Kruger National Park, South Africa. *Journal of Ecology*, **102**, 566-575.

Cumming, G.S., and Allen, C.R. (2017). Protected areas as social-ecological systems: perspectives from resilience and social-ecological systems theory. *Ecological Applications*, **27**, 1709-1717.

Daskin, J.H., Stalmans, M., and Pringle, R.M. (2016). Ecological legacies of civil war: 35-year increase in savanna tree cover following wholesale large-mammal declines. *Journal of Ecology*, **104**, 79-89.

Deinet, S., Ieronymidou, C., McRae, L., et al. (2013). *Wildlife comeback in Europe: the recovery of selected mammal and bird species. Final report to Rewilding Europe by ZSL, Birdlife International and the European Bird Census Council*. London: Zoological Society of London.

Dirzo, R., Young, H.S., Galetti, M., Ceballos, G., Isaac, N.J.B., and Collen, B. (2014). Defaunation in the Anthropocene. *Science*, **345**, 401-406.

Donlan, C.J., Berger, J., Bock, C.E., et al. (2006). Pleistocene rewilding: an optimistic agenda for twenty-first century conservation. *The American Naturalist*, **168**, 660-681.

Donlan, J., Green, H.W., Berger, J., et al. (2005). Re-wilding North America. *Nature*, **436**, 913-914.

Doughty, C.E., Roman, J., Faurby, S., et al. (2016a). Global nutrient transport in a world of giants. *Proceedings of the National Academy of Sciences of the United States of America*, **113**, 868-873.

Doughty, C.E., Wolf, A., Morueta-Holme, N., et al. (2016b). Megafauna extinction, tree species range reduction, and carbon storage in Amazonian forests. *Ecography*, **39**, 194-203.

Ellis, E.C., Kaplan, J.O., Fuller, D.Q., Vavrus, S.J., Goldewijk, K.K., and Verburg, P.H. (2013). Used planet: a global history. *Proceedings of the National Academy of Sciences of the United States of America*, **110**, 7978-7985.

Estes, J.A., Terborgh, J., Brashares, J.S., et al. (2011). Trophic downgrading of Planet Earth. *Science*, **333**, 301-306.

Faurby, S., and Svenning, J.-C. (2015). Historic and prehistoric human-driven extinctions have reshaped global mammal diversity patterns. *Diversity and Distributions*, **21**, 1155-1166.

Faurby, S., and Svenning, J.-C. (2016). Resurrection of the island rule: human-driven extinctions have obscured a basic evolutionary pattern. *The American Naturalist*, **187**, 812-820.

Fløjgaard, C., Bladt, J., and Ejrnæs, R. (2017a). *Naturpleje og arealstørrelser med særligt fokus på Natura 2000 områderne. Videnskabelig rapport fra DCE – Nationalt Center for Miljø og Energi nr. 228*. Aarhus: Aarhus University.

Fløjgaard, C., De Barba, M., Taberlet, P., and Ejrnæs, R. (2017b). Body condition, diet and ecosystem function of red deer (*Cervus elaphus*) in a fenced nature reserve. *Global Ecology and Conservation*, **11**, 312-323.

Ford, A.T., Goheen, J.R., Otieno, T.O., et al. (2014). Large carnivores make savanna tree communities less thorny. *Science*, **346**, 346-349.

Fragoso, J.M.V. (1997). Tapir-generated seed shadows: scale-dependent patchiness in the Amazon rain forest. *Journal of Ecology*, **85**, 519-529.

Franklin, C.M.A., and Harper, K.A. (2016). Moose browsing, understorey structure and plant species composition across spruce budworm-induced forest edges. *Journal of Vegetation Science*, **27**, 524-534.

Galetti, M. (2004). Parks of the Pleistocene: recreating the Cerrado and the Pantanal with megafauna. *Natureza & conservaçao*, **2**, 93-100.

Galetti, M., Pires, A.S., Brancalion, P.H.S., and Fernandez, F.A.S. (2017). Reversing defaunation by trophic rewilding in empty forests. *Biotropica*, **49**, 5-8.

Galetti, M., Moleón, M., Jordano, P., et al. (2018). Ecological and evolutionary legacy of megafauna extinctions. *Biological Reviews*, **93**, 845-862.

Gibbs, J.P., Hunter, E.A., Shoemaker, K.T., Tapia, W.H., and Cayot, L.J. (2014). Demographic outcomes and ecosystem implications of giant tortoise reintroduction to Española Island, Galapagos. *PLoS ONE*, **9**, e110742.

Gill, J.L. (2014). Ecological impacts of the late Quaternary megaherbivore extinctions. *New Phytologist*, **201**, 1163-1169.

Griffiths, C.J., Hansen, D.M., Jones, C.G., Zuël, N., and Harris, S. (2011). Resurrecting extinct interactions with extant substitutes. *Current Biology*, **21**, 762-765.

Guldemond, R.A.R., Purdon, A., and van Aarde, R.J. (2017). A systematic review of elephant impact across Africa. *PLoS ONE*, **12**, e0178935.

Hansen, D.M. (2015). Non-native megaherbivores: the case for novel function to manage plant invasions on islands. *AoB Plants*, **7**, plv085.

Hansen, D.M., Donlan, C.J., Griffiths, C.J., and Campbell, K.J. (2010). Ecological history and latent conservation potential: large and giant tortoises as a model for taxon substitutions. *Ecography*, **33**, 272-284.

Haynes, G. (2012). Elephants (and extinct relatives) as earth-movers and ecosystem engineers. *Geomorphology*, **157-158**, 99-107.

Higgs, E., Falk, D.A., Guerrini, A., et al. (2014). The changing role of history in restoration ecology. *Frontiers in Ecology and the Environment*, **12**, 499-506.

Hobbs, R.J., and Cramer, V.A. (2008) Restoration ecology: interventionist approaches for restoring and maintaining ecosystem function in the face of rapid environmental change. *Annual Review of Environment and Resources*, **33**, 39-61.

Hobbs, R.J., Arico, S., Aronson, J., et al. (2006). Novel ecosystems: theoretical and management aspects of the new ecological world order. *Global Ecology and Biogeography*, **15**, 1-7.

Hoegh-Guldberg, O., Hughes, L., McIntyre, S., et al. (2008). Assisted colonization and rapid climate change. *Science*, **321**, 345-346.

Hopcraft, J.G.C., Olff, H., and Sinclair, A.R.E. (2010). Herbivores, resources and risks: alternating regulation along primary environmental gradients in savannas. *Trends in Ecology & Evolution*, **25**, 119-128.

Hovick, T.J., Elmore, R.D., and Fuhlendorf, S.D. (2014). Structural heterogeneity increases diversity of non-breeding grassland birds. *Ecosphere*, **5**, 1-13.

Howison, R.A., Olff, H., van de Koppel, J., and Smit, C. (2017). Biotically driven vegetation mosaics in grazing ecosystems: the battle between bioturbation and biocompaction. *Ecological Monographs*, **87**, 363-378.

Hughes, F.M.R., Stroh, P.A., Adams, W.M., Kirby, K.J., Mountford, J.O., and Warrington, S. (2011). Monitoring and evaluating large-scale, 'open-ended' habitat creation projects: a journey rather than a destination. *Journal for Nature Conservation*, **19**, 245-253.

Hughes, F.M.R., Adams, W.M., and Stroh, P.A. (2012). When is open-endedness desirable in restoration projects? *Restoration Ecology*, **20**, 291-295.

Jepson, P. (2016). A rewilding agenda for Europe: creating a network of experimental reserves. *Ecography*, **39**, 117-124.

Jepson, P., and Schepers, F. (2016). *Policy brief: making space for rewilding: creating an enabling policy environment*. Oxford/Nijmegen:

Rewilding Europe, University of Oxford.

Johnson, C.N., Alroy, J., Beeton, N.J., et al. (2016a). What caused extinction of the Pleistocene megafauna of Sahul? *Proceedings of the Royal Society of London B: Biological Sciences*, **283**(1824).

Johnson, C.N., Rule, S., Haberle, S.G., Kershaw, A.P., McKenzie, G.M., and Brook, B.W. (2016b). Geographic variation in the ecological effects of extinction of Australia's Pleistocene megafauna. *Ecography*, **39**, 109-116.

Johnstone, J.F., Allen, C.D., Franklin, J.F., et al. (2016). Changing disturbance regimes, ecological memory, and forest resilience. *Frontiers in Ecology and the Environment*, **14**, 369-378.

Jones, C.G., Lawton, J.H., and Shachak, M. (1994). Organisms as ecosystem engineers. *Oikos*, **69**, 373-386.

Jorge, M.L.S.P., Galetti, M., Ribeiro, M.C., and Ferraz, K.M.P.M.B. (2013). Mammal defaunation as surrogate of trophic cascades in a biodiversity hotspot. *Biological Conservation*, **163**, 49-57.

Klaver, I., Keulartz, J., and Van Den Belt, H. (2002). Born to be wild. *Environmental Ethics*, **24**, 3-21.

Knapp, A.K., Blair, J.M., Briggs, J.M., et al. (1999). The keystone role of bison in North American tallgrass prairie – bison increase habitat heterogeneity and alter a broad array of plant, community, and ecosystem processes. *BioScience*, **49**, 39-50.

Law, A., McLean, F., and Willby, N.J. (2016). Habitat engineering by beaver benefits aquatic biodiversity and ecosystem processes in agricultural streams. *Freshwater Biology*, **61**, 486-499.

Law, A., Gaywood, M.J., Jones, K.C., Ramsay, P., and Willby, N.J. (2017). Using ecosystem engineers as tools in habitat restoration and rewilding: beaver and wetlands. *Science of the Total Environment*, **605-606**, 1021-1030.

Lee, W.G., Wood, J.R., and Rogers, G.M. (2010). Legacy of avian-dominated plant-herbivore systems in New Zealand. *New Zealand Journal of Ecology*, **34**, 28-47.

Lundgren, E.J., Ramp, D., Ripple, W.J., and Wallach, A.D. (2018). Introduced megafauna are rewilding the Anthropocene. *Ecography*, **41**, 857-863.

Marshall, K.N., Cooper, D.J., and Hobbs, N.T. (2014). Interactions among herbivory, climate, topography and plant age shape riparian willow dynamics in northern Yellowstone National Park, USA. *Journal of Ecology*, **102**, 667-677.

Martin, P.S. (1967). Pleistocene overkill. *Natural History*, December, 32-38.

Mech, L.D. (2012). Is science in danger of sanctifying the wolf? *Biological Conservation*, **150**, 143-149.

Miller, S.M., Harper, C.K., Bloomer, P., Hofmeyr, J., and Funston, P.J. (2015). Fenced and fragmented: conservation value of managed metapopulations. *PLoS ONE*, **10**, e0144605.

Mills, L.S., Soulé, M.E., and Doak, D.F. (1993). The keystone-species concept in ecology and conservation. *BioScience*, **43**, 219-224.

Navarro, L.M., and Pereira, H.M. (2012). Rewilding abandoned landscapes in Europe. *Ecosystems*, **15**, 900-912.

Nenzén, H.K., Montoya, D., and Varela, S. (2014). The impact of 850,000 years of climate changes on the structure and dynamics of mammal food webs. *PLoS ONE*, **9**, e106651.

Nogués-Bravo, D., Simberloff, D., Rahbek, C., and Sanders, N.J. (2016). Rewilding is the new Pandora's box in conservation. *Current Biology*, **26**, R87-R91.

Ordonez, A., Williams, J.W., and Svenning, J.-C. (2016). Mapping climatic mechanisms likely to favour the emergence of novel communities. *Nature Climate Change*, **6**, 1104-1109.

Owen-Smith, R.N. (1988). *Megaherbivores: the influence of very large body size on ecology*. Cambridge: Cambridge University Press.

Ozinga, W.A., Römermann, C., Bekker, R.M., et al. (2009). Dispersal failure contributes to plant losses in NW Europe. *Ecology Letters*, **12**, 66–74.

Paine, R.T. (1980). Food webs: linkage, interaction strength and community infrastructure. *Journal of Animal Ecology*, **49**, 667–685.

Pérez-Méndez, N., Jordano, P., García, C., and Valido, A. (2016). The signatures of Anthropocene defaunation: cascading effects of the seed dispersal collapse. *Scientific Reports*, **6**, 24820.

Pooley, S., Barua, M., Beinart, W., et al. (2017). An interdisciplinary review of current and future approaches to improving human–predator relations. *Conservation Biology*, **31**, 513–523.

Poschlod, P., and Bonn, S. (1998). Changing dispersal processes in the central European landscape since the last ice age: an explanation for the actual decrease of plant species richness in different habitats? *Acta Botanica Neerlandica*, **47**, 27–44.

Pringle, R.M., Goheen, J.R., Palmer, T.M., et al. (2014). Low functional redundancy among mammalian browsers in regulating an encroaching shrub (*Solanum campylacanthum*) in African savannah. *Proceedings of the Royal Society of London B: Biological Sciences*, **281**(1785).

Ripple, W.J., and Beschta, R.L. (2004). Wolves and the ecology of fear: can predation risk structure ecosystems? *BioScience*, **54**, 755–766.

Ripple, W.J., and Beschta, R.L. (2012). Trophic cascades in Yellowstone: the first 15 years after wolf reintroduction. *Biological Conservation*, **145**, 205–213.

Ripple, W.J., Beschta, R.L., Fortin, J.K., and Robbins, C.T. (2014a). Trophic cascades from wolves to grizzly bears in Yellowstone. *Journal of Animal Ecology*, **83**, 223–233.

Ripple, W.J., Estes, J.A., Beschta, R.L., et al. (2014b). Status and ecological effects of the world's largest carnivores. *Science*, **343** (6167), 1241484.

Root-Bernstein, M., Guerrero-Gatica, M., Piña, L., Bonacic, C., Svenning, J.-C., and Jaksic, F.M. (2017). Rewilding-inspired transhumance for the restoration of semiarid silvopastoral systems in Chile. *Regional Environmental Change*, **17**, 1381–1396.

Rubenstein, D.R., Rubenstein, D.I., Sherman, P.W., and Gavin, T.A. (2006). Pleistocene Park: does re-wilding North America represent sound conservation for the 21st century? *Biological Conservation*, **132**, 232–238.

Rull, V. (2008). Speciation timing and neotropical biodiversity: the Tertiary–Quaternary debate in the light of molecular phylogenetic evidence. *Molecular Ecology*, **17**, 2722–2729.

Sandom, C., Faurby, S., Sandel, B., and Svenning, J.-C. (2014a). Global late Quaternary megafauna extinctions linked to humans, not climate change. *Proceedings of the Royal Society of London B: Biological Sciences*, **281**, 20133254.

Sandom, C.J., Ejrnæs, R., Hansen, M.D.D., and Svenning, J.-C. (2014b). High herbivore density associated with vegetation diversity in interglacial ecosystems. *Proceedings of the National Academy of Sciences of the United States of America*, **111**, 4162–4167.

Seddon, P.J. (2017). The ecology of de-extinction. *Functional Ecology*, **31**, 992–995.

Sinclair, A. (2003). The role of mammals as ecosystem landscapers. *Alces*, **39**, 161–176.

Smith, F.A., Boyer, A.G., Brown, J.H., et al. (2010). The evolution of maximum body size of terrestrial mammals. *Science*, **330**, 1216–1219.

Stegner, M.A., and Holmes, M. (2013). Using palaeontological data to assess mammalian community structure: potential aid in conservation planning. *Palaeogeography, Palaeoclimatology, Palaeoecology*, **372**, 138–146.

Stein, A., Gerstner, K., and Kreft, H. (2014). Environmental heterogeneity as a universal driver of species richness across taxa, biomes and spatial scales. *Ecology Letters*, **17**, 866-880.

Svenning, J.-C., Pedersen, P.B.M., Donlan, C.J., et al. (2016). Science for a wilder Anthropocene: synthesis and future directions for trophic rewilding research. *Proceedings of the National Academy of Sciences of the United States of America*, **113**, 898-906.

Terborgh, J., Davenport, L.C., Niangadouma, R., et al. (2016). Megafaunal influences on tree recruitment in African equatorial forests. *Ecography*, **39**, 180-186.

Ubilla, M., Rinderknecht, A., Corona, A., and Perea, D. (2018). Mammals in last 30 to 7ka interval (late Pleistocene-early Holocene) in Southern Uruguay (Santa Lucía River Basin): last occurrences, climate, and biogeography. *Journal of Mammalian Evolution*, **25**, 291-300.

van der Plas, F., Howison, R.A., Mpanza, N., Cromsigt, J.P.G.M., and Olff, H. (2016). Different-sized grazers have distinctive effects on plant functional composition of an African savannah. *Journal of Ecology*, **104**, 864-875.

Vavra, M., Parks, C.G., and Wisdom, W.J. (2007). Biodiversity, exotic plant species, and herbivory: the good, the bad, and the ungulate. *Forest Ecology and Management*, **246**, 66-72.

Vera, F.W.M. (2009). Large-scale nature development – the Oostvaardersplassen. *British Wildlife*, **20**, 28-36.

Weinstein, S., Titcomb, G., Agwanda, B., Riginos, C., and Young, H. (2017). Parasite responses to large mammal loss in an African savanna. *Ecology*, **98**, 1839-1848.

Wilder, B.T., Betancourt, J.L., Epps, C.W., Crowhurst, R.S., Mead, J.I., and Ezcurra, E. (2014). Local extinction and unintentional rewilding of bighorn sheep (*Ovis canadensis*) on a desert island. *PLoS ONE*, **9**, e91358.

Wright, N.A., Steadman, D.W., and Witt, C.C. (2016). Predictable evolution toward flightlessness in volant island birds. *Proceedings of the National Academy of Sciences of the United States of America*, **113**, 4765-4770.

Young, H.S., McCauley, D.J., Helgen, K.M., et al. (2013). Effects of mammalian herbivore declines on plant communities: observations and experiments in an African savanna. *Journal of Ecology*, **101**, 1030-1041.

Zimov, S.A. (2005). Pleistocene park: return of the mammoth's ecosystem. *Science*, **308**, 796-798.

Rewilding through land abandonment

STEVE CARVER
University of Leeds

Throughout human history, the overwhelming trajectory of landscape change across much of the world has been the reduction in wild and unmanaged ecosystems. This has come about through the appropriation and modification of terrestrial ecosystems and their conversion to some form of human-dominated land use. The extent and level of landscape change over time has roughly mirrored that of human population growth. The early long, slow clearance of natural ecosystems (e.g. forest) and conversion to agriculture has been followed by a few millennia of much more intensive disturbance of nature to support increasingly large settlements (Diamond, 1997). These patterns of human population and land use have been mapped from 6000 years BP to current day (Klein Goldewijk et al., 2011; Ellis et al., 2013), showing exactly how we have altered the global terrestrial biome to support our burgeoning population and ever-increasing demand for greater wealth and a better standard of living. These studies show how croplands and pasture have increased exponentially, with rapid expansion in the amount of land given over to agricultural production from the 1500s onwards. Despite the Malthusian theory of limits to growth, agricultural production has managed to keep pace with demand for food such that we are still able to support a global population in excess of 7 billion people. This has partly been achieved by land-take (i.e. appropriation of wild lands into agricultural production) and agricultural intensification permitted by technological advances such as mechanisation, the widespread use of chemicals and genetic modification, as well as global distribution networks. Nonetheless, the interplay between demand for better living standards and intensification has, ironically, led over the last few decades to some former agricultural lands being abandoned (Rudel et al., 2005; Hindmarch and Pienkowski, 2008). This has, in turn, allowed nature to reclaim the unused land through a process known as 'passive rewilding'.

This chapter will discuss the rewilding of abandoned landscapes, with an emphasis on Europe, through passive management approaches. Abandonment is seen both as a cultural and ecological phenomenon, considering in turn its implications for both human and natural systems. The debate is complicated by the diverse cultural traditions of scientists which inevitably influence the ways in which abandonment is studied and understood with some characterising it as a threat and others as an opportunity (Otero et al., 2015). Abandonment and passive rewilding through non-intervention management are classified according to spatial and temporal scales together with associated causes and drivers. The chapter will address key questions relating to the ongoing debates about passive rewilding over more active approaches, such as does passive rewilding replace management for nature, with management by nature? In other words, is passive rewilding, as opposed to active rewilding, more about process-based objectives rather than desired compositional outcomes? With passive rewilding are we trying to (re)create a former or future nature in the spaces and opportunities that abandoned lands provide, and how do we do that? The chapter concludes with some thoughts on the likely future of abandoned farmland and its possible reclamation as concerns grow over further population growth, climate change, and food security at a more global level.

Abandonment and passive rewilding: drivers and definitions
Horror vacui, 'nature abhors a vacuum'

Although the origin of this Latin phrase is attributed to Aristotle and has its origins in physics, it is often applied to modern ecological thinking about niche-filling: the processes of in-migration and colonisation where space is abandoned by one species (e.g. through death, disturbance, or out-migration) and taken up by another. Abandonment is the term used to refer to the removal of any form of human land use whereupon the land is left in an 'unused' state. Farmland abandonment occurs when agricultural land becomes too difficult and/or costly to farm, resulting in cessation of farming activity, leaving the land without crops or grazing animals (Navarro and Pereira, 2015).

Because nature abhors a vacuum, abandoned land rarely remains unused for long as various species of flora and fauna soon move in to colonise the space left behind by cessation of agricultural use. Over time, the mix of species may well change through early to late successional stages depending on species in-migration rates and any human interference. While not all the species colonising abandoned land may be considered wild or native, in the absence of direct human management, the process of colonisation of abandoned land is in itself natural, although this is open to debate because it will be affected by human activities and influence (Peterken, 1993, 1996; Pearce, 2015). This may be termed 'passive rewilding', as the rewilding process happens unaided and without direct human intervention or influence.

The outcome of land abandonment under rewilding scenarios is defined here as the spontaneous restoration of mainly native vegetation (acknowledging that non-natives will inevitably be part of the mix) that takes place in the absence of domestic livestock grazing and other forms of human intervention, together with a concomitant return of full trophic processes through the in-migration of insects, birds, reptiles, amphibians, and native mammals (including top predators) exploiting the new niches along with increasing decomposition processes. This is the 'non-intervention' approach that is widely practised to achieve favourable conservation status in areas of primary (i.e. wilderness) habitat and in strictly protected core areas of National Parks and nature reserves. Thus, passive rewilding in the absence of domestic plants and animals may be regarded as the purest form of rewilding because human intervention, influence, and management is effectively zero.

Drivers

The causes of farmland abandonment are complex, but include a series of well-known economic, social, cultural, political, and environmental drivers. Economic drivers forcing land abandonment may be allied to external markets and pricing of agricultural produce that includes crops (food and fodder) and livestock products (meat, milk, eggs, wool, etc.) together with internal costs of production (labour, fuel, land, etc.). The demand for cheap food together with the buying power of large supermarket chains, processing costs, and the globalisation of food markets has driven down farm-gate prices resulting in increasingly marginal profits for farmers on all but the best agricultural land (Appleby et al., 2003). When costs exceed profits, continuing production quickly becomes unsustainable, and without external subsidies some form of land-use change is required. In these situations, farmers often respond in one of three ways: intensification, extensification, or afforestation (Navarro and Pereira, 2015). Where possible, farmers may choose to intensify production by upgrading production using modern farming methods and so be able to produce more per unit of land more efficiently and at a lower cost. This can also mean changing from one crop type to another (e.g. from arable to horticultural) or moving to added-value systems such as organic farming (Gabriel et al., 2013; Gadanakis et al., 2015). The alternative is extensification, whereby the farmer may decide to increase the size of the farm, reduce stocking rates and/or diversify into multiple uses, thus maintaining a level of output while keeping costs as low as possible. A third option is afforestation, which offers long-term returns from timber sales with minimal management overheads. This is especially common where land capability is poor and agriculture is therefore marginal. When none of these options is economically feasible, the situation may arise where agricultural production is abandoned entirely and the land left to nature.

Social, cultural, and political drivers of land abandonment vary between countries and regions and even between farms depending on the exact set of prevailing circumstances. The last few decades have seen a rise in globalisation and the realisation of regional disparities in lifestyles and wealth. Rural depopulation has had notable effects on patterns of agricultural production and abandonment over the centuries. The industrial revolution saw a huge migration of people from the countryside to find work in factories and cities (Mantoux, 2013), but the effects on the reduction of the agricultural workforce were largely offset by associated mechanisation and scientific improvements in agricultural practices. More recently, rural depopulation has consisted of younger people migrating from the countryside in search of better-paid work in towns and cities, because farming is often seen as a less-attractive option involving long hours of hard work with little financial reward (MacDonald et al., 2000). This is perhaps most marked in marginal and remote agricultural communities where intensification through mechanisation is impossible due to physical limitations of soil, topography, or climate. As a result, we often see ageing populations in some remote rural communities where land has been gradually abandoned as younger people move away and older farmers progressively stop working the land (Navarro and Pereira, 2015).

Rapidly changing political fortunes can result in regionwide land abandonment. Baumann and colleagues (2011) provide an example of how economic, social, and political drivers have combined to force large-scale land abandonment in traditional agricultural production systems in western Ukraine. Here, as in many other former-Soviet bloc counties, the collapse of communism in the early 1990s led to a shift away from a centrally planned economy and systems of state support and guaranteed prices. This radical restructuring of the agricultural sector introduced foreign competition, liberalisation, and budget cuts, resulting in rural out-migration and widespread land abandonment with up to 56 per cent of farmland being abandoned in the study region (Baumann et al., 2011). This pattern has been repeated across much of the former Soviet Union including Estonia, Latvia, Czech Republic, and Romania, with several studies citing similar sets of drivers (Peterson and Aunap, 1998; Nikodemus et al., 2005; Kuemmerle et al., 2008; Václavik and Rogan, 2009).

Similar patterns are also seen across much of western Europe, focusing largely on mountainous and semi-arid regions of the Pyrenees and Mediterranean (García-Ruiz and Lana-Renault, 2011), where either spontaneous or induced abandonment has resulted in some areas going out of production. Environmental drivers, together with the gradual or rapid collapse of traditional farming communities, can result in spontaneous abandonment as land becomes too difficult to farm because of deterioration in physical factors such as climate and soils. In some cases, political and economic intervention can cause induced abandonment, such as with the European Common Agricultural Policy (CAP) on set-aside, where land was taken out of production to reduce food surpluses and limit the financial burden of agricultural subsidies (Hine et al., 2016). This has, however, tended to be temporary rather than permanent abandonment, with fields lying fallow rather than rewilding over longer time periods.

Mapping abandonment

Patterns in geography are best represented as maps, and farmland abandonment is no exception. However, it is not an especially easy phenomenon to map as it is both highly space–time variant and presents itself in diverse ways. There are two main approaches: statistical indices of abandonment rates or potential, and direct monitoring using remote-sensing techniques. The former attempts to predict where abandonment may happen or provide general statistics on how much land is being abandoned over large spatial scales, while the latter monitors patterns of abandonment by direct measurement and tends towards local studies.

Attempts have been made to map and model the potential for land abandonment across Europe using various statistical indices. The EU has itself created an agri-environment indicator showing the risk of land abandonment which is available at NUTS2 level (Eurostat, 2013). This has been estimated through a statistical analysis of key indicators including those showing weak land markets, low farm income, lack of farm investment, ageing farmer populations, lack of farmer education, low farm size, remoteness, low population density, and lack of buy-in to farm support schemes. The model shows a higher risk of farm abandonment in southern EU Member States (Portugal, Spain, Italy, Greece, and Romania), the Baltic States (Latvia, Estonia, Lithuania, but also including parts of northern Finland and Sweden) and north-western Ireland. Agricultural areas in these regions with a high dependency on grazing livestock are shown to be particularly at risk, although the resolution of the NUTS2 data makes further interpretation difficult. Verburg and Overmars (2009) use a variation of the Dyna-CLUE model to predict natural vegetation regrowth, or passive rewilding, on recently abandoned land based on a range of scenarios taken from the EU-RURALIS2 project. The model is based on spatial allocation requiring the estimation of demand and how it satisfies estimates across defined regions. Navarro and Pereira (2015) map hotspots across Europe, where areas categorised as 'agriculture' in 2000 are projected to become rewilded or afforested in 2030 and that are common to the scenarios of the CLUE model and how they respond to globalisation of agricultural markets covering the range of socioeconomic conditions across the axes of regionalisation versus globalisation, and willingness versus reluctance against sustainable lifestyle changes at the societal level (Verburg and Overmars, 2009). These hotspots are expressed as a percentage of each 100-km^2 grid cell. These models are taken by Ceauşu and colleagues (2015) and mapped alongside wilderness quality indices for Europe based on remoteness from roads and settlements, absence of light pollution, difference from potential

104   S. CARVER

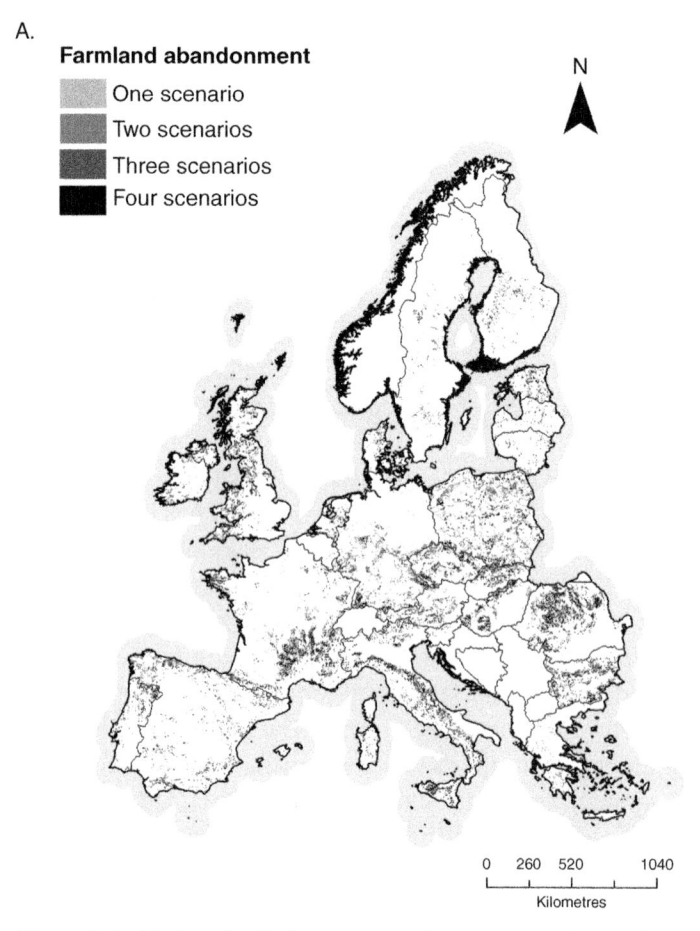

A.
Farmland abandonment
- One scenario
- Two scenarios
- Three scenarios
- Four scenarios

Figure 6.1. (A) Farmland abandonment in Europe projected for the year 2040 under different scenarios by the Dyna-CLUE model (after Ceaușu et al., 2015). (B) Wilderness quality index for Europe (after Kuiters et al., 2013). (A black and white version of this figure will appear in some formats. For the colour version, please refer to the plate section.)

natural vegetation, and proportion of net primary productivity harvested by humans. These maps show recognisable patterns in wilderness quality with strong latitudinal and altitudinal trends and overlap with areas of high megafauna species richness. Critically, the analysis shows areas of high potential farmland abandonment in areas of intermediate wilderness quality, indicating that such areas once abandoned could add to Europe's wilderness areas through passive rewilding (Ceaușu et al., 2015). This pattern is shown in Figure 6.1.

The most comprehensive mapping to date is provided by Kuemmerle and colleagues (2016) in mapping hotspots of land-use change across Europe between 1990 and 2006 using high-resolution spatial indicators of land-use

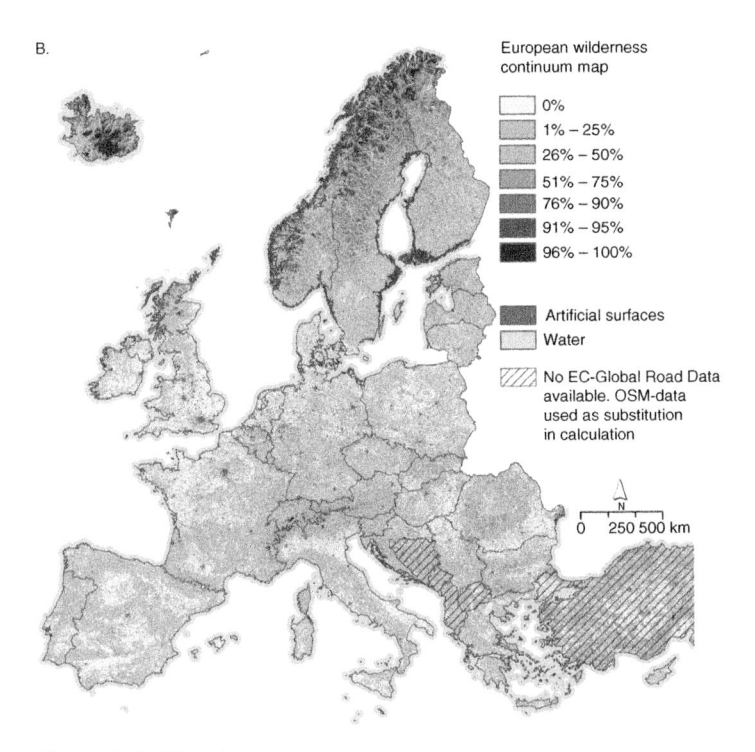

Figure 6.1. (Cont.)

change. These include input metrics relating to production inputs (e.g. fertiliser application) and output metrics relating to yield (e.g. biomass removed by grazing). Absolute land-use changes were calculated for each indicator and hotspots showing areas of both increase and decrease mapped.

The second approach to mapping farmland abandonment is to use temporal analysis of high-resolution remotely sensed imagery. Here, satellite images from before and after a period of abandonment are compared to map the areas affected. This is a task for which remote sensing and image analysis is particularly well suited and many studies have been carried out to provide robust estimates of abandonment rates and provide detailed maps of abandonment (Baumann et al., 2011; Prishchepov et al., 2012; Alcantara et al., 2013; Estel et al., 2015). However, some difficulties distinguishing between pasture and abandoned arable land exist, because the reflection signatures between these two land covers are remarkably similar before a natural tree cover develops (Lunetta and Lyon, 2004).

Combined approaches that use spatially explicit statistical indices to correct, validate, or co-rectify classifications based on remotely sensed imagery are becoming increasingly popular (e.g. Weissteiner et al., 2011). Other methods based on human 'by eye' calibration of classified imagery are also available. These rely on campaigns to collect volunteered geographic information

(VGI) on land cover classifications and associated degrees of human impact/intervention using Geo-Wiki tools (Perger et al., 2012; See et al., 2015).

Impacts of abandonment

The EU describes farmland abandonment as the 'cessation of agricultural activities on a given surface of land which leads to undesirable changes in biodiversity and ecosystem services' (Eurostat, 2013). This implies that abandonment is seen very much in a negative light and a failure of EC rural social policies. However, depending on perspective, abandonment and associated rewilding can also be seen has having benefits as well as costs.

Costs

Abandonment is often used as a pejorative term, certainly in farming circles within which the general and widely held view is that both abandonment and rewilding are viewed as backward and retrograde steps. Here the most obvious impact is the loss of potentially productive farmland that may have taken generations of farmers to clear, improve, and bring into production. To abandon such land, however marginal it may be, is likely to be viewed negatively by most farmers because of threats to livelihoods, communities, and the social/cultural fabric of those regions in question (Ruskule et al., 2013).

There are EU policies and regulations that are designed specifically to keep such land in what is called 'Good Agricultural and Environmental Condition' (GAEC) and subsidies are paid to farmers to maintain the land in this state. This has had knock-on implications for former farmland that is no longer actively farmed but is managed by wildlife conservation groups in that it can effectively mean that a farming pressure (e.g. through 'conservation grazing') is incentivised whether it is farmed or not if payments are to be maintained (Plachter and Hampicke, 2010).

Biodiversity loss is often cited as a cost of land abandonment (MacDonald et al., 2000). Here the implication is that farmland, particularly that which is managed using traditional (i.e. old-fashioned and non-industrialised) farming techniques and methods, is home to valued assemblages of species that have become associated with these landscapes. Examples include farmland birds, insects, small mammals, and amphibians that are associated with features in the farming landscape such as hedgerows, ponds, and meadows (Benton et al., 2003). Many of these traditionally farmed landscapes have been designated as High Nature Value (HNV) farming areas. These are areas of Europe where traditional agriculture supports, and is associated with, farmland biodiversity, and as such are valued for their wildlife alongside cultural heritage, high-quality artisanal farm produce, and much-needed rural employment. Many HNV farming areas coincide with areas at higher levels of risk of abandonment, leading to worries about biodiversity loss as well as socioeconomic impacts (Henle et al., 2008; Terres et al., 2015).

On the cessation of farming, abandoned land will generally go through a series of successional stages before reaching its potential natural vegetation (PNV) associated with the concurrent edaphic conditions as determined by climate, soil, drainage, and topography. Early stage succession usually involves a phase of low shrubby and often thorny vegetation commonly referred to as 'scrubbing up' that may well compete with and crowd out species associated with HNV (Benjamin et al., 2005; O'Rourke, 2016). The increase of shrubby vegetation and associated dead/dry biomass on the ground may also present an elevated fire risk, particularly in semi-arid landscapes, although it should be recognised that many such ecosystems are, more often than not, already human-modified cultural landscapes. Management by cutting, controlled burning, and grazing are techniques that are often used to manage risk in areas prone to wildfire (Botelho and Fernandes, 1999; Brown, 2002; Moritz et al., 2014), which then halt abandonment and succession. This is true of the heather moorland landscapes of upland Britain, which are actively managed by cutting, burning, and draining for the benefit of gamebirds (red grouse *Lagopus lagopus scotica*) and driven shoots (Tharme et al., 2001).

Benefits

Allowing abandoned farmland to rewild naturally using passive or non-intervention approaches can provide a range of benefits that may well offset the loss of food production and other costs. These include biodiversity gains through improved wildlife habitats, reductions in soil erosion, improved water quality, maintenance of base flows, enhanced carbon sinks, and reduced flood risk further down the drainage basin.

Biodiversity gains arising from farmland abandonment may seem to go contrary to claims that abandonment causes biodiversity loss. The answer to this apparent contradiction lies in considerations about exactly what kind of biodiversity we are talking about, as the species mix inhabiting an ecosystem will vary according to the land management systems involved (Queiroz et al., 2014). Results from a meta-analysis of studies on biodiversity in abandoned farmland in the Mediterranean region by Plieninger and colleagues (2014) indicate a slight increase in overall biodiversity, although the picture is far from clear or uniform across all studies with differences between taxa, spatio-temporal scales, land uses, landforms, and climate. It ought to be clear that there will be both winners and losers when land is abandoned. Abandoned land that is allowed to rewild by passive means without the limiting pressure from domestic grazing will obviously favour more woodland species as shrubs and trees return. Studies in north-west Spain have shown that bird species associated with shrubland and woodland show a significant increase, while those associated with open, farmland habitats have declined (Plieninger et al., 2014; Regos et al., 2016). Deciding what species assemblages we want will determine conservation approaches here, but as far as rewilding is concerned, management should favour native species and habitats over those dependent on human intervention and niches (Carver, 2014).

Soil erosion has long been associated with agriculture (Morgan, 2009) and abandonment has been shown to reduce soil erosion (Cerdà, 1997; García-Ruiz and Lana-Renault, 2011). Rewilding following abandonment of Mediterranean farmland means the expanding and permanent vegetation cover protects soils from rain-splash erosion and surface run-off, improving infiltration rates and soil characteristics leading to reductions in soil loss and sediment delivery. This in turn can affect stream morphology and ecology and reduces sedimentation in reservoirs (García-Ruiz and Lana-Renault, 2011). Improvements in soil properties and reductions in sediment yields also have benefits in terms of water quality for supply and have benefits for river ecology. Similar effects will be seen in other landscapes such as alpine pastures, where reduced stocking rates and increased forest cover also results in increased infiltration rates, reduced run-off, and less soil and gully erosion (Tasser et al., 2003) and reductions in downstream flood events, while afforestation also has benefits in terms of avalanche protection (Olschewski et al., 2012).

Carbon storage and sequestration can also benefit from abandonment as more carbon is locked up in woody vegetation and in the soil/peat carbon store with correspondingly less being lost in run-off or the atmosphere through cropping, grazing, and land tillage (Schierhorn et al., 2013). Cessation of burning practices has been shown to increase carbon sequestration and reduce losses as dissolved organic carbon as well as having benefits to stream biodiversity and potentially reducing peak flows and downstream flooding (Brown et al., 2015).

Approaches to passive rewilding

Passive rewilding is based on a 'leave it to nature' philosophy where human intervention is kept to an absolute minimum or even zero. As such, this form of rewilding relies almost entirely on natural processes to determine which species colonise abandoned land, although it is often moderated by unintended human influences. Some limited intervention may be required to remove domestic species and prevent or minimise external impacts, such as grazing by stray domestic livestock or colonisation by non-native/exotic invasive species (e.g. sika deer in parts of the UK). However, the management of non-native/exotic invasive species is often intensive, expensive, and requiring aggressive treatments over extended periods that may be argued to be incompatible with passive rewilding (Manchester and Bullock, 2000). As such, these must be considered on a case-by-case basis.

Non-intervention management

On the face of it, passive rewilding of abandoned land through the application on non-intervention management principles seems like an elegant solution; it can be done at 'no extra cost' to government, NGOs, or land owners; it can deliver economic benefits associated with ecosystem services; and it can deliver new natural landscapes and habitats that benefit both nature and people alike. Stepping back from abandoned land and letting nature decide is easy enough. However, it might not always deliver the expected outcomes, although much depends on the contextual setting of the landscape in question.

Deary and Warren (2017) studied visions and understanding of rewilding among land managers and owners of upland estates in Scotland that were engaged in some level of rewilding. Results show a wide variety of views on rewilding, but the term itself is problematic, with many respondents citing the 're-' as indicating attempts to turn back the ecological clock to a time before widespread human settlement and intervention, together with the feeling that it sounded more political than practical. The study is also interesting because it brings to the fore a series of wider concerns about processes that are relevant to abandonment, passive rewilding, and non-intervention. The 'unnatural' ecological starting point of many rewilding projects within an already ecologically damaged and impoverished landscape was quoted as an issue that could create problems for passive rewilding, as natural conditions may need to be restored through active intervention before ecosystems can function naturally and autonomously. This refers to the problems of missing natural elements found across most, if not all, human-modified ecosystems. These elements can include both species and processes, as well as ecological habitats and niches.

A key problem affecting the ecology of the Scottish Highlands is red deer (*Cervus elaphus*) numbers which have grown to high densities due in part to estate management practices and the lack of a natural predator (e.g. the wolf *Canis lupus*). Deer are managed in Scotland by stalking and culling (human predation), and by deer fences which mimic behavioural traits driven by predators by excluding deer from certain areas to allow woodland regeneration (Clements, 2016). Without the missing trophic driver of the wolf, management interventions are required to prevent overgrazing by deer (Beschta and Ripple, 2016). In this context, passive rewilding through non-intervention management will not provide the landscape with the conditions required for creating a return to nature that is faithful to its true ecological potential without the reintroduction of missing predators (i.e. species) that drive trophic cascades (i.e. process) controlling deer numbers and behaviour (Nilsen et al., 2007; Arts et al., 2016). Without all the necessary ecological building blocks in place, or at least nearby from where they can migrate into recently abandoned land, it is likely that passive rewilding will (at least in the short term) result in some unexpected results. In the case of the Scottish Highlands, ceasing deer management operations might just result in even more deer and even more overgrazing and further suppression of natural tree regeneration.

In smaller landscape units enclosed by human-modified landscapes, non-native invasive species can be a problem. Invasive ruderal weeds such as Japanese knotweed (*Fallopia japonica*), Giant hogweed (*Heracleum mantegazzianum*), and Himalayan balsam (*Impatiens glandulifera*) have become a problem in many abandoned, brownfield, or 'feral' sites, where they vigorously outcompete native species and spread rapidly to other sites (Gallardo et al., 2016). Both spatial adjacency (connectivity to sources of non-native invasive species) and isolation (lack of connectivity to sources of native species) therefore becomes a problem for smaller abandoned land parcels, suggesting that passive rewilding might not necessarily be the most appropriate management policy across all landscape scales, at least in the short term. The spread of invasive non-native plants has been facilitated and accelerated by 'feral' lands along transport infrastructure (e.g. railway and motorway embankments, river and canal sides) with clear implications for native species and habitats.

Towards a classification of rewilding approaches

Rewilding, whether passive or active, is a process. It takes place over both space and time. It is entirely possible, perhaps even normal, for a rewilding project to start out as either passive or active and at some point in time become the other. The change may be sudden (e.g. resulting from a change in policy and management) or it may be gradual (e.g. as active interventions are scaled back). Rewilding actions might not be the same across a whole project area, with different or distinct approaches being pursued in different areas of the target site depending on requirements. As such, a classification of abandonment and passive/non-intervention management might usefully be set out along a space–time continuum along with more active management approaches.

In spatial ecology, space can vary from patch through mosaics to landscape scales. Rewilding can also mirror this spatial continuum. It is theoretically possible to rewild the grass lawn around your house by not mowing it (Kun, personal communication). In practice, however, it is unlikely to achieve much other than a mess of weeds and matted tall grasses because of its small scale and lack of in-migration. However, rewilding works better at larger scales that encompass multiple ecotones, groups, and mosaics (Scott et al., 1999; Freemark et al., 2002; Pereira and Navarro, 2015). Ultimately, whole landscape rewilding can create the greatest ecological benefit, although the scope for such large-scale rewilding is limited by multiple constraints of land availability and social, political, and cultural barriers. Rewilding can be classified according to spatial extent as: small-scale (patch), medium-scale (mosaic or group), and large-scale (landscape). Naturally, the thresholds between these approaches are fuzzy rather than discrete, depending on the ecological and human landscape context or setting.

The causes of land abandonment and subsequent rewilding have a considerable influence on spatial scale and configuration (Munroe et al., 2013). Land abandonment, whether deliberate, circumstantial, or accidental, takes place over space and time and within a matrix of contemporaneous land use, both rural and urban. How this spatial land-use matrix is configured strongly affects landscape and ecological connectivity. This is a key aspect of rewilding as outlined in the original three C's or 'cores, corridors and carnivores' model (Soulé and Noss, 1998; Worboys et al., 2010). Here, the emphasis is on creating large landscape-scale ecological networks through the connecting of existing protected areas. This can occur in multiple ways: through expansion of existing core areas (e.g. rewilding around core area edges) and connecting these using linear (e.g. riparian) and landscape corridors (e.g.

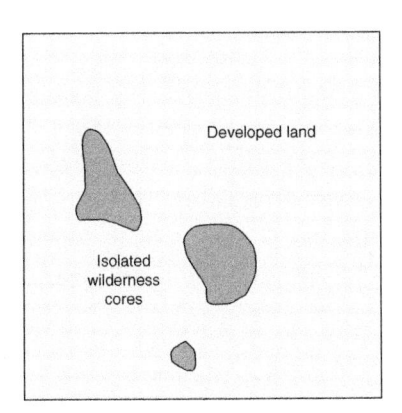

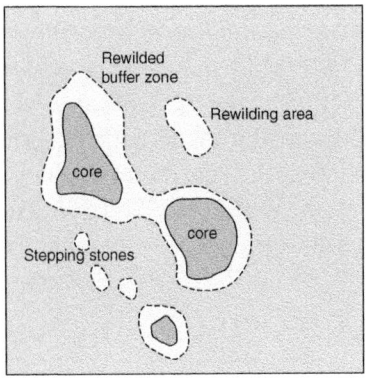

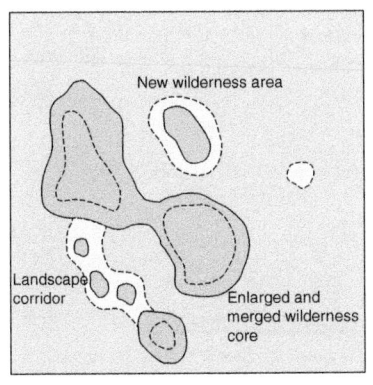

Figure 6.2. Restoration of wilderness via rewilding in the three C's model (after Carver and Fritz, 2016). (A black and white version of this figure will appear in some formats. For the colour version, please refer to the plate section.)

112   S. CARVER

wildlife friendly farming corridors) and stepping stones of small, yet new protected areas that provide temporary refuges to species moving through the landscape between larger core areas (Figure 6.2).

Time is also a key aspect of any rewilding project. Rewilding does not happen overnight. Many rewilding projects do not actually have defined endpoints, preferring instead to explicitly acknowledge that it is a process in achieving spontaneous replacement and perpetuation of natural processes (Browning and Yanik, 2004). This may take only a few years for vascular plants, birds, and mammals, but can take decades for trees, and hundreds of years to develop a mature woodland. Passive rewilding after land abandonment implies a non-intervention process that continues over time. Nonetheless, some intervention is usually employed even if only to implement a programme of scientific monitoring or to control access to the site from visitors or domestic livestock from adjacent land. The latter usually involves some form of impermeable barrier, such as a wall or fence. Analysis of monitoring data may reveal problems that require further intervention to rectify or adjust depending on site plans or changes in the wider circumstances or site setting. Thus, it may be that passive rewilding plans will change in response over time. This temporal flux between approaches over time means that, over time and depending on progression and management decisions, the style of rewilding can flip between passive and active and back again (Figure 6.3).

Simple land abandonment with no human intervention 'flat lines' over time, implying that the ecosystem that develops therein is determined largely by natural processes and may be regarded as passive rewilding in its purest sense (the red line in Figure 6.3), accepting that any rewilding site will be

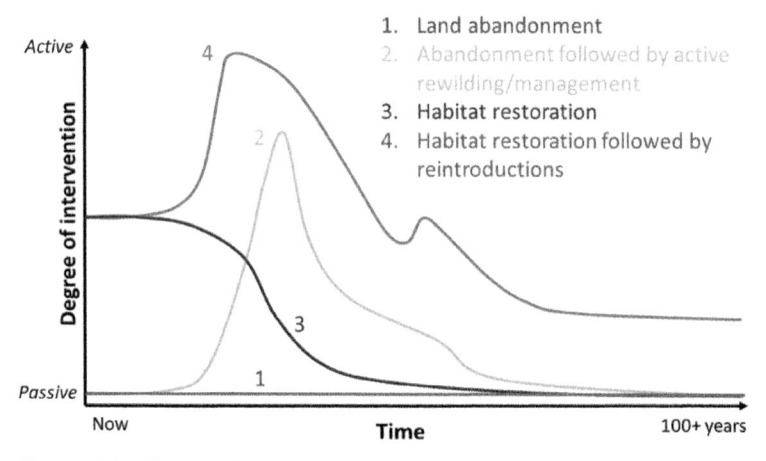

Figure 6.3. Changes in types and categories of rewilding over time. (A black and white version of this figure will appear in some formats. For the colour version, please refer to the plate section.)

influenced to a greater or lesser degree by human activities in adjacent land. Land which is abandoned and starts to rewild on its own but is then managed, albeit as a rewilding site, by some later intervention for conservation goals to direct or reset successional trajectories follows a passive–active–passive pattern (the yellow line in Figure 6.3). Examples where this might be expected include abandoned land affected by non-native invasive species that require management to prevent them dominating the ecology of the site, or where the abandoned land is too isolated for native species to appear spontaneously so requiring some management intervention such as assisted regeneration of native woodland. Active rewilding of abandoned land through habitat restoration might well occur from day 1 where it is recognised that habitats need to be actively restored through human intervention to replace missing species and process. Once re-established, natural habitats and ecosystems should prove to be self-sustaining (especially on a landscape scale) and management intervention scaled back to passive rewilding, so following an active–passive trajectory (the blue line in Figure 6.3). Habitat restoration is often a prerequisite for the reintroduction of higher order predators and herbivores to re-establish trophic cascades within a rewilding site (the green line in Figure 6.3). Again, active rewilding is often necessary to establish the correct vegetation patterns and ecological niches before missing species can be returned to the landscape, after which management can be scaled back to monitoring programmes and the occasional intervention to control or maintain species numbers by culling or top-up introductions from outside the area. Whatever the original causes of abandonment or the initial level of management intervention, the overall trajectory of rewilding ought to be towards a passive system wherein nature and ecological processes are able and allowed to take care of themselves in the absence of domestic plant and animal species wherever possible.

Examples of passive rewilding

How the process of rewilding progresses over time after initial abandonment as shown in Figure 6.3 is perhaps best illustrated using real-world examples. Here a series of short case studies are used to show how the causes, type, and progression of rewilding may vary over space and time.

The Cuningar Loop is a meander loop in the River Clyde as it flows through Glasgow, Scotland. Originally the location of several reservoirs used to store water pumped from the river in the early 1800s, the land fell into disuse after the construction of an aqueduct bringing clean, unpolluted water from Loch Katrine in the 1850s. At around 15 ha in size the Cuningar Loop is an example of a small-scale urban 'wasteland' or brownfield site which has been allowed to rewild with little human intervention, resulting in passive regeneration of native woodland. The site has subsequently been taken over by Forestry Commission Scotland (FCS) and remodelled during the 2013/14 Commonwealth Games into a 'woodland park' with redesigned and built features including a path network, adventure play facilities, BMX park, outdoor bouldering/climbing area, and riverside walk. While it may be argued that this has increased the public appeal and footfall in the loop it has, in fact, reversed many of the ecological gains from over 150 years of passive rewilding (Sinnett et al., 2017). As such, the site may be regarded as a good example of the abandonment of a small-scale industrial site followed by passive rewilding and recent active remodelling and management to meet local social needs for outdoor recreation and enjoyment.

Scar Close is a nature reserve in the Yorkshire Dales National Park, northern England, consisting mainly of limestone pavement and native grassland, woodland, and heath. The 60-ha reserve is owned and managed by Natural England and, critically, has had grazing by domestic livestock excluded from the area since being made stock-proof in 1974. Beyond the encircling fence and walls very little management has taken place on the reserve except for monitoring and some limited cutting/removal of non-native species (mainly sycamore trees). The site is a classic example of opportunistic, small-scale abandonment followed by passive rewilding with minimal intervention (sycamore control) and stock exclusion. The site is particularly interesting because of the neighbouring Southerscales nature reserve, which is managed by the Yorkshire Wildlife Trust under a Higher Level Scheme (HLS) agreement that stipulates a minimum level of 'conservation' grazing by sheep and cattle to maintain Biodiversity Action Plan (BAP) habitats (mainly calcareous grassland, limestone pavement, blanket bog, and upland flushes). Both sites are roughly the same size, at the same altitude, and both are limestone pavements. While Southerscales maintains a floristic biodiversity of around 50 principal species, mainly in inaccessible 'grykes' (i.e. deep fissures where grazing animals cannot reach) in the limestone pavement, the adjacent Scar Close reserve has nearly 250 species (Fisher, 2011; Newlands, personal communication), some of which are supported by and contribute to a gradually spreading horizontal growing medium covering the formerly bare limestone 'clints'. Not only is the floristic biodiversity higher at the ungrazed Scar Close, but the biomass is far greater and exhibits more natural forms unconstrained by heavy grazing by domestic livestock (sheep and cattle). Furthermore, species of particular conservation interest such as globeflower (*Trollius europaeus*) which are susceptible to overgrazing are absent on Southerscales, but locally common on Scar Close.

Carrifran Wildwood is a community habitat restoration project in the southern uplands of Scotland near Moffat. The site comprises a single 650-ha valley formerly grazed by sheep that was bought by a community group in 2000 with the express intention of returning the land to a native woodland. The project has the stated aims to recreate 'an extensive tract of mainly forested wilderness with most of the rich diversity of native species present in the area before human activities became dominant. The woodland will not be exploited commercially and human impact will be carefully managed. Access will be open to all, and it is hoped that the Wildwood will be used throughout the next millennium as an inspiration and an educational resource' (Newton and Ashmole, 2000). Volunteers have now planted over half a million trees based on locally sourced seed using ecological principles to design planting patterns that best suit local edaphic conditions (Ashmole and Ashmole, 2009). The project has recently expanded to encompass the neighbouring Talla and Gamehope valleys, roughly trebling the size of the project area. Carrifran Wildwood is an example of an intentional, medium-scale habitat restoration project, and as such may be described as abandonment of agriculture (grazing) with active rewilding (assisted tree regeneration) with the long-term intention of moving to a passive rewilding phase.

The Portuguese island of Madeira is a subtropical island in the Atlantic Ocean about 800 km west of the African coast. The island's agriculture was focused initially on grain and later sugar cane and wine for export together with fruit and vegetables. Production is limited by the extremely steep terrain and availability of irrigation water for which a complex landscape of terraces and irrigation canals (levadas) was constructed. Many of the former agricultural areas have been abandoned as the island's economy has shifted more towards tourism and agriculture in the more difficult terrain has become uneconomic (Cruz et al., 2009). This has led to the recovery of native vegetation on many former terraced fields and gardens mixed in with feral domesticated plants. This is an example of a circumstantial, medium-scale passive rewilding following farmland abandonment.

Out-migration of the largely rural population of Norway to the Americas during the mid to late nineteenth century and early twentieth century has led to many rural areas in south-west Norway becoming largely abandoned. These former sheep-grazing areas have since reforested as a result of reductions in domestic livestock grazing in circumstances where there was no inward migration of surplus wild grazers to supress tree regrowth. This is a good example of human out-migration across large-scale landscapes leading to land abandonment and subsequent passive rewilding of native forest. The example is interesting in respect to the earlier example of deer management in Scotland, as it adequately demonstrates how the landscape of Scotland might look if deer and sheep numbers are reduced to allow native tree regeneration at a landscape scale (Halley, 2017).

The meltdown of nuclear reactor number 4 in April 1986 at the Chernobyl Nuclear Power Plant in the Ukraine is perhaps the most well-known example of land abandonment at a large, landscape scale. The disaster resulted in large areas of the surrounding countryside and nearby city of Pripyat being contaminated by high levels of radioactive fallout. A 30-km exclusion zone was declared and tens of thousands of people evacuated. As a result, despite early impacts from direct radiation such as forest dieback, in the absence of humans the area has become a haven for wildlife, with significant populations of moose, wolf, lynx, and bear with natural habitats continuing to reclaim both urban and agricultural land (Webster et al., 2016). The Chernobyl disaster remains to date the largest example of accidental passive rewilding resulting from widespread abandonment on a landscape scale. The area immediately over the border in Belarus has been designated as the 1300-km^2 Polesie State Radioecological Reserve; the biggest in Belarus and one of the biggest in Europe.

Futures?

The future of rewilding, whether passive or active, or whether resulting from land abandonment or other causes, is uncertain. Nevertheless, it presents a more positive and proactive approach to conservation aimed at halting and perhaps reversing biodiversity loss across a range of landscape scales and situations. This chapter has focused mainly on passive rewilding following the abandonment of agricultural land, and the costs and benefits, drivers and barriers to this kind of rewilding have been described and assessed.

The 'elephant in the room' in most discussions about the protection of wild lands and the creation of more through the process of rewilding is inevitably that of human population growth. The total global human population recently passed 7 billion people and is projected to pass 8 billion in or around 2023. While longer-term forecasts predict a levelling-off of global population, this still creates a huge pressure on food production and distribution such that global food security issues may well override rewilding gains over the coming decades. Together with other global uncertainties such as geopolitics, water resources, and climate change, this may lead to reversals in land abandonment as historically farmed landscapes are brought back into production as it becomes profitable and necessary to do so. However, it should be clear that there are careful trade-offs to be made between global food security and biodiversity loss (Cramer et al., 2017). Ecosystem services from natural processes and rewilded landscapes may well override the limited potential for food production in marginal lands. Indeed, carefully rewilded landscapes, especially in more marginal upstream areas, may prove essential in protecting the most productive lowland farms from flooding, soil erosion, landslides, drought, and extreme weather events. Long-term protection of rewilding sites is necessary to protect these areas from reappropriation if they are to have any degree of permanency.

There are more cost-effective, ecologically sensitive, and beneficial ways to enhance human well-being than encroaching on wilderness or reversing rewilding sites. These include the efficiencies in agricultural production to be gained from a move towards wasting less food, eating less meat, and better planning of land-sparing and land-sharing initiatives (Phalan et al., 2011). Add these to the benefits from ecosystem service delivery and it ought to be possible to devise optimum land partition models that allocate nature/agriculture land mixes that meet the needs of 'planet, people, profit' in equal measure (Kristenkas, 2014).

The bottom line about land abandonment and passive rewilding concerns the nature of nature itself. We perhaps need to accept that nature can survive without our benevolent guiding hand and accept what she gives us in return. It might not necessarily be what we expect or what we might want (and it may be influenced by our own activities before and outside of the rewilding target area), but the process of that rewilding will be natural nonetheless. Everything depends on whether we are prepared to step back and see what happens. Such an approach maintains that abandonment and passive rewilding is about process-based objectives rather than specified compositional outcomes. And what about the use or acceptance of non-native species within that process? Some definitions of passive rewilding allow for the use of domestic herbivores as 'analogues' for native species (Nogués-Bravo et al., 2016) while it has been argued here and elsewhere (Carver, 2014) that true rewilding ought to be based on native fauna only which, critically, covers all trophic levels including top predators. There is perhaps a deeper philosophical argument to be made here, for while it is recognised that the disturbance to ecological succession created by large grazing herbivores is essential to the creation and maintenance of varied ecological mosaics and their associated biodiversity (Hodder and Bullock, 2009), the role of domestic analogues in the absence of a predatory pressure is limited in many ways to maintaining a farming pressure on the land. The application of the term 'rewilding' to the extensification of agriculture, regardless of the benefits in terms of biodiversity gains, is at best a kind of 'rewilding-lite' (Carver, 2014). Passive rewilding with domestic analogues is an example where strict rewilding principles are being traded-off against social, political, and cultural norms. This is somehow missing the point and is, in the final analysis, perhaps not true rewilding, or as Anthony Sinclair so succinctly puts it: 'In the end if a term, either restoration or rewilding, applies to everything, it also means nothing' (Sinclair, 2017).

References

Alcantara, C., Kuemmerle, T., Baumann, M., et al. (2013). Mapping the extent of abandoned farmland in Central and Eastern Europe using MODIS time series satellite data. *Environmental Research Letters*, **8**, 035035.

Appleby, M.C., Cutler, N., Gazzard, J., et al. (2003). What price cheap food? *Journal of Agricultural and Environmental Ethics*, **16**, 395–408.

Arts, K., Fischer, A., and van der Wal, R. (2016). Boundaries of the wolf and the wild: a conceptual examination of the relationship between rewilding and animal reintroduction. *Restoration Ecology*, **24**, 27–34.

Ashmole, M., and Ashmole, P. (2009). *The Carrifran Wildwood story: ecological restoration from the grass roots*. Jedburgh: Borders Forest Trust.

Baumann, M., Kuemmerle, T., Elbakidze, M., et al. (2011). Patterns and drivers of post-socialist farmland abandonment in Western Ukraine. *Land Use Policy*, **28**, 552–562.

Benjamin, K., Domon, G., and Bouchard, A. (2005). Vegetation composition and succession of abandoned farmland: effects of ecological, historical and spatial factors. *Landscape Ecology*, **20**, 627–647.

Benton, T.G., Vickery, J.A., and Wilson, J.D. (2003). Farmland biodiversity: is habitat heterogeneity the key? *Trends in Ecology & Evolution*, **18**, 182–188.

Beschta, R.L., and Ripple, W.J. (2016). Riparian vegetation recovery in Yellowstone: the first two decades after wolf reintroduction. *Biological Conservation*, **198**, 93–103.

Botelho, H., and Fernandes, P.M. (1999). Controlled burning in the Mediterranean countries of Europe. In Eftichidis, G., Balabanis, P., and Ghazi, A. (Eds.), *Wildfire management* (pp. 163–170). European Commission.

Brown, L.E., Holden, J., Palmer, S.M., Johnston, K., Ramchunder, S.J., and Grayson, R. (2015). Effects of fire on the hydrology, biogeochemistry, and ecology of peatland river systems. *Freshwater Science*, **34**, 1406–1425.

Brown, T. (2002). Minimizing wildfire risk with grazing. *Rangelands Archives*, **24**, 17–18.

Browning, G., and Yanik, R. (2004). Wild Ennerdale: letting nature loose. *ECOS*, **24**, 34–38.

Carver, S. (2014). Making real space for nature: a continuum approach to UK conservation. *ECOS*, **35**, 4–14.

Carver, S.J., and Fritz, S. (2016). *Mapping wilderness*. Dordrecht: Springer.

Ceauşu, S., Hofmann, M., Navarro, L.M., Carver, S., Verburg, P.H., and Pereira, H.M. (2015). Mapping opportunities and challenges for rewilding in Europe. *Conservation Biology*, **29**(4), 1017–1027.

Cerdà, A. (1997). Soil erosion after land abandonment in a semiarid environment of southeastern Spain. *Arid Land Research and Management*, **11**, 163–176.

Clements, V. (2016). Native woodlands inside and outside the main red deer range in Scotland. *Scottish Forestry*, **70**, 24–29.

Cramer, W., Egea, E., Fischer, J., et al. (2017). Biodiversity and food security: from trade-offs to synergies. *Regional Environmental Change*, **17**, 1257–1259.

Cruz, M.J., Aguiar, R., Correia, A., Tavares, T., Pereira, J.S., and Santos, F.D. (2009). Impacts of climate change on the terrestrial ecosystems of Madeira. *International Journal of Design & Nature and Ecodynamics*, **4**, 413–422.

Deary, H., and Warren, C.R. (2017). Divergent visions of wildness and naturalness in a storied landscape: practices and discourses of rewilding in Scotland's wild places. *Journal of Rural Studies*, **54**, 211–222.

Diamond, J. (1997). *Guns, germs and steel: a short history of everybody for the last 13,000 years*. New York, NY: W.W. Norton.

Ellis, E.C., Kaplan, J.O., Fuller, D.Q., Vavrus, S., Goldewijk, K.K., and Verburg, P.H. (2013).

Used planet: a global history. *Proceedings of the National Academy of Sciences of the United States of America*, **110**, 7978-7985.

Estel, S., Kuemmerle, T., Alcántara, C., Levers, C., Prishchepov, A., and Hostert, P. (2015). Mapping farmland abandonment and recultivation across Europe using MODIS NDVI time series. *Remote Sensing of Environment*, **163**, 312-325.

Eurostat. (2013). Agri-environment indicator – risk of land abandonment. Eurostat Statistics Explained, March 2013. http://ec.europa.eu/eurostat/statistics-explained/index.php/Agri-environmental_indicator_-_risk_of_land_abandonment#Indicator_definition

Fisher, M. (2011). Restoring wildlife: ecological concepts and practical applications. *Restoration Ecology*, **19**, 292-293.

Freemark, K., Bert, D., and Villard, M.A. (2002). Patch-, landscape-, and regional-scale effects on biota. In Gutzwiller, K.J. (Ed.), *Applying landscape ecology in biological conservation* (pp. 58-83). New York, NY: Springer.

Gabriel, D., Sait, S.M., Kunin, W.E., and Benton, T.G. (2013). Food production vs. biodiversity: comparing organic and conventional agriculture. *Journal of Applied Ecology*, **50**, 355-364.

Gadanakis, Y., Bennett, R., Park, J., and Areal, F.J. (2015). Evaluating the sustainable intensification of arable farms. *Journal of Environmental Management*, **150**, 288-298.

Gallardo, B., Zieritz, A., Adriaens, T., et al. (2016). Trans-national horizon scanning for invasive non-native species: a case study in western Europe. *Biological Invasions*, **18**, 17-30.

García-Ruiz, J.M., and Lana-Renault, N. (2011). Hydrological and erosive consequences of farmland abandonment in Europe, with special reference to the Mediterranean region – a review. *Agriculture, Ecosystems and Environment*, **140**, 317-338.

Halley, D. (2017). Land use in Norway and Scotland. NINA. www.nina.no/english/News/News-article/ArticleId/3941

Henle, K., Alard, D., Clitherow, J., et al. (2008). Identifying and managing the conflicts between agriculture and biodiversity conservation in Europe – a review. *Agriculture, Ecosystems and Environment*, **124**, 60-71.

Hindmarch, C., and Pienkowski, M.W. (2008). *Land management: the hidden costs*. Chichester: John Wiley & Sons.

Hine, R.C., Ingersent, K.A., and Rayner, A.J. (2016). *The reform of the Common Agricultural Policy*. New York, NY: Springer.

Hodder, K.H., and Bullock, J.M. (2009). Really wild? Naturalistic grazing in modern landscapes. *British Wildlife*, **20**, 37-43.

Kistenkas, F.H. (2014). Innovating European nature conservation law by introducing ecosystem services. *GAIA – Ecological Perspectives for Science and Society*, **23**, 88-92.

Klein Goldewijk, K., Beusen, A., Van Drecht, G., and De Vos, M. (2011). The HYDE 3.1 spatially explicit database of human-induced global land-use change over the past 12,000 years. *Global Ecology and Biogeography*, **20**, 73-86.

Kuemmerle, T., Hostert, P., Radeloff, V.C., van der Linden, S., Perzanowski, K., and Kruhlov, I. (2008). Cross-border comparison of post-socialist farmland abandonment in the Carpathians. *Ecosystems*, **11**, 614.

Kuemmerle, T., Levers, C., Erb, K., et al. (2016). Hotspots of land use change in Europe. *Environmental Research Letters*, **11**(6), 064020.

Kuiters, A.T., van Eupen, M., Carver, S., Fisher, M., Kun, Z., and Vancura, V. (2013). Wilderness register and indicator for Europe final report (EEA Contract No: 07.0307/2011/610387/SER/B. 3).

Lunetta, R.S., and Lyon, J.G. (2004). *Remote sensing and GIS accuracy assessment*. Boca Raton, FL: CRC Press.

MacDonald, D., Crabtree, J.R., Wiesinger, G., et al. (2000). Agricultural abandonment in mountain areas of Europe: environmental consequences and policy response. *Journal of Environmental Management*, **59**, 47-69.

Manchester, S.J., and Bullock, J.M. (2000). The impacts of non-native species on UK biodiversity and the effectiveness of control. *Journal of Applied Ecology*, **37**, 845-864.

Mantoux, P. (2013). *The industrial revolution in the eighteenth century: an outline of the beginnings of the modern factory system in England*. Abingdon: Routledge.

Morgan, R.P.C. (2009). *Soil erosion and conservation*. New York, NY: John Wiley & Sons.

Moritz, M.A., Batllori, E., Bradstock, R.A., et al. (2014). Learning to coexist with wildfire. *Nature*, **515**, 58-66.

Munroe, D.K., van Berkel, D.B., Verburg, P.H., and Olson, J.L. (2013). Alternative trajectories of land abandonment: causes, consequences and research challenges. *Current Opinion in Environmental Sustainability*, **5**, 471-476.

Navarro, L.M., and Pereira, H.M. (2015). Rewilding abandoned landscapes in Europe. In Pereira, H.M. and Navarro, L.M. (Eds.), *Rewilding European landscapes* (pp. 3-23). New York, NY: Springer.

Newton, A., and Ashmole, P. (2000). *Carrifran Wildwood Project: native woodland restoration in the Southern Uplands of Scotland. Management Plan*. Jedburgh: Borders Forest Trust.

Nikodemus, O., Bell, S., Grīne, I., and Liepiņš, I. (2005). The impact of economic, social and political factors on the landscape structure of the Vidzeme Uplands in Latvia. *Landscape and Urban Planning*, **70**, 57-67.

Nilsen, E.B., Milner-Gulland, E.J., Schofield, L., Mysterud, A., Stenseth, N.C., and Coulson, T. (2007). Wolf reintroduction to Scotland: public attitudes and consequences for red deer management. *Proceedings of the Royal Society of London B: Biological Sciences*, **274**, 995-1003.

Nogués-Bravo, D., Simberloff, D., Rahbek, C., and Sanders, N.J. (2016). Rewilding is the new Pandora's box in conservation. *Current Biology*, **26**(3), R87-R91.

O'Rourke, E. (2016). Landscape values: high nature value farming on the Iveragh Peninsula. In Collins, T., Kindermann, G., Newman, C., and Cronin, N. (Eds.), *Landscape values: place and praxis* (pp. 248-253). Galway: Centre for Landscape Studies, NUI Galway.

Olschewski, R., Bebi, P., Teich, M., Hayek, U.W., and Grêt-Regamey, A. (2012). Avalanche protection by forests – a choice experiment in the Swiss Alps. *Forest Policy and Economics*, **17**, 19-24.

Otero, I., Marull, J., Tello, E., et al. (2015). Land abandonment, landscape, and biodiversity: questioning the restorative character of the forest transition in the Mediterranean. *Ecology and Society*, **20**, 7.

Pearce, F. (2015). *The new wild: why invasive species will be nature's salvation*. London: Icon Books.

Pereira, H.M., and Navarro, L.M. (2015). *Rewilding European landscapes*. New York, NY: Springer.

Perger, C., Fritz, S., See, L., et al. (2012). A campaign to collect volunteered geographic information on land cover and human impact. In Jekel, T., Car, A., Strobl, J., and Griesebner, G. (Eds.), *GI_Forum 2012: Geovisualization, Society and Learning* (pp. 83-91). Berlin: Herbert Wichmann.

Peterken, G.F. (1993). *Woodland conservation and management*. London: Chapman and Hall.

Peterken, G.F. (1996). *Natural woodland: ecology and conservation in northern temperate regions*. Cambridge: Cambridge University Press.

Peterson, U., and Aunap, R. (1998). Changes in agricultural land use in Estonia in the 1990s detected with multitemporal Landsat MSS imagery. *Landscape and Urban Planning*, **41**, 193-201.

Phalan, B., Onial, M., Balmford, A., and Green, R.E. (2011). Reconciling food production and biodiversity conservation: land sharing and land sparing compared. *Science*, **333**, 1289-1291.

Plachter, H., and Hampicke, U. (2010). *Large-scale livestock grazing: a management tool for nature conservation*. Berlin: Springer Science & Business Media.

Plieninger, T., Hui, C., Gaertner, M., and Huntsinger, L. (2014). The impact of land abandonment on species richness and abundance in the Mediterranean Basin: a meta-analysis. *PLoS ONE*, **9**(5), e98355.

Prishchepov, A.V., Radeloff, V.C., Baumann, M., Kuemmerle, T., and Müller, D. (2012). Effects of institutional changes on land use: agricultural land abandonment during the transition from state-command to market-driven economies in post-Soviet Eastern Europe. *Environmental Research Letters*, **7**(2), 024021.

Queiroz, C., Beilin, R., Folke, C., and Lindborg, R. (2014). Farmland abandonment: threat or opportunity for biodiversity conservation? A global review. *Frontiers in Ecology and the Environment*, **12**, 288-296.

Regos, A., Domínguez, J., Gil-Tena, A., Brotons, L., Ninyerola, M., and Pons, X. (2016). Rural abandoned landscapes and bird assemblages: winners and losers in the rewilding of a marginal mountain area (NW Spain). *Regional Environmental Change*, **16**, 199-211.

Rudel, T.K., Coomes, O.T., Moran, E., et al. (2005). Forest transitions: towards a global understanding of land use change. *Global Environmental Change*, **15**, 23-31.

Ruskule, A., Nikodemus, O., Kasparinskis, R., Bell, S., and Urtane, I. (2013). The perception of abandoned farmland by local people and experts: landscape value and perspectives on future land use. *Landscape and Urban Planning*, **115**, 49-61.

Schierhorn, F., Müller, D., Beringer, T., Prishchepov, A.V., Kuemmerle, T., and Balmann, A. (2013). Post-Soviet cropland abandonment and carbon sequestration in European Russia, Ukraine, and Belarus. *Global Biogeochemical Cycles*, **27**, 1175-1185.

Scott, J.M., Norse, E.A., Arita, H., et al. (1999). The issue of scale in selecting and designing biological reserves. In Soulé, M.E. and Terborgh, J. (Eds.), *Continental conservation. Scientific foundations of regional reserve networks* (pp. 19-37). London: Kogan Page.

See, L., Perger, C., Hofer, M., Weichselbaum, J., Dresel, C., and Fritz, S. (2015). LACO-WIKI: an open access online portal for land cover validation. *ISPRS Annals of Photogrammetry, Remote Sensing and Spatial Information Sciences*, **2**, 167-171.

Sinclair, A. (2017). The future of conservation: lessons from the past and the need for rewilding of ecosystems. www.youtube.com/watch?v=hAGYP9-cZ7Q&feature=youtu.be

Sinnett, D., Calvert, T., Martyn, N., et al. (2017). *Green infrastructure: how is green infrastructure research translated into practice outside the UK?* Technical Report. Bristol: University of the West of England.

Soulé, M., and Noss, R. (1998). Rewilding and biodiversity: complementary goals for continental conservation. *Wild Earth*, **8**, 18-28.

Tasser, E., Mader, M., and Tappeiner, U. (2003). Effects of land use in alpine grasslands on the probability of landslides. *Basic and Applied Ecology*, **4**, 271-280.

Terres, J.M., Scacchiafichi, L.N., Wania, A., et al. (2015). Farmland abandonment in Europe: identification of drivers and indicators, and development of a composite indicator of risk. *Land Use Policy*, **49**, 20-34.

Tharme, A.P., Green, R.E., Baines, D., Bainbridge, I.P., and O'Brien, M. (2001). The effect of management for red grouse shooting on the population density of breeding birds on heather-dominated moorland. *Journal of Applied Ecology*, **38**, 439-457.

Václavík, T., and Rogan, J. (2009). Identifying trends in land use/land cover changes in the context of post-socialist transformation in central Europe: a case study of the greater Olomouc region, Czech Republic. *GIScience & Remote Sensing*, **46**, 54-76.

Verburg, P.H., and Overmars, K.P. (2009). Combining top-down and bottom-up dynamics in land use modeling: exploring the future of abandoned farmlands in Europe with the Dyna-CLUE model. *Landscape Ecology*, **24**, 1167.

Webster, S.C., Byrne, M.E., Lance, S.L., et al. (2016). Where the wild things are: influence of radiation on the distribution of four mammalian species within the Chernobyl Exclusion Zone. *Frontiers in Ecology and the Environment*, **14**, 185-190.

Weissteiner, C.J., Boschetti, M., Böttcher, K., Carrara, P., Bordogna, G., and Brivio, P.A. (2011). Spatial explicit assessment of rural land abandonment in the Mediterranean area. *Global and Planetary Change*, **79**(1-2), 20-36.

Worboys, G., Francis, W.L., and Lockwood, M. (2010). *Connectivity conservation management: a global guide (with particular reference to mountain connectivity conservation)*. London: Earthscan.

CHAPTER SEVEN

Rewilding and restoration

JAMES R. MILLER
University of Illinois Urbana-Champaign
RICHARD J. HOBBS
University of Western Australia

There is a preponderance of verbs with the prefix 're-' in the conservation lexicon: reintroduce, remediate, reconnect, reforest, remediate, recover, reintroduce, restore, and so on (Corlett, 2016). A relatively recent addition to this list (and unlike the other verbs, a newly minted word) is the focus of this book – rewild.

The term 'rewild' was born two decades ago, as Michael Soulé and Reed Noss (1998) advocated for the restoration and protection of extensive wilderness areas and the keystone role played by wide-ranging large animals, especially carnivores. This was in keeping with the mission of the Wildlands Network (formerly the Wildlands Project), which Soulé founded, and its early emphasis on connecting extensive blocks of federal and state land in the western United States (Soulé and Terborgh, 1999). The basic premise here is that the elimination of these keystone species triggers a cascade of changes that eventually results in ecosystem degradation and extirpation of other species. Rewilding was viewed as a key step in reversing this degradation and restoring self-regulating ecological communities (Soulé and Noss, 1998), a thread that runs through all of the various forms of rewilding that have followed. This North American framework focused on extant native species whose ranges had been greatly reduced by human land use, such as the grizzly bear (*Ursus arctos horribilis*), grey wolf (*Canis lupus*), and mountain lion (*Puma concolor*). Meanwhile, a somewhat different, more audacious approach to rewilding was being proffered on the other side of the world.

Australia is similar to North America with regard to impacts on native species and their habitats since European colonisation, as well as the rapid loss of many medium- and large-bodied species soon after initial human contact (Bradshaw and Ehrlich, 2015). Although the scenario played out in Australia at least 40,000 years earlier (Johnson, 2006), these waves of extinction resulted in many imbalances and vacant niches on both continents. Tim Flannery, a scientist with the Australian Museum, suggests that one of the most important of these imbalances stems from the lack of large carnivores (Flannery, 1994). Around the same time that Soulé and his colleagues were considering the situation in western North America, Flannery suggested strategies for addressing this imbalance in Australia, including the introduction of the closest living relative of the continent's extinct gigantic carnivorous goannas, the Komodo dragon (*Varanus komodoensis*). Weighing up to 100 kg, this largest living species of lizard could help bring about trophic stability in the tropics of the Northern Territory and Queensland, according to Flannery, by dampening environmental degradation wrought by population spikes of introduced placental herbivores and by reducing populations of introduced carnivores. At the same time, he lamented the fact that the Komodo dragon is one of only a few species suitable for such an introduction because all of the large herbivorous and carnivorous marsupials that once dominated Australia's megafauna are now extinct. With a hint of envy, Flannery (1994) compared this situation to that of North America, observing that there the Pleistocene megafauna could largely be recreated by importing extant conspecifics from Asia, Africa, and South America.

It is difficult to know if anyone in North America paid much attention to Flannery's suggestion. Yet a strikingly similar proposal was made just a decade later by Donlan and colleagues (2005, 2006), calling for functional restoration of the megafauna that existed on the continent 13,000 years ago by introducing horses, camelids, cheetahs, elephants, lions, and the like from Asia and Africa. They labelled this process 'Pleistocene rewilding', and their proposal has generated a fair measure of debate ever since (Rubenstein et al., 2006; Caro, 2007; Oliveira-Santos and Fernández, 2009; Toledo et al., 2011; Nogués-Bravo et al., 2016; Rubenstein and Rubenstein, 2016).

The approach proposed by Caro and his colleagues is considered a separate form of rewilding by some (Corlett, 2016; Nogués-Bravo et al., 2016). Others see it as a specific form of trophic rewilding, defined as 'an ecological restoration strategy that uses species introductions to restore top-down trophic interactions and associated trophic cascades to promote self-regulating biodiverse ecosystems' (Svenning et al., 2016). In this sense, Pleistocene rewilding is true to the original vision of rewilding offered by Soulé and Noss (1998), although the argument for ecological replacement of long-extinct species is a radical departure.

An alternative to trophic rewilding is ecological rewilding, or the passive management of ecological succession, typically on agricultural land that has been abandoned (Navarro and Pereira, 2012). Whereas trophic rewilding involves top-down processes, ecological rewilding (also known as passive rewilding, although again, some consider these two strategies distinct for reasons that escape us; see Corlett, 2016) is more of a blend. Here, the emphasis is on managing succession to restore fundamental ecosystem processes, but the key role of large-bodied herbivores and carnivores is also recognised (Navarro and Pereira, 2012; Pereira and Navarro, 2015). As with trophic rewilding, the goal is still restoring self-sustaining ecosystems and biodiversity protection, but this goal has been broadened to include other ecosystem services.

At the foundation of rewilding is ecological restoration, as evidenced by the language used by proponents of the former to describe it. Some go so far as to argue that the objectives of rewilding now so closely mimic those of restoration (Nogués-Bravo et al., 2016) that the distinction between the two is unclear. The most commonly used definition of restoration comes from the Society for Ecological Restoration's *Primer on Ecological Restoration*: 'Ecological restoration is the process of assisting the recovery of an ecosystem that has been degraded or destroyed' (SER Science and Policy Working Group, 2004), and once more the parallels with rewilding are evident. What of the methodologies employed? Again, the linkages between restoration and ecological rewilding appear fairly strong, given the latter's focus on abandoned agricultural land (Cramer and Hobbs, 2007) and manipulating ecological succession (Walker et al., 2007). The connections between trophic rewilding and restoration are perhaps more tenuous, as restoration has historically placed a much greater emphasis on plants than animals (McAlpine et al., 2016). In the sections that follow, we explore the relationship between rewilding and ecological restoration in greater detail, with an emphasis on how each could benefit from greater integration of these two endeavours.

Trophic rewilding and ecological restoration

The conceptual framework for trophic rewilding, as initially envisioned by Soulé and Noss (1998), has received considerable attention in the ensuing decades (Terborgh and Estes, 2010; Estes et al., 2011; Ripple et al., 2014). Empirical support, however, is scant and somewhat mixed. Perhaps the most well-known example of trophic rewilding that involved the reintroduction of a regionally extirpated species is the restoration of wolves to Yellowstone National Park. The reintroduction resulted in both direct effects on the wolves' prey and indirect effects on other wildlife and environmental features (Ripple and Beschta, 2012). When species are globally extinct, however, the only means of trophic rewilding is the introduction of functionally similar non-indigenous species, or ecological replacements (Seddon et al., 2014).

One of the most successful instances of trophic rewilding using ecological replacements is the introduction of exotic tortoises (*Astrochelys radiata*, *Aldabrachelys gigantea*) on Mauritian offshore islands as substitutes for extinct giant tortoises (*Cylindraspis* spp.) (Griffiths et al., 2010, 2011, 2013). These introductions have highlighted the role of giant tortoises as ecological engineers in these island systems, resulting in improved seed dispersal, establishment of native trees, and suppression of invasive plants (Griffiths et al., 2010, 2011). The apparent success of these ecological replacements motivated similar introductions of exotic tortoises to reinstate processes lost with the extinction of endemic giant tortoises in numerous other island systems (Hansen et al., 2010). Results have been mixed, however, with desirable effects on vegetation structure and composition in some cases, but either no effect or undesirable effects in others, such as the dispersal by tortoises of invasive plants (Hansen et al., 2010). Clearly, the challenge is to identify replacement species that will perform the desired ecosystem function, and the more time that has passed since the extinction of the original species, the greater this challenge will be (Seddon et al., 2014).

The difficulty in identifying suitable replacement species posed by time lags appears to hold for reintroductions of extirpated species. There is little support for reintroducing species extirpated > 5000 years ago based on a review of the literature (Svenning et al., 2016), and support was even weaker for the effectiveness of ecological replacements. This review is hampered by a geographical bias towards North America, Europe, and oceanic islands, but the real constraint is a lack of empirical research (Svenning et al., 2016). One thing that critics and proponents alike agree on is the need to rectify this knowledge gap by gathering empirical data on carefully designed introductions (Seddon et al., 2014; Nogués-Bravo et al., 2016; Rubenstein and Rubenstein, 2016; Svenning et al., 2016; Fernández et al., 2017), ideally using species that can be easily monitored and managed (Corlett, 2013). Only then will we be able to sort out the factors underlying the considerable variability in the magnitude of trophic impacts in such efforts to date (Boitani and Linnell, 2015).

Even though trophic rewilding has not garnered much empirical support to date, the lack of attention paid to this strategy in restoration circles is, on the face of it, surprising. The evidence at hand suggests that trophic rewilding can be an effective restoration strategy in at least some circumstances, and after all, the track record of ecological restoration is replete with uncertain outcomes (Lockwood and Pimm, 1999; Hobbs, 2009; Suding, 2011). One factor contributing to the exclusion of trophic rewilding from the major currents of ecological restoration is likely the emphasis on animals as agents of recovery.

Ecological restoration has historically focused on vegetation and associated soil and geomorphological characteristics, going back to its antecedents in mine reclamation and its roots in the recovery of tallgrass prairies in the Midwestern USA (Allison, 2012). Indeed, a recent edited volume on restoration (Allison and Murphy, 2017) contains relatively few references to fauna and does not consider faunal roles in restoration at all. The same can be said of a recent set of international standards proposed for ecological restoration, published by the Society for Ecological Restoration (McDonald et al., 2016a). Here, when animals are considered, it is usually with regard to plant–animal interactions, especially ways that various species such as grazing or browsing animals might affect restoration plantings. There is also a tacit assumption that restoring vegetation will facilitate the colonisation of the system by fauna of all shapes and sizes.

Passive recolonisation of restored sites by wildlife, which is a cornerstone of ecological rewilding, plays an equally prominent role in restoration. The so-called 'Field of Dreams' hypothesis is premised on the notion that 'if you build it, they will come'. The presumption here is that habitat enhancement or creation will be followed by the redevelopment of faunal communities, or creating structure will beget biotic recovery (Palmer et al., 1997; Sudduth et al., 2011). The Field of Dreams hypothesis is referred to as one of the myths of restoration ecology (Hilderbrand et al., 2005). The concept of a self-assembling ecosystem is appealing, especially in cases where budgets and time are limited – that is, most cases – and the process of community assembly has clear relevance in restoration (Temperton et al., 2004). Yet the dynamic nature of community assembly and the role of stochasticity in this process have hampered the search for generalisable assembly rules (Temperton and Hobbs, 2004). While Field of Dreams may be an attractive metaphor, the process of actually restoring habitat and ensuring its use by target species is much more complex. Effective restoration will require shifting from a simplistic conceptualisation of habitat as land cover to a more nuanced consideration of key resources and the timing of their availability, species interactions, mobility, and landscape characteristics, to name a few (Miller and Hobbs, 2007). As the pool of target species and spatial scale of restoration expands, this complexity increases exponentially.

The complexity of broad-scale habitat restoration goes beyond the requirements of the target species. Intensive restoration is expensive and real-world constraints are imposed by the strategies of funding sources, which often prefer to manage risk by spreading their resources across multiple projects (Johnson, 2017). This effectively restricts restoration in any one project to a relatively small area. It would seem that restoration is at an impasse, given the formidable barriers to effective habitat restoration on the one hand and the limitations of passive restoration on the other. At the same time, a variety of species that include wide-ranging herbivores and carnivores are indeed recolonising Europe in impressive numbers (Deinet et al., 2013; Boitani and Linnell, 2015) and this same phenomenon is occurring elsewhere (Chapron et al., 2014). These range expansions should not, however, be construed solely as a response to habitat restoration or rewilding, although reforestation does play a role. Europe's dramatically increasing populations of large carnivores, for example, largely exist outside of protected areas in multiple use, human-dominated landscapes (Chapron et al., 2014).

Overall, we believe that the patterns described above offer several lessons for trophic rewilding and ecological restoration. Proponents of rewilding, particularly ecological rewilding, would do well to recognise the limitations of depending on passive recolonisation and simply excluding human actions. A blend of approaches will be required to facilitate the recovery of a broad spectrum of biodiversity. More intensive restoration efforts focused on key resources required by particular species could occur in patches embedded in landscapes dominated by less-intensive management. Ongoing management will be required for some ecosystem types. For instance, there is uncertainty about the ability of herbivores to maintain open habitats in the absence of restoration actions such as burning or clearing (Kirby, 2009; Kerley et al., 2012). On the other hand, restoration practitioners would do well to recognise that wildlife not only respond to restoration, but are also agents of this process. As with most things in ecology, a given strategy may work well in some situations and not others, and successful outcomes will depend on a range of factors relating to both the area being restored and the fauna being targeted.

Ecological rewilding and ecological restoration

Ecological rewilding, like ecological restoration, takes a vegetation-centric approach that focuses on succession (Navarro and Pereira, 2012), although the role of fauna in shaping successional processes is also recognised. The idea of restoration largely being a series of interventions aimed at directing ecological succession has been around for a while (Luken, 1990; Walker et al., 2007). Ecosystems gradually recover from disturbance through a successional sequence involving species arriving, establishing and growing at different rates, development of species interactions, and modification of the local environment. Hence, left to themselves, damaged ecosystems will eventually develop some form of vegetation cover and the recovering system will mediate ecosystem processes. Sometimes, however, recovery is very slow or not in a direction deemed desirable, and in these instances, restoration efforts aim to speed up and direct the successional process. Undoubtedly, simply allowing successional processes to operate is likely to be a less-costly option than intervening, especially if this involves expensive soil works, plant establishment, and the like (Prach and Hobbs, 2008). Legacies of past land use, however, are a major determinant of the direction that succession will take and these legacies may persist for decades or longer (Fraterrigo, 2012). Work on abandoned farmland indicates that land-use legacies result in situations where successional processes either stall or lead to alternative assemblages that may or may not be desirable in terms of their composition and provision of ecosystem services (Cramer and Hobbs, 2007; Cramer et al., 2008).

Given that a major rewilding target, particularly in Europe, is abandoned pastoral and agricultural lands, lessons learned from old field succession are particularly relevant. Rewilding frequently assumes that successional processes will gradually take predominantly short-stature herbaceous communities through to more tree-dominated communities. This, in turn, is expected to result in the enhanced provision of a range of ecosystem services – primarily supporting (e.g. habitat for biodiversity), cultural (e.g. recreation), and regulating services (e.g. carbon sequestration). Provisioning services will necessarily be reduced (Navarro and Pereira, 2012).

This assumed trajectory of rewilding is likely to work well in situations where vegetation succession is relatively rapid – where, for instance, there is a ready source of propagules, either in the seed bank or in nearby patches, and where there are no inherent barriers to their establishment. It is less likely to work in situations where species arrival and establishment is not guaranteed or where this is an exceedingly slow process, as is often the case in landscapes characterised by low productivity. In these cases, active restoration will likely be required, and not only in the early phases of the process, as some proponents of rewilding suggest (Navarro and Pereira, 2012). Rewilded areas are sometimes compared to previous uses of the land using a snapshot of the ecosystem services that are provided (Navarro and Pereira, 2012). Yet even if the trajectory of rewilding that is often assumed is indeed the one that transpires, the magnitude and type of ecosystem services provided will vary over time. In the early phases of rewilding, the system may be perceived as degraded compared to what was there prior to abandonment. As rewilding proceeds, there is also the potential for succession to divert from the expected trajectory, or for the system to shift to alternative states that prevent further successional development (Figure 7.1). In these cases, progressive diminution of ecosystem services is likely.

The potential for these scenarios to unfold will necessitate a set of trigger points for deciding whether alternative states and trajectories are desirable, and whether further management intervention is required (e.g. Grant, 2006). Are those involved in making these decisions willing to adopt a strategy that accepts the journey and doesn't worry so much about the destination – i.e. open-ended restoration (Hughes et al., 2011, 2012) – or do they see the achievement of a particular type of 'wildness' as more important? And what of public perception? These questions get to the heart of what is actually meant by ecological rewilding and what role humans have in achieving it (Hobbs et al., 2010).

Such issues are front and centre in one form of ecological rewilding that can be seen in many areas of Europe. Take Scotland as an example. Pine woodland is re-establishing in upland areas after centuries of management that maintained the vegetation as low heathland (Hobbs, 2008). There and elsewhere in Europe, debate continues regarding the desirable extent of closed woodland or forest versus open areas or areas with scattered trees. In addition, the more

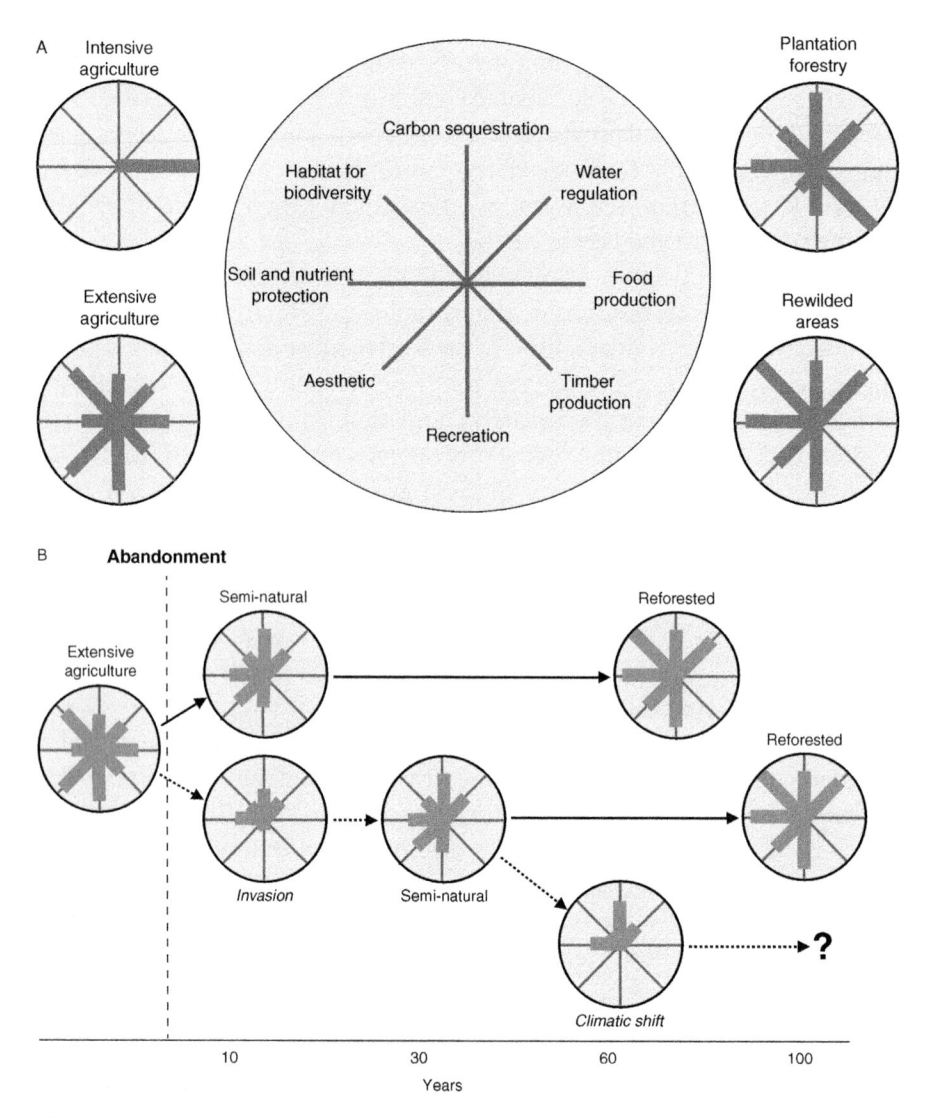

Figure 7.1. Panel A depicts proposed differences in the provision of ecosystem services by rewilded areas compared to land uses typically proposed for rewilding (*sensu* Navarro and Pereira, 2012). Panel B indicates how ecosystem services in areas dedicated to extensive agriculture prior to abandonment might change over time as rewilding proceeds. Solid lines indicate the trajectory of succession preferred by proponents of rewilding (e.g. Navarro and Pereira, 2012). Dotted lines indicate the succession as affected by two key sources of environmental uncertainty – plant invasions and climate change. The progression from invaded to semi-natural presumes an intervention that would likely involve clearing and replanting.

open agricultural and pastoral landscapes are often associated with deeply held cultural values and provide habitat for a range of species that do not occur in more wooded areas. Examples from Europe include alpine meadows characterised by high floristic diversity, and arable land-dependent bird species such as the skylark (*Alauda arvensis*) and reed bunting (*Emberiza schoeniclus*). The persistence of these systems depends on continued interventions of the sort that have been ongoing for centuries or millennia, such as the management of dehasas, cork oak savannas, and the like (Bugalho et al., 2011). Conserving or restoring such systems will involve the continuation or reinstatement of traditional management practices rather than their abandonment to rewilding. Thus, in cultural landscapes, there will need to be a balance of approaches that mesh the ambitions of rewilding with the requirements of species that inhabit managed systems and the cultural values that may be lost. This is clearly pertinent in many European landscapes, but also has to be considered more broadly in the context of lands managed (currently or previously) by indigenous peoples in the Americas, Australia, and elsewhere (e.g. Mann, 2005; Gammage, 2011).

This is also relevant more broadly, where ambitious global restoration goals need to be approached strategically and acted on in ways that do not compromise existing values. For instance, Veldman and colleagues (2015) showed that the *Atlas of Forest and Landscape Restoration Opportunities* (World Resources Institute, 2014) misclassified 9 million km^2 of grassy biomes globally as 'deforested' or 'degraded' and therefore providing 'opportunities' for forest restoration. If these areas were to be 'restored' by planting trees, significant biodiversity losses would undoubtedly ensue because of the damage to diverse grassland ecosystems (Bond, 2016). Clearly, care is required in interpreting calls for restoration or rewilding that might involve significant modification of existing ecosystems and the loss of their biodiversity and services. A tension therefore exists between the goals of rewilding, which essentially aims to remove human influence from the landscape, and broader goals of maintaining cultural landscapes that involve keeping humans *in* landscapes as keystone ecological species. This tension also extends to restoration. While traditional restoration aims to 'kick-start' ecosystem recovery with the ultimate goal of the process proceeding unaided, there is increasing recognition that ongoing management will be required in many restoration areas, particularly where a narrowly defined target system has been identified.

Ecological rewilding has a vision of broad-scale landscape change, yet in many places it may be necessary to adapt this to take account of the factors discussed above. Current landscapes consist of mosaics of patches of different land uses, degrees of modification, and the like. Even if the overall goal is to achieve ecological rewilding over the entire landscape, it is likely that this will involve a range of different actions in different places – succession or passive restoration may work in some areas, while more active restoration may be necessary in others if, for instance, invasive weeds are perceived as a problem. This reflects the more general observation that effective and responsible interventions will require a pluralistic approach drawn from several existing practices (Hobbs et al., 2011). Different objectives might be set for different parts of the landscape, requiring a broad portfolio-based approach that takes into account the range of ecological conditions, conservation goals, and likely ecosystem services provided in different areas (Hobbs et al., 2014, 2017). A portfolio approach considers the range of ecological conditions present in the focal area and the array of management options available, seeking the most effective mix of interventions (including non-intervention) to achieve agreed-upon conservation and social goals. Ecological rewilding may indeed be the most appropriate approach in some cases but prove inappropriate or ineffective in others, and could theoretically be carried out on a relatively small scale. Rewilding for a butterfly is likely different than rewilding for a bison.

The place of history, revisited

Ecological restoration has traditionally been driven by historically based goals (Higgs et al., 2014). More recently, there has been much debate about the degree to which restoration targets based on historical ecosystems still make sense (Higgs et al., 2014). These deliberations are motivated by a growing awareness of the extent and rapidity of environmental change. Although restorationists have been aware of these changes for some time, other concerns have taken precedence. For example, the *SER Primer on Ecological Restoration* (2004) offers much in the way of guidance, but invasive species and climate change are hardly mentioned. Around this same time, Harris and his colleagues (2006) delivered a wake-up call to the restoration community by asserting that the scope and pace of climate change will force a reconsideration of goals for restoration projects across the board. There is now broad acknowledgement within this community that ecosystems and species distributions are changing, resulting in new species assemblages and altered ecosystem processes over relatively short time spans (Hobbs et al., 2009, 2018). There is also growing recognition that with these changes, it will not be possible to fully restore all ecosystems, despite continuing calls for full restoration to be the accepted goal in most situations (McDonald et al., 2016a, 2016b). In response, others have appealed for a more inclusive approach that recognises the profound degree of alteration in many systems and the suite of different goals that are possible (Higgs et al., 2014; Miller and Bestelmeyer, 2016).

A rear-view mirror perspective also characterised early rewilding efforts, epitomised by the Pleistocene version. That proposal goes far beyond the emphasis on historical goals in restoration by targeting the prehistorical.

Critics have seized upon the many differences between the Pleistocene and today, including a wholly different climate, existence of a fauna that was not present at that time, and the ubiquitous presence of people and prevalence of human land use (Rubenstein et al., 2006; Oliveira-Santos and Fernandez, 2009; Toledo et al., 2011).

Moving forward from the fixation on the way back in Pleistocene rewilding, trophic rewilding has adopted multiple perspectives on the role of history. These range from looking back several millennia in the case of ecological replacements and the past century or two in the case of reintroductions, to looking ahead in the case of assisted colonisation (Seddon et al., 2014). In ecological rewilding, historical conditions are viewed as a source of inspiration and a knowledge base to enhance understanding of ecological functioning, but the trajectory of ecosystem change that is anticipated is not rooted in history (Pereira and Navarro, 2015). Rather, the uncertainty posed by environmental change is acknowledged and the potential for novel combinations of plant and animal communities to develop is met not with alarm but rather pragmatic acceptance. The notion of ecological communities characterised by novelty has been received somewhat differently among restoration ecologists.

Novel ecosystems

Over the last decade or so, the concept of novel ecosystems has vaulted from obscurity to being referenced in hundreds of scientific articles and serving as the central theme of more than a few symposia and book chapters, as well as the focus of an edited volume (Miller and Bestelmeyer, 2016). Along the way, it has generated no small amount of controversy, particularly among restoration ecologists (Hobbs et al., 2009; Murcia et al., 2014; Miller and Bestelmeyer, 2016). Numerous definitions of the term 'novel ecosystem' have been proposed (Hobbs et al., 2013; Morse et al., 2014; Truitt et al., 2015). Although there are differences among these renderings, several common threads emerge: as a function of human activities, the components of novel ecosystems and their interactions differ from those that prevailed historically; these systems tend to maintain their novel character without further human intervention; and a novel system cannot be restored to its historical state, whether due to ecological or practical limitations.

The novel ecosystems concept has recently met with strong opposition from at least a few critics (Aronson et al., 2014; Murcia et al., 2014; Simberloff et al., 2015). Two possible explanations for this resistance are (1) that some restorationists equate the idea that full restoration is not feasible in some cases with a kind of surrender to degradation, and (2) that recognising novel ecosystems will lead to the replacement of restoration with environmental engineering (Allison, 2017). Yet being cognisant of the growing emergence of novel ecosystems is not the same as actively managing for novelty, although the latter may be required in instances where out of necessity restoration shifts from compositional goals to functional goals (Perring et al., 2015).

It is the emphasis on functional goals that is central to some forms of rewilding, particularly in the reliance on replacement species. These species, introduced to fill key roles that are vacant due to extinction, are necessarily novel elements. Clearly, there is risk involved here as the history of species introductions is one fraught with unexpected consequences. At the same time, human-wrought environmental change has increased the element of risk in many strategies that have long been part of the conservation toolbox (Lawler et al., 2010). It follows that this same amplification of risk stemming

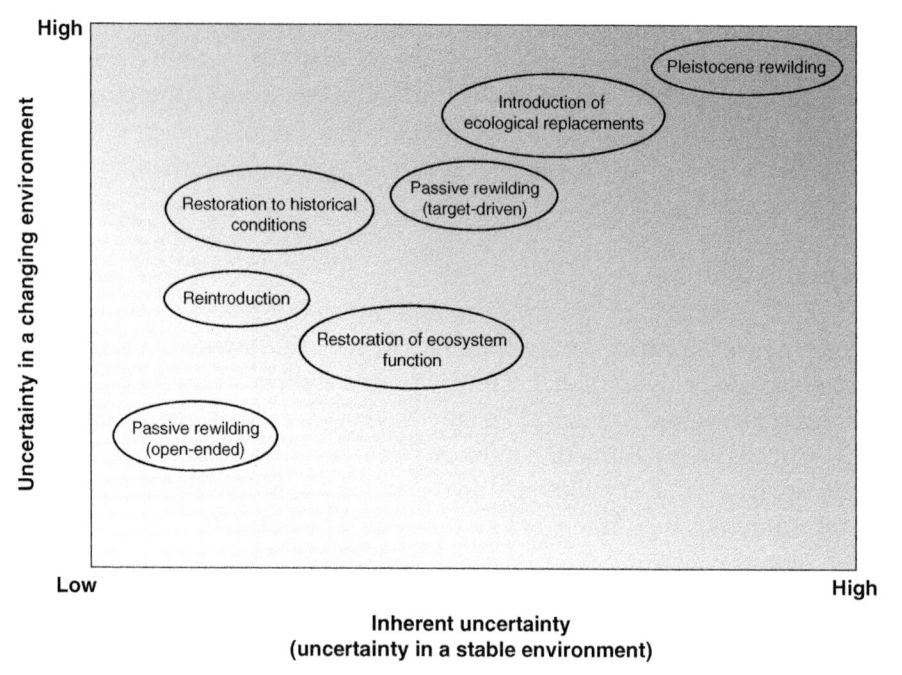

Figure 7.2. Management strategies typically employed in rewilding and ecological restoration plotted with respect to the relative degree of uncertainty in their effectiveness. The x-axis represents uncertainty associated with a management strategy under relatively stable environmental conditions. The y-axis represents uncertainty as a function of rapidly changing environmental conditions due to climate change, modifications in land use and land cover, and other human actions. Thus, strategies represented at the bottom of the plot are more robust to uncertainty associated with rapidly changing environmental conditions, whereas those at the top of the plot are associated with a higher degree of risk under such conditions (*sensu* Lawler et al., 2010).

from human activities will attend restoration actions and rewilding approaches, some more than others (Figure 7.2). Therefore, part of the challenge is to manage risk and wherever possible to select approaches where it is minimised. It is likely that the level of risk deemed tolerable in a given situation will be commensurate with the severity of environmental change and the risk of inaction.

Proponents of rewilding seem more comfortable with such risks than many engaged in ecological restoration. If rewilding has the goal of 'letting nature do its own thing', even allowing for some initial intervention, novel assemblages are virtually guaranteed by altered environmental conditions and new species mixes (Hobbs et al., 2018). Will proponents of rewilding be willing to accept the results of an uncertain and unpredictable assembly process? Again, to some extent, this comes down to the broader question of whether humans pull back from nature and accept the consequences or actively intervene and manage to achieve particular desired assemblages.

This question becomes even more moot where humans actively design ecosystems for particular purposes. Higgs (2017) has examined in detail the differences between designed and novel ecosystems, which can be largely discussed in the context of different intents. Designed ecosystems are, from the start, developed with particular ecological services or functions in mind and often require ongoing intervention, whereas novel ecosystems develop once direct human intervention ceases and in the absence of intent. However, the distinction is not perfect. A designed ecosystem may exhibit its own inherent dynamics, despite human direction and even with active management. For instance, a plantation forest may be colonised by a range of understorey species as well as fauna that utilise such habitats. On the other hand, a novel ecosystem can be managed to guide it towards particular goals.

Designed ecosystems would appear to be an odd place to look for rewilding. Yet one of the most commonly cited examples of rewilding, the Oostvaardersplassen in the Netherlands, was intentionally created by pushing back the sea with a dyke, draining a polder, and actively introducing various plant and animal species to colonise the former seabed. There is no link at all to the historical ecosystem at that site. It was *designed* to be a completely unique terrestrial ecosystem – where a marine ecosystem had previously existed – that would come to self-organise and require only minimal intervention thereafter. Cities abound with examples of areas that were entirely designed to begin with, but then started a process of autonomous assembly when their previous use ceased. The New York High Line is an excellent case in point. This elevated railway in Manhattan ceased operation in the 1980s and developed a rich vegetation comprising native and non-native species, and is now managed as a heavily used urban park (David and Hammond, 2011). Indeed, Marris (2011) suggests that places such as abandoned urban lots represent examples of the 'new wild' where nature actually has a chance to go its own way. Admittedly, this is not classical restoration or rewilding of the forms that we have considered, but we think that there are tenets of each endeavour that may be relevant in such circumstances.

Conclusions

We have focused on the relationship between two of the 're-' words that we listed at the beginning of this chapter – rewilding and restoration. Of course, other terms in the list are subsumed in these two activities, and by virtue of the prefix, all imply a sense of returning to some prior condition. As we have noted above, the notion of a return to the past is problematic in both restoration and rewilding, but inherent in both endeavours is consideration of the past. Going forward, both will almost certainly benefit from an ongoing dialogue about what such considerations might entail.

Other elements of rewilding frameworks can inform restoration, and vice versa. As ecological restoration has matured, the definition of success has expanded from one based largely on ecological fidelity to include other considerations, such as cultural revitalisation, community participation, and social justice (Higgs, 1997, 2003). Some of this thinking is evident in passive rewilding (Navarro and Pereira, 2012), but there needs to be greater recognition that at the root of all conservation actions are human values. As noted above, the idea of simply allowing succession to proceed in the absence of human interference is fine until it takes an unexpected turn that confers less value than the system that was abandoned. One can imagine relying on an adaptive approach in rewilding that comprises both scientific and social frameworks, as is increasingly adopted in restoration (Allison, 2017) and in conservation more generally. There are many stakeholders in rewilding projects, whether they are acknowledged or not, whose opinions will affect the process and its outcome. These voices need to be heard.

A salient point about ecological restoration that has contributed to the incredible growth in its popularity over the last few decades is the range of spatial scales over which it can be implemented. There are the megarestoration projects of the scale that rewilding aspires to, ranging from Florida's Everglades to Gondwana Link in south-western Australia (www.gondwanalink.org/) to Brazil's Atlantic Forest restoration. In other cases, restoration is undertaken as one element of landscape-scale conservation frameworks as well as on individual farmsteads and in urban neighbourhoods. Could there be parallel tracks for rewilding across a gradient of spatial scales? What, for example, are the opportunities for rewilding in the 'new wild' of urban areas? Demonstrating the value of rewilding at finer scales in places where many people are exposed to it on a regular basis has the potential to grow support for more extensive projects (Miller and Hobbs, 2002).

Reframing rewilding in a way that can be implemented across a gradient of spatial scales will facilitate a portfolio approach that aims for appropriate actions across a range of different situations. Trophic rewilding can work well where there are large areas of habitat and where herbivores have become over-abundant and are wrecking the system. Reintroductions of large animals can kick-start or facilitate broader restoration. Ecological rewilding can work where systems are likely to self-regenerate and where important cultural values are not likely to be lost. Both of these types of rewilding can also work in concert with each other. Finally, ecological restoration activities that intervene to accelerate or change the direction of successional development are going to be essential in many areas where natural recovery processes are compromised. Rather than seeking to compartmentalise and differentiate among these various approaches, we suggest that they should all be seen as complementary tools to be used to promote conservation and ecosystem service goals in an increasingly humanised world. The future shape of both the human and natural world largely depends on how well we accomplish the juggling act of deciding which tools are appropriate where and when.

References

Allison, S.K. (2012). *Ecological restoration and environmental change: renewing damaged ecosystems*. London: Routledge.

Allison, S.K. (2017). Ecological restoration and environmental change. In Allison, S.K., and Murphy, S.D. (Eds.), *The Routledge handbook of ecological restoration* (pp. 485–495). London: Routledge.

Allison, S.K., and Murphy, S.D. (2017). *The Routledge handbook of ecological restoration*. London: Routledge.

Aronson, J., Murcia, C., Kattan, G.H., Moreon-Mateosa, D., Kixon, K., and Simberloff, D. (2014). The road to confusion is paved with novel ecosystem labels: a reply to Hobbs et al. *Trends in Ecology & Evolution*, **29**, 646–647.

Boitaini, L., and Linnell, J.D.C. (2015). Bringing large mammals back: large carnivores in Europe. In Pereira, H.M., and Navarro, L.M. (Eds.), *Rewilding European landscapes* (pp. 67–84). Heidelberg: Springer.

Bond, W.J. (2016). Ancient grasslands at risk. Highly biodiverse tropical grasslands are at risk from forest-planting efforts. *Science*, **351**, 120–122.

Bradshaw, C.J.A., and Ehrlich, P.R. (2015). *Killing the koala and poisoning the prairie*. Chicago, IL: University of Chicago Press.

Bugalho, M.N., Caldeira, M.C., Pereira, J.S., Aronson, J., and Pausas, J.G. (2011). Mediterranean cork oak savannas require human use to sustain biodiversity and ecosystem services. *Frontiers in Ecology and the Environment*, **9**, 278–286.

Caro, T. (2007). The Pleistocene re-wilding gambit. *Trends in Ecology & Evolution*, **22**, 281–283.

Chapron, G., Kaczensky, P., Linnell, J.D.C., et al. (2014). Recovery of large carnivores in Europe's modern human-dominated landscapes. *Science*, **346**, 1517–1519

Corlett, R.T. (2013). The shifted baseline: prehistoric defaunation in the tropics and its consequences for biodiversity conservation. *Biological Conservation*, **163**, 13–21.

Corlett, R.T. (2016). Restoration, reintroduction, and rewilding in a changing world. *Trends in Ecology & Evolution*, **31**, 453–462.

Cramer, V.A., and Hobbs, R.J. (2007). *Old fields: dynamics and restoration of abandoned farmland*. Washington, DC: Island Press.

Cramer, V.A., Hobbs, R.J., & Standish, R.J. (2008). What's new about old fields? Land abandonment and ecosystem assembly. *Trends in Ecology & Evolution*, **23**, 104–112.

David, J., and Hammond, R. (2011). *High Line: the inside story of New York City's park in the sky*. New York, NY:Farrar, Straus and Giroux.

Deinet, S., Ieronymidou, C., McRae, L., et al. (2013). Wildlife comeback in Europe: The recovery of selected mammal and bird species. Final Report to Rewilding Europe by ZSL, Birdlife International and the European Bird Census Council, Zoological Society of London, UK.

Donlan, C.J., Greene, H.W., Berger, J., et al. (2005). Re-wilding North America. *Nature*, **436**, 913–914.

Donlan, C.J., Berger, J., Bock, C.E., et al. (2006). Pleistocene rewilding – an optimistic agenda for twenty-first century conservation. *The American Naturalist*, **168**, 660–681.

Estes, J.A., Terborgh, J., Brashares, J.S., et al. (2011). Trophic downgrading of planet Earth. *Science*, **333**, 301–306.

Fernández, N., Navarro, L.M., and Pereira, H.M. (2017). Rewilding: a call for boosting ecological complexity in conservation. *Conservation Letters*, **10**, 276–278.

Flannery, T. (1994). *The future eaters*. Kew, Victoria: Reed Books.

Fraterrigo, J.M. (2012). Landscape legacies. In Levin, S.A. (Ed.), *Encyclopedia of biodiversity*, 2nd edition (chapter 388). San Diego, CA: Academic Press.

Gammage, B. (2011). *The biggest estate on Earth: how Aborigines made Australia*. Crows Nest, New South Wales: Allen & Unwin.

Grant, C.D. (2006). State-and-transition successional model for bauxite mining rehabilitation in the jarrah forest of Western Australia. *Restoration Ecology*, **14**, 28–37.

Griffiths, C.J., Jones, C.G., Hansen, D.M., et al. (2010). The use of extant non-indigenous tortoises as a restoration tool to replace extinct ecosystem engineers. *Restoration Ecology*, **18**, 1–7.

Griffiths, C.J., Hansen, D.M., Jones, C.G., Zuël, N., and Harris, S. (2011). Resurrecting extinct interactions with extant substitutes. *Current Biology*, **21**, 762–765.

Griffiths, C.J., Zuë, L.N., Jones, C.G., Ahamud, Z., and Harris, S. (2013). Assessing the potential to restore historic grazing ecosystems with tortoise ecological replacements. *Conservation Biology*, **27**, 690–700.

Hansen, D.M., Donlan, C.J., Griffiths, C.J., and Campbell, K.J. (2010). Ecological history and latent conservation potential: large and giant tortoises as a model for taxon substitutions. *Ecography*, **33**, 272–284.

Harris, J.A., Hobbs, R.J., Higgs, E., and Aronson, J. (2006). Ecological restoration and global climate change. *Restoration Ecology*, **14**, 170–176.

Higgs, E. (1997). What good is ecological restoration? *Conservation Biology*, **11**, 338–348.

Higgs, E. (2003). *Nature by design: people, natural process, and ecological restoration*. Cambridge, MA: MIT Press.

Higgs, E. (2017). Novel and designed ecosystems. *Restoration Ecology*, **25**, 8–13.

Higgs, E., Falk, D.A., Guerrini, A., et al. (2014). The changing role of history in restoration ecology. *Frontiers in Ecology and the Environment*, **12**, 499–506.

Hilderbrand, R.H., Watts, A.C., and Randle, A.M. (2005). The myths of restoration ecology. *Ecology and Society*, 10, 19 [online].

Hobbs, R.J. (2008). Woodland restoration in Scotland: ecology, history, culture, economics, politics and change. *Journal of Environmental Management*, **90**, 2857–2865.

Hobbs, R.J. (2009). Looking for the silver lining: making the most of failure. *Restoration Ecology*, **17**, 1–3.

Hobbs, R.J. (2017). Where to from here? Challenges for restoration and revegetation in a fast-changing world. *The Rangeland Journal*, **39**, 563–566.

Hobbs, R.J., Higgs, E., and Harris, J.A. (2009). Novel ecosystems: implications for conservation and restoration. *Trends in Ecology & Evolution*, **24**, 599–605.

Hobbs, R.J., Cole, D.N., Yung, L., et al. (2010). Guiding concepts for park and wilderness stewardship in an era of global environmental change. *Frontiers in Ecology and the Environment*, **8**, 483–490.

Hobbs, R.J., Hallett, L.M., Ehrlich, P.R., and Mooney, H.A. (2011). Intervention ecology: applying ecological science in the 21st century. *BioScience*, **61**, 442–450.

Hobbs R.J., Higgs, E.S., and Hall, C.M. (2013). Defining novel ecosystems. In Hobbs, R.J., Higgs, E.S., and Hall, C.M. (Eds.), *Novel ecosystems: intervening in the new ecological world order* (pp. 58–60). Oxford: Wiley-Blackwell.

Hobbs, R.J., Higgs, E., Hall, C.M., et al. (2014). Managing the whole landscape: historical, hybrid, and novel ecosystems. *Frontiers in Ecology and the Environment*, **12**, 557–564.

Hobbs, R.J., Higgs, E.S., and Hall, C.M. (2017). Expanding the portfolio: Conserving nature's masterpieces in a changing world. *BioScience*, **67**, 568–575.

Hobbs, R.J., Valentine, L.E., Standish, R.J., and Jackson, S.T. (2018). Movers and stayers: novel assemblages in changing environments. *Trends in Ecology & Evolution*, **33**, 116–128.

Hughes, F.M.R., Stroh, P.A., Adams, W.M., Kirby, K.J., Mountford, J.O., and Warrington, S. (2011). Monitoring and evaluating large-scale, 'open-ended' habitat creation projects: a journey rather than a destination. *Journal for Nature Conservation*, **19**, 245–253.

Hughes, F.M.R., Adams, W.M., and Stroh, P.A. (2012). When is open-endedness desirable in restoration projects? *Restoration Ecology*, **20**, 291–295.

Johnson, C.N. (2006). *Australia's mammal extinctions: a 50,000 year history*. Cambridge: Cambridge University Press.

Johnson, J. (2017). Scaling up restoration in a time of change – observations from Western Australia. *SER News*, **31**, 9–15.

Kerley, G.I.H., Kowalczyk, R., and Cromsigt, J.P.G.M. (2012). Conservation implications of the refugee species concept and the European bison: king of the forest or refugee in a marginal habitat? *Ecography*, **35**, 519–529.

Kirby, K.J. (2009). Policy in or for wilderness. *British Wildlife*, **20**, 59–62.

Lawler, J.J., Tear, T.H., Pyke, C., et al. (2010). Resource management in a changing and uncertain climate. *Frontiers in Ecology and the Environment*, **8**, 35–43.

Lockwood, J.L., and Pimm, S.L. (1999). When does restoration succeed? In Weiher, E., and Keddy, P. (Eds.), *Ecological assembly rules: perspectives, advances, retreats* (pp. 363–392). Cambridge: Cambridge University Press.

Luken, J.O. (1990). *Directing ecological succession*. New York, NY: Chapman and Hall.

Mann, C.C. (2005). *1491: new revelations of the Americas before Columbus*. New York, NY: Alfred A. Knopf.

Marris, E. (2011). *Rambunctious garden: saving nature in a post-wild world*. New York, NY: Bloomsbury.

McAlpine, C., Catterall, C.P., Mac Nally, R., et al. (2016). Integrating plant- and animal-based perspectives for more effective restoration of biodiversity. *Frontiers in Ecology and the Environment*, **14**, 37–45.

McDonald, T., Gann, G.D., Jonson, J., & Dixon, K.W. (2016a). *International standards for the practice of ecological restoration – including principles and key concepts*. Washington, DC: Society for Ecological Restoration.

McDonald, T., Jonson, J., & Dixon, K.W. (2016b). National standards for the practice of ecological restoration in Australia. *Restoration Ecology*, **24**, S4–S32.

Miller, J.R., and Bestelmeyer, B.T. (2016). What's wrong with novel ecosystems, really? *Restoration Ecology*, **24**, 577–582.

Miller, J.R., and Bestelmeyer, B.T. (2017). What the novel ecosystems concept provides: a reply to Kattan et al. *Restoration Ecology*, **25**, 488–490.

Miller, J.R., and Hobbs, R.J. (2002). Conservation where people live and work. *Conservation Biology*, **16**, 330–337.

Miller, J.R., and Hobbs, R.J. (2007). Habitat restoration – do we know what we're doing? *Restoration Ecology*, **15**, 382–390.

Morse, N.B., Pellissier, P.A., Cianciola, E.N., et al. (2014). Novel ecosystems in the Anthropocene: a revision of the novel ecosystem conceprt for pragmatic applications. *Ecology and Society*, **19**, 2.

Murcia, C., Aronson, J., Kattan, G.H., Moreno-Mateos, D., Dixon, K., and Simberloff, D. (2014). A critique of the 'novel ecosystem' concept. *Trends in Ecology & Evolution*, **29**, 548–553.

Navarro, L.M., and Pereira, H.M. (2012). Rewilding abandoned landscapes in Europe. *Ecosystems*, **15**, 900–912.

Nogués-Bravo, D., Simberloff, D., Rahbek, C., and Sanders, N.J. (2016). Rewilding is the new Pandora's box in conservation. *Current Biology Magazine*, **26**, R83–R101.

Oliveira-Santos, L.G.R., and Fernández, F.A.S. (2009). Pleistocene rewilding, Frankenstein ecosystems, and an alternative conservation agenda. *Conservation Biology*, **24**, 4–5.

Palmer, M.A., Ambrose, R.F., and Poff, N.L. (1997). Ecological theory and community restoration ecology. *Restoration Ecology*, **5**, 291–300.

Pereira, H.M., and Navarro L.M. (Eds) (2015). Preface. In *Rewilding European landscapes* (pp. v–x). Heidelberg: Springer.

Perring, M.P., Standish, R.J., Price, J.N., et al. (2015). Advances in restoration ecology: rising to the challenges of the coming decades. *Ecosphere*, **6**, 131.

Prach, K., and Hobbs, R.J. (2008). Spontaneous succession vs. technical reclamation in the restoration of disturbed sites. *Restoration Ecology*, **16**, 363–366.

Ripple, W.J., and Beschta, R.L. (2012). Trophic cascades in Yellowstone: the first 15 years after wolf reintroduction. *Biological Conservation*, **145**, 205–213.

Ripple, W.J., Estes, J.A., Beschta, R.L., et al. (2014). Status and ecological effects of the world's largest carnivores. *Science*, **343**, 1241484.

Rubenstein, D.R., and Rubenstein, D.I. (2016). From Pleistocene to trophic rewilding – a wolf in sheep's clothing. *Proceedings of the National Academy of Sciences of the United States of America*, **113**, E1.

Rubenstein, D.R., Rubenstein, D.I., Sherman, P.W., and Gavin, T.A. (2006). Pleistocene Park – does re-wilding North America represent sound conservation for the 21st century? *Biological Conservation*, **132**, 232–238.

Seddon, P.J., Griffiths, C.J., Soorae, P.S., and Armstrong, D.P. (2014). Reversing defaunation – restoring species in a changing world. *Science*, **345**, 407–412.

SER Science and Policy Working Group. (2004). *The SER primer on ecological restoration*. www.ser.org and Society for Ecological Restoration, Tucson, Arizona.

Simberloff, D., Murcia, C., and Aronson, J. (2015). 'Novel ecosystems' are a Trojan horse for conservation. http://ensia.com/voices/novel-ecosystems-are-a-trojanhorse-for-conservation/ (accessed 6 March 2015).

Soulé, M., and Noss, R. (1998). Rewilding and biodiversity. *Wild Earth*, **8**, 2–11.

Soulé, M., and Terborgh, J. (1999). *Continental conservation: scientific foundations of regional reserve networks*. Washington, DC: Island Press.

Sudduth, E.B., Hassett, B.A., Cada, P., and Bernhardt, E.S. (2011). Testing the field of dreams hypothesis: functional responses to urbanization and restoration in stream ecosystems. *Ecological Applications*, **21**, 1972–1988.

Suding, K.N. (2011). Toward an era of restoration ecology: successes, failures, and opportunities ahead. *Annual Review of Ecology, Evolution, and Systematics*, **42**, 465–487.

Svenning, J.C., Pedersen, P.B.M., Donlan, C.J., et al. (2016). Science for a wilder Anthropocene – synthesis and future directions for trophic rewilding research. *Proceedings of the National Academy of Sciences of the United States of America*, **113**, 898–906.

Temperton, V.M., and Hobbs, R.J. (2004). The search for ecological assembly rules and its relevance to restoration ecology. In Temperton, V.M., Hobbs, R.J., Nuttle, T., and Halle, S. (Eds.), *Assembly rules and restoration ecology: bridging the gap between theory and practice* (pp. 34–54). Washington, DC: Island Press.

Temperton, V.M., Hobbs, R.J., Nuttle, T., and Halle, S. (Eds.) (2004). *Assembly rules and restoration ecology: bridging the gap between theory and practice*. Washington, DC: Island Press.

Terborgh, J., and Estes, J.A. (2010). *Trophic cascades: predators, prey, and the changing dynamics of nature*. Washington, DC: Island Press.

Toledo, D., Agudelo, M.S., and Bentley, A.L. (2011). The shifting of ecological restoration benchmarks and their social impacts: digging deeper into Pleistocene re-wilding. *Restoration Ecology*, **19**, 565–568.

Truitt, A.M., Granek, E.F., Duveneck, M.J., Goldsmith, K.A., Jordan, M.P., and Yazzie, K.C. (2015). What is novel about novel ecosystems: managing change in an ever-changing world. *Environmental Management*, **55**, 1217–1226.

Veldman, J.W., Overbeck, G.E., Negreiros, D., et al. (2015). Where tree planting and forest expansion are bad for biodiversity and ecosystem services. *BioScience*, **65**, 1011–1018.

Walker, L.R., Walker, J., and Hobbs, R.J. (2007). *Linking restoration and ecological succession*. New York, NY: Springer.

World Resources Institute. (2014). *Atlas of forest and landscape restoration opportunities*. Washington, DC: World Resources Institute. www.wri.org/resources/maps/

CHAPTER EIGHT

Understanding the factors shaping the attitudes towards wilderness and rewilding

NICOLE BAUER

Swiss Federal Institute for Forest, Snow and Landscape Research WSL, Economics and Social Sciences, Social Sciences in Landscape Research Group

ALINE VON ATZIGEN

University of Zurich and Ethnographic Museum; Swiss Federal Institute for Forest, Snow and Landscape Research WSL, Economics and Social Sciences, Social Sciences in Landscape Research Group

Only few wilderness areas, representing an invaluable part of the world's natural heritage, remain in the highly developed regions of the world. Therefore, the rewilding of landscapes is increasingly being discussed as a management option to combat biodiversity loss and reshape landscapes in regions where the need for agricultural and forest land is decreasing. As many of the regions where rewilding is being discussed are densely populated, potential wilderness would be located in the vicinity of inhabited areas, leading to a high probability of conflict between people in favour of rewilding and those opposed to it.

Rewilding projects – in addition to their ecological use and value – have to be in accordance with the values and needs of the local population and should not interfere with their plans for the region. Hence, the acceptance and support by those who live around potential wilderness areas is very important and before taking a decision about rewilding an area or not, it is crucial to learn more about the attitudes of the general public towards nature and towards the expansion of wilderness. Therefore, the present chapter reviews current knowledge on public attitudes towards nature, wilderness, and rewilding. It takes a special focus on empirical studies from the field of environmental psychology and environmental sociology, assessing the attitudes of the general population by directly asking them questions on their relation to the environment, while in other chapters (in particular Chapters 2 and 3) the historical, cultural, and political origins of wilderness were discussed.

Background to attitudinal research and its relevance to understanding social responses to nature

Attitudes cannot be directly observed and must be inferred from self-report. They represent a tendency to evaluate an entity such as an object or an idea in a positive or negative way (Eagly and Chaiken, 1993) and are often described in terms of three components: the affective component indicating a person's feelings about the attitude object, the behavioural or conative component describing the way the attitude influences a person's behaviour, and the cognitive component, a person's belief/knowledge about an attitude object (Table 8.1). Although some attitudes may arise at least partly from genetic sources, most attitudes are primarily learned (Baron and Byrne, 1994) and attitudes formed through direct experiences are stronger than those formed from listening to others or observing others (Fazio et al., 1982). There is usually only a weak correlation between attitudes and behaviour, but specific attitudes

Table 8.1. *Overview of the terms used in attitudinal research related to nature.*

Term	Definition
Attitudes (Eagly and Chaiken, 1993)	Tendency to evaluate an entity such as an object or an idea in a positive or negative way; there are three components of attitudes: – the affective component indicating a person's feelings about the attitude object, – the behavioural component describing the way the attitude influences a person's behaviour, and – the cognitive component, a person's belief/knowledge about an attitude object
Value orientations (Schwartz, 1992; Schwarz and Boehnke, 2004)	Criteria people use to select and justify actions and to evaluate people and events. The term refers to a more general concept than attitudes, and value orientations are expected to influence attitudes and actions
Human–nature relationship (Kellert, 1980, 1993)	The term is generally used as a synonym for 'nature concern'. It is supposed to be described by different dimensions (e.g. biocentric, ecocentric, biophobic), and influences attitudes
Visions of Nature (van den Born et al., 2001)	This is an umbrella concept consisting of three elements: (a) the images of nature that identify the types of natural environments people distinguish, (b) the values of nature that determine why nature is seen as important, and (c) the images reflecting the relationship that describe how people see the appropriate relationship between humans and nature (e.g. humans as exploiters of nature, or humans as participants in nature)

are known to be much better predictors of behaviour than general ones. In contrast to these findings, recent research shows a strong connection between individuals' relationship with nature and their environmental behaviour and decision-making (Braito et al., 2017; Muhar et al., 2017).

The research on the way people relate to the environment uses different concepts: value orientations, human–nature relationships, and visions of nature are the words most often encountered in the literature focusing on the relation between humans and the environment. These concepts inform and influence more specific attitudes towards nature and the use of this limited number of concepts facilitates the comparison of results between different studies. Nevertheless, the concepts used in attitudinal research on nature are not to be used synonymously with each other as they address different perspectives on the ways people relate to the environment. Some of these concepts are derived from theory, while others have an empirical basis or have been refined by large surveys. The empirical studies related to the different concepts follow both deductive and inductive reasoning, use both quantitative and qualitative methods, and lead to typologies that differ in respect of types, positionality of humans and nature to each other as well as content (e.g. the feeling towards nature or the quality of the relationship). Most of the empirical studies focus on European countries and North America, providing a good foundation of understanding about the way the public in these countries relates to nature. There is, however, a knowledge gap on how humans relate to nature for other parts of the world.

Value orientations

In psychology, values are defined as the criteria people use to select and justify actions and to evaluate people and events. They are more general than attitudes and have an influence on the more concrete attitudes and actions in different domains.

The theory of basic human values postulates 10 fundamental values that are differentiated in all societies: power, achievement, hedonism, stimulation, self-direction, universalism, benevolence, tradition, conformity, and security (Schwartz, 1992). 'Nature concern' has been identified as a subtype of the 'universalism' type by Schwarz and Boehnke (2004) and can be characterised by the items 'unity with nature', 'protecting the environment', and 'world of beauty'.

In the context of environmental decision-making, the term value describes an ideal that provides reasons for protecting the environment or for setting aside land (e.g. ecocentric value orientation, respect for nature) and is often used as a synonym for 'attitudes', although it is known to be more general than attitudes (for more details see Bauer, 2016).

Wilson (1993) describes the two theoretical concepts biophilia – the love of all that lives – and its antithesis, biophobia – the tendency for people to be afraid of nature as the two elementary value orientations that focus on the feelings of humans towards nature and that can be found in many empirical studies. This classification of value orientations has been refined by Stern and Dietz (1994), who define three basic environmental value sets (the egoistic, the altruistic, and the biospheric value orientations) and by Thompson and Barton (1994), who propose an anthropocentric value orientation, strongly connected with a preference for farmlands, an ecocentric value orientation that correlates with a preference for wild lands (Kaltenborn and Bjerke, 2002), and an environmentally apathetic orientation.

Human–nature relationship

Mounting environmental problems have led to the reconsideration of the human–nature relationship and to an intensification of its empirical exploration in recent years. In the empirical literature 'human–nature relationship' is often used as a synonym for 'nature concern' or other environmental values. The human–nature relationship with its different dimensions (e.g. biocentric, ecocentric, biophobic) influences the more concrete attitudes towards wilderness.

At the beginning of the definition of the human–nature relationship is the work of Kellert (1980, 1993). He established, on the basis of a survey, a typology of the human–nature relationship with nine types: (1) the utilitarian dimension (the material value of nature is important), (2) the naturalistic dimension (nature is regarded with wonder and awe), (3) the ecological scientific dimension (nature is investigated systematically), (4) the aesthetic dimension (the beauty of nature is paramount), (5) the symbolic dimension (nature is a source of symbols), (6) the humanistic dimension (humans feel emotionally connected with nature), (7) the moralistic dimension (humans feel responsible for nature), (8) the dominating dimension (humans master the natural world), and (9) the negative dimension (humans view nature with fear and aversion). The human–nature relationship of a person can be described by more than one of these dimensions; each dimension illustrates which values a person considers as being especially important. In the dimensions described by Kellert, the concepts of biophilia and biophobia are clearly apparent (Wilson, 1993), although not all of the dimensions can be assigned to the two concepts.

Visions of nature

In the previous section, we have seen that the relationship of people with nature is expressed in terms of values and beliefs that are considered as conceptually different from each other. In the Netherlands the relationship is often analysed in a more holistic way (van den Born et al., 2001) and is – similarly to the values and the human–nature relationship – seen as having an influence on more concrete attitudes. The 'visions of nature' is an umbrella concept originating in the Netherlands and consists of three elements: (a) the images of nature that identify the types of natural environments people distinguish, (b) the values of nature that determine why nature is seen as important (intrinsic and instrumental values are distinguished as in Lockwood, 1999), and (c) the images reflecting the relationship that describe how people see the appropriate relationship between humans and nature (e.g. humans as exploiters of nature, or humans as participants in nature). In comparison to the approaches focusing on one single dimension, e.g. values or beliefs, this approach provides more information on the human–nature relationship and the interrelations between the three elements.

People and nature

There are a multitude of studies assessing the relationship of humans with nature using one of the above-mentioned approaches. At the first glance these seem to have a different focus than those looking more specifically at wilderness and rewilding. However, we know from social science studies that the public doesn't perceive nature and wilderness as strictly different concepts (e.g. van den Born et al., 2001), and that the relationship with nature has a strong influence on the attitudes towards wilderness and rewilding and potential decisions on supporting or being opposed to management actions in favour of rewilding (e.g. Mondini and Hunziker, 2018).

Perception of nature

Many studies on the visions of nature from the Netherlands identify five or six images of nature: e.g. Buijs (2009) found five ideal types of images of nature: the wilderness image, the autonomy image, the inclusive image, the aesthetic image, and the functional image, all with different implications for natural resource management. The wilderness image was characterised by an underlying ecocentric value orientation, combined with the perception of nature as being resilient and with adherence to a hands-off approach to management. The wilderness image was found in other studies in the Netherlands as well (van den Born et al., 2001; de Groot and van den Born, 2003) and wild nature was usually perceived as more natural than the arcadian nature, which is strongly influenced by humans.

In the studies on the perception of nature, Buijs et al. (2009) found cultural differences: they identified a set of three nature images in a study comparing the human–nature relationship of immigrants from Islamic countries and native Dutch people: (a) the wilderness image, focusing on ecocentric values and on the independence of nature; (b) the functional image, a concept with an anthropocentric value orientation, in line with the intensive management of nature; and (c) the inclusive image, with an ecocentric value orientation and defined by an intimate relationship between humans and nature. In this study, 44 per cent of the immigrant respondents endorsed the functional image while the majority of the native Dutch respondents supported the wilderness image (see also a review of literature from France and the Netherlands: Buijs et al., 2006). The results of a study in Poland with local people (Hunka et al., 2009), however, found the presence of the wilderness image and a wide recognition of intrinsic (ecocentric) values (Lockwood, 1999) very similar to the studies in the Netherlands with local Dutch people.

The wilderness image, the arcadian image, and the functional image seem to be key images across multiple studies. The presence of the wilderness image in the general population in many of the studies taking place in, for example, the Netherlands, Poland, and Switzerland (van den Born et al., 2001; Buijs et al., 2006, 2009; Bauer et al., 2009; Hunka et al., 2009) leads to the hypothesis that the idea of unmanaged nature is both present and generally appreciated among the general public in Western societies.

Relation with nature

As we know that the human–nature relationship is predictive for the attitudes towards rewilding and phenomena related to rewilding, we review here some of the studies on the relation of humans with nature. We focus on the handful of studies that explicitly assess the human–nature relationship and the attitudes towards wilderness in the same survey.

One of these studies took place in Switzerland, where Bauer and colleagues (2009) identified a typology describing four general attitudes towards nature: (a) nature lovers; (b) nature sympathisers; (c) nature-connected users; and (d) nature controllers. Nature sympathisers have an emotionally distanced attitude towards nature and show aspects of an environmentally apathetic orientation (Thompson and Barton, 1994). Nature controllers can be considered to have an anthropocentric value orientation, nature lovers have an ecocentric value orientation, and nature-connected users a biocentric value orientation with anthropocentric aspects (Stenmark, 2002). The majority of respondents showed clear biophiliac tendencies, with nature lovers accounting for 31 per cent of the sample and nature-connected users accounting for 27 per cent. The four types were found to have different attitudes towards wilderness and rewilding (see below).

This study has been replicated in the south-western Carpathians in Romania (Bauer et al., 2018), where it revealed two human–nature relationship types, 'progressive nature friends' and 'traditional nature users', which differ in their feelings towards wilderness, attitudes towards protected areas, and perceptions of development opportunities of the regions. For the 'traditional nature users' accounting for 64 per cent of the sample, the utilitarian aspect of nature is especially important, while the 'progressive nature friends' accounting for 36 per cent of the sample have a more ecocentric value orientation.

Many of the studies on the human–nature relationship found differences between people with a different cultural background (Buijs et al., 2009; Hunka et al., 2009; Kloek et al., 2017). In the study of Bauer and colleagues (2009) there were differences in the perception and assessment of nature for the members of different language regions within Switzerland: in the cluster of the 'nature controllers', the proportion of Italian-speaking and the amount of French-speaking people is especially high, while the German-speaking people are especially present in the cluster of the 'nature lovers'.

Similarly, Kloek and colleagues (2017) found differences between native Dutch, Chinese, and Turkish immigrants in the assessment of the naturalness of four types of nature, and Fox and Xu's (2017) comparison of the nature values of British and Chinese respondents revealed that the British reported more biophilic tendencies than the Chinese. These studies hint at a cultural variation in the human–nature relationship and are in accordance with studies indicating that childhood experiences are decisive for the perception of nature and the human–nature relationship during adulthood (Kals et al., 1999; Ward Thompson et al., 2008).

Although there are cultural differences concerning the perception of nature and the value orientations which are not yet well documented or understood, the dominance of the ecocentric value orientation in the general population from, for example, the Netherlands, Switzerland, and Poland is striking.

As well as cultural factors, environmental differences influence attitudes towards nature: people from rural areas are reported to have a more utilitarian approach to nature (Bauer, 2005; Bauer et al., 2018) and to define nature differently (Kloek et al., 2017). In a review of literature from France and the Netherlands, Buijs and colleagues (2006) found three main images of nature: the arcadian image, representing the idyllically pastoral landscape; the wilderness image, related to uninfluenced nature; and the functional image, related to agricultural use of land. A shift from the functional image to the wilderness image and arcadian image was reported to be influenced by urbanisation.

Furthermore, different social groups (e.g. farmers, anglers, conservationists, livestock owners, land owners, etc.) with different functional ties to nature (e.g. dependent on the use of nature for a living or not) were found to have a different relationship with nature (Buijs et al., 2006; Bauer et al., 2009, 2018); social groups depending on the use of nature for a living (e.g. famers, anglers, livestock owners, etc.) were found to have a more utilitaristic relationship with nature and to be more attached to an arcadian image of nature, while the groups using nature for leisure adhered more to a wilderness image. Additionally, people favouring an arcadian image were found to live more often in the countryside, while the wilderness image was more present among the urban population. Other influencing factors have been found to be the age of the respondents (a more utilitarian value orientation is associated with elderly people, e.g. Bauer, 2005) and religious beliefs (Muslims and Christians gave ecocentric reasons and non-religious people gave anthropocentric reasons for nature conservation; Kloek et al., 2017). Most of the research done to date is based on the assumption that the human–nature relationship is stable over different situations; however, an important insight from more recent research is that people hold multiple human–nature relationships that are context-sensitive and that individuals' environmental decision-making is connected to their relationship with nature (Braito et al., 2017; Muhar et al., 2017). This means that there are many context variables that influence environmental decision-making, e.g. to opt for or against non-interventionist forest management strategies, or repressive measures for predator management.

Although the research on the human–nature relationship used different quantitative and qualitative methods, included different samples, took place in different western European countries, and compared different ethnic groups, there is still a surprisingly important overlap in the nature images and in the underlying value orientations identified in these studies. The wilderness image, the arcadian image, and the functional image seem to be the key images and the anthropocentric, ecocentric, biocentric/biophiliac, and biophobic value orientations seem to be quantitatively the most dominant value orientations within the cultural contexts examined. In most of the studies on the human–nature relationship we could detect a strong nature-friendly relationship, expressed in the quantitative dominance of an ecocentric, biocentric, or biophiliac value orientation in the samples.

People and rewilding
Wilderness, wildness, and rewilding in social science studies

While there is some social science research on wild nature and rewilding from European countries and North America, there is almost none from other regions of the world. This is surprising, as much of Europe lacks existing wilderness and wilderness areas that form the core of most North American rewilding initiatives have been established long ago. The differences in the presence of wilderness reflect Europe's higher human population density and longer history of intensive human land use (Corlett, 2016). While in North America the aim of rewilding was initially to restore the pre-Columbian wilderness, in Europe the aim has been to create 'wildness' (characterised by spontaneity and absence of human control) in areas that have been managed for millions of years (Hall, 2014).

The differences concerning the presence of wilderness have implications for the definition of wilderness as perceived by the public: in the USA and Canada wilderness is usually seen as nature with distinct ecological features that is protected in designated wilderness areas. In Europe where wilderness is almost non-existent, the situation is less clear. The public thus does not have a clearly defined concept of wilderness used by natural scientists and hence there are no easy definitions of wilderness that can be used in social science studies (Bauer, 2005). Some studies indicate that the word wilderness has negative connotations for some people, e.g. in Switzerland and neighbouring countries (Stremlow and Sidler, 2002). It often remains unclear what the public means exactly by the word wilderness and in some studies the term is apparently used for nature in general. Due to the lack of a definition that is well understood by the public and the potentially negative connotations of the word wilderness, the word 'wilderness' is generally avoided in questionnaires or interviews and paraphrased (e.g. 'areas in which nature is not or is no longer influenced') and the participants of surveys are asked to indicate the characteristics of this kind of nature either in open-ended questions or by giving them a predefined set of alternatives to choose from (Bauer et al., 2009).

According to some social science studies, 'wilderness' as defined by the survey participants can be found in wetlands, large forests, and remote mountain areas, and apart from natural features (e.g. dense vegetation), few human traces, little infrastructure, and few persons using an area are the key criteria found in surveys (Bauer, 2005; Lupp et al., 2011).

Rewilding has been discussed as a nature conservation strategy for some years now (Jørgensen, 2015) and there are many different types of rewilding (Corlett, 2016). The passive rewilding that is characterised by little or no human interference (Schnitzler, 2014; Corlett, 2016) is the concept generally used in social science studies, in which rewilding is usually defined as a process in which a formerly cultivated landscape develops without human control while excluding land that has actively been changed to initiate a return to near-natural conditions, e.g. as in many river renaturalisation projects (Höchtl et al., 2007, Bauer et al., 2009).

Attitudes towards wilderness and rewilding

In this section we are going to review the literature on (a) the assessment of long-existing wilderness and wilderness areas (in the USA and Canada) and the reactions to planned changes such as the designation of new wilderness areas, as we expect that rewilding in other parts of the world will lead to the designation of new protected areas with similar consequences to those in North America. Furthermore, we include studies (b) on wilderness defined as nature that is no longer influenced and on rewilding as the process of getting wild(er), and (c) on wilderness-related phenomena, such as spontaneous reforestation, dead wood, and the spontaneous resurgence of animals such as large carnivores (i.e. animals that have not been actively introduced). We include these studies because spontaneous reforestation and dead wood are visible signs of a lack of management and cultivation that result in a public response and because the resurgence or spread of wild animals is often considered a symbol of rewilding.

Most of the studies on the attitudes towards wilderness originate in the USA or Canada where wilderness protection has long been a culturally relevant topic. Lutz and colleagues (1999) compared rural and urban attitudes towards wilderness in British Columbia, Canada. An urban and a rural sample, including 75 participants each, with similar sociodemographic variables differed with regard to the number of visits to wilderness, with rural participants visiting wilderness more often than urban dwellers. The study found that both groups have positive attitudes towards wilderness but have different concepts of wilderness. The urbanites often regarded areas as wilderness notwithstanding the evidence of human intrusion (e.g. signs of logging activity, roads, and hydroelectric dams); the rural respondents had generally a higher threshold for what they considered wilderness and considered areas with any such activities as non-wilderness.

Positive attitudes towards wilderness are also reported by Cordell and colleagues (1998), who investigated general attitudes towards wilderness in the USA on a sample of 1900 people: of these, 44 per cent acknowledged the existence of protected wilderness areas and the majority of the respondents had a positive attitude towards wilderness areas: 56 per cent would prefer more wilderness to be protected. In a follow-up study comparing the national survey from 1994 with data collected in the year 2000 using the same approach but a smaller random sample of inhabitants of the USA, Cordell and colleagues (2003) found that only 52 per cent considered the size of the wilderness areas to be too small; urban residents, people living in eastern states, young and white residents opted for a larger size of the protected wilderness.

Rudzitis and Johansen (1991) investigated the wilderness attitude of a random sample of residents of 11 counties with federally designated wilderness areas and a significant population growth. They found that the attitude towards wilderness was more positive among people that had recently moved to the region (85 per cent) than among long-term local residents (75 per cent). The study of Durrant and Shumway (2004) based on the same questions as Rudzitis and Johansen (1991) used a random sample of residents from six very sparsely inhabited south-eastern Utah counties. The government proposed additional wilderness areas on state-owned land in these counties and this had been leading to polarising debates over wilderness areas. Fourteen per cent of the respondents reported that existing wilderness in the county was a reason to move and stay there and 34 per cent were in favour of additional wilderness areas. In this survey negative attitudes towards wilderness areas were found to be mainly linked to a wider debate on the wilderness designation.

Another example of differentiated perceptions to wilderness within the local population was found in a study in the south-western Carpathians in Romania (Bauer et al., 2018): while local people's feelings towards wild nature were generally positive, the planned wilderness areas (located in the core areas of long-existing National Parks) were assessed far more critically as people incorrectly associated these plans with further restrictions on the future use of the forest. These study results underline the importance of a timely communication of potential restrictions for the outcomes of conservation projects.

Although wilderness is relatively rare in European countries, the idea of protection of wild nature has been the focus of increased attention in recent years. In a quantitative survey of the Swiss population on the human–nature relationship and the attitudes towards wilderness and rewilding Bauer and colleagues (2009) found four different types of human–nature relationship with different attitudes towards wilderness and rewilding leading to a roughly equal classification of the population into wilderness opponents (51 per cent) and wilderness proponents (49 per cent). While the 'nature-connected users' and the 'nature controllers' evaluate wilderness negatively, the 'nature sympathisers' and the 'nature lovers' evaluate it in a positive way. The wilderness opponents have different reasons for their attitude towards rewilding: the 'nature-connected users' feel threatened by the rewilding of nature and the 'nature controllers' regard the perceived spread of wilderness with contempt.

In the third nationwide survey on nature awareness in Germany, the Federal Agency for Nature Conservation (BMUB/BfN, 2014) investigated the general attitudes towards wilderness using a sample of 2000 randomly selected participants from all regions of Germany. In this survey 65 per cent of the respondents preferred nature to be more wild; this preference was especially marked in younger people (below 30 years) and well-educated people. Similarly, Höchtl and colleagues (2005) analysed the attitudes towards the wilderness in the Val Grande National Park in Italy and noticed that the wild nature in the protected areas was assessed positively while the rewilding of the landscape outside the Park was perceived differently. The local population from the region of the Val Grande Park perceived the (unintended) rewilding of formerly cultivated areas as negative (see also Höchtl et al., 2007). The loss of the familiar landscape of former times and the attachment to this landscape was hypothesised by the authors to be the main reason for the negative evaluation of this unintended landscape change.

Some studies have investigated the attitudes towards important and easily perceptible attributes of passive rewilding developments that are associated with the rewilding of landscapes, e.g. the spontaneous regrowth of forests (Hunziker et al., 2008) and dead wood (e.g. Stelzig, 2000). For example, Hunziker and colleagues (2008) found that reforestation in the Alps was rated much more positively by tourists and a random population sample of Swiss residents than by local inhabitants. One of the few studies from outside North America and western Europe is the work of Stelling and colleagues (2017), who investigated stakeholders' attitudes towards regrowth in an abandoned agricultural landscape in south-eastern Australia. The participants viewed the regrowth through one of three frames: 'Control', a negative interpretation of the regrowth; 'Accept', a positive interpretation; and 'Ambivalent', a combination of the 'Control' and 'Accept' frames.

In a recent study, Gundersen and colleagues (2017) found that survey participants rated photographs of forest settings with dead wood digitally removed substantially higher than the corresponding original non-edited photographs. Respondents' familiarity with the ecological role of dead wood or an additional text containing information about dead wood's ecological benefits lead to a more positive assessment. The relevance of ecological knowledge has been found to be relevant for the positive evaluation of dead wood in other studies as well (e.g. Stelzig, 2000). Other influencing factors are the social status and the age: the lower the status and the higher the age of the respondents, the more negative were attitudes towards non-interventionist forest management strategies (Müller and Job, 2009). In addition, researchers found an association between the human–nature relationship and the assessment of dead wood: the more biocentric the attitude of the participants, the more pronounced was the opposition against the removal of dead wood in a National Park (Lupp and Konold, 2008; Lupp et al., 2011; Sacher et al., 2017).

The resurgence of wild animals is often seen as a symbol of wilderness and rewilding and there are a multitude of surveys on the attitudes towards wildlife and changes in these attitudes in Western societies. In a replica of the survey of Stephen Kellert on attitudes towards animals, Kelly et al. (2016) found the greatest differences over time for historically stigmatised species (e.g. vultures, wolves, and coyotes): the attitudes in 2014 were significantly more positive for these species than in 1978. However, Treves and colleagues (2013) measured the attitudes towards wolves in Wisconsin in 2001 and 2009 and noticed an increased agreement with statements reflecting fear of wolves, and the inclination to poach a wolf. Neither the time span over which respondents reported exposure to wolves locally nor self-reported losses of domestic animals to wolves correlated with changes in attitude. As the abundances of grey wolves in Wisconsin were at their highest level in 60 years at the time of the second survey, the authors assumed that the mere presence of more wolves was decisive for the more negative attitudes compared to 2001.

Zimmermann and colleagues (2001) reviewed a variety of predator acceptance surveys in Norway and proposed a model explaining people's attitudes and fears as a function of the distance between predators and humans, and as a function of the length of experience of living in the same area. The proportion of people with a negative attitude increased to a maximum with the arrival of large carnivores, and decreased with experience over time. Both before and after carnivore arrival, many people that reported being afraid of carnivores still did not have a negative attitude towards large carnivores. In this study, level of fear did not seem to provide a direct measure of predator acceptance. In line with the model of Zimmermann et al. (2001) and the expectation of a decrease of fear/increase in acceptance over time are the results of Mkonyi and colleagues (2017): while 79 per cent of the respondents of their study in Northern Tanzania held negative attitudes towards large carnivores, the years at residence and the knowledge of carnivores were positively associated with positive attitudes towards the animals. Interestingly, in a recent study by Mondini and Hunziker (2018) in two mountainous regions in Switzerland, fear, perceived damage caused by bears, and value orientations were strong and significant predictors for attitudes towards bears, which were, in turn, strong predictors for attitudes towards repressive measures for bear management. Neither knowledge about bears nor affectedness was associated with attitudes towards bears or towards repressive measures, respectively.

The association between the human–nature relationship or values towards nature with the acceptance of predators and wild animals has been found in other studies as well. In their study of 1200 secondary school pupils, Hermann and Menzel (2013) found that wildlife value orientations were useful predictors of an intention to support the return of wolves as a re-emerging, but still rare, species in Germany. The wildlife value orientations (WVO) are a rather new development and are defined as patterns of basic beliefs that give direction and meaning to fundamental values in the context of wildlife and that can be linked to specific attitudes (Fulton et al., 1996). Empirical research has identified two main wildlife value orientations: utilitarism and mutualism (Teel et al., 2005), or – in other studies – domination and mutualism (Manfredo et al., 2009). Individuals with a mutualistic view of wildlife attribute rights and care to wildlife while domination-orientated individuals prioritise human benefits over wildlife interest (Hermann and Menzel, 2013). The parallels of the WVO to the dominating dimension (humans as master of the natural world) of Kellert (1980, 1993) and the anthropocentric and ecocentric value orientation of Thompson and Barton (1994) are evident.

What can we learn from the studies on wilderness, rewilding, and wilderness-related phenomena?

Public attitudes towards wilderness, rewilding, and wilderness-related phenomena are less positive than attitudes about a wider human–nature relationship. In many cases the assessment of the free development of nature is ambivalent or even negative and reasons for being opposed include aesthetic reasons, feelings of fear, and loss of control. Wilderness and wilderness-related phenomena are assessed differently if they are related to an uninhabited protected area compared to people's everyday landscape. Negative attitudes towards the rewilding of landscapes have been most marked outside of protected areas (Höchtl et al., 2005). However, the intentionality of the rewilding may affect this differentiation in attitudes: within a protected area without significant human habitation rewilding is usually a central part of a strategy, whereas in the landscape where people are living it may be perceived as a side effect of land abandonment. Local people hence experience or fear the change of the landscape they are attached to and are therefore opposed to such developments, while tourists and the general public from other regions are much more positive about rewilding developments (Hunziker et al., 2008).

In western Europe and the USA, the attitudes of the public not experiencing rewilding of nature in their home region are neutral (half of the population in favour and half of it opposed: Bauer et al., 2009) or even positive (Lutz et al., 1999; BMUB/BfN, 2014) and the attitudes towards long-term wilderness areas were reported to be positive (Rudzitis and Johansen, 1991; Cordell et al., 1998). Plans to establish new wilderness areas or to increase the size of long-term wilderness areas has often met with the opposition of local residents and may lead to negative attitudes towards the wilderness areas themselves (Durrant and Shumway, 2004; Bauer et al., 2018). Interestingly, negative attitudes were more pronounced for those living for a long time in the area and were thus attached to a certain state of the landscape than for those who had recently migrated, suggesting that this reaction may be linked to the fear of the landscape changes associated with rewilding or to the perception that new/larger wilderness areas would lead to new restrictions in use.

Similarly, in western Europe and the USA, younger people with a higher socioeconomic status and higher education are more in favour of the establishment of new/larger wilderness areas, wilderness, rewilding, and wilderness-related phenomena such as dead wood and resurgence of wild animals. Those in favour of wilderness are mostly urbanites and hence are less affected by these developments than are people living in the countryside and therefore less dependent on the use of nature for a living. Older people or those growing up in a culture with the ideal of an arcadian nature are more opposed to the free development of nature and its consequences (Bauer et al., 2009; Buijs et al., 2009). The studies on the assessment of dead wood (e.g. Gundersen et al.,

2017) provided additional insights into the role that knowledge can play in influencing attitudes, where people became more positive towards the presence of dead wood when they were provided information about its ecological function. However, in relation with the resurgence of wild animals, the knowledge of the species has not consistently been identified as a predictor of the attitude towards the animal (Mkonyi et al., 2017; Mondini and Hunziker, 2018). Similarly, the association between the time of residency and time of exposure with attitudes can be inconsistent (Zimmermann et al., 2001; Treves et al., 2013; Mkonyi et al., 2017). In some studies fear or concern about safety is seen as a direct predictor of the attitudes towards wild animals, while in other studies it is not (Zimmermann et al., 2001; Roskaft et al., 2007). The main reason for the contradictory results concerning the attitudes towards wild animals can be seen in the relevance of certain characteristics such as the population density of wild animals and the kind of animal considered (wolves, bears, lynxes, cougars, etc.); these make it difficult to directly compare the studies on the attitudes towards wild animals and to compare them with the studies on other wilderness-related phenomena.

Conclusions

Attitudes towards wilderness, rewilding, and wilderness-related phenomena are strongly interrelated with the human–nature relationship. In Western contexts, these are also strongly related to specific situations, and differences in the attitudes towards wilderness can be partially ascribed to a different probability and degree of being directly afflicted by landscape changes, the establishment of new protected areas, and the restrictions associated with these. The human–nature relationship is associated with multiple demographic variables including age, place of residence, education, as well as the nature-relatedness of the professional background. In the Western context, a biophilic or ecocentric value orientation is associated with a younger, well-educated urban population with few functional ties to nature and is strongly associated with positive attitudes towards wilderness, rewilding, and wilderness-related phenomena. Another important aspect is the existence of differences in the relation between humans and nature associated with cultural provenance and religion. Furthermore, the human–nature relationship of young immigrants spending most of their life in a country is reported still to be strongly influenced by their culture of origin.

Limitations and implications for future research

Before discussing practical implications of the insights gained from the literature review, we would like to point out some limitations of the existing empirical evidence and identify some important research areas for the future.

The most obvious limitation of the research on the relation of humans towards nature and the assessment of wilderness and rewilding reviewed above is the strong focus on North America and Europe and the lack of information on attitudes towards nature, wilderness, and rewilding for many other regions of the world. There is a need for more research (e.g. in Africa, Asia, and South America), leading to a better understanding of the human–nature relation and the attitudes towards rewilding in these regions for a more accurate mapping of attitudes and to inform conservation management decisions.

The finding of cultural differences in the human–nature relationship and the finding that young immigrants spending most of their life in a country are reported to still be strongly influenced by their culture of origin raises the question of the importance of early childhood experiences on the relationship with nature and calls for more studies to investigate the role of childhood in forming environmental attitudes and to find ways to influence the human–nature relationship to foster more sustainable relationships (see also Kals et al., 1999; Ward Thompson et al., 2008). These findings also reveal relevant gaps in knowledge about the human–nature relationship of different cultures, and indicate the need for more research on the human–nature relationship in developing countries and more research on people with an immigrant background in Western societies, as the cultural differences in the human–nature relationship might influence decisions about rewilding and the way of proceeding in planned rewilding projects.

Moreover, the role of knowledge about natural processes in wilderness and their influence on biodiversity has until now been widely ignored in studies of the general attitudes of the public towards wilderness and rewilding. Further research on the impact of ecological knowledge on the attitudes towards wilderness should be a priority, as these can help to inform managers and decision-makers about the impact of different environmental education measures on the attitudes and potentially help to find innovative ways to positively influence attitudes towards rewilding, if the decision for rewilding is considered a socially sustainable solution for a certain region.

Another important finding of the review is that local, site-specific attitudes to rewilding are less positive than those revealed by more general studies of the wider public's attitudes to nature. This phenomenon is often referred to as the NIMBY (Not In My Backyard) effect and is well known in many fields of environmental research, e.g. the discussion about renewable energy and the placement of wind turbines (Devine-Wright and Howes, 2010) or in relation with nature protection projects (Byrka et al., 2017). Some other studies identify the risk perception of local people in relation with the potential consequences of the planned projects as motives for being opposed to these and describe perceived environmental injustice, fairness of the process, and personal commitment to others as the crucial factors in the perceived risk perceptions (Wolsink and Devilee, 2009). For rewilding projects this means that the motives for being opposed to rewilding would have to be studied with a more detailed and specific assessment including the human–nature relationship, the attitudes towards rewilding, the perceived risks related to the rewilding project as well as the context factor. Such a more differentiated assessment would be advantageous as well to obtain more diverse and in-depth information on the finding that rewilding outside of protected areas are viewed more negatively than rewilding in a protected area.

In most studies on attitudes towards predators and the resurgence of different wild animals as a symbol of rewilding the value orientations are not assessed, even though these are known to be good predictors for the more concrete attitudes towards wildlife and knowledge about the human–nature relationship can help elucidate values towards wildlife (Buijs, 2009; de Groot et al., 2011). This stresses the need to integrate questions on the human–nature relationship in social science research on wild animals.

An important insight from the most recent research is that individual and collective understandings of the human–nature relationship are especially relevant because the human–nature relationship concepts work in tandem with situational factors to induce or constrain behaviour (e.g. potential decisions on supporting or being opposed to management actions in favour of rewilding, on non-interventionist forest management strategies; Flint et al., 2013; Braito et al., 2017). The situational factors that must be considered are as follows. (1) The thematic focus of a decision process, i.e. the kind of natural resource that may be affected by human activities. For example, the research of van den Born (2006) suggested that farmers who own both cropland and forests tend to act according to a 'stewardship of nature' orientation in forests, but guided by 'mastery of nature' orientations in relation to the cropland (Yoshida et al., 2017). (2) Attitudes and emotions, as these are relatively sensitive to situational contexts. For example, Muhar and Böck (2017) reported that in the case of attitudes towards flood control, personal involvement in recent extreme events can activate concepts of control rather than concepts of partnership with nature. Additionally, (3) the culture (e.g. norms, traditions) and social structures of societies (e.g. race, gender, ethnicity, class; Flint et al., 2013) and (4) the governance regime, for example, which incorporates the organisational setup of stakeholder involvement in decision-making, are important considerations. As it is still unclear how individual and collective psychological processes interact with each other to influence behaviour formation, more research on the role of context is needed.

Rewilding has been a nature conservation strategy for some years now, leading to the question as to whether the rewilding initiatives and the rewilding of the landscapes have influenced public attitudes towards wilderness and rewilding. The results of the studies analysing attitude changes over time seem to be mixed: there are some indications that wilderness is evaluated more positively, or perceived as more important, than it used to be, and in most of the studies on the human–nature relationship we could detect a strong nature-friendly relationship. This leads us to the hypothesis that the human–nature relationship might have shifted slightly towards a more ecocentric value orientation in the western European countries and in North America and that the attitudes towards rewilding might have slightly improved over the last 20 years. However, in order to make reliable statements about the change of attitudes towards nature and wilderness, a monitoring of wilderness attitudes would be needed. No such studies have yet been conducted and there is a need to address this research gap in order to evaluate the impact of the rewilding of landscapes on public attitudes.

While there is a growing interest on the influence of nature contact on human health and well-being, there is still very little knowledge about the effect of wild nature on humans. One of the few studies looking at differently managed forests found that a walk through a 'tended' urban forest had a more positive effect on short-term well-being than a walk through a 'wild' forest (Martens et al., 2011). Reasons for this were hypothesised to be the denser vegetation of the wild forest, inducing fears in the visitors, and the fact that – at the time of the experiment – people were relatively unfamiliar with forests with a high amount of dead wood, leading the participants to think about the state of the visited forest, thoughts that interfered with the restoration process of the participants. However, the long-term effects of contact with wild nature are still widely unknown.

Implications for practice

As many of the regions where rewilding is discussed are densely populated, potential wilderness areas would be located in the vicinity of inhabited areas. As the decisions about rewilding projects should take into account the views and attitudes of those who live around potential wilderness areas, every rewilding project should start with a detailed and specific assessment of the human–nature relationship, the attitudes towards rewilding of the local population, and other relevant variables. The existence of multiple human–nature relationships within one person underlines the importance of considering the context variables of the planned project during this assessment (Flint et al., 2013), while other studies call for the integration of, for example, the risks that are perceived by the local population in relation with the planned rewilding project (Wolsink and Devilee, 2009). This broad assessment will help to identify well-defined target groups from the data, thus allowing the different attitudes to be mapped and helping to customise specific interventions and management actions.

In a next step in connection with the decision about rewilding, all stakeholders, and especially the inhabitants of the nearest communities, should be included in a participatory process. In this phase the active involvement of the population is crucial for the further acceptance and support of potential wilderness areas and can only be achieved by overcoming social barriers that prevent people from taking part in participatory landscape development (Schenk et al., 2007).

Once the decision about the rewilding has been taken (ideally by the vote of the inhabitants of the nearest communities) there are different ways to positively influence the attitudes of the different groups. Information campaigns often used by environmental NGOs that communicate the need and the benefit of wilderness areas could be used to strengthen positive attitudes of those in favour of wilderness. However, from former research we know that information campaigns would be inappropriate for influencing the attitudes of people opposed to wilderness (Petty and Cacioppo, 1986). The attitudes of those having a negative attitude towards wilderness could best be changed to be more wilderness-accepting by using role models (well-known politicians, actors) who communicate their dedication to environmental protection and wilderness.

In conclusion, the literature review on the attitudes towards nature and wilderness presented here provides a valuable starting point, as it indicates those topics that may be important to consider in more detail during rewilding processes. However, these studies cannot replace a more detailed and specific assessment of the human–nature relationship and the attitudes towards rewilding that is needed in regions with (prospective) rewilding initiatives.

References

Baron, R.A., and Byrne, D. (1994). *Social psychology: understanding human interaction.* Needham Heights, MA: Allyn and Bacon.

Bauer, N. (2005). Für und wider Wildnis – Soziale Dimensionen einer aktuellen gesellschaftlichen Debatte. Zürich, Bristol-Stiftung; Bern, Stuttgart, Wien, Haupt. 185 S.

Bauer, N. (2016). Social values of wilderness in Europe. In Bastmeijer, K. (Ed.), *Wilderness protection in Europe. The role of international, European and national law* (pp. 94–113). Cambridge: Cambridge University Press.

Bauer, N., Wallner, A., and Hunziker, M. (2009). The change of European landscapes: human–nature relationships, public attitudes towards rewilding, and the implications for landscape management. *Journal of Environmental Management*, **90**, 2910–2920.

Bauer, N., Vasile, M., and Mondini, M. (2018). Attitudes towards nature, wilderness and protected areas: a way to sustainable stewardship in the South-Western Carpathians. *Journal of Environmental Planning and Management*, **61**, 857–877.

BMUB/BfN Bundesministerium für Umwelt, Naturschutz, Bau und Reaktorsicherheit und Bundesamt für Naturschutz. (2014).

2013 Nature awareness study: Population survey on nature and biological diversity. www.bfn.de/fileadmin/BfN/gesellschaft/Dokumente/nature-awareness-study-2013.pdf (accessed 27 October 2017).

Braito, M., Böck, K., Flint, C., Muhar, A., Muhar, S., and Penker. M. (2017). Human-nature relationships and the complexity of environmental behaviour. *Environmental Values*, **26**, 365–389.

Buijs, A.E. (2009). Public natures. In *Social representations of nature and land practices*. Wageningen: Wageningen University.

Buijs, A.E., Pedroli, B., and Luginbühl, Y. (2006). From hiking through farmland to farming in a leisure landscape: changing social perceptions of the European landscape. *Landscape Ecology*, **21**, 375–389.

Buijs, A.E., Elands, B.H.M., and Langers, F. (2009). No wilderness for immigrants: cultural differences in images of nature and landscape preferences. *Landscape and Urban Planning*, **91**, 113–123.

Byrka, K., Kaiser, F.G., and Olko, J. (2017). Understanding the acceptance of nature-preservation-related restrictions as the result of the compensatory effects of environmental attitude and behavioral costs. *Environment and Behavior*, **49**, 487–508.

Cordell, H.K., Tarrant, M.A., McDonald, B L., and Bergstrom, J.C. (1998). How the public views wilderness: more results from the USA survey on recreation and the environment. *International Journal of Wilderness*, **4**, 28–31.

Cordell, H.K., Tarrant, M.A., and Green, G.T. (2003). Is the public viewpoint of wilderness shifting? *International Journal of Wilderness*, **9**, 27–32.

Corlett, R.T. (2016). Restoration, reintroduction, and rewilding in a changing world. *Trends in Ecology & Evolution*, **31**, 453–462.

de Groot, W.T., and van den Born, R.J.G. (2003). Visions of nature and landscape type preferences: an exploration in the Netherlands. *Landscape and Urban Planning*, **63**, 127–138.

de Groot, M., Drenthen, M., and de Groot, W.T. (2011). Public visions on the human/nature relationship and their implications for environmental ethics. *Environmental Ethics*, **33**, 25–44.

Devine-Wright, P., and Howes, Y. (2010). Disruption to place attachment and the protection of restorative environments: a wind energy case study. *Journal of Environmental Psychology*, **30**, 271–280.

Durrant, J.O., and Shumway, J.M. (2004). Attitudes toward wilderness study areas: a survey of six southeastern Utah counties. *Environmental Management*, **33**, 271–283.

Eagly, A.H., and Chaiken, S. (1993). *The psychology of attitudes*. Fort Worth, TX: Harcourt, Brace, & Janovich.

Fazio, R.H., Chen, J., McDonel, E.C., and Sherman, S.J. (1982). Attitude accessibility, attitude–behavior consistency, and the strength of the object-evaluation association. *Journal of Experimental Social Psychology*, **18**, 339–357.

Flint, C.G., Kunze, I., Muhar, A., Yoshida, Y., and Penker, M. (2013). Exploring empirical typologies of human–nature relationships and linkages to the ecosystem services concept. *Landscape and Urban Planning*, **120**, 208–217.

Fox, D., and Xu, F. (2017). Evolutionary and socio-cultural influences on feelings and attitudes towards nature: a cross-cultural study. *Asia Pacific Journal of Tourism Research*, **22**, 187–199.

Fulton, D.C., Manfredo, M.J., and Lipscomb, J. (1996). Wildlife value orientations: a conceptual and measurement approach. *Human Dimensions of Wildlife*, **1**, 24–47.

Gundersen, V., Stange, E.E., Kaltenborn, B.P., and Vistad, O.I. (2017). Public visual preferences for dead wood in natural boreal forests: the effects of added information. *Landscape and Urban Planning*, **158**, 12–24.

Hall, M. (2014). Extracting culture or injecting nature? Rewilding in transatlantic perspective. In Drenthen, M. and

Keulartz, J. (Eds.), *Old World and New World perspectives in environmental philosophy* (pp. 17-35). Cham: Springer.

Hermann, N., and Menzel, S. (2013). Predicting the intention to support the return of wolves: a quantitative study with teenagers. *Journal of Environmental Psychology*, **36**, 153-161.

Höchtl, F., Lehringer, S., and Konold, W. (2005). *Kulturlandschaft oder Wildnis in den Alpen? Fallstudien im Val Grande-Nationalpark und im Stronatal (Piemont/Italien)*. Zürich: Haupt Verlag.

Höchtl, F., Lehringer, S., and Konold, W. (2007). Wilderness: what it means when it becomes a reality – a case study from the southwestern Alps. *Landscape and Urban Planning*, **70**, 85-95.

Hunka, A.D., De Groot, W.T., and Biela, A. (2009). Visions of nature in Eastern Europe: a Polish example. *Environmental Values*, **18**, 429-452.

Hunziker, M., Felber, P., Gehring, K., Buchecker, M., Bauer, N., and Kienast, F. (2008). Evaluation of landscape change by different social groups. Results of two empirical studies in Switzerland. *Mountain Research and Development*, **28**, 140-147.

Jørgensen, D. (2015). Rethinking rewilding. *Geoforum*, **65**, 482-488.

Kals, E., Becker, R., and Rieder, D. (1999). Förderung natur- und umweltschützenden Handelns bei Kindern und Jugendlichen. In Linneweber, V. and Kals, E. (Eds.), *Umweltgerechtes Handeln: Barrieren und Brücken* (pp. 190-209). Heidelberg: Springer.

Kaltenborn, B.P., and Bjerke, T. (2002). The relationship of general life values to attitudes toward large carnivores. *Human Ecology Review*, **9**, 55-61.

Kellert, S.R. (1980). Contemporary values of wildlife in American Society. Institutional Series Report Nr. 1. In Shaw, W.W. and Zube, E.H. (Eds.), *Wildlife values* (pp. 241-267). Tucson, AZ: Center for Assessment of Noncommodity Natural Resource Values, University of Arizona.

Kellert, S.R. (1993). The biological basis for human values of nature. In Kellert, S.R. and Wilson, O. (Eds.), *The biophilia hypothesis* (pp. 42-69). Washington, DC: Island Press.

Kelly, A.G, Slagle, K.M., Wilson, R.S., Moeller, S.J., and Bruskotter, J.T. (2016). Changes in attitudes toward animals in the United States from 1978 to 2014. *Biological Conservation*, **201**, 237-242.

Kloek, M.E., Buijs, A.E., Boersema, J.J., and Schouten, M.G.C. (2017). Cultural echoes in Dutch immigrants' and non-immigrants' understandings and values of nature. *Journal of Environmental Planning and Management*, **61**, 818-840.

Lockwood, M. (1999). Humans valuing nature: synthesising insights from philosophy, psychology and economics. *Environmental Values*, **8**, 381-401.

Lupp, G., and Konold, W. (2008). Landscape perceptions and preferences of both residents and tourists: a case study in the Müritz National Park (Germany). In Siegrist, D., Clivaz, C., Hunziker, M., and Iten, S. (Eds.), *Visitor management in nature-based tourism – strategies and success factors for parks and recreational areas* (pp. 47-58). Series of the Institute of Landscape and Open Space. HSR University of Applied Sciences Rapperswil 2.

Lupp, G., Höchtl, F., and Wende, W. (2011). Wilderness – a designation for Central European landscapes? *Land Use Policy*, **28**, 594-603.

Lutz, A.R., Simpson-Housley, P., and De Man, A.F. (1999). Wilderness: rural and urban attitudes and perceptions. *Environment and Behaviour*, **31**, 259-266.

Manfredo, M.J., Teel, T.L., and Henry, K.L. (2009). Linking society and environment: a multilevel model of shifting wildlife value orientations in the Western United States. *Social Science Quarterly*, **90**, 407–427.

Martens, D., Gutscher, H., and Bauer, N. (2011). Walking in 'wild' and 'tended' urban forests: the impact on psychological well-being. *Journal of Environmental Psychology*, **31**, 36–44.

Mkonyi, F.J., Estes, A.B., Msuha, M.J., Lichtenfeld, LL., and Durant, S.M. (2017). Local attitudes and perceptions toward large carnivores in a human-dominated landscape of northern Tanzania. *Human Dimensions of Wildlife*, **22**, 314–330.

Mondini, M., and Hunziker, M. (2018). Psychological factors influencing human attitudes towards brown bear: a case study in the Swiss Alps. *Umweltpsychologie*, in press.

Muhar, A., and Böck, K. (2017). Mastery over nature as a paradox: societally implemented but individually rejected. *Journal of Environmental Planning and Management*, **61**, 994–1010.

Muhar, A., Raymond, C.M., van den Born, R.J.G., et al. (2017). A model integrating social-cultural concepts of nature into frameworks of interaction between social and natural systems. *Journal of Environmental Planning and Management*, **61**, 756–777.

Müller, M., and Job, H. (2009). Managing natural disturbance in protected areas: tourists' attitudes towards the bark beetle in a German national park. *Biological Conservation*, **142**, 375–383.

Petty, R.E., and Cacioppo, J.T. (1986). *Communication and persuasion – central and peripheral routes to attitude change*. New York, NY: Academic Press.

Roskaft, E., Handel, B., Bjerke, T., and Kaltenborn, B.P. (2007). Human attitudes towards large carnivores in Norway. *Wildlife Biology*, **13**, 172–185.

Rudzitis, G., and Johansen, H.E. (1991). How important is wilderness? Results from a United States survey. *Environmental Management*, **15**, 227–233.

Sacher, P., Kaufmann, S., and Mayer, M. (2017). Wahrnehmung der natürlichen Waldentwicklung im Nationalpark Harz durch Besucher. *Naturschutz und Landschaftsplanung*, **49**(9), 291–299.

Schenk, A., Hunziker, M., and Kienast, F. (2007). Factors influencing the acceptance of nature conservation measures – a qualitative study in Switzerland. *Journal of Environmental Management*, **83**, 66–79.

Schnitzler, A. (2014). Towards a new European wilderness: embracing unmanaged forest growth and the decolonisation of nature. *Landscape and Urban Planning*, **126**, 74–80.

Schwartz, S.H. (1992). Universals in the content and structure of values: theory and empirical tests in 20 countries. In Zanna, M.P. (Ed.), *Advances in experimental social psychology* (pp. 1–65). New York, NY: Academic Press.

Schwartz, S.H., and Boehnke, K. (2004). Evaluating the structure of human values with confirmatory factor analysis. *Journal of Research in Personality*, **38**, 230–255.

Stelling, F., Allan, C., and Thwaites, R. (2017). Nature strikes back or nature heals? Can perceptions of regrowth in a post-agricultural landscape in South-eastern Australia be used in management interventions for biodiversity outcomes? *Landscape and Urban Planning*, **158**, 202–210.

Stelzig, I. (2000). Zur Akzeptanz von Totholz in deutschen Wald-Nationalparken. In Trommer, G. and Stelzig, I. (Eds.), *Naturbildung und Naturakzeptanz. Beiträge zur biologischen Forschung* (pp. 117–123). Frankfurt: Shaker-Verlag.

Stenmark, M. (2002). *Environmental ethics and policy-making*. Aldershot: Ashgate.

Stern, P.C., and Dietz, T. (1994). The value basis of environmental concern. *Journal of Social Issues*, **50**, 65–84.

Stremlow, M., and Sidler, C. (2002). *Schreibzüge durch die Wildnis*. Bern: Haupt-Verlag.

Teel, T., Dayer, A., Manfredo, M.J., and Bright, A. (2005). *Regional results from the research project entitled 'Wildlife Values in the West' (Project Rep. No. 58)*. Project Report for the Western Association of Fish and Wildlife Agencies. Fort Collins, CO: Colorado State University,

Human Dimensions in Natural Resources Unit.

Thompson, S.C.G., and Barton, M.A. (1994). Ecocentric and anthropocentric attitudes toward the environment. *Journal of Environmental Psychology*, **14**, 149-157.

Treves, A., Naughton-Treves, L., and Shelley, V. (2013). Longitudinal analysis of attitudes toward wolves. *Conservation Biology*, **27**, 315-323.

van den Born, R.J.G. (2006). Implicit philosophy – images of relationships between humans and nature in the Dutch population. In van den Born, R.J.G., Lenders, R.H.J., and de Groot, W.R. (Eds.), *Visions of nature. A scientific exploration of people's implicit philosophies regarding nature in Germany, the Netherlands and the United Kingdom* (pp. 63-83). Berlin: LIT Verlag.

van den Born, R.J.G., Lenders, R.H.J., de Groot, W.T., and Huijsman, E. (2001). The new biophilia: an exploration of visions of nature in Western countries. *Environmental Conservation*, **28**, 65-75.

Ward Thompson C., Aspinall, P., and Montarzino, A. (2008). The childhood factor: adult visits to green places and the significance of childhood experience. *Environment and Behavior*, **40**, 111-143.

Wilson, E.O. (1993). Biophilia and the conservation ethic. In Kellert, S. and Wilson, E.O. (Eds.), *The biophilia hypothesis* (pp. 31-41). Washington, DC: Island Press.

Wolsink, M., and Devilee, J. (2009). The motives for accepting or rejecting waste infrastructure facilities. Shifting the focus from the planners' perspective to fairness and community commitment. *Journal of Environmental Planning and Management*, **52**, 217-236.

Yoshida, Y., Flint, C.G., and Dolan, M. (2017). Farming between love and money: US Midwestern farmers' human-nature relationships and impacts on watershed conservation. *Journal of Environmental Planning and Management*, **61**, 1033-1050.

Zimmermann, B., Wabakken, P., and Dötterer, M. (2001). Human-carnivore interactions in Norway: how does the re-appearance of large carnivores affect people's attitudes and levels of fear? *Forest Snow and Landscape Research*, **76**, 137-153.

CHAPTER NINE

Health and social benefits of living with 'wild' nature

CECILY MALLER, LAURA MUMAW,
and BENJAMIN COOKE
Centre for Urban Research, RMIT University

As nature protection and rewilding initiatives gain momentum around the globe there is renewed focus on the potential health and social benefits of interactions with nature for individual people and communities. Among other benefits, 'rewilding'[1] has been promoted as a means of reconnecting people with nature, addressing the so-called 'nature deficit disorder' (Louv, 2008) and 'ecological boredom' thought to characterise modern life, and living in cities in particular (Monbiot, 2013). In this chapter we focus on the health and social benefits of nature protection and rewilding in urban as opposed to other contexts for three reasons.

First, cities are the most dominant form of human settlement, with the majority of people around the world now living in urban contexts (United Nations General Assembly, 2016). Second, concern for the health and wellbeing of people living in cities is evident in many urban national and international policy agendas, for example the UN's New Urban Agenda (United Nations General Assembly, 2016). Signed at Habitat III, the United Nations Conference on Housing and Sustainable Urban Development (Quito, Ecuador in October 2016), Principle 100 of the New Urban Agenda aims to 'support the provision of well-designed networks of safe, inclusive for all inhabitants, accessible, green, and quality public spaces and streets ... promoting walkability and cycling towards improving health and well-being' (United Nations General Assembly, 2016, p. 4). Third, although often thought to have disrupted and fragmented landscapes, cities have been recognised as biodiversity 'hotspots' and key sites for threatened species (Garrard and Bekessy, 2014; Ives et al., 2016). Momentum is building to protect nature in cities and reconceptualise urban environments as valued forms of habitat (Low, 2003; Hinchliffe and Whatmore, 2006; Puppim de Oliveira et al., 2011). This stance does not seek to gloss over the environmental impact of cities, or override the significance of

---

[1] We acknowledge that rewilding is a contested term, as other chapters in this volume attest. It was beyond the scope of this chapter to discuss its limitations.

'intact' or more fully functioning ecosystems, conservation zones, and National Parks. The point is that urban environments present a unique context for thinking about nature and health because of the impacts and changes brought about by people and urbanisation processes on ecosystems, and because the human-centric character of urban environments detracts from how animal and plant species, albeit living side-by-side with humans, are conceived. Human–non-human species interactions in urban environments have the potential to affect outcomes for human health and nature protection in myriad ways. With its focus on the health and social benefits of rewilding in cities, this chapter complements the case studies of urban rewilding discussed in Chapter 14.

To set the scene, the first section of the chapter introduces our framing of urban rewilding in the context of understanding human–nature relationships. The next section provides an overview of the research on the human health and social benefits of contact with nature. The following section turns to the specific health and social benefits from rewilding programmes and projects as experienced in urban environments. The subsequent section situates rewilding in contemporary debates around the social dimensions of greening cities. The final section discusses some of the challenges, potential harms, and conflicts that may arise from urban rewilding projects. We conclude the chapter by highlighting directions for further research.

Urban rewilding and human–nature relationships

In contrast to the more traditional forms of rewilding taking place in peri-urban or rural areas (see Chapters 4–7), urban rewilding is different for a number of reasons, mostly due to the presence and activities of people (Diemer et al., 2003). Ideas of 'wilderness' that position nature and humanity as inherently separate can undervalue the richness of nature in urban areas simply because humans live in them (Rink and Herbst, 2011; Threlfall and Kendal, 2018), overlook cities as loci for rewilding activities, and discourage the nurturing of human appreciation and stewardship of non-human life forms (Braun, 2005; Beatley and Bekoff, 2013). We define urban rewilding in this chapter as: *any initiative or programme that seeks to encourage biodiversity, ecosystem function, and the persistence of native species in a range of urban settings, including on private and public land; and both human-supported and 'natural' colonisation of urban environments by native species.*

The first part of our definition means that a range of human-led activities to improve the biodiversity of urban environments at a number of scales are included, from the restoration of parkland and degraded remnants on the fringes of cities to gardening to encourage wildlife in residents' backyards. The last part of our definition allows for both deliberate, human-controlled, and 'accidental' forms of rewilding to be included. There are many examples of unmanaged colonisation by native species of abandoned or neglected areas of cities or in human created habitats such as gardens. In the eastern parts of the USA, white-tailed deer (*Odocoileus virginianus*), a native species once hunted by humans to low numbers, has returned with vigour thanks to the proliferation of suburban gardens that provide food and shelter from predators and hunters (DeNicola et al., 2010). In Brisbane, Australia, brush turkeys (*Alectura lathami*, a large megapode) have returned to the city since the early 1970s due to local households planting subtropical rainforest plants in their gardens. Undeterred by the expansion of houses, domestic cats, and increased traffic, the number of brush turkeys in Brisbane has dramatically increased since the 1980s (Jones et al., 2004). This broad definition allows us to explore a range of contexts and situations where people may benefit from increased contact with native animals and plants in cities.

Overview of the health and social benefits of interacting with nature

Beginning with the parks movement in the USA and Europe in the nineteenth century, and now backed up by global evidence from diverse disciplines including environmental psychology, epidemiology, and health promotion (Maller et al., 2008; Egorov et al., 2016), access to and contact with nature is increasingly regarded as a key dimension of what makes cities liveable and healthy (Frumkin, 2003; Hinchliffe and Whatmore, 2006; Maller et al., 2010; Pellegrini and Baudry, 2014; Lowe et al., 2015). Benefits provided by nature in cities include environmental or ecosystem services like heat mitigation, pollution reduction, drinking water, and stormwater protection (McDonald, 2015). Other health and social benefits derive specifically from personal contact with animals and plants (Jones, 2018), or experiencing nature as part of a community, for example through restoration activities (Townsend, 2006; Mumaw et al., 2017).

To understand the diversity of benefits derived from different ways of experiencing nature, Keniger and colleagues (2013, p. 917) suggest a typology of nature interactions: indirect (experiencing nature while not being present in it); incidental (encountering nature while performing another activity); and intentional (being in nature through direct intention, from hiking to environmental volunteering). The idea of encouraging humans to engage with nature in cities for conservation and environmental education goals has existed historically since the 1980s (Adams, 2005). Early recommended approaches focused on planning of green spaces for recreation, conservation, and environmental education in cities, but also suggested establishing wildlife sanctuaries on public and private land, and incorporating wildlife habitat in backyards (Adams and Leedy, 1987).

All types of interactions with nature have been shown to have one or more of a range of benefits, including: cognitive, psychological, physiological, social, and spiritual (Keniger et al., 2013). For example, cognitive and psychological benefits from having access to nature include stress reduction and increased capacity for attention (Ulrich et al., 1991; Kaplan, 1995; Maller et al., 2006). The physical health benefits include improved immune function, increased physical activity, reduced cardiovascular morbidity, and improved pregnancy outcomes (Egorov et al., 2016). There is evidence that contact with nature provides opportunities for social connection and improved emotional health (Coley et al., 1997; Townsend, 2006; Maller 2009; Soulsbury and White, 2016) and higher life satisfaction (Honold et al., 2016).

Experiencing and interacting with nature can also contribute to a person's sense of control and security, inspiration and imagination, feelings of connection and belonging, learning and developing skills, and shaping identity – all dimensions of subjective well-being (Russell et al., 2013; Fish et al., 2016; Mumaw et al., 2017). Actively taking care of nature, from environmental volunteering to ecological restoration, provides additional forms of subjective and social well-being related to 'giving back' or making a contribution not only for nature, but for future generations (Warburton and Gooch, 2007; Mumaw, 2017), and/or one's own place and community (Husk et al., 2016). Chan and colleagues (2016) make a cogent case for how human relationships with nature and relational values for nature underpin the foundations of social well-being and cultural codes of peoples across the globe. This body of knowledge provides a rich background from which the health and social benefits of urban rewilding can be understood.

The health and social benefits of rewilding cities

We now turn to the potential health and social benefits arising from different ways of engaging with rewilding in cities, ranging from indirectly receiving the benefits of rewilding projects, for example having access to urban green space, to actively participating in rewilding activities. As urban rewilding is a new and developing area, we draw on a broad range of literature relevant to our definition of urban rewilding as including *any initiative or programme that seeks to encourage biodiversity, ecosystem function and the persistence of native species in a range of urban settings, including on private and public land; and both human-supported and 'natural' colonisation of urban environments by native species.*

To understand how different health and social benefits may be generated by different types of human–nature interactions, we have devised a simple categorisation tool building on the typology of Keniger and colleagues (2013) (Figure 9.1). This tool consists of quadrants defined by a two-dimensional matrix. The vertical axis of the matrix represents the type of human–nature interaction, in a gradient ranging from passive or incidental interactions at one end, to deliberative or intentional interactions intended to foster biodiversity, for example conservation and purposive rewilding, at the other. The

HEALTH AND SOCIAL BENEFITS   169

Legend: examples of different interactions
1. Viewing trees from an inner city window
2. Playing in sports park
3. Walking dog through small reserve
4. Vegetable gardening
5. Nature hike
6. Gardening for wildlife
7. Restoring creekside habitat

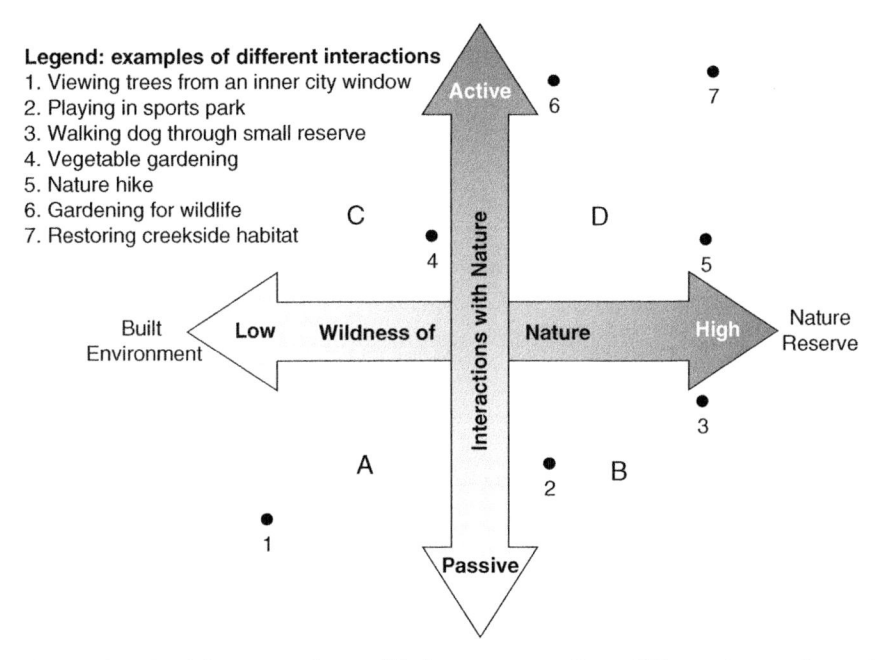

Figure 9.1. Tool for comparing well-being outcomes from different interactions with urban nature (adapted from Keniger et al., 2013). Quadrants signify passive to active fostering of nature (y-axis), in low to high wildness of nature (x-axis). (A black and white version of this figure will appear in some formats. For the colour version, please refer to the plate section.)

horizontal axis represents a 'wildness' gradient of the environment, ranging from human-built environments for human purposes at one end to reserves for nature at the other. In presenting this tool, we recognise that the categorisation will be subjective and not always definitive, and that the quadrants are not mutually exclusive. We position a variety of cases for illustration and describe them in the legend for Figure 9.1. What distinguishes this framework from that of Keniger and colleagues (2013) is the integration of their typologies of urban nature with typologies of *interacting with* those forms of nature. This facilitates a more nuanced comparative exploration of the well-being benefits of urban rewilding activities in different forms of urban nature, for example comparing benefits from vegetable gardening, gardening to improve wildlife habitat, and habitat restoration. We use the tool here to discuss the human benefits of urban rewilding programmes.

Given our definition of rewilding, the 'wildness of nature' continuum of Figure 9.1 includes the colonisation or retention of native species populations in small urban green spaces like gardens, through to reserves rich with ecological communities of the native biodiversity of an area. While we draw some distinctions between benefits in different quadrants, we highlight that knowledge about the range of health and social benefits associated with interacting with nature and rewilding or biodiversity, including the mechanisms by which they occur, is still being developed (Dallimer et al., 2012; Pett et al., 2016; Flies et al., 2017; Mills et al., 2017). Although generally positive, there are variations and inconsistencies highlighted in the literature. Literature reviews of studies on well-being from nature have cautioned about publication bias and inadequate controls, particularly in relation to quantitative studies (Capaldi et al., 2015; Husk et al., 2016). We also point out that literature on the benefits of interacting with nature in general is substantially skewed to studies from Europe, North America, and Western societies (Keniger et al., 2013), and extending this focus to other societies and regions is required.

The remainder of this section reviews the health and social outcomes from different types of interactions with nature in cities, by discussing each quadrant in turn.

Quadrant A: passive or incidental interaction with less wild nature

This first quadrant (Figure 9.1) captures simple, incidental encounters with plants and animals in built environments that may occur while people are doing something else, such as commuting to work. The encounters with plants and animals in these scenarios may or may not be with native species. Simply viewing natural but not necessarily wild environments from a window is reported to assist with stress recovery (Ulrich et al., 1991; Kaplan, 2001). For example, in a review of the relationship between urban green space and human health and well-being, Tzoulas and colleagues (2007) report on experimental studies that show that views of trees and grass from apartment buildings improve adult residents' ability to cope with major life issues and reduce their mental fatigue, and improve attention capacity in children. More recently, Honold and colleagues (2016) found that residents in Berlin who had views of high amounts of diverse kinds of vegetation from their homes had significantly lower cortisol levels. They also found that participants who regularly walked along a vegetated trail by a canal in their neighbourhood or had a commute that involved viewing nature had significantly lower cortisol levels and reported significantly higher life satisfaction than infrequent users (Honold et al., 2016). Rewilding projects that restore habitat and enhance the presence of green space, vegetation, and the visibility of wildlife in urban areas, from bees to birds, are also likely to provide these passive well-being benefits as people go about their daily routines.

Quadrant B: passive or incidental interaction with wilder nature

The second quadrant in Figure 9.1 captures simple, incidental encounters with native plants and animals that occur in green spaces and parks in local neighbourhoods, particularly those with higher biodiversity. People may not go to these areas to expressly interact with plants and animals, but may incidentally encounter them while they are there. Some studies have shown a relationship between residents' expressed well-being and the level of biodiversity in their neighbourhoods (Luck et al., 2011; Botzat et al., 2016). The effects vary with biodiversity scale (for example, ecosystems versus species) and type (for example, trees versus insects), and are often mixed (Botzat et al., 2016). At least one study shows that well-being benefits from a park visit, including ability to reflect and regain perspective, developing emotional bonds to the park, and feeling a stronger sense of identity, increase with its species richness (Fuller et al., 2007). In terms of urban nature more generally, research shows that the amount of neighbourhood green space is correlated with stronger social ties reported among neighbours and greater prosocial activity in the neighbourhood (Capaldi et al., 2015). Capaldi and colleagues (2015) report on a number of studies that show even brief contacts with urban nature, for example taking walks through green spaces, improve mood and emotional state. Interestingly, people's stated motivations for using an urban park may underestimate the rich array of outcomes they report experiencing after interacting with that environment (Irvine et al., 2013). A survey of 312 park users in 13 separate urban parks in Sheffield, UK, showed that their motivations for using a park were primarily about undertaking physical pursuits, principally walking (as part of a route, walking the dog, or exercise) followed by enjoying the park's qualities. By contrast, they reported effects after the activity principally related to physical restoration (relaxation and refreshment); positive emotions (happiness and wonder); place-related attachment to or appreciation of the place; and mental restoration (tranquillity or serenity). This indicates 'near nature' that is part of everyday encounters is of value to urban rewilding, and simultaneously, contributes to a variety of health and well-being benefits (Cox et al., 2017).

Possibly related to some of the positive health and well-being outcomes reported in the literature is the notion that there is a connection between human microbiomes and urban biodiversity levels, or the environmental microbiome (Flies et al., 2017; Mills et al., 2017). Flies and colleagues (2017, p. 2) suggest interactions with biodiverse urban green spaces positively influence the human microbiome 'which acts in the long term to suppress inflammation and reduce chronic immunological diseases'. In fact, Flies and colleagues (2017, p. 6) suggest one potential explanation for discrepancies observed in the literature on the benefits of nature in cities could be 'attributed to the unmeasured microbial diversity of the green spaces'. Interest is progressing in this regard, with a number of studies recently being published on the relationship between the human microbiome and urban biodiversity (for example, Hanski et al., 2012; Rook, 2013; Flies et al., 2017; Mills et al., 2017). Mills and colleagues (2017, p. 2) propose the 'microbiome rewilding hypothesis' that argues increasing biodiversity in urban green spaces 'can rewild the environmental microbiome to a state that benefits human health by primary prevention as an ecosystem service'. As well as influencing passive encounters, this potential outcome, along with other health and social benefits, is also likely to be associated with intentional interactions with nature as depicted in quadrants C and D (Figure 9.1).

Quadrant C: intentional interactions with less wild nature

The third quadrant, Quadrant C (Figure 9.1), refers to intentional interactions in urban green spaces that are perceived by the actors as for humans rather than wild nature, although native plants and animals may live there. Many urban gardens fall into this category, being seen as places for people (Davies and Webber, 2004) and/or growing non-native fruits, vegetables, and flowers. Clayton (2007, p. 223) notes that 'the garden seems to be seen as part of the domestic world ... rather than as part of wild nature'. Bhatti and Church (2004) found that gardens are valued most for making 'a house a home', and least for 'where you can care for the planet' (although they did find that 20–25 per cent of UK householders valued their gardens as places to encourage wildlife – see Quadrant D). Nonetheless, gardeners report benefits of observing nature and its cycles, relaxation, and self-expression (Clayton, 2007). People perceive their gardening and relationships with their garden as distinct from interactions with nature in parks (Bernardini and Irvine, 2007). Bernardini and Irvine (2007) found that householders derive self-esteem and self-efficacy from shaping the garden with their own hands, and learning through challenges and experimentation. This creates strong attachments to place, and contributes to their sense of identity (Bernardini and Irvine, 2007). Torres and colleagues (2017) report that French community gardeners express personal and social benefits from their involvement, including sensory interactions with nature, reflection and relaxation, place-based learning, new friendships, and opportunities to share their knowledge and discoveries with others. Gardeners can feel a 'sense of wilderness' in the garden, which is highly valued 'because of the associated symbolic meanings such as the perceived rhythm of life' (Bernardini and Irvine, 2007). Several authors highlight that opportunities for environmental stewardship, through activities such as community gardening, can strengthen community resilience, build ecological knowledge, and instil an ethic of caring for nature that can be a foundation for fostering urban biodiversity (Ernstson et al., 2010; Colding and Barthel, 2013; Mumaw, 2017).

Quadrant D: intentional interactions with wilder nature

The fourth quadrant (Figure 9.1) includes intentional actions with wilder nature. Although we acknowledge there are other ways to intentionally interact with nature (e.g. bird feeding, wildlife care), in our discussion we focus on the fostering of native species and communities in the urban landscape on public or private land (at the upper end of the quadrant). These activities connect more clearly with urban rewilding initiatives compared to the previous three quadrants, although all four have relevance in different ways as already discussed. To the right of the quadrant, studies on the motivations for, and benefits of, environmental volunteering on tracts of urban public wildlife habitats show that participants receive many of the well-being benefits described in other quadrants, including: learning about nature and self-expression (Bruyere and Rappe, 2007; Asah and Blahna, 2012); peace of mind and a sense of community (Grese et al., 2000); positive emotions and being able to socialise (Asah et al., 2014); and attachments to local nature and the places they care for (Ryan and Grese, 2005; Husk et al., 2016). The environmental volunteering activities reviewed in all these studies fit our definition of urban rewilding and include intentional interactions in cities such as ecological restoration of stream and terrestrial habitats (Grese et al., 2000; Ryan and Grese, 2005), conservation and land management (Bruyere and Rappe, 2007), or environmental enhancement of watersheds, habitats, or re-greening of waste sites (Husk et al., 2016). Notably, and not surprisingly, participants in these studies expressed additional well-being benefits unique to this quadrant, for example: 'causing good things to happen', 'protecting natural places from disappearing', and 'making the world better for others' (Grese et al., 2000).

To the left of the quadrant, we have positioned the example of wildlife gardening. Wildlife gardening activities include 'removing environmental weeds, planting and protecting indigenous vegetation and vegetative structure, and providing habitat for indigenous wildlife' (Mumaw, 2017, p. 94). Similar well-being benefits to those described by environmental volunteers on public land above are expressed by participants engaged in municipal wildlife gardening, that is, gardening to complement their local government's management of public land to restore and protect native species and communities (Mumaw, 2017; Mumaw et al., 2017). These benefits include strengthened connections with nature, place, and community; rejuvenation, wonder, and 'quality of life' from experiencing nature and its cycles; learning and sharing new knowledge and skills; and a sense of achievement from improving the environment (Mumaw et al., 2017). Some participants also expressed strong feelings of hope and motivation from their involvement: 'it makes you feel good about your neighbourhood' and 'you see the willingness of people to make a difference ... I come back refreshed again and feeling more positive' (Mumaw et al., 2017). Feelings of well-being come from participants seeing their gardens as a continuum with the wild landscape, endorsement that they are contributing to fostering indigenous flora and fauna, and visible involvement of local government and other community members (Mumaw, 2017; Mumaw et al., 2017). Those feelings of well-being associated with personal development and living a meaningful life have been found to independently contribute to quality of life (Tay and Diener, 2011).

Challenges of urban rewilding

There are normative assumptions about rewilding being beneficial for humans that must be acknowledged and unpacked if rewilding is to deliver positive outcomes to urban communities more broadly. Little research has critically engaged with how the human residents of cities perceive and experience urban rewilding interventions and the potentially increasing numbers, or changing types, of animals and plants sharing and comprising urban neighbourhoods (although see Rupprecht, 2017). This part of the chapter considers some of the practicalities and challenges of going about implementing urban rewilding programmes and initiatives, as these will impact health and social outcomes, both positive and negative.

As highlighted in the first part of this chapter, rewilding cities as places of dense human habitation has different implications than rewilding in landscapes that are less densely populated. The qualities of rewilded ecologies themselves have the potential to impact negatively on urban human communities; allergies caused by pollen, tree roots as trip hazards, falling branches that can kill or injure, leaves that accumulate and clog drains, bird droppings on washing, and bites and stings are just some examples of 'ecosystem disservices' (Lyytimäki et al., 2008). Multispecies interactions, as encouraged through rewilding, are not always nice or pleasant. Just as traditional restoration ecology thinking has had to reconcile the need for human–environment coexistence in urban ecology, conceptions of urban rewilding must consider the challenges of coexistence (Dearborn and Kark, 2010). We suggest that to live with, and in, urban rewilding in ways that are attentive to benefits and harms necessitates an emphasis on the democratic processes and experimental practices of environmental management (Lorimer and Driessen, 2014). In this sense we must paint a more nuanced picture of rewilding into which health and social benefits can be positioned.

The emergence of a critical discussion on equity and environmental justice in urban greening offers an entry point for considering both the distribution of benefits and the complexities of human–environment interactions for urban rewilding (Rutt and Gulsrud, 2016; Gould and Lewis, 2017). We argue that to progress a rewilding agenda that is equitable and beneficial means an honest engagement with the social and environmental harms that could also be generated alongside the health and well-being benefits.

The greening of cities is inherently bound up with urban development and regeneration projects that seek to change neighbourhood character, attract private investment, and increase property prices (Wolch et al., 2014). Projects perceived to be imposed on communities from outside that do not take account of local interests can cause resentment and resistance (Curran and Hamilton, 2012; Kabisch and Haase, 2014). Greening activities are also not evenly distributed and the benefits not evenly experienced among people of differing socioeconomic status, with wealthy parts of cities often the greenest (Heynen et al., 2006; Perkins, 2011). Rewilding as a means for increasing health and social benefits through nature interactions and access must also be alert to the potential for projects to be captured by private interests or enclosed through private property, despite promising public good benefits (Perkins, 2011). Rewilding that displaces communities by raising property prices or displacing industry in favour of residential development represents a form of 'green gentrification' (Gould and Lewis, 2017). Left unchecked, 'green gentrification' will exacerbate inequity and compromise any contribution that rewilding can make to health and social benefits. The types of interactional benefits that rewilding could facilitate noted in Figure 9.1 must be realised as part of a broader urban sustainability agenda that is cognisant of social equity considerations (Gould and Lewis, 2017).

Entwined with considerations around equitable rewilding outcomes are the processes by which rewilding is pursued. If human residents do not accept or warm to rewilding for biodiversity, the success of such programmes is thrown into doubt because of the human-centric mandate of cities. If cities are to be places where wildlife is welcomed, popular support, and indeed active engagement, is crucial. Building on relationships that care for nature, like forms of gardening from private to community and wildlife gardening, and environmental volunteering, show promise for engaging urban residents. Indeed, as discussed in the previous section (Figure 9.1, Quadrant D), some health and social benefits derive specifically from active participation in rewilding efforts, and when situated in a collaboration between community members and local authorities (Mumaw, 2017). Borrowing from wider discussion about conservation in the Anthropocene, we also suggest that urban rewilding is best performed as a collaborative experiment, where interested and affected urban residents have the opportunity to play a role in realising the health and social benefits that might emerge from rewilding efforts (Lorimer and Driessen, 2014). Transformative environmental outcomes can be achieved in municipal areas, while building urban social and institutional resilience, through governance that links active citizens with networks and local authorities jointly responding to local social–ecological context (Buijs et al., 2016). With the dynamic and evolving socioecology of the Anthropocene city, collaborative and experimental rewilding is a clear avenue for bringing people into the remaking of urban ecologies.

A collaborative ethos for rewilding would also help to push back against any attempts at co-option by private or governmental interests as a top-down managerial project for delivering economic or development benefits (Gabriel, 2016). It also provides a way of insulating against the potential for rewilding projects and policy to simply be transferred from one city to the next, independent of local conditions, social context, and community collaboration (Peck, 2011). A de-centred urban rewilding is also worth pursuing as a way of pushing back against the ordered, manicured, formal parks and gardens that are generally considered as being the spaces that provide well-being benefits (Rupprecht et al., 2015). A collaborative rewilding could invoke more informal, messy, and creative spaces that allow for a wider trajectory of ecological flourishing (Threlfall and Kendal, 2018), where the potential for ecosystem disservices can be more clearly and transparently addressed and where more diverse forms of human–environment interaction can take root (Pellegrini and Baudry, 2014). Embracing 'informal' and 'wild' spaces also helps us to see, preserve, and build on the myriad wild and unruly spaces in cities that already exist and are home to diverse ecological assemblages and human–environment interactions (Rupprecht and Byrne, 2014; Threlfall and Kendal, 2018). Drawing different perspectives and voices into rewilding has the potential to reframe how nature is viewed and experienced in urban environments and the potential for a range of health and social benefits to be achieved.

Conclusions

Set in the context of urban rewilding, this chapter has discussed the health and social benefits that can arise from fostering and protecting nature in cities. A range of benefits were discussed that varied according to whether the human–nature interactions were passive or more intentional. In doing so, we devised a conceptual tool, extending Keniger et al.'s (2013) framework, comprised of four quadrants to depict how different forms of urban nature and different intensities of interactions can produce different outcomes for the people and communities involved. Rather than just being relevant to urban rewilding, this tool could potentially assist in exploring a wide range of human–nature interactions. Each type of interaction with nature has the potential to create positive outcomes. Even simple everyday activities such as work commutes present the opportunity for passive opportunities for people to experience animals and plants in their city.

Studies to date suggest there may be a relationship between higher levels of biodiversity in urban areas and increased health and social outcomes, and that fostering and supporting biodiversity generates additional quality of life outcomes related to living a meaningful life. Interactions associated with social activity appear to generate socially related benefits additional to personal benefits. Overall, however, the evidence is still inconclusive about what specific qualities of nature, including biodiversity, what forms of interaction, and what particular contexts are important for stimulating different health and well-being benefits. More specifically, understanding of the mechanisms through which the health and social benefits associated with interacting with nature, rewilding, or enhanced biodiversity is still developmental and we urge further research in this regard. There is also a gap in understanding the different social outcomes from urban rewilding, and the health benefits of contact with nature more generally, in different parts of the world, particularly the Global South. As the world becomes more urbanised and rewilding projects multiply, the need to understand the human dimensions and how health and well-being may differ by culture, geographic region, or demographic group is becoming critical. There is also a bias towards measuring, quantifying, and reporting the benefits of contact with nature while the potential harms or disservices that might arise are sometimes ignored. To progress an urban rewilding agenda that is equitable and beneficial means an honest engagement with the social and environmental harms that could also be generated alongside potential health and well-being benefits. It is therefore essential that the human and social dimensions of urban rewilding share equal billing with biodiversity and ecological imperatives.

Acknowledgements

The authors thank Paula Arcari who provided essential background research for this chapter.

References

Adams, L.W. (2005). Urban wildlife ecology and conservation: a brief history of the discipline. *Urban Ecosystems*, **8**, 139-156.

Adams, L.W., and Leedy, D.L. (1987). *Integrating Man and Nature in the Metropolitan Environment: Proceedings of a National Symposium on Urban Wildlife*, 4-7 November 1986, National Institute for Urban Wildlife, Chevy Chase, Maryland, USA.

Asah, S.T., and Blahna, D.J. (2012). Motivational functionalism and urban conservation stewardship: implications for volunteer involvement. *Conservation Letters*, **5**, 470-477.

Asah, S.T., Lenentine, M.M., and Blahna, D.J. (2014). Benefits of urban landscape eco-volunteerism: mixed methods segmentation analysis and implications for volunteer retention. *Landscape and Urban Planning*, **123**, 108-113.

Beatley, T., and Bekoff, M. (2013). City planning and animals: expanding our urban compassion footprint. In Basta, C. and Moroni, S. (Eds.), *Ethics, design and planning of the built environment*, Urban and Landscape Perspectives 12 (pp. 185-195). Dordrecht: Springer.

Bernardini, C., and Irvine, K.N. (2007). The nature of urban sustainability: private or public greenspaces?' In Kungolos, A.G., Brebbia, C.A., and Beriatos, E. (Eds.), *Sustainable development and planning III* (Vol. II, pp. 661-674). Southampton: WIT Press.

Bhatti, M., and Church, A. (2004). Home, the culture of nature and meanings of gardens in late modernity. *Housing Studies*, **19**, 37-51.

Botzat, A., Fischer, L.K., and Kowarik, I. (2016). Unexploited opportunities in understanding liveable and biodiverse cities. A review on urban biodiversity perception and valuation. *Global Environmental Change*, **39**, 220-233.

Braun, B. (2005). Environmental issues: writing a more-than-human urban geography. *Progress in Human Geography*, **29**, 635-650.

Bruyere, B., and Rappe, S. (2007). Identifying the motivations of environmental volunteers. *Journal of Environmental Planning and Management*, **50**, 503-516.

Buijs, A.E., Mattijssen, T.J.M., Van der Jagt, A.P.N., et al. (2016). Active citizenship for urban green infrastructure: fostering the diversity and dynamics of citizen contributions through mosaic governance. *Current Opinion in Environmental Sustainability*, **22**, 1-6.

Capaldi, C.A., Passmore, H.-A., Nisbet, E.K., Zelenski, J.M., and Dopko, R.L. (2015). Flourishing in nature: a review of the benefits of connecting with nature and its application as a wellbeing intervention. *International Journal of Wellbeing*, **5**, 1-16.

Chan, K.M.A., Balvanera, P., Benessaiah, K., et al. (2016). Opinion: why protect nature? Rethinking values and the environment. *Proceedings of the National Academy of Sciences of the United States of America*, **113**, 1462-1465.

Clayton, S. (2007). Domesticated nature: motivations for gardening and perceptions of environmental impact. *Journal of Environmental Psychology*, **27**, 215-224.

Colding, J., and Barthel, S. (2013). The potential of 'Urban Green Commons' in the resilience building of cities. *Ecological Economics*, **86**, 156-166.

Coley, R.L., Kuo, F.E., and Sullivan, W.C. (1997). Where does community grow? The social context created by nature in urban public housing. *Environment and Behavior*, **29**, 468-495.

Cox, D.T.C., Shanahan, D.F., Hudson, H.L., et al. (2017). Doses of neighborhood nature: the benefits for mental health of living with nature. *BioScience*, **67**, 147-155.

Curran, W., and Hamilton, T. (2012). Just green enough: contesting environmental gentrification in Greenpoint, Brooklyn. *Local Environment*, **17**, 1027-1042.

Dallimer, M., Irvine, K.N., Skinner, A.M.J., et al. (2012). Biodiversity and the feel-good factor: understanding associations between self-reported human well-being and species richness. *BioScience*, **62**, 47-55.

Davies, R., and Webber, L. (2004). Enjoying our backyard buddies: social research informing the practice of mainstream community education for the conservation of urban wildlife. *Australian Journal of Environmental Education*, **20**, 77-87.

Dearborn, D.C., and Kark, S. (2010). Motivations for conserving urban biodiversity. *Conservation Biology*, **24**, 432-440.

DeNicola, A.J., VerCauteren, K.C., Curtis, P.D., and Hyngstrom, S.E. (2010). *Managing white-tailed deer in suburban environments: a technical guide*. Ithaca, NY: Cornell University Press.

Diemer, M., Held, M., and Hofmeister, S. (2003). Urban wilderness in Central Europe: rewilding at the urban fringe. *International Journal of Wilderness*, **9**, 7-11.

Egorov, A.I., Mudu, P., Braubach, M., and Martuzzi, M. (2016). *Urban green spaces and health*. Copenhagen: WHO Regional Office for Europe.

Ernstson, H., Barthel, S., Andersson, E., and Borgström, S.T. (2010). Scale-crossing brokers and network governance of urban ecosystem services: the case of Stockholm. *Ecology and Society*, **15**, 1-25.

Fish, R., Church, A., and Winter, M. (2016). Conceptualising cultural ecosystem services: a novel framework for research and critical engagement. *Ecosystem Services*, **21**, 208–217.

Flies, E.J., Skelly, C., Negi, S.S., et al. (2017). Biodiverse green spaces: a prescription for global urban health. *Frontiers in Ecology and the Environment*, **15**, 510–516.

Frumkin, H. (2003). Healthy places: exploring the evidence. *American Journal of Public Health*, **93**, 1451–1454.

Fuller, R.A., Irvine, K.N., Devine-Wright, P., Warren, P.H., and Gaston, K.J. (2007). Psychological benefits of greenspace increase with biodiversity. *Biology Letters*, **3**, 390–394.

Gabriel, N. (2016). No place for wilderness: urban parks and the assembling of neoliberal urban environmental governance. *Urban Forestry & Urban Greening*, **19**, 278–284.

Garrard, G.E., and Bekessy, S.A. (2014). Land use and land management. In Byrne, J., Sipe, N., and J. Dodson (Eds.), *Australian environmental planning: challenges and future prospects* (pp. 61–72). Abingdon: Routledge.

Gould, K.A., and Lewis, T.L. (2017). *Green gentrification: urban sustainability and the struggle for environmental justice.* Abingdon: Routledge.

Grese, R.E., Kaplan, R., Ryan, R.L., and Buxton, J. (2000). Psychological benefits of volunteering in stewardship programs. In Gobster, P.H. and Hull, R.B. (Eds.), *Restoring nature: perspectives from the social sciences and humanities* (pp. 265–279). Washington, DC: Island Press.

Hanski, I, von Hertzen, L, Fyhrquist, N, et al. (2012). Environmental biodiversity, human microbiota, and allergy are interrelated. *Proceedings of the National Academy of Sciences of the United States of America*, **109**, 8334–8339.

Heynen, N., Kaika, M., and Swyngedouw, E. (2006). *In the nature of cities: urban political ecology and the politics of urban metabolism.* London: Routledge.

Hinchliffe, S., and Whatmore, S. (2006). Living cities: towards a politics of conviviality. *Science as Culture*, **15**, 123–138.

Honold, J., Lakes, T., Beyer, R., and van der Meer, E. (2016). Restoration in urban spaces. *Environment and Behavior*, **48**, 796–825.

Husk, K., Lovell, R., Cooper, C., Stahl-Timmins, W., and Garside, R. (2016). Participation in environmental enhancement and conservation activities for health and well-being in adults: a review of quantitative and qualitative evidence. *Cochrane Database of Systematic Reviews*, **5**.

Irvine, K.N., Warber, S.L., Devine-Wright, P., and Gaston, K.J. (2013). Understanding urban green space as a health resource: a qualitative comparison of visit motivation and derived effects among park users in Sheffield, UK. *International Journal of Environmental Research and Public Health*, **10**, 417–442.

Ives, C.D., Lentini, P.E., Threlfall, C.G., et al. (2016). Cities are hotspots for threatened species. *Global Ecology and Biogeography*, **25**, 117–126.

Jones, D.N. (2018). *The birds at my table: why we feed wild birds and why it matters.* Ithaca, NY: Cornell University Press.

Jones, D.N., Sonnenburg, R., and Sinden, K.E. (2004). Presence and distribution of Australian brushturkeys in the greater Brisbane region. *Sunbird: Journal of the Queensland Ornithological Society*, **34**, 1–9.

Kabisch, N., and Haase, D. (2014). Green justice or just green? Provision of urban green spaces in Berlin, Germany. *Landscape and Urban Planning*, **122**, 129–139.

Kaplan, R. (2001). The nature of the view from home. *Environment and Behavior*, **33**, 507–542.

Kaplan, S. (1995). The restorative benefits of nature: toward an integrative framework. *Journal of Environmental Psychology*, **15**, 169–182.

Keniger, L., Gaston, K., Irvine, K., and Fuller, R. (2013). What are the benefits of interacting with nature? *International Journal of Environmental Research and Public Health*, **10**, 913.

Lorimer, J., and Driessen, C. (2014). Wild experiments at the Oostvaardersplassen:

rethinking environmentalism in the Anthropocene. *Transactions of the Institute of British Geographers*, **39**, 169–181.

Louv, R. (2008). *Last child in the woods: saving our children from nature-deficit disorder*. Chapel Hill, NC: Algonquin Books.

Low, T. (2003). *The new nature*. Camberwell, Victoria: Penguin.

Lowe, M., Whitzman, C., Badland, H., et al. (2015). Planning healthy, liveable and sustainable cities: how can indicators inform policy? *Urban Policy and Research*, **33**, 131–144.

Luck, G.W., Davidson, P., Boxall, D., and Smallbone, L. (2011). Relations between urban bird and plant communities and human well-being and connection to nature. *Conservation Biology*, **25**, 816–826.

Lyytimäki, J., Petersen, L.K., Normander, B., and Bezák, P. (2008). Nature as a nuisance? Ecosystem services and disservices to urban lifestyle. *Environmental Sciences*, **5**, 161–172.

Maller, C. (2009). Promoting children's mental, emotional and social health through contact with nature: a model. *Health Education*, **109**, 522–543.

Maller, C., Townsend, M., St Leger, L., et al. (2008). *Healthy parks, healthy people. The health benefits of contact with nature in a park context: a review of relevant literature*. Melbourne: School of Health and Social Development, Faculty of Health, Medicine, Nursing and Behavioural Sciences, Deakin University & Parks Victoria.

Maller, C.J., Townsend, M., Pryor, A., Brown, P.B., and St Leger, L. (2006). Healthy parks healthy people: contact with nature as an upstream health promotion intervention for populations. *Health Promotion International*, **21**, 45–54.

Maller, C.J., Henderson-Wilson, C., and Townsend, M. (2010). Re-discovering nature in everyday settings: or how to create healthy environments and healthy people. *EcoHealth*, **6**, 553–556.

McDonald, R.I. (2015). *Conservation for cities: how to plan and build natural infrastructure*. Washington, DC: Island Press.

Mills, J.G., Weinstein, P., Gellie, N.J.C., Weyrich, L.S., Lowe, A.J., and Breed, M.F. (2017). Urban habitat restoration provides a human health benefit through microbiome rewilding: the Microbiome Rewilding Hypothesis. *Restoration Ecology*, **25**, 866–872.

Monbiot, G. (2013). *Feral: searching for enchantment on the frontiers of rewilding*. London: Penguin.

Mumaw, L. (2017). Transforming urban gardeners into land stewards. *Journal of Environmental Psychology*, **52**, 92–103.

Mumaw, L.M., Maller, C., and Bekessy, S. (2017). Strengthening wellbeing in urban communities through wildlife gardening. *Cities and the Environment (CATE)*, **10**(1), article 6, 1–18.

Peck, J. (2011). Geographies of policy: from transfer-diffusion to mobility-mutation. *Progress in Human Geography*, **35**, 773–797.

Pellegrini, P., and Baudry, S. (2014). Streets as new places to bring together both humans and plants: examples from Paris and Montpellier (France). *Social and Cultural Geography*, **15**, 871–900.

Perkins, H.A. (2011). Gramsci in green: neoliberal hegemony through urban forestry and the potential for a political ecology of praxis. *Geoforum*, **42**, 558–566.

Pett, T.J., Shwartz, A., Irvine, K.N., Dallimer, M., and Davies, Z.G. (2016). Unpacking the people–biodiversity paradox: a conceptual framework. *BioScience*, **66**, 576–583.

Puppim de Oliveira, J.A., Balaban, O., Doll, C.N.H., Moreno-Peñaranda, R., Gasparatos, A., Iossifova, D., and Suwa, A. (2011). Cities and biodiversity: perspectives and governance challenges for implementing the Convention on Biological Diversity (CBD) at the city level. *Biological Conservation*, **144**, 1302–1313.

Rink, D., and Herbst, H. (2011). From wasteland to wilderness – aspects of a new form of urban nature. In Richter, M. and Weiland, U. (Eds.), *Applied urban ecology: a global framework* (pp. 82–92). Chichester: John Wiley & Sons.

Rook, G.A. (2013). Regulation of the immune system by biodiversity from the natural environment: an ecosystem service essential to health. *Proceedings of the National Acadamy of Sciences USA*, **110**, 18360-18367.

Rupprecht, C.D.D. (2017). Ready for more-than-human? Measuring urban residents' willingness to coexist with animals. *Fennia*, **195**, 142-160.

Rupprecht, C. and Byrne, J. (2014). Informal urban green-space: comparison of quantity and characteristics in Brisbane, Australia and Sapporo, Japan. *PLoS ONE*, **9** (6), e99784.

Rupprecht, C.D.D., Byrne, J.A., Ueda, H., and Lo, A.Y. (2015). 'It's real, not fake like a park': residents' perception and use of informal urban green-space in Brisbane, Australia and Sapporo, Japan. *Landscape and Urban Planning*, **143**, 205-218.

Russell, R., Guerry, A.D., Balvanera, P., et al. (2013). Humans and nature: how knowing and experiencing nature affect well-being. *Annual Review of Environment and Resources*, **38**, 473-502.

Rutt, R.L., and Gulsrud, N.M. (2016). Green justice in the city: a new agenda for urban green space research in Europe. *Urban Forestry & Urban Greening*, **19**, 123-127.

Ryan, R.L., and Grese, R.E. (2005). Urban volunteers and the environment: forest and prairie restoration. In Barlett, P.F. (Ed.), *Urban place: reconnecting with the natural world* (pp. 173-188). Cambridge, MA: MIT Press.

Soulsbury, C.D., and White, P.C.L. (2016). Human–wildlife interactions in urban areas: a review of conflicts, benefits and opportunities. *Wildlife Research*, **42**, 541-553.

Tay, L., and Diener, E. (2011). Needs and subjective well-being around the world. *Journal of Personality and Social Psychology*, **101**, 354-365.

Threlfall, C.G., and Kendal, D. (2018). The distinct ecological and social roles that wild spaces play in urban ecosystems. *Urban Forestry & Urban Greening*, **29**, 348-356.

Torres, A., Nadot, S., and Prevot, A.-C. (2017). Specificities of French community gardens as environmental stewardships. *Ecology and Society*, **22**, 1-13.

Townsend, M.A. (2006). Feel blue? Touch green! Participation in forest/woodland management as a treatment for depression. *Urban Forestry & Urban Greening*, **5**, 111-120.

Tzoulas, K., Korpela, K., Venn, S., et al. (2007). Promoting ecosystem and human health in urban areas using Green Infrastructure: a literature review. *Landscape and Urban Planning*, **81**, 167-178.

Ulrich, R.S., Simons, R.F., Losito, B.D., Fiorito, E., Miles, M.A., and Zelson, M. (1991). Stress recovery during exposure to natural and urban environments. *Journal of Environmental Psychology*, **11**, 231-248.

United Nations General Assembly. (2016). The New Urban Agenda Explainer. Paper presented to United Nations Conference on Housing and Sustainable Urban Development (Habitat III), 17-20 October, Quito, Ecuador. http://resilientneighbors.com/wp-content/uploads/2016/12/NUA-explained.pdf.

Warburton, J., and Gooch, M. (2007). Stewardship volunteering by older Australians: the generative response. *Local Environment*, **12**, 43-55.

Wolch, J.R., Byrne, J., and Newell, J.P. (2014). Urban green space, public health, and environmental justice: the challenge of making cities 'just green enough'. *Landscape and Urban Planning*, **125**, 234-244.

CHAPTER TEN

The psychology of rewilding

SUSAN CLAYTON
College of Wooster

How will people respond to rewilded landscapes, or to the idea of rewilding in a specific context? The success of efforts to introduce rewilded areas will depend in part on public acceptance of these practices. This chapter aims not to predict people's responses to rewilding, but to discuss factors that might influence those responses. The discussion is rooted in the fact that attitudinal responses are not entirely (and perhaps sometimes not at all) guided by rational calculations about the utility of landscapes. People's attitudes, rather, are likely to be affected by the symbolic meanings of landscapes. To imagine what those meanings might be, it is worth considering human perceptions, values, and most important, sense of relationship in regard to the natural world. However, symbolic meanings are malleable and socially mediated; they change in response to a changing culture, and they can be altered by the way in which a landscape or a policy is framed. Even the meaning and value of nature is culturally specific. Thus there is no definitive answer to the question of human preferences.

This chapter begins by exploring human conceptions of wildness, and examining some individual differences in those conceptions. It will then review some previous research on attitudes towards specific landscape management practices, as well as the values that underlie those attitudes. I discuss the values and attitudes of conservation professionals in particular, and present the results of a survey about their attitudes towards a variety of conservation practices. Because attitudes towards rewilding are related to perceptions of the relationship between humans and the natural world, I argue that we need to be mindful of the ways in which that relationship is changing. After commenting on the methods that are used to obtain information about people's preferences, I conclude by encouraging proponents of rewilding to include humans in their plans.

Understandings of 'wildness'

Both the idea of wildness, and nature in general, are things that people value. But what do they mean? As has been noted elsewhere, people typically consider nature to be something that is separate from human experience and human interventions (Rozin et al., 2012; Bauer, 2016). Historian Roderick Nash (e.g. in Hausdoerffer, 2017) has defined it as 'land that is not controlled'. Anthropologist Kay Milton has described the fundamental qualities of nature – the qualities that nature protection focuses on – as including beauty, diversity, personhood, and wildness. She defines wildness primarily as freedom from human influence, and says that it is implied in 'the very concept of nature' (Milton, 2002, p. 112).

Some systematic data on perceptions of wildness and naturalness have been collected. For example, Watson and colleagues (2015) asked visitors to a natural wilderness area in California about the ways in which they characterised wilderness. The researchers looked at perceptions that the wilderness was wild (little evidence of human activity), natural (contained no non-native species), and unconfined (lacking restrictions on human behaviour), based on a factor analysis of statements generated by wilderness managers. 'Wildness' emerged as the most important feature of wilderness. It is worth noting, however, that visitors to a wilderness area may have attitudes about wilderness that differ from those of the general public. Rozin and colleagues (2012) looked at attitudes towards naturalness in a representative sample of European and American residents, specifically with regard to food. They found an overall positive evaluation of the concept of 'natural', accompanied by a preference for food products that had not been interfered with even when their composition was identical to food that had been subject to human intervention. In other words, the process of human intervention or non-intervention was more important than the outcome in assessing naturalness. Applied to landscapes, this might suggest that human interventions would be seen as compromising the naturalness of a setting, even if the goal was to restore an ecosystem to a more pristine or healthy condition.

In general, people do prefer landscapes that appear free from human intervention (Milton, 2002; Bauer et al., 2009) – although landscapes that are unmanaged are not always preferred. For example, Khew and colleagues (2014) found that people in Singapore had a higher preference for seeing manicured landscapes rather than naturalistic ones, apparently because the former had higher aesthetic value. When unmanaged landscapes are preferred, it is not simply an aesthetic preference based on objective characteristics. In an experimental study, McMahan and colleagues (2016) showed participants photographs of outdoor landscapes and obtained ratings for naturalness, aesthetic quality, desire to visit the place depicted, and the appropriateness of designating the place as a protected area. The researchers told half the participants that the areas depicted had been unaltered, and the other half that the *identical images* depicted places that had been altered by humans. People gave higher ratings on every variable to the places that they believed had not been altered. The perceived naturalness of the landscape was the direct predictor of the other ratings: that is, places that were perceived as natural were rated as higher in aesthetic quality, appropriateness for protection, and desirability as a place to visit. Apparently, it's not only that unaltered landscapes are preferred because of the way they look; independent of their appearance, the fact of being unaltered adds value to a place and makes it seem more worthy of being protected. This preference for non-intervention has been described as ideational: that is, based on beliefs about the value of naturalness rather than on a functional or even perceptual difference between the landscapes (McMahan et al., 2016).

As suggested by these results, laypeople's judgements about the extent to which a landscape is natural or even healthy are not always accurate. In a random sample of Australian respondents, Williams and Cary (2002) found that perceived naturalness was valued, but that there was no relationship between preference ratings and the actual ecological health of the landscapes. Interestingly, Ford et al. (2014) found a strong positive correlation between people's evaluations of the beauty of a landscape under a particular management practice and their rating that the practice will have a positive impact on the natural environment. Ribe (2002) has suggested that perceptions of attractiveness can be used as a proxy for perceptions of environmental health.

To fully grasp the ways in which people think about wildness, it is also relevant to examine some of the more metaphorical or symbolic ways in which the term 'rewilding' has been used. George Monbiot's (2013) book, *Feral*, discusses the need to rewild ourselves as well as our landscapes, finding ways to increase our connection with nature. Similarly, ecologist Marc Bekoff (2014) wrote a book, *Rewilding our Hearts*, in which he talked about 'undoing the unwilding' of human beings that is associated with modern culture. The modern Western lifestyle, with its heavy use of technology and an educational system that largely occurs indoors, has, according to Bekoff, diminished people's knowledge about the natural world in ways that are harmful to humans as well as to their support for conservation. This argument has also been made elsewhere; Soga and Gaston (2016), for example, discuss the reduction of human experiences of nature and the potential harmful impacts for humans. With an even closer focus on human well-being, psychologist Patricia Hasbach (2014) has discussed rewilding as a therapeutic approach: a way of emphasising direct sensory experiences with the natural world in order to promote psychological health among those, especially Westerners, who are increasingly distant from the natural world.

For the purposes of this chapter, these descriptions of rewilding from a human perspective do two important things: they implicitly or explicitly affirm the significance of a nature that is free from human control and from the impacts of modern culture; and they argue that this freedom from human control is important for human beings to experience. Hasbach (2014) describes the benefits as follows: 'Wildness offers us the opportunity to experience ourselves embedded in the natural world and reminds us of our place in the order of things. It invites feelings of awe ... and a sense of humility' (p. 3). Ironically, human attempts to experience a landscape free from human influence increase the human influence on those landscapes, threatening the very wildness that is valued.

Individual differences

The value people hold for wilderness, or for nature in general, is a significant determinant of attitudes. It is important to recognise, however, that the value for 'wilderness' may be culturally specific. In previous historical eras and in Western societies, wilderness was often considered to be frightening, and wilderness areas were seen as something that would increase in value if subjected to human management (Cronon, 1996). In the modern era, ethnic minority groups in both the USA and European countries seem to prefer more managed and less natural landscapes compared to the white majority in those regions. In one study comparing Dutch immigrants and native Dutch residents, the immigrants' preference for managed landscapes was partly explained by a lower value for wilderness (Buijs et al., 2009). The very concept of wilderness has been criticised as suggesting a Eurocentric perspective in its emphasis on freedom from human influence (Cronon, 1996). Indigenous peoples have traditionally lived in contact with nature, experiencing human and non-human life as fundamentally intertwined (Sabloff, 2001); some indigenous languages do not have a word for 'wild' (Salmon, 2017). The idea that nature should be free from human influence may be not only naïve and outdated but reflective of colonialist attitudes that have the potential to marginalise and harm native traditions. Indeed, at times people have been removed from their native lands because their presence conflicted with the dominant Western ideology of uninhabited wilderness (Bauer et al., 2009; Dowie, 2009).

The fact that indigenous or immigrant perspectives might differ from those of the dominant culture is a reminder that not everyone values unmanaged nature to the same extent. Bauer (2016), for example, reports a study in which tourists were more positive about a wilderness strategy (reforestation) for land development than were the local inhabitants, and states that in general 'people who are not affected by wilderness ... are more positive about it' than those who interact with it more closely (pp. 103–104). It is worth considering why people might not like wilderness. For many people, particularly those who have less experience with nature, it can represent a source of danger (Bixler and Floyd, 1997; Koole and Van den Berg, 2005). Indeed, Koole and Van den Berg (2005) found that people were more likely to think about death in 'wild' natural settings than in managed settings or in the city. Wilderness and unrestrained nature pose a potential threat to human safety and security that can lead those who feel unprepared to deal with those perceived dangers to respond with fear and dislike.

One individual difference that has been repeatedly found in attitudes towards wilderness is that people who perceive nature as having intrinsic value also assign greater value to wilderness (Buijs et al., 2009). Koole and Van den Berg (2005), for example, found a positive correlation between eco-centric values (valuing nature for its own sake) and a preference for views of wilderness over managed or urban landscapes, while Khew found that conservationists preferred more naturalistic landscapes over more obviously managed landscapes. Research by Tang and colleagues (2015) showed that those with a higher sense of connectedness to nature also showed greater preference for images of more wild landscapes.

Importantly, Tang et al.'s (2015) research also uncovered a mechanism underlying this preference. Hypothesising that people who are more connected to nature actually perceive the landscape differently than those who feel more disconnected, they found indeed that people higher in connectedness rated forests as higher in mystery (defined as an aspect of a setting that encourages further exploration) and legibility (the perception that one would be able to find one's way around a landscape), suggesting that these people anticipated the potential for more positive interactions in those settings. They also rated the forests as more restorative. Statistical analyses showed that these perceptions of the forests, in turn, accounted for much of the difference in preferences.

Connection to nature may be a significant influence on attitudes towards wildness, particularly in cultures where such a connection is limited. Connection to nature is a multidimensional construct, referring to an emotional response to nature as well as a cognitive perception of oneself as interdependent with the natural world. It can be a transitory feeling, but a stable tendency to perceive oneself as connected, which I have labelled environmental identity (Clayton, 2003, 2012), may be developed on the basis of repeated significant experiences in nature (Soga et al., 2016; Prévot et al., 2018). Environmental identity (EID) essentially defines the personal relationship with nature, which varies along a spectrum from separation and domination (low EID) to more equal and intimate (high EID).

Having a strong EID should, consistent with the research by Tang, affect reactions to wild landscape, leading to a perception that the landscape is less strange and 'alien' and that one will be able to effectively thrive in that landscape. It is not clear, though, that a strong EID would predict support for rewilding. Bauer and colleagues (2009) found that one's relationship with nature affected attitudes towards rewilding, but people who they defined as 'nature-connected' were opposed to rewilding. Although we do not have data on this, it is likely that at least some of the experiences with nature that lead to the development of EID take place in landscapes that are managed by humans, because these important experiences are likely to take place in early to middle childhood; so a connection to nature does not necessarily mean a connection to wild nature.

In addition to connection, knowledge is also likely to make a difference in attitudes towards wildness. For one thing, knowledge about a landscape is likely to make it seem less threatening (Bixler and Floyd, 1997). For another thing, knowledge about a landscape may include more accurate understanding of its state of health. Given that people's preferences are not always based on an accurate perception of the landscape, what are the attitudes of people who have a deeper understanding of ecosystem functioning – the conservation professionals? Hagerman and Satterfield (2014) surveyed about 140 people whom they describe as 'a globally representative sample of conservation experts' (p. 550) including both academic and NGO expertise to explore their attitudes towards conservation goals. The researchers found that 'maintaining ecological processes' was the conservation goal most strongly endorsed; 'protecting wilderness with minimum direct human impacts' was second. Thus, a preference for wilderness also exists among experts.

Attitudes towards landscape management practices

Rewilding falls within a spectrum of ways in which people engage with natural species and landscapes in order to achieve specific goals. Attitudes towards other types of interventions are also relevant to understanding the public response to rewilding. Vining and colleagues (2000), for example, looked at public attitudes towards ecological restoration in Chicago, following controversy over an existing project designed to restore a native prairie. Based on analysis of public documents, the researchers found evidence that both proponents and opponents valued nature, but they were defining it differently: opponents of restoration objected to the clearing of mature trees and felt that nature should be allowed to 'take its course'. However, implications for human well-being were considered equally important with those for ecological well-being. When the researchers asked for evaluations of a hypothetical restoration attempt, respondents were more positive than negative about restoration, and were neutral about both the removal of non-native species and about the desirability of human intervention. The strongest preference was for allowing for public involvement in the decision.

Public attitudes towards species reintroduction have been inconsistent. For the most part, there is a high degree of public support (e.g. Watson et al., 2015), but a high degree of variability among different species as well as among individuals with different characteristics. This reminds us that, rather than applying consistent general principles, people approach specific conservation practices based on values that are relevant to them in a specific situation. Approval of reintroductions is affected by aesthetic liking for the species, as well as a perception of whether one's own interests are implicated. Thus, people who fear predation by wolves, for example, tend to be less supportive of reintroduction efforts (Dressel et al., 2015).

Attitudes towards specific practices are also likely to be affected by knowledge level. A review of research on attitudes towards geoengineering, for example, showed that knowledge levels were fairly low, and that attitudes were affected not only by values and perceptions of the relationship between humans and nature, but also by both knowledge and the way in which the issue was framed (Scheer and Renn, 2014). Attitudes towards urban wetlands, in another study, were responsive to information provided about the value of wetlands, an effect that was mediated by the wildlife value orientation of the respondent (Straka et al., 2016). Similarly, Pauwels (2013) found a wide range of attitudes towards synthetic biology, and concluded that the state of knowledge was one of the most important determinants of attitudes (although knowledge did not necessarily increase support, tending rather to increase perception of risks); in addition, the specific application that was highlighted (flu vaccine, which was viewed positively by a majority, versus accelerated animal growth, which was viewed negatively) had a large impact on support. Attitudes towards rewilding are also likely to be malleable in response to the information that people receive about the topic. Notably, attitudes will also be affected by factors external to the practice, such as trust in the people and organisations who are responsible for designing and implementing it (Vining et al., 2000; Ford et al., 2014).

With specific reference to the need to adapt to climate change, Hagerman and Satterfield (2014) asked experts about attitudes towards a variety of conservation strategies. Although there was a high degree of agreement that climate change might require the adoption of non-traditional policies such as revised guidelines for prioritising conservation investments, the more interventionist practices, such as assisted migration and allowing nonnative species to become established in protected areas, received little agreement. Similar results were found by Watson and colleagues (2015), who found support for management interventions to restore natural conditions (reintroduce native species, remove invasive species), but not for interventions to promote adaptation to climate change (moving plants and animals to new habitats, introducing new genetic material). They also found that people who more highly valued 'naturalness' as an aspect of wilderness were more in favour of the first set of interventions, and less in favour of the second set.

Wildlife management has been described as 'Purposefully influencing interactions among and between people, wildlife, and habitats' (Decker et al., 2009, p. 315). This definition is significant not only in its inclusion of people, but also in that it emphasises interactions as opposed to static states of being. Indeed, an influential (and controversial) paper by Kareiva and Marvier (2012) stated as core postulates that 'the fate of nature and that of people are deeply intertwined' (p. 965) and 'conservation must occur within human-altered landscapes' (p. 966). However, this principle is not universally accepted. Reactions to specific conservation practices are likely to depend in part on the degree to which this interdependence of humans and the rest of nature is accepted and/or prioritised (see e.g. Robbins and Moore, 2013). In addition, as described by Holmes and colleagues (2017), reactions may depend on underlying values that assign anthropocentric or biocentric values to nature.

Values and goals

I have mentioned the relevance of values to attitudes towards wilderness, a topic Nicole Bauer has explored (Chapter 8). Values are preferred end-states or ways of being. They affect attitudes towards land management through a combination with beliefs: a person's policy evaluation stems largely from their beliefs about how that policy will affect the things that he or she values. Thus the attitude has both a cognitive (belief) component and an affective (value) component. Researchers have documented a variety of possible values for nature, and the extent to which people endorse these values will partially determine their attitudes towards wilderness as well as its management (Bauer et al., 2009; Ford et al., 2014). Goals follow directly from values: valuing a particular view of nature, such as its wildness, should lead to a goal of preserving natural areas that are free from human interventions, whereas valuing the economic potential of natural resources would lead to a goal, and a preference, for effectively obtaining those natural resources (Ribe, 2002).

Values for nature include not only the belief that it has intrinsic value – also described as biocentric values – but also valuing it for utilitarian, aesthetic, moralistic, cultural/symbolic, and scientific values, among others. The existence of multiple values for nature, often held by the same individuals, is important for understanding the reactions to different types of nature management, including rewilding. People may object to a practice from a perspective that emphasises nature's existence value or cultural value, and yet find it more compelling when they consider its utilitarian or scientific values. As described above, Watson and colleagues (2015) found that people supported reintroduction of species in order to make a wilderness area more 'natural', but were less supportive of human interventions that were aimed at helping a wilderness area adapt to the impacts of climate change.

As with values towards nature in general, values underlying conservation are multiple. What valued goals are conservation practices, including rewilding, trying to achieve? Utilitarian goals of preserving species that may provide services to humans exist alongside, but sometimes in tension with, biocentric values that emphasise the preservation of natural species for their own sake. Indeed, conservation professionals are capable of endorsing different sets of values at the same time. Sandbrook and colleagues (2010) surveyed young conservation professionals at an international conference and found not only that divergent values were represented, but also that these values were associated with different attitudes towards conservation goals. The most common values were biocentric, emphasising the intrinsic value of nature. However, a preservationist attitude seemed to stress the importance of avoiding extinctions as a primary goal. Interestingly, this perspective put less emphasis on maintaining natural areas that excluded human involvement. Other values that were represented were an emphasis on biodiversity – that is, the number of species maintained was more important than the preservation of species as an end goal – and a more utilitarian focus on the value that biodiversity provides to humans.

Rudd (2011) used a larger sample and different methodology to examine the conservation values and management attitudes among conservation professionals. In an interesting demonstration that multiple values coexist, the two values that had the highest rank as well as the highest level of agreement were 'conservation planning needs to understand how people and nature interact' and 'biological diversity should be conserved because it sustains ecosystem function' (p. 1169). The first of these acknowledges human interdependence with nature, while the second could reflect more biocentric values. But the two statements that received the lowest rankings were ones that either prioritised human needs ('the value of biological diversity depends on its usefulness to people') or overrode them ('long-term residents should be displaced from protected areas if conservation needs warrant').

More recently, Holmes and colleagues (2017) interviewed 30 professionals attending the 2014 International Conference on Conservation Biology in order to develop a detailed description of values and beliefs about the relationship between humans and nature. They found evidence for three different attitude profiles: one that emphasised human involvement with nature, and the associated goal of reducing humans' emotional separation from nature; one that stressed biocentric values and emphasised the conservation of biodiversity and ecosystems as a goal; and one that took a more pragmatic approach, suggesting that economic motivations could be harnessed to achieve conservation goals. Mace (2014) has argued that a shifting understanding or conceptualisation of the relationship between humans and nature is leading to a shift in conservation values, but this shift is still in progress rather than complete.

Conservation professionals

Conservation professionals are at the front lines with regard to landscape management, both in terms of their knowledge about current and potential practices, and in terms of their commitment to conserving biodiversity. However, their values, and their understandings of the relationship between humans and nature, are complex. To add to our knowledge about attitudes towards conservation practices among conservation professionals, I conducted an online survey. The principal research questions were:

- What are the attitudes towards a variety of wildlife management practices, including rewilding?
- Can these attitudes be related to other attitudes, values, and personal characteristics, particularly beliefs about the relationship between humans and nature?

Sample

Using a snowball sample technique, the request to participate was disseminated online in 2014 using professional listservs (the Society for Conservation Biology) and Facebook pages (the Nature Conservancy, Wildlife Conservation Society, Flora and Fauna International, the Zoological Society of London) and encouraging people in the conservation community to share it with colleagues. Participants were targeted as 'conservation scientists and professionals', but there were no screening mechanisms in place and anyone who wanted to complete the survey was allowed to do so. The recruitment message explained that the survey was designed to assess attitudes towards a variety of issues regarding the protection and conservation of nature. Three hundred and thirty-four people completed the online survey. About 60 per cent were female. The largest proportion of respondents (44 per cent) was between 25 and 34 years of age, but the age range was substantial, and close to 6 per cent were over 64. All responses were anonymous. The range of degrees held was also broad: 32 per cent had a doctorate, 41 per cent a master's, 24 per cent a bachelor's degree, and 3 per cent said they did not have a college degree.

Measures

Conservation practices. The survey asked respondents to rate their attitudes towards six conservation practices, on a 1–5 scale. The practices are listed in Table 10.1.

Other variables. The survey also measured environmental identity, perceptions of scientists' responsibility for protecting healthy ecosystems, perceptions of the relationship between humans and nature (as helpful or harmful), choice among various principles for conserving species (i.e. all species should be protected, or they should be prioritised for several different reasons including utility to humans, chance of survival, importance to ecosystem, or use of least resources), endorsement of the intrinsic or utilitarian value of nature, and personal motivations for doing conservation work (environmental conservation, intellectual curiosity, career development, concern about local communities, and concern about the future). Finally, I included a single item assessing a general sense of optimism or pessimism about the future.

Results

As can be seen in Table 10.1, rewilding was rated right below the middle of the scale, which is 'sometimes a good idea'. It was rated more positively than synthetic biology, which is arguably the least 'natural' approach. The highest ratings were given to the practices that intervened only in human markets and not directly with species or ecological processes: selling naming rights and establishing a market for ecosystem services. In general, ratings of all the practices were positively intercorrelated (although the correlation between rewilding attitudes and attitudes towards selling naming rights was not significant). Thus, ratings seemed to reflect an overall attitude towards conservation practices rather than clear preferences for some more than others.

There were no gender differences in attitudes towards rewilding. There was a significant relationship between attitudes and highest degree ($F[3, 317] = 2.7$, $p = .05$), but it was non-linear. Only 10 people reported having less than a bachelor's degree, and they tended to be more negative about rewilding; if those people were removed, however, there was a linear relationship such that more education was associated with more negative attitudes about rewilding ($F[2, 308] = 3.88$, $p = .02$; Figure 10.1). There was also a significant difference associated with discipline: biologists were more

Table 10.1. *Mean ratings of six conservation practices. Brief definitions for each practice were provided; in addition, participants were given a 'don't know/no opinion' option.*

Practice	Mean	SD
Rewilding	2.79	1.00
Assisted migration	2.78	0.92
Synthetic biology	1.94	0.86
Selling the naming rights for newly discovered species	3.09	1.33
Culling individual animals	2.90	0.94
Creating markets for ecosystem services	3.92	0.96

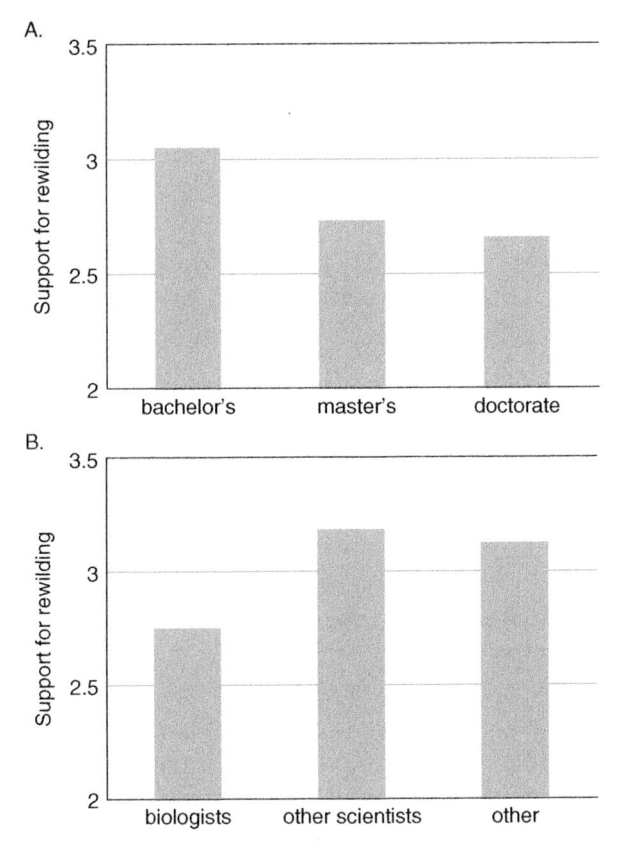

Figure 10.1. Significant differences in attitudes towards rewilding. (A) Differences associated with education; (B) differences associated with discipline.

negative about rewilding than other scientists and non-scientists, who did not differ ($F[2,220] = 3.3$, $p = .04$).

Most respondents saw humans as both helpful and harmful to the conservation of nature ($N = 185$), but they were much more likely to see humans as harmful ($N = 145$) than helpful ($N = 5$). They were most likely to endorse both an intrinsic value for nature and the relevance of human needs ('the intrinsic value of nature is important, but human needs are also relevant', $N = 141$), but were more likely to say nature's intrinsic value was most important ('it's important to protect and conserve nature for its own sake', $N = 117$), than to say human needs were most important ($N = 7$). Surprisingly, there was no relationship between attitudes towards rewilding and perceptions of humans as harmful to nature, perceptions of scientists as responsible for protecting healthy ecosystems, or environmental identity. Similarly, there was no association between attitudes towards rewilding and intrinsic value for nature versus consideration of utilitarian values.

As a principle for prioritising species, only two of the five possible answers received a significant number of responses: 'all species should be protected, without exception', and 'we should prioritize the species that are most important to ecosystem functioning'. People who endorsed the latter principle were more supportive of rewilding ($t[297] = 2.1$, $p = .04$); they were also significantly lower in EID ($t[305] = 2.4$, $p = .02$), suggesting that this perspective reflected a view of humans as more separate from nature.

Finally, attitudes towards rewilding were significantly correlated with only one of the personal motives: concern about the future ($r = .16$, $p = .005$), which was also correlated with EID ($r = .23$, $p < .001$). They were uncorrelated with optimism.

In order to see whether the pattern of associations with attitudes towards rewilding was distinctive, I looked for similar associations with attitudes towards the other interventionist strategies: assisted migration, culling individual animals, and synthetic biology. Intrinsic value for nature was not correlated with attitudes towards any of the practices, nor were these values statements associated with EID. People who endorsed the prioritisation of species were also more supportive of synthetic biology and of culling individual animals. Concern about the future was associated with support for assisted migration ($r = .16$). Support for synthetic biology was associated with intellectual curiosity ($r = .14$) and with concern about one's career development ($r = .20$). There were no differences in attitudes towards these practices associated with discipline or with the amount of education. EID did not correlate with attitudes towards any of the practices, although it was related to perceptions of scientists as responsible ($r = .20$, $p < .001$).

Discussion

In this survey of conservation professionals, attitudes towards rewilding (and towards other practices) were not associated with biocentric values, a sense of environmental identity, or a general perception of the relationship between humans and the natural world. The fact that these core beliefs were not associated with any of the practices described in the survey may indicate that conservation professionals have a somewhat pragmatic basis for their attitudes towards these practices, considering only whether they are effective and practical. (The survey did not ask for those kinds of assessments, so I can only speculate.) The fact that biologists were more negative than other professionals does support the idea that relevant knowledge might be more important than values in determining these attitudes. Similarly, those with a PhD were more negative than those with only a bachelor's degree. Interestingly, these relationships with degree and discipline were not seen in attitudes towards the other practices, perhaps because the other practices are more familiar and accepted within the scientific community.

Attitudes towards rewilding were related to a belief that ecosystem functioning should be prioritised when selecting species to protect, and to concern for the future. The concern for the future suggests a broad tendency to think about the long term, and the prioritisation of ecosystem functioning suggests, again, a more pragmatic approach to conservation. Interestingly, EID was negatively related to the prioritisation of ecosystem functioning, and positively related to concern for the future; this demonstrates, perhaps, that attitudes towards rewilding may not be directly associated with perceptions of the relationship between humans and nature: attitudes towards specific practices are complex, incorporating more than one core idea.

Acknowledging change

Interactions between humans and the natural world, important even in a pre-technological era, have become even more powerful. Human behaviour has affected every aspect of the natural world (Crutzen, 2002) and, because of climate change, could potentially lead to devastating impacts on species and ecosystems. It is widely recognised and acknowledged that people have impaired the ability of landscapes to look after themselves: in the absence of human management, landscapes will not necessarily return to a more 'natural' form. In addition, what is 'natural' will change along with a changing climate. As historian and former park naturalist William Tweed wrote: '"Natural" processes cannot lead reliably to "natural" results in a world where anthropogenic climate change, global pollution, and habitat fragmentation have changed the operating rules, and where society's very definition of *nature* is no longer clear' (2010, p. 189). To put it succinctly: naturalness (well-functioning natural systems and processes) and wildness (freedom from human intervention) may become incompatible (Watson et al., 2015). In other words, humans and nature are, more clearly than ever, interrelated.

What does this mean for rewilding? It takes a difficult and perhaps even paradoxical position on the relationship between humans and nature. On the one hand, it is based on the assumption that human intervention is needed to achieve the optimal environmental conditions. On the other hand, the optimal conditions towards which it strives often seem to maintain a conception of humans and nature as separate (Deary and Warren, 2017): that ecosystems will, once the desired species have been introduced, return to a baseline in which humans were not yet present (as perhaps suggested by the negative relationship between EID and prioritisation of ecosystem functioning in the results presented above). This paradoxical position is behind at least some of the controversy over rewilding as a practice. It also means that attitudes towards rewilding are hard to predict.

Changing patterns of interaction between humans and nature have affected humans as well as the ecosystem (Soga and Gaston, 2016; Clayton et al., 2017).

As Kareiva and Marvier (2012) describe, conservation practices have outcomes for people as well as for biodiversity. One concern that has been repeatedly raised is the possibility that, as humans have less interaction with the natural world, they will find less in it to value. Indeed, research suggests that experiences with nature are a predictor of environmental concern (Zaradic et al., 2009). In a downward spiral, people who are kept apart from nature may consider it to be less important and show less concern about protecting it (Soga and Gaston, 2016). However, people are increasingly losing the types of knowledge that may enable them to feel comfortable in 'wild' landscapes (Clayton et al., 2017). Rewilded landscapes that find ways to symbolically and emotionally incorporate interactions with humans, perhaps while simultaneously restricting physical interactions, are more likely to be politically successful in a context in which wildlife depends on human support and protection. Such interactions, which could include virtual or remote access, merit further exploration and study.

Conclusions

The research suggests several conclusions. One is that many people like the idea of landscapes that are free from human intervention, and some believe that it may encourage humility. A second is that many people endorse the intrinsic value of nature. Those people are more likely to support wilderness, as are people who feel more connected to nature. A third is that the concept of wilderness may be rejected as being too exclusive of human experience. Attitudes towards a practice may be affected by a number of things, including the perceived consequences of that practice, knowledge about that practice, and socially mediated understandings of the relationship between humans and nature. It is possible to give people information that influences these perceptions and attitudes. Although responses to landscapes and landscape practices are conceptually related to values and perceptions of the relationship between humans and the natural world, in a specific case the attitudes may be more strongly affected by more specific knowledge and experiences.

It is important to recognise the variety of methods that have been used to examine people's preferences for landscapes and for conservation practices, and the possible implications for limits on reliability and validity of the results. First, the sample that is studied may be a convenience sample; a nationally representative sample; or a sample of people who use an area in a particular way (e.g. forest owners; wilderness visitors). None of these samples is worthless, as even a convenience sample can provide valuable information about the relationship between personality variables and preferences; but none is perfect, because as described above, attitudes are likely to be affected by a range of factors including national culture and one's relationship to a geographic area.

Almost all of the samples are restricted to people from technologically developed cultures. Second, preferences may be elicited in response to photographs and/or to written descriptions of a landscape management practice. These methods may give both too little information – lacking the sensory detail that comes from actually interacting with an environment – and too much information – explicitly describing practices that a person might fail to recognise, or overlook, when actually perceiving a landscape. Finally, the information gained is limited to the questions that were asked. Knowing that people find a landscape attractive, for example, is not the same as knowing that they will support paying to maintain it; knowing that a particular management practice is disliked does not mean that it would be considered unacceptable. There is typically a great gulf between the findings of any single study and the implications for the practices in a particular landscape.

Rewilding initiatives exist within an ecological context in which there is great concern about declining levels of biodiversity, as well as a social context in which human experiences of nature are diminished. Mace (2014) traced the evolution of conservation from an emphasis on 'nature for itself', with a focus on wilderness and the preservation of natural spaces without humans, to what she called the 'people and nature' approach. Such an approach has to include the role of cultural institutions in influencing – and developing – sustainable interactions between humans and the natural world. (See also the results of the studies by Sandbrook et al., 2010.) Rewilding practices, as examples of these institutions, should consider the ways in which humans continue to interact with natural systems and not just the ways in which humans intervene to create natural systems.

Because landscapes and landscape management have symbolic meaning, rewilding and other practices could have unintended consequences by affecting the ways in which people respond to the landscape. 'Wild' landscapes have been described as encouraging humility, whereas rewilding might encourage hubris. In an example of unintended consequences, the practice of culling individual wolves has been linked to a drop in the wolf population, which researchers believe to result from increased poaching (Chapron and Treves, 2016). They argue that culling or delisting the wolves is perceived by potential hunters as an indication that illegal hunting will be tolerated. Similarly, successful rewilding initiatives could increase people's confidence in human ability to control and positively affect natural landscapes, and thus decrease caution about protecting existing species and habitats.

Recognition of human involvement in nature entails obligations as well as privileges for humans. For example, Salmon (2017) has called for us to think about our own place in landscapes, how we are connected to ecosystems and what we owe to them. Robbins and Moore (2013), describing rewilding as 'experimental conservation theatre' (p. 14), call for it to be recognised as something that occurs within a sociopolitical context as well as an ecological one. Accepting the fact that human and ecological systems are fundamentally intermingled, rather than trying to separate human spheres of activity from ecosystems, or human priorities from biocentric ones, may be the most honest and ultimately the most practical way to protect biodiversity in a human-dominated world.

References

Bauer, N. (2016). Social values of wilderness in Europe. In Bastmeijer, K. (Ed.), *Wilderness protection in Europe* (pp. 94–113). Cambridge: Cambridge University Press.

Bauer, N., Wallner, A., and Hunziker, M. (2009). The change of European landscapes: human-nature relationships, public attitudes towards rewilding, and the implications for landscape management in Switzerland. *Journal of Environmental Management*, **90**, 2910–2920.

Bekoff, M. (2014). *Rewilding our hearts*. Novato, CA: New World Library.

Bixler, R., and Floyd, M. (1997). Nature is scary, disgusting, and uncomfortable. *Environment and Behavior*, **5**, 202–247.

Buijs, A.E., Elands, B.H., and Langers, F. (2009). No wilderness for immigrants: cultural differences in images of nature and landscape preferences. *Landscape and Urban Planning*, **91**, 113–123.

Chapron, G., and Treves, A. (2016). Blood does not buy goodwill: allowing culling increases poaching of a large carnivore. *Proceedings of the Royal Society of London B: Biological Sciences*, **283**, doi:10.1098/rspb.2015.2939

Clayton, S. (2003). Environmental identity: a conceptual and an operational definition. In Clayton, S. and Opotow, S. (Eds.), *Identity and the natural environment* (pp. 45–65). Cambridge, MA: MIT Press.

Clayton, S. (Ed.) (2012). Environment and identity. In *Oxford Handbook of Environmental and Conservation Psychology* (pp. 164–180). New York, NY: Oxford.

Clayton, S., Colléony, A., Conversy, P., et al. (2017). Transformation of experience: toward a new relationship with nature. *Conservation Letters*, **10**, 645–651.

Cronon, W. (1996). The trouble with wilderness: or, getting back to the wrong nature. *Environmental History*, **1**, 7–28.

Crutzen, P.J. (2002). Geology of mankind. *Nature*, **415**, 23–23.

Deary, H., and Warren, C.R. (2017). Divergent visions of wildness and naturalness in a storied landscape: practices and discourses of rewilding in Scotland's wild places. *Journal of Rural Studies*, **54**, 211–222.

Decker, D., Siemer, W., Leong, K., Riley, S., Rudolph, B., and Carpenter, L. (2009). Conclusion: what is wildlife management? In Manfredo, M., Vaske, J., Brown, P., Decker, D., and Duke, E. (Eds.), *Wildlife and society: the science of human dimensions* (pp. 315–328). Washington, DC: Island Press.

Dowie, M. (2009). *Conservation refugees: the hundred-year conflict between global conservation and native peoples*. Cambridge, MA: MIT Press.

Dressel, S., Sandström, C., and Ericsson, G. (2015). A meta-analysis of studies on attitudes toward bears and wolves across Europe 1976–2012. *Conservation Biology*, **29**, 565–574.

Ford, R.M., Williams, K.J., Smith, E.L., and Bishop, I.D. (2014). Beauty, belief, and trust: toward a model of psychological processes in public acceptance of forest management. *Environment and Behavior*, **46**, 476–506.

Hagerman, S.M., and Satterfield, T. (2014). Agreed but not preferred: expert views on taboo options for biodiversity conservation, given climate change.

Ecological Applications, **24**, 548-559. doi:10.1890/13-0400.1

Hasbach, P. (2014). Rewilding for human flourishing. *The Counselor* (Newsletter of the Oregon Counseling Association), 2-4.

Hausdoerffer, J. (2017). Epilogue: a conversation with Roderick Nash. In Van Horn, G. and Hausdoerffer, J. (Eds.), *Wildness: relations of people and place* (pp. 243-254). Chicago, IL: University of Chicago Press.

Holmes, G., Sandbrook, C., and Fisher, J. (2017). Understanding conservationists' perspective on the new-conservation debate. *Conservation Biology*, **31**, 353-363.

Kareiva, P., and Marvier, M. (2012). What is conservation science? *BioScience*, **62**, 962-969.

Khew, J.Y.T., Yokohari, M., and Tanaka, T. (2014). Public perceptions of nature and landscape preference in Singapore. *Human Ecology*, **42**, 979-988.

Koole, S L., and Van den Berg, A.E. (2005). Lost in the wilderness: terror management, action orientation, and nature evaluation. *Journal of Personality and Social Psychology*, **88**, 1014.

Mace, G. (2014). Whose conservation? *Science*, **345**, 1558-1560.

McMahan, E.A., Cloud, J.M., Josh, P., and Scott, M. (2016). Nature with a human touch: human-induced alteration negatively impacts perceived naturalness and preferences for natural environments. *Ecopsychology*, **8**, 54-63.

Milton, K. (2002). *Loving nature*. New York, NY: Routledge.

Monbiot, G. (2013). *Feral: searching for enchantment on the frontiers of rewilding*. London: Penguin Books.

Pauwels, E. (2013). Public understanding of synthetic biology. *BioScience*, **63**, 79-89.

Prévot, A.-C., Clayton, S., and Mathevet, R. (2018). The relationship of childhood upbringing and university degree program to environmental identity: experience in nature matters. *Environmental Education Research*, **24**, 263-279.

Ribe, R.G. (2002). Is scenic beauty a proxy for acceptable management? The influence of environmental attitudes on landscape perceptions. *Environment and Behavior*, **34**, 757-780. doi:10.1177/001391602237245

Robbins, P., and Moore, S. (2013). Ecological anxiety disorder: diagnosing the politics of the Anthropocene. *Cultural Geographies*, **20**, 3-19.

Rozin, P., Fischler, C., and Shields-Argelès, C. (2012). European and American perspectives on the meaning of natural. *Appetite*, **59**, 448-455.

Rudd, M. (2011). Scientists' opinions on the global status and management of biological diversity. *Conservation Biology*, **25**, 1165-1175.

Sabloff, A. (2001). *Reordering the natural world*. Toronto: University of Toronto Press.

Salmon, E. (2017). No word. In Van Horn, G. and Hausdoerffer, J. (Eds.), *Wildness: relations of people and place* (pp. 24-32). Chicago, IL: University of Chicago Press.

Sandbrook, C., Scales, I., Vira, B., and Adams, W. (2010). Value plurality among conservation professionals. *Conservation Biology*, **25**, 285-294.

Scheer, D., and Renn, O. (2014). Public perceptions of geoengineering and its consequences for public debate. *Climatic Change*, **125**, 305-314. https://doi.org/10.1007/s10584-014-1177-1

Soga, M., and Gaston, K.J. (2016). Extinction of experience: the loss of human–nature interactions. *Frontiers in Ecology and the Environment*, **14**, 94-101.

Soga, M., Gaston, K.J., Koyanagi, T.F., Kurisu, K., and Hanaki, K. (2016). Urban residents' perceptions of neighbourhood nature: does the extinction of experience matter? *Biological Conservation*, **203**, 143-150.

Straka, T.M., Kendal, D., and van der Ree, R. (2016). When ecological information meets high wildlife value orientations: influencing preferences of nearby residents for urban wetlands. *Human Dimensions of Wildlife*, **21**, 538-554.

Tang, I., Sullivan, W.C., and Chang, C. (2015). Perceptual evaluation of natural landscapes: the role of the individual connection to nature. *Environment and Behavior*, **47**, 595–617. doi:10.1177/0013916513520604

Tweed, W. (2010). *Uncertain path*. Oakland, CA: University of California Press.

Vining, J., Tyler, E., and Kweon, B.S. (2000). Public values, opinions, and emotions in restoration controversies. In Gobster, P. and Hull, R.B. (Eds.), *Restoring nature: perspectives from the social sciences and humanities* (pp. 143–162). Washington, DC: Island Press.

Watson, A., Martin, S., Christensen, N., Fauth, G., and Williams, D. (2015). The relationship between perceptions of wilderness character and attitudes toward management intervention to adapt biophysical resources to a changing climate and nature restoration at Sequoia and Kings Canyon National Parks. *Environmental Management*, **56**, 653–663.

Williams, K.J.H., and Cary, J. (2002). Landscape preferences, ecological quality, and biodiversity protection. *Environment and Behavior*, **34**, 257–274.

Zaradic, P.A., Pergams, O.R., and Kareiva, P. (2009). The impact of nature experience on willingness to support conservation. *PLoS ONE*, **4**(10), e7367.

CHAPTER ELEVEN

The high art of rewilding: lessons from curating Earth art

MARCUS HALL
University of Zurich

Creating high art

'Spiral Jetty' is an enigmatic ribbon of rocks stretching out from a desolate shore of the Great Salt Lake, flanked by a sagebrush desert, and accessible only by a bumpy journey on a long dirt road. The jetty itself is wide enough to walk on, leading out 300 metres to curve upon itself, and curve again, repeating this form to finally reach a dead end. As high art constructed in 1970 by a New York artist who became enthralled with the site and leased it from the state, the jetty is the fruit of a bulldozer, dump truck, and loader working in tandem to heap rocks and sand into a new installment of what was called *Earth art*, a rising artistic expression of the day. Acknowledging the value of the jetty, the State of Utah recently designated it an 'Official State Work of Art' and has set it aside to be protected for generations.

However, Spiral Jetty is also disintegrating, sinking into the mud, losing its distinct forms as it disappears from view. Some art appreciators are calling for its restoration. Others feel that it should be left to melt back into the landscape, letting nature do what it will do to human ambition. The artist who constructed the jetty, Robert Smithson, celebrated entropy, that tendency of materials to spontaneously fall apart, so presumably his jetty should be allowed to weather the elements on its own, subject to shifting sands and fluctuating water levels. Still, curation is needed: *curating* the monument requires measures of preservation and restoration, as well as an appreciation for wild nature. Decisions must be taken about the degree and extent to which this work should be protected. Spiral Jetty is no Machine in the Garden, but Art in the Wilderness (Figure 11.1).

In what follows, I argue that rewilding is very much like curating. Instilling greater wildness in a place or object or organism means that we make decisions about which kinds of the wild we admire and promote. If we admire a former wild, then we set out to bring some of it back; if we admire an intrinsic wild, then we do our best to unbind and unleash it. Although wildness is that state of nature beyond human control, once humans enter the scene, they cannot help

Figure 11.1. Robert Smithson's Spiral Jetty on the Great Salt Lake, USA, 2005. Wikimedia Commons. (A black and white version of this figure will appear in some formats. For the colour version, please refer to the plate section.)

but modify it, if only by changing their perceptions of it. When wild nature is no longer out of sight and out of mind, human entry into it rewilds it for one's own purposes and needs. Curators do their best to exhibit an artist's intent, so that curating wild nature is the project of promoting and displaying our vision of optimal wildness, which is the goal of rewilding. Although rewilders may not see themselves as nature's curators, I believe that curating is the pursuit they have set out for themselves. Below I outline key differences between preserving, restoring, and rewilding, before suggesting how the notion of curating may encompass them all. In the next pages, I invite rewilders to consider for a moment how, if they were put in charge of Spiral Jetty, they would choose to care for this disintegrating work of Earth art. Artists and rewilders need to better appreciate each other's work, for there is much logic and analysis in producing art, just as there is much art and intuition in rewilding.

Interpreting the jetty

As if to commemorate Spiral Jetty's newly protected status, the *New York Times Magazine* published a full feature article about one family's pilgrimage to visit the jetty and record their thoughts about its creation and preservation. The article by Heidi Julavits (2017), 'The Art at the End of the World', is deeply celebratory, invoking the mysteries of the American Desert, the circuitous journey required to arrive there, and the feelings produced after finally viewing Spiral Jetty in salt and rock. Following an afternoon at the site, exploring its curving shapes, trudging along the muddy shore, and taking in the longer view from the hill, Julavits wrote that she had been 'thinking a lot about interior landscapes, those uninhabited places inside of us that cannot be contained (or explained) by any map. Interior landscapes are shaped by all kinds of forces: geographic or familial or cultural or genetic.' Although Smithson's jetty is a human creation, it clearly invokes more primal feelings, ones that the author cherished even as she wondered about the future of the jetty. How will rising numbers of visitors impair the site, at the same time that it sinks further in the mud? What would Smithson, who died a few years after designing the jetty, have wanted to happen to his artwork, Julavits asked. 'Would he want it restored? Would his championing of entropic thinking deem the opposite?' Proper curation, she understood, must endeavour to display the intent of the artist and the spirit of the artwork, even as its walls crumble and its outlines disappear, with the larger jetty returning by degrees into the wild (Julavits, 2017).

The *Times* article proved popular, at least judging by the number of readers who chose to write in commentaries and express their feelings about this essay and the art object it describes. In fact, Spiral Jetty shows itself to be a controversial subject, one that fascinates as well as exasperates depending on interpreter. For many, the article provided a pause for reflection about a man with a vision on an arid landscape who directed heavy-equipment operators to haul 6650 tons of rock out into a vast salty lake, the largest in the Western hemisphere, for the purpose of placing designs on the Earth. But where some see art, others see hubris. Still others simply acknowledge this curiosity that, because it draws tourists, deserves our attention and care. For many, the jetty is becoming a kind of natural monument or even sacred site, like Ayers Rock or Devils Tower, except this site was created by the human hand on top of natural forces. Much like the various feelings and emotions elicited by a wild landscape that vary according to observer, Spiral Jetty incubates new meanings of nature and the human place in it.

Indeed, we might go further, and seek to classify the range of online comments received about this description of the jetty, as such comments offer insights about how and why the jetty should be curated. It is a fascinating exercise to read through these readers' comments, taking them one at a time, for they provide a window into the world of art appreciation, while revealing snippets about human relationships with the natural world. Below I categorise some 300 comments into five general categories while reporting representative quotations. It seems that everyone has an opinion about how to curate Smithson's jetty. Commenters variously felt:

(1) That Spiral Jetty is a true masterpiece: 'Since I may not get to Utah to see this in person, I am more than grateful for this so sensitively written essay which enabled me, like the crows, to feel a certain sameness with the world, to silently sing along the same song.'

(2) That coming from a big city made it more impressive. Said a reader from London: 'Spiral Jetty is hands down the greatest work of art ever conceived and created by the hands of man.' Said a reader from Alaska: 'I have read and reread this article, trying to see what I may have missed. This family would benefit from getting out of the city from time to time.'

(3) That high art needs more grounding: 'And then there's the fact that Spiral Jetty just isn't – that – isolated. Look it up on Google Maps and you'll see a couple cars in the parking lot and another a few meters down the road. It's clearly a tourist attraction, if one that's relatively hard to access.'

(4) That the jetty elicits new interpretations: 'How about this: Spiral Jetty is a metaphor for American mental landscape ... The jetty stem heads and points across a body of water to a distant shore – the journey across the ocean to reach "America", an extroverted achievement subject to exogenous conditions. On the jetty, the traveler never reaches that distant shore but, following the line, is turned inward.'

(5) That the jetty evokes our day's widening cultural schism: 'Wow. Just reading the comments here, I have a new understanding of the corrosive impact of social media on creativity and the courage it takes to put your thoughts and ideas out there. I loved the article. I read it to the end, in a context in which, so often, time is too short for that. As with any piece, some aspects spoke to me more than others ... But does that mean I must berate the author? Belittle her? Trash her present and her past? Make assumptions about her relationships? Her motives? ... Must the tall poppies always be cut down? To the author: it's a beautiful piece.'

There is one more point about Spiral Jetty that needs mentioning: this artwork is intimately linked to climate change and human water use. Originally constructed as a design of rocks projecting a half-metre above the lake level, the jetty went on to be submerged or re-exposed according to weather trends and water consumption. Because the Great Salt Lake has no outlet, like an enormous Dead Sea, it rises on wet years to inundate the jetty. And then on dry, hot years, the lake's level falls to reveal Smithson's artwork. Lake levels drop further still when irrigation and water consumption increase in the upstream metropolis. After sinking from view just a few years after its construction, and staying hidden for decades, the jetty definitively reappeared in 2002, and continues to march higher today as the shoreline recedes further. In a real sense, Spiral Jetty is a barometer of humanity's global folly. An atmosphere overloading with carbon coupled with upstream city growth is measured by a rising jetty surrounded by a few drying puddles. The historic record of the lake's water levels demonstrates that our century is becoming hotter, drier, and more consumptive than ever before (Figures 11.2 and 11.3).

THE HIGH ART OF REWILDING   205

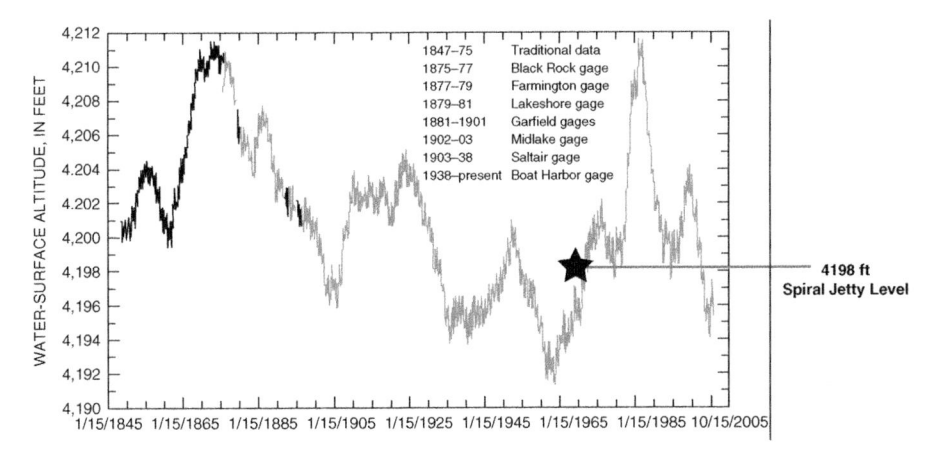

Figure 11.2. Historic levels of Great Salt Lake (1845–2005), showing level of Spiral Jetty. US Geological Survey data. https://ut.water.usgs.gov/greatsaltlake/elevations/. (A black and white version of this figure will appear in some formats. For the colour version, please refer to the plate section.)

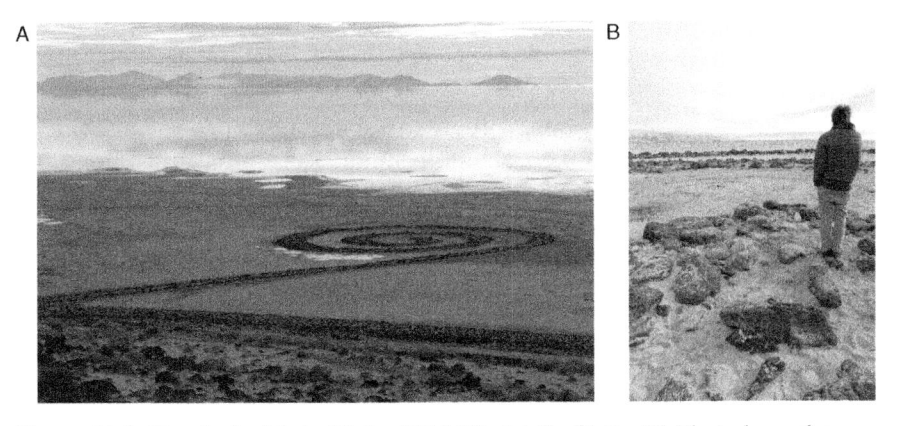

Figure 11.3. Dry-docked Spiral Jetty, 2018 (A); detail of jetty (B). Photo by author.

Given these many interpretations of Spiral Jetty, the project of curating it requires that we plan our activities and make decisions, not only about whether we manage the materials that compose the jetty, as by tidying up the rock walls and dredging the mud beside them, but also about whether we manage the surrounding water levels, as by searching for larger cures that address a shrinking lake. Robert Smithson's own curatorial advice would suggest that his creation be left to its own devices, winding down on its own. However, as the jetty becomes more evident, in our mind and on the land, rising above the receding waters, then Smithson's louder message is that we must be cautious. A dry-docked Spiral Jetty is a warning that we need to set out on a different path, because curating the jetty is also curating the Earth.

Clearly, Spiral Jetty requires more than simple restoration, as one must consider the intent of the artist, the experience of the viewer, and the spontaneous effects of time on the monument. Like a rewilder who considers the best options for managing a site, a curator's strategies for taking care of this monument go to the core of justifying human activities that are passive or active, hidden or overt, analytical or intuitive. What are the differences between restoring art and restoring nature? Should one seek to restore the original or an updated version of the original? More generally, what are the differences between restoring and rewilding, and does rewilding represent a wholly new form of nature management? The following sections take up these questions in turn.

Restoring art versus restoring nature

There is an old debate in the restoration profession about whether ecosystem restorers can be compared to art restorers. The 'Sistine Chapel' debate served to highlight key differences between restoring a static artwork created all at once by a single master, and restoring a dynamic, constantly evolving ecosystem populated by myriad species and processes. It was generally agreed that restoring artworks and ecosystems both present enormous challenges, with decisions needing to be taken about techniques, timing, and intent. Yet because artworks are human creations, restoring them seemed to represent a very different challenge than that of restoring natural creations. Some argued that attempting to restore nature, unlike restoring art, is conceptually flawed because untouched nature can never arise from retouching it (Katz, 1992); others felt that restoring nature is akin to producing a forgery (Elliot, 1982). Still others pointed out that natural works include more moving parts than artworks, and also embody deeper pasts (Hall, 2005). There was also the suggestion that nature runs by its own processes and laws apart from the human world, whereas art is centrally a human creation (Losin, 1988).

However, extending the Sistine Chapel debate to include Earth art complicates the distinction between restoring art and restoring nature, as there is more overlap between these two pursuits if the artwork is composed of natural materials that are left to change through natural processes. Because Smithson's Spiral Jetty is constructed out of rock and dirt, the project of restoring it is more like the pursuit of an ecological restorer who replaces soils and situates them in appropriate positions. Just as an ecologist may use nearby undisturbed reference sites as models to follow in recreating more natural conditions, so too will a curator use references found in the Earth artist's notes and conversations as guides to renewing this art. Restorers of both projects also share the goal of reinstating 'moving targets', as part of Spiral Jetty's fascination rests on how it ages with time. In fact, the jetty is very much a living work, termed *process art*, which varies not only according to lake levels, but also according to wind and wave directions, angles of the sun, salt concentrations, algae varieties that produce different hues, and itinerant bird and insect species that light on its rocks encrusting with microorganisms. Every day and every weather pattern produce a new jetty experience, much as a natural site grows and evolves by month and year. Thus, unlike a painting or a statue sheltered in a museum, Spiral Jetty is exposed to the elements, reflecting different colours and eliciting different moods depending on the time it is visited. Composed initially of inert elements, the jetty now forms habitat for living organisms as it cycles through the seasons. Like good ecological restoration, faithful Earth art curation seeks to incorporate nature's own contributions. Nature's organic creativity is celebrated and displayed in both kinds of restoration projects.

Given Smithson's admiration for entropy in fashioning his art, the challenge of restoring it requires special handling, because one can argue that his works should remain free of any intervention. Unlike a programme of *active* management by the human hand, Spiral Jetty might be left to *passive* management whereby nature's spontaneous forces are given free rein to modify the jetty. As art critic Wayne Anderson declares after considering a range of possible interventions to the jetty, 'We will just have to be patient and make no rash moves.' In this cautious approach, strategies of curating Earth art are not very far from those of restoring nature, especially if the human hand can be deemed a natural element. As Anderson explains, 'if we humans are embedded in nature, as Smithson believed ... then human intervention would be an aspect of the ever changing natural process that Spiral Jetty is undergoing'. It turns out that Smithson, himself, called at various times for active management of the jetty, reportedly noting that should water levels rise, 'he would place more rocks on it to build it up so it would remain out of the water more of the time' (Anderson, 2004). Perhaps the best management plan for this famous example of Earth art would be *guided entropy*, whereby humans lightly intervene alongside spontaneous processes.

So just as an ecological restorer hopes to remove undesirable elements from a site, such as alien invasive species or contaminated soils, Smithson's curators may likewise ensure the long-term survival of Spiral Jetty, as by adding or removing rocks or modifying some other aspect of the site. There is talk of removing the thick sediments that have deposited between the jetty's rings, for example, and some visitors are already realigning stones at the site (Eppich et al., 2011; Julavits, 2017). As another enthusiast explained after visiting the jetty, 'somehow we can't stop wanting to pause time's effects' (Chianese, 2013). Just as Smithson's curators cannot help but decide which elements need adding or which conditions represent acceptable entropic states, ecological restorers cannot help but insert their own human values into a site, as by favouring this or that species or this or that former condition. Because the human hand is directed by a subjective mind, curators like restorers cannot help but place human values onto their objects that go beyond what initially existed.

Restoring versus rewilding

In distinguishing between restoring and rewilding, one might suspect that rewilding is largely an upgraded version of restoration, or Restoration v2.0. However, rewilding is more than a modernised version of restoring: rewilding promotes very different conceptual assumptions than its noble cousin even when updated. The main differences between these two environmental healing practices centre on historical sensitivity and target conditions: restorers, unlike rewilders, are traditionally bound by an objective past. Bringing back faithful or authentic former states, after all, has long been the highest goal of restorers who see themselves 'assisting the recovery of an ecosystem that has been degraded, damaged or destroyed' (SER, 2004). As a result, *historical fidelity* has been much trumpeted by those who help such ecosystems to recover; before restorers ever act on their chosen site or object, their first challenge has been to describe as accurately as possible the conditions that once existed because this former state is considered less damaged than its modern-day counterpart. Archival records, old maps, former species lists, early oral descriptions, historic photographs, and even scientific proxy data, such as ancient pollen records taken from sediment cores, all provide hard evidence that can be used to objectively describe a former state that is the restoration goal. The supposed existence of a better past has spawned enormous interest in historical ecology, the science that aims to describe earlier natural states and the changes they underwent to arrive at the present (Egan and Howell, 2001; Pzabó, 2015; Beller et al., 2017).

Yet beyond the enormous difficulty in identifying relevant historical records that might be used in sketching the goal of a restoration project, there is also the challenge of accurately interpreting such records. An old map, even a detailed one, might show river courses and basic vegetation types, for example, but it can never show all former natural elements or species present, nor can an early photograph, with inevitable missing information being filled in by studying reference sites, extrapolating natural processes, or else relying on intuition, hunch, or other more subjective methods. As any historian knows, the project of describing the past produces a different story depending on the values and biases of the historians telling the story, even when they are presented with identical archival sources. Moreover, the same historian may recount different stories from one occasion to the next, as the pursuit of historical truth has generally disappeared as a goal of the historical profession (Novick, 1988). Restorers, as applied historians, find themselves confronting the impossibility of producing error-free descriptions of former places, objects, or organisms (Alagona et al., 2012; Hall, 2014b). Owing to the past's elusive objectivity, restorers are now promoting sensitivities to nature's many pasts, and the fact that natural states undergo many kinds of historical changes (Marris, 2011; Higgs et al., 2014).

There is the additional challenge of agreeing to the precise date of an optimal past. 'Snapshots' of former systems therefore become especially suspect, since such images represent only a single blink in a range of acceptable pasts. For Americans, 1492 has always been a special date, after which their ecosystems became increasingly connected to, and modified by, those of the Old World as globalisation progressed. Yet other American restorers counter that 13,000 years ago is a better goal, as humans with all their ecosystem-changing activities had not yet ventured across the Bering Land Bridge to populate the Americas. Similarly for New Zealanders who rely on such criteria, the eighteenth century is considered a key moment after which their islands became connected to the rest of the world. But so too could an eleventh-century New Zealand ecosystem be judged as the crucial target because that was the time when these islands first saw humans arriving from Polynesia. Europeans for their part may point to the agricultural or else industrial revolutions as being crucial ecological turning points, and so form the basis of the goals of restoration, as natural states before these times were presumably less corrupted by the human hand. Yet even if sensitive to a site's many former snapshots, restorers such as Balaguer and colleagues insist that 'the restoration practitioner guides or steers the restoration process towards an historically-grounded configuration' (Balaguer et al., 2014).

With greater awareness of the ambiguity presented by restoration's many conceivable historic end-points, a few ecologically minded propose that one should reinstate former 'moving targets' rather than former static states (Clewell and Aronson, 2013). Yet by acknowledging that ecosystems are dynamic interactions of living and non-living components that continually undergo structural and compositional changes involving evolutionary and successional processes, then restorers' goals become even more difficult to define. The project of recreating moving targets, even more than that of recreating static states, promotes still greater reliance on subjectivity because more information is needed to describe accurate dynamic conditions. In the time that follows a preferred moving target, the fact that climates kept changing, species kept invading, and humans kept disrupting, means that there is little possibility to return to that target, however defined, because these former dynamic conditions could not exist under the constraints of today's altered world. Richard Hobbs and co-authors (2011) provided the signal call that Restoration v1.0 had ended by advocating a new restorative pursuit, calling it *intervention ecology*. 'Restoration,' they declared, 'evokes for the layperson the increasingly untenable notion that an ecosystem can be returned to some previous state.' David Lowenthal (2015) waxes more eloquently that 'In nature restoration, human agency is the passion that dare not speak its name, human mastery omnipresent but unseen.'

An important advance for recreating moving targets relies on the activities of animals to do the restoring, especially animals similar to ones that have gone extinct. In this scenario, humans no longer carry out the day-to-day activities themselves for restoring ecosystems – whether clipping, weeding, planting, digging – but instead rely on the restorative actions of an extinct organism's modern-day analogue or 'proxy' – such as grazing or predation. Although environmental *restoration* has long been characterised as an active, hands-on process that distinguishes it from spontaneous, hands-off environmental *regeneration* (Bradshaw, 1996), restoring with analogue species further removes the human hand from the restorative process. Thus, even if Josh Donlan and colleagues label their project as 'rewilding', their objective continues to be one of restoring to an ideal past, which in their North American case occurred during the Pleistocene. As they explain, 'Pleistocene history and taxon substitutions can provide us with new benchmarks for restoration' (Donlan et al., 2006).

One realises that as conceived, *Pleistocene rewilding* is still a form of restoration, albeit to a deeper historic baseline, and one utilising activities and functions of analogue animals instead of people (see Chapter 4). Yet Pleistocene rewilders' proposed introduction of analogue species, such as elephants and lions – which are similar but not identical to extinct counterparts – can never reproduce precise ecological activities carried out by the Pleistocene's actual megafauna (Corlett, 2016). Rewilding with the Pleistocene in mind is still an act of restoration, one that may get closer to reproducing former ecological conditions, but one that can never replicate those historic conditions, even in the long run. Time's arrow moves ever forward, altering objects and processes that feed back to alter these objects and processes. It seems that until one can invent a time machine to travel back over the decades or millennia to re-establish endangered flora and fauna and then track these organisms and their activities up to the present, then ecological managers must describe their work as *intervention* more than *restoration*.

Rewildness versus rewilderness

At the most basic level, rewilders can be distinguished from restorers by the fact that their ultimate goal is favouring wild conditions rather than past conditions. Neither real nor mythic pasts are crucial for rewilders, because

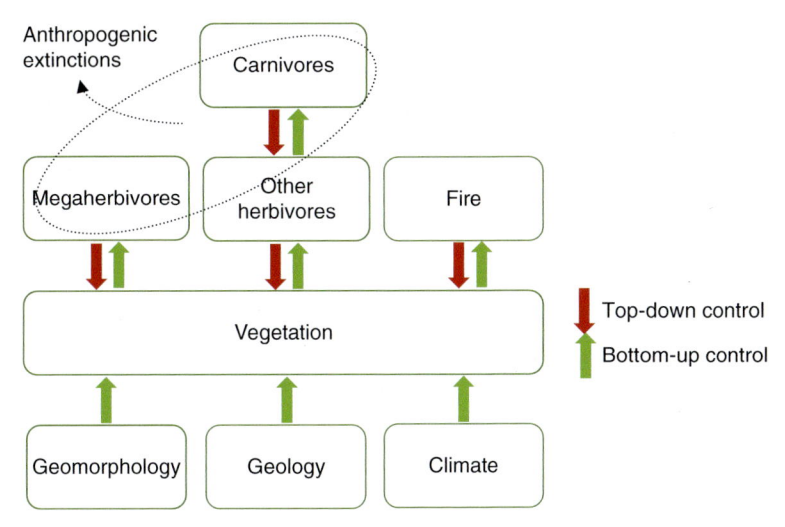

Figure 4.1. A simplified diagram of the main components of a terrestrial ecosystem showing where top-down and bottom-up controls operate, and where anthropogenic extinctions in the late Pleistocene (and ongoing) have removed a major component of the control system. Note that Pleistocene megacarnivores might have regulated megaherbivore populations through predation on juveniles (Van Valkenburgh et al., 2016), but humans are the only predators known to influence the dynamics of megaherbivore populations. (A black and white version of this figure will appear in some formats.)

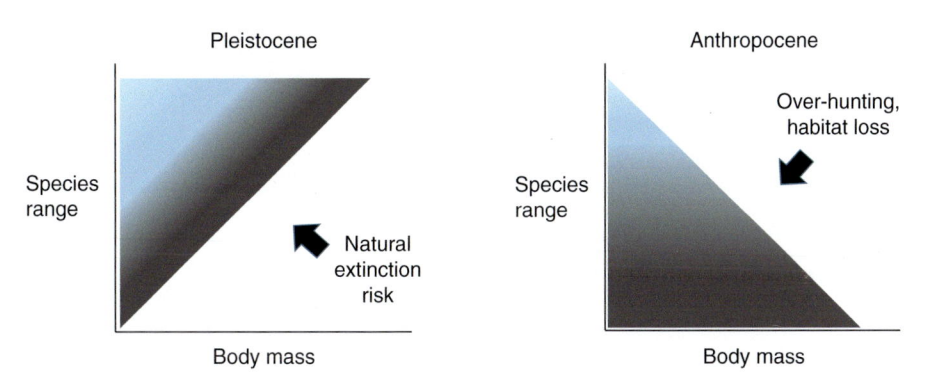

Figure 4.2. Conceptual representations of the relationship between species geographic range and body mass (logarithmic) for a continental assemblage of terrestrial mammals in Pleistocene and Anthropocene epochs. (A black and white version of this figure will appear in some formats.)

REAL-WORLD TROPHIC REWILDING PROJECTS

Figure 5.1. Real-world trophic rewilding projects. (A,B) Yellowstone National Park (USA) – large-scale (~9000 km^2) project with wolf (*Canis lupus*) reintroduction, with large herbivore populations and strong vegetation impacts continuing, but perhaps modulated by restored predation (2011). (C) Experimental beaver (*Castor fiber*) reintroduction in Scotland, with strong vegetation impacts (Knapdale, ~44 km^2, 2013). (D) Free-ranging beaver (*C. fiber*) reintroduction in western Jutland (Denmark), also with strong vegetation impacts (2010). Small to moderately sized fenced projects in densely occupied European regions, with semi-wild horses (*Equus ferus*) and cattle (*Bos primigenius*), and sometimes also other species, in (E,F) Mols Bjerge (120 ha, Denmark, 2017), (G) Knepp Estate (1400 ha, England, 2017), (H) Kasted Mose (62 ha, Denmark, 2017), (I,J) Sydlangeland (120+ 25 ha, Denmark, 2017), and (K,L) Gelderse Poort (700 ha, Netherlands, 2016). (M) Large-scale project with introduction of tapirs (*Tapirus terrestris*) and other historically extirpated species in a subtropical savanna setting (Iberá, ~1500 km^2, Argentina), with emerging restoration of seed dispersal and herbivory (2017). (N) Fenced project with Eurasian elk (moose, *Alces alces*), extirpated since mid-Holocene, to restore browsing in raised bog ecosystem (Lille Vildmose, 2100 ha, Denmark, 2017). (O) Use of functional analogue (*Bubalus bubalis*) of Pleistocene extinct water buffalo (*B. murrensis*) in small fenced project (Kasted Mose, 62 ha, Denmark, 2017). Photos: J.-C. Svenning. (A black and white version of this figure will appear in some formats.)

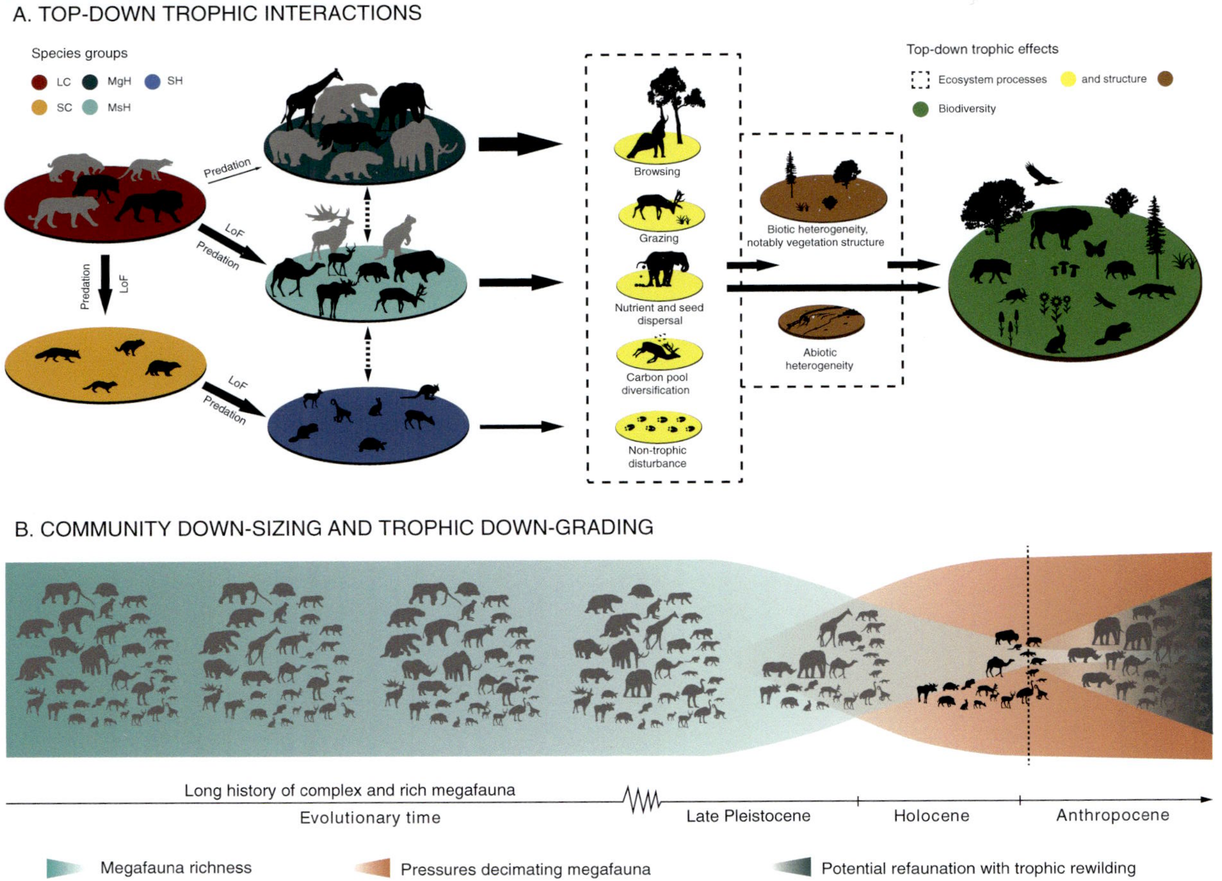

A. TOP-DOWN TROPHIC INTERACTIONS

B. COMMUNITY DOWN-SIZING AND TROPHIC DOWN-GRADING

Figure 5.2. (A) Outline of how trophic rewilding may promote biodiversity via effects on ecosystem processes and structure. Species groups: large carnivores (LC), small carnivores (SC), megaherbivores (MgH), mesoherbivores (MsH), and small herbivores (SH), with representatives of extinct (grey) and extant (black) species are shown with their top-down trophic interactions. LoF, landscape of fear. (B) Species-rich and complex megafaunas have been the norm of the evolutionary timescales (transparent figurines), where current (dotted line and black figurines) species diversity evolved, but have undergone strong decimation during the last 50,000 years, with potential recovery with trophic rewilding as one possible Anthropocene scenario and continuing pressures and losses as another. (A black and white version of this figure will appear in some formats.)

A.

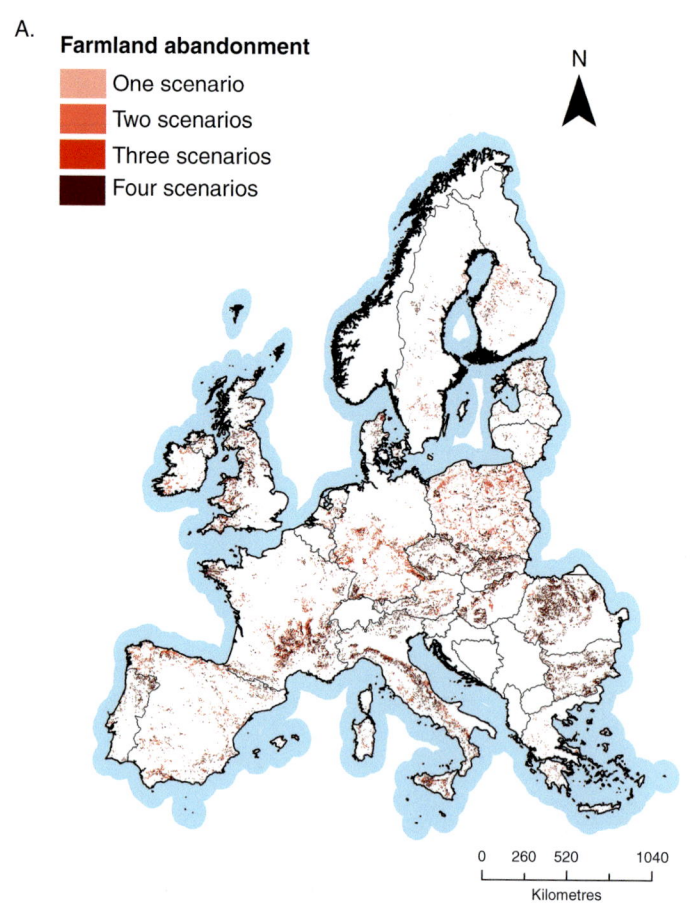

Farmland abandonment
- One scenario
- Two scenarios
- Three scenarios
- Four scenarios

Figure 6.1. (A) Farmland abandonment in Europe projected for the year 2040 under different scenarios by the Dyna-CLUE model (after Ceauşu et al., 2015). (B) Wilderness quality index for Europe (after Kuiters et al., 2013). (A black and white version of this figure will appear in some formats.)

B.

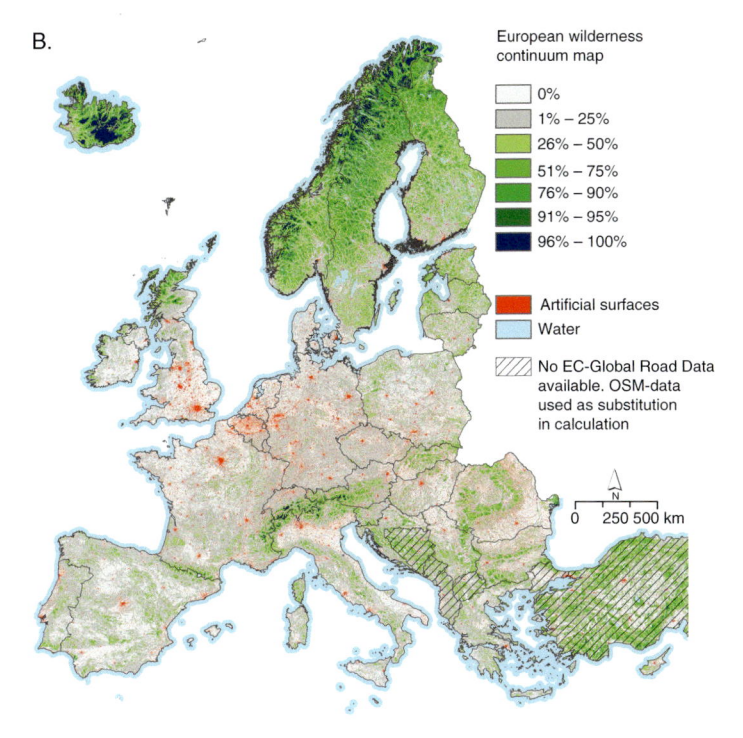

European wilderness continuum map
- 0%
- 1% – 25%
- 26% – 50%
- 51% – 75%
- 76% – 90%
- 91% – 95%
- 96% – 100%

Artificial surfaces
Water
No EC-Global Road Data available. OSM-data used as substitution in calculation

0 250 500 km

Figure 6.1. (Cont.)

Figure 4.3. It is standard practice to reduce woody cover on rangelands in the western USA using mechanical tools, such as the 'bull-hog mulcher' (left), as an artificial substitution for the ecosystem function once performed by Pleistocene megaherbivores in the same way as elephants in Africa today (right). Photo credits: Fecon Inc. (left) and Stein Moe (right). (A black and white version of this figure will appear in some formats.)

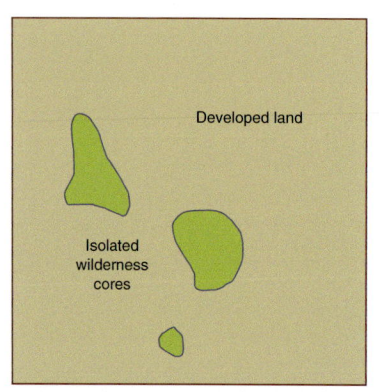

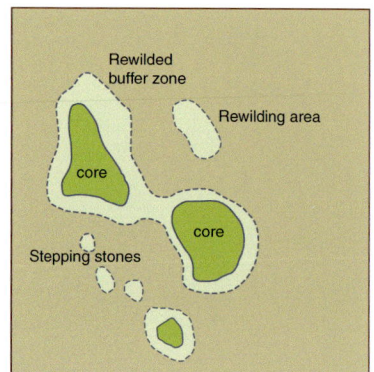

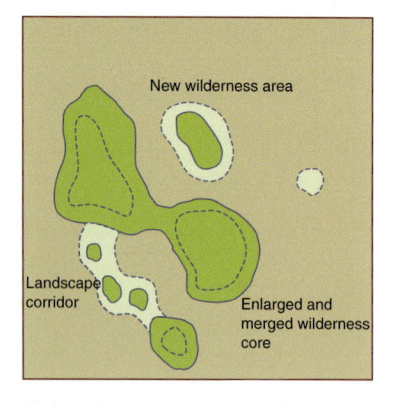

Figure 6.2. Restoration of wilderness via rewilding in the three C's model (after Carver and Fritz, 2016). (A black and white version of this figure will appear in some formats.)

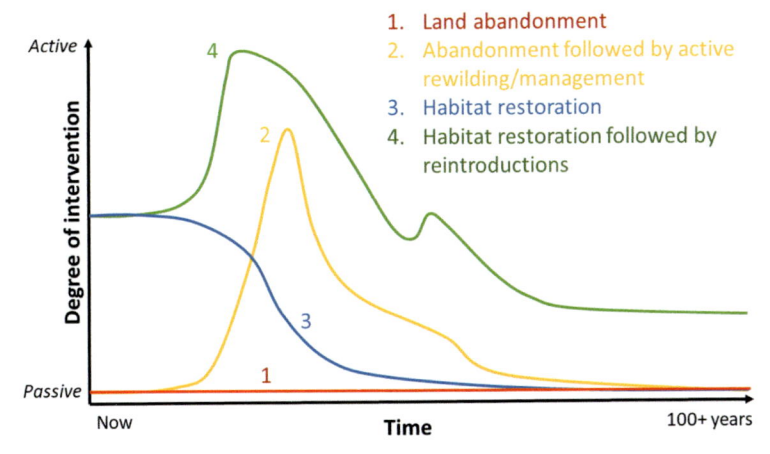

Figure 6.3. Changes in types and categories of rewilding over time. (A black and white version of this figure will appear in some formats.)

1. Land abandonment
2. Abandonment followed by active rewilding/management
3. Habitat restoration
4. Habitat restoration followed by reintroductions

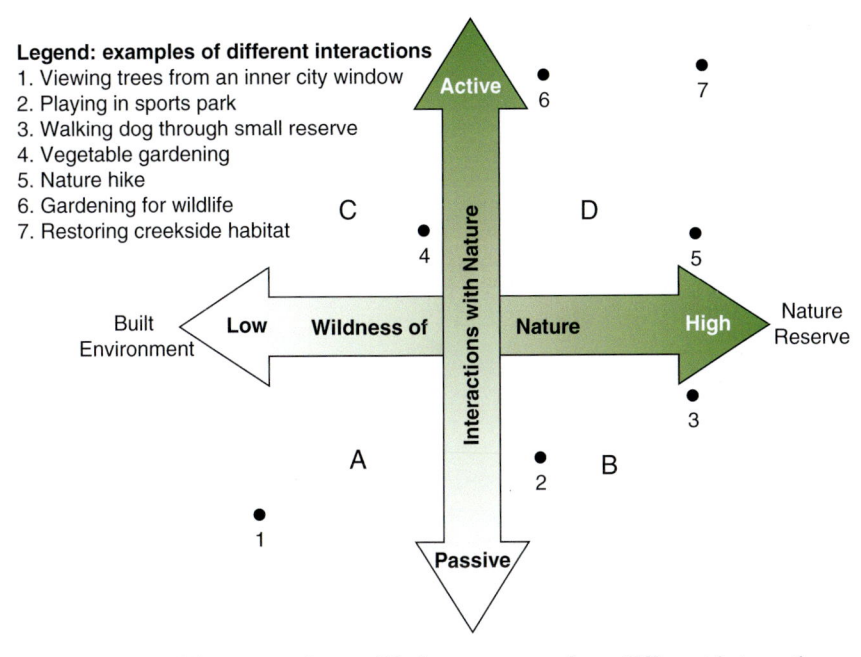

Legend: examples of different interactions
1. Viewing trees from an inner city window
2. Playing in sports park
3. Walking dog through small reserve
4. Vegetable gardening
5. Nature hike
6. Gardening for wildlife
7. Restoring creekside habitat

Figure 9.1. Tool for comparing well-being outcomes from different interactions with urban nature (adapted from Keniger et al., 2013). Quadrants signify passive to active fostering of nature (y-axis), in low to high wildness of nature (x-axis). (A black and white version of this figure will appear in some formats.)

Figure 11.1. Robert Smithson's Spiral Jetty on the Great Salt Lake, USA, 2005. Wikimedia Commons. (A black and white version of this figure will appear in someformats.)

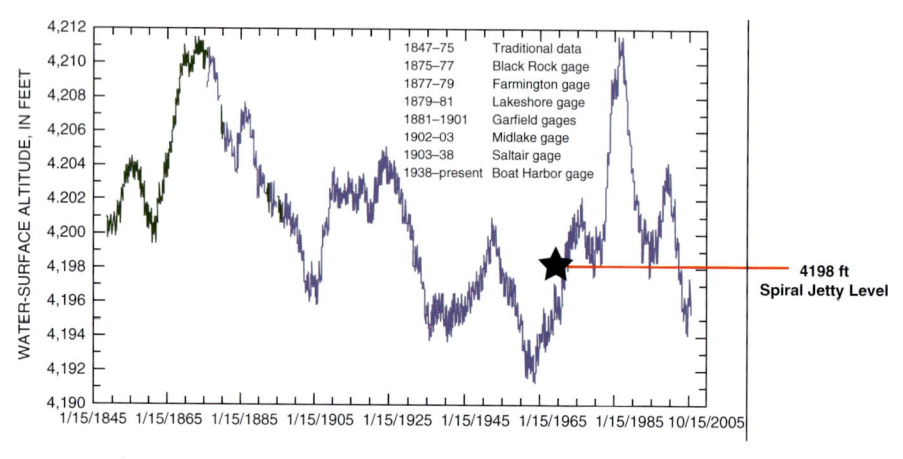

Figure 11.2. Historic levels of Great Salt Lake (1845–2005), showing height of Spiral Jetty. US Geological Survey data. https://ut.water.usgs.gov/greatsaltlake/elevations/. (A black and white version of this figure will appear in some formats.)

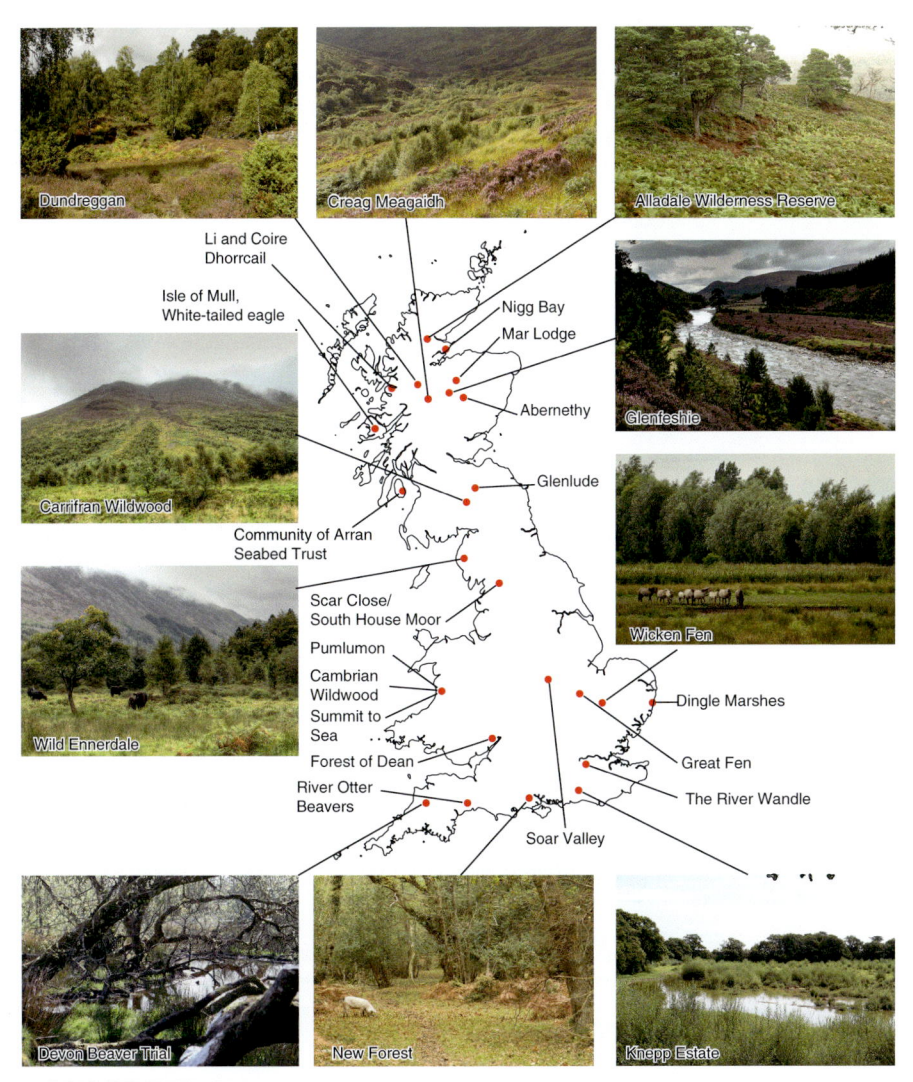

Figure 12.1. Rewilding-associated projects across Britain as identified by Rewilding Britain, Trees for Life, Mark Fisher, and the authors. Photos C. Sandom. (A black and white version of this figure will appear in some formats.)

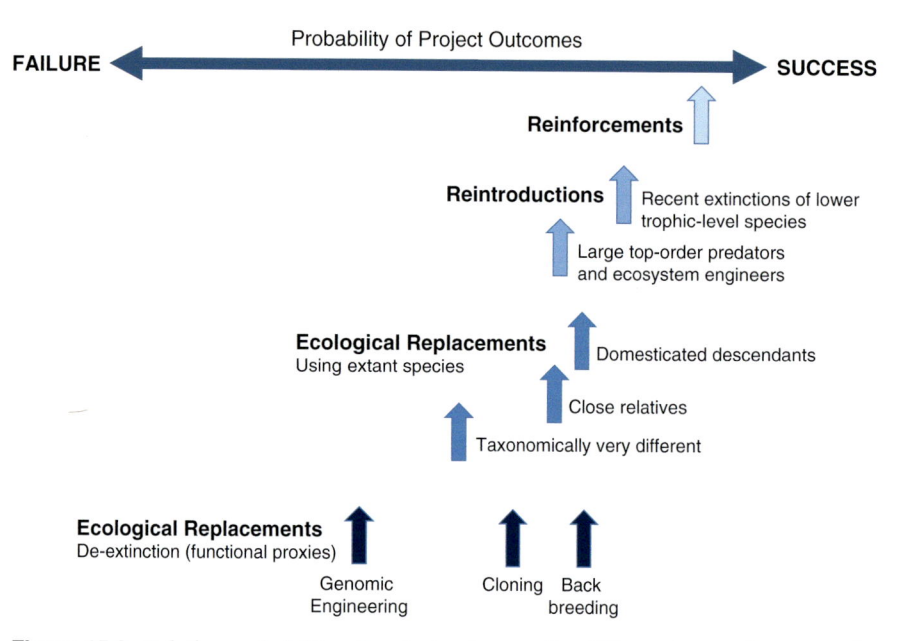

Figure 15.2. Relative probability of project outcomes for different types of conservation translocation that could be undertaken for rewilding: reinforcements, reintroductions, and ecological replacements. Reintroductions are separated into recent extinctions of lower-trophic-level species and reintroductions of large top-order predators or ecosystem engineers. Ecological replacement projects can use either extant species, or apply genetic manipulations to create functional proxies of an extinct species; for the latter, the risks and uncertainties in the technological creation of founder groups of proxy species are not taken into account. Outcome probability takes into account risk and uncertainty, including perceived probability of negative impacts on ecosystems, and negative social attitudes. The placement of arrows within translocation type indicates approximate relative probability of outcomes, all other things being equal. Vertical alignment of arrows between translocation types is less reliably indicative of relative probabilities of outcomes due to likely marked variation in risk factors and uncertainties between the different types of projects. (A black and white version of this figure will appear in some formats.)

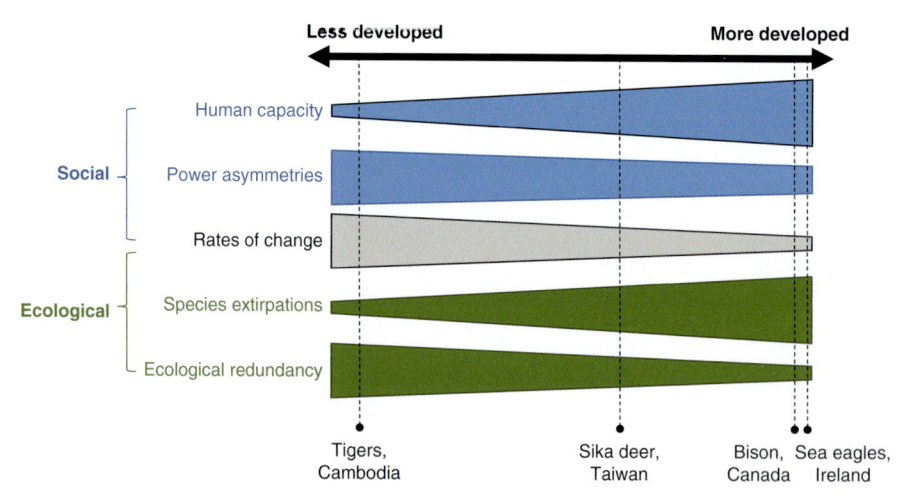

Figure 19.1. A heuristic of the relative differences in the social and ecological contexts for rewilding, on a continuum between less and more developed countries and regions, informed by Walker (1992), Armitage and Johnson (2006), Ensor (2011), and Butler and colleagues (2014). The indicative positions of the four rewilding case studies considered in this chapter are also shown. (A black and white version of this figure will appear in some formats.)

Figure 16.1. Key top-down regulating species discussed in the text: (A) wolves and their effect on Yellowstone's (B) elk population (note bark stripping on the aspen); and (C) dingoes in Australia. © (A) Mark Kent (CC BY-SA2), (B) Kristal Kraft CC BY-SA2, and (C) Paul Balfe CC BY2. (A black and white version of this figure will appear in some formats.)

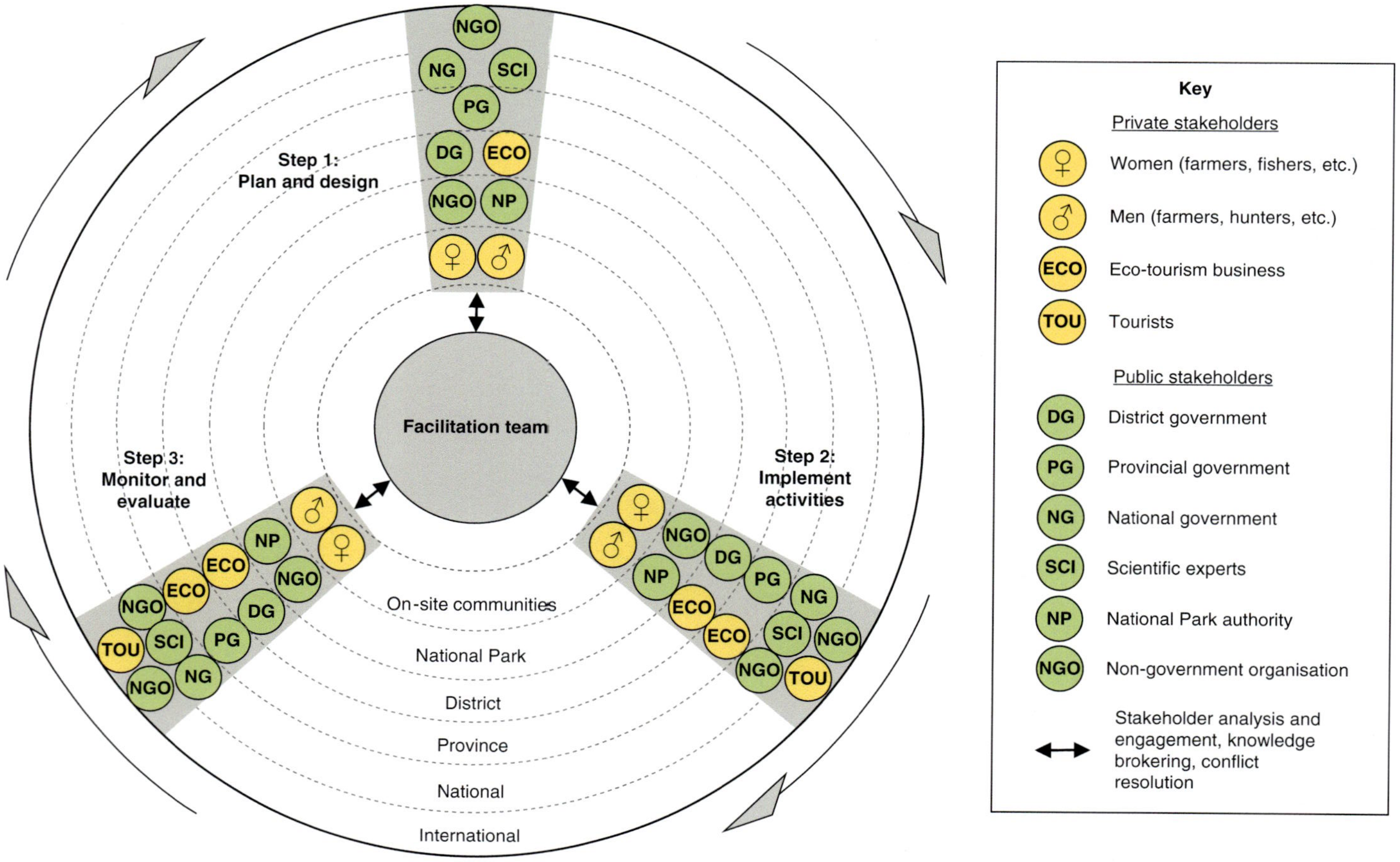

Figure 19.2. The ACM structure and process for the governance of rewilding. Likely stakeholders in a rewilding initiative are differentiated as private or public, and by their scale, from on-site to international. (A black and white version of this figure will appear in some formats.)

the wild is their primary concern, and elements of the wild can be found in both pristine and domestic places, as well as in present and past states. Both the tree growing in the forest and the tree growing in the backyard can be considered wild at some level, as both depend on the same physiologic processes and both are subject to the same pathogens, for example, and so both can be cherished by rewilders. If the wild is nature-apart-from-culture, elements of wildness can be found everywhere, from sap flowing in a tree to vibrating cellulose molecules (Cronon, 1995). The wild, like beauty, is found in the eye of the beholder, but it is also found where one has decided one will find it.

Such considerations of the wild lead to a key distinction of rewilding: whereas some rewilders aim to reinstate *wildness*, others hope to reinstate *wilderness*. Wildness, the quality, can be independent of historic references and targeted baselines; wilderness, the place, is the state of nature free from people. The signal moment in the history of rewilding may have occurred when domestic and semi-domestic grazers, rather than analogue species, began being used to instil wildness into ecosystems. It was found, for example, that judicious domestic grazers could preserve and maintain basic levels of biodiversity in grasslands (Metera et al., 2010). Modern and primitive breeds of sheep, cows, and horses, as animals whose form, behaviour, and function has been modified through several generations of human husbandry, were selected as 'natural grazers' (or 'naturalistic grazers') to direct ecosystems towards wilder end-points (Hancock et al., 2010; Hodder and Bullock, 2010). In the case of Heck cattle utilised in many rewilding projects, these seemingly extinct animals 'brought back to life' were actually the product of a meticulous breeding programme carried out in 1930s Germany for combining various rare and hardy breeds from across Europe (Lorimer and Driessen, 2016). By utilising such grazers to create more wild systems, rewilders are implicitly agreeing that humans can play a positive role in shaping these systems. The wildness produced by Heck cattle includes a human component (Vermeulen, 2015; Jepson, 2016).

Thus, instead of utilising species as similar as possible in form and function to the ones that existed before humans drove them to extinction, rewilders who depend on domestic or semi-domestic grazers are assuming that human designs have a legitimate role in helping produce self-willed systems. In the example of the Netherlands' Oostvaardersplassen, by favouring grazers but not predators in managing this reserve for *Naturentwicklung* – nature development – rewilders are demonstrating that they believe humanity can play a positive role in curating the wild. And when Oostvaardersplassen's rewilders cull sick or excessive grazers in the reserve, they are showing still greater faith in the human ability to design wildness on the land (Van Wieren and Bruinderink, 2010). Employing semi-domestic ungulates to clip brush and graze meadows for reinstating wilder systems is a declaration that nature and culture both contribute to the rewilding endeavour. Sensitive curators of Spiral Jetty share this belief that nature can be steered by culture for the benefit of both.

In fact, the Dutch project of promoting 'nature development' implies that a rewilder is actually more interested in future states than past states. Spontaneous processes are favoured by letting them loose, releasing them so that their final products are anticipated but cannot be predicted. There may even be distinct transatlantic interpretations of rewilding, with Europeans aiming to reinstate *wildness* and Americans favouring *wilderness* (Hall, 2014a). Such continental preferences are reflected in dominant founding myths, as North and South Americans more deeply embrace a view of their land as formerly 'pristine' which, if not completely human-free, was inhabited by peoples felt to be not completely human. It merits repeating that restoring wilderness, in the vision of the Pleistocene rewilders, is really a project of restoration, because such wilderness can exist only before civilisation arrived. Yet this kind of wilderness can never be a realistic goal for rewilders, because wilderness becomes humanised the instant one seeks to enter, manage, and reinstate it. Logically consistent rewilding is therefore the project of promoting wildness, not wilderness – reinstating wild qualities, not wild places. As a result, rewilders need not be preoccupied with describing former wild places. Rewilders must instead be concerned with identifying activities and processes that promote wild qualities. Rewilding, like curating, assumes knowledgeable and enlightened human overseers.

Earth art in Europe

A year after Robert Smithson constructed his jetty on the edge of Great Salt Lake, he travelled across the Atlantic, accepting an invitation to develop a sample of his art in the Netherlands. In searching for appropriate *non-sites*, as he called them, he eventually settled on an old flooded quarry near the northern village of Emmen, producing his only installment of Earth art outside the USA. There he directed a crane to scoop sand and gravel into an arching dike along the shoreline to form two half-circles together with an overlying mound with a spiralling pathway. 'Broken Circle/Spiral Hill' has become another shrine that requires its curators to make hard decisions about the best way to manage a piece of earth that is also art (Figure 11.4).

However, in this Dutch case, Smithson's enigmatic shapes have been restored to their full forms, so that nearly five decades after its construction, this artwork displays its original outlines while undergoing occasional reshaping, clipping, and brushing. In the intervening years, the surrounding lake level likewise fluctuated, serving to inundate and heavily erode the work, leaving its sandy shapes in disarray. There was also a struggle over ownership of the site, coupled with growing public concern over the disappearance of

Figure 11.4. Broken Circle/Spiral Hill near Emmen, the Netherlands. Wikimedia Commons.

Smithson's creation. The result of these events and concerns is that this artwork has been reborn, with the Dutch government stepping in to bring it back. Yet some Smithson aficionados point out that this restoration went beyond the intent of the artist, as this artwork's edges have been shored up with wooden planks, making it even more permanent and giving it a look that is 'distinctly artificial and unlike Smithson's organic integration of rock and water' (Boettger, 2006, p. 227). The Dutch curators apparently considered it their obligation to improve on this *objet d'art* by allowing their own judgements to supersede those of the artist. In the quarry near Emmen, Smithson's managers have apparently become *gardeners* more than *restorers* in deciding to ignore cautions against full restoration. As a local admirer explained, 'What was meant as temporary art to be taken over by the elements must now be nurtured' (Marijnissen, 2011).

There is still the possibility that Smithson would have applauded the intensive restoration of his Dutch creation. Even if Smithson had not originally intended for Broken Circle/Spiral Hill to be permanent, as it had been commissioned as a 'temporary public artwork' (Tuchman, 2017), he still gave advice on its management after his departure. When pressed by the site's owners about what should be done with the artwork, Smithson responded that they should weed its flat surfaces and shore up its sides with wooden planks (Isola, 1996). So if nature was the initial curator of this Earth art, humans-apart-from-nature became its final curator. It seems that in this densely populated, intensively managed country, where more than a quarter of the land has been wrestled from the sea, people do not hesitate to take charge where they see nature becoming too invasive. One realises that across the Atlantic, Spiral Jetty continues to wax and wane under the vagaries of the elements, whereas at Broken Circle/Spiral Hill humans are working to combat entropy. In the Dutch style of curating earth that is art, people are less hesitant about putting their mark on nature.

If one then travels across this small country to compare the curation of Broken Circle/Spiral Hill with the rewilding of Oostvaardersplassen, one witnesses first-hand how these seemingly disparate pursuits show a great deal of overlap. Both projects are situated on artificial, reclaimed landscapes and rely on human ingenuity to create more interesting natural forms. Both the project of rewilding biodiversity in a wetland and that of curating sandy shapes in a quarry can be considered high art because human will is being manifested on the land. The rewilders decide on which grazers to introduce or cull, in which proportions at which time of the year; the curators decide on which interventions to use with which materials at which frequency. Both rewilding and curating can be passive, hands-off processes applied over large scales, but both of these Dutch projects are active, hands-on processes focused on relatively small areas. Rewilders and curators are both looking towards the future, as they consider ways to usher current systems towards more desirable states. If Broken Circle/Spiral Hill exhibits lower levels of biodiversity than Oostvaardersplassen, it also displays more interesting forms and shapes. Managers at both sites are curating deeply humanised places, accounting for nature's spontaneity while manifesting culture's wishes.

The hubris of rewilding

Proponents of rewilding are of course no strangers to criticism. Just as curating Smithson's artworks invites a heated discussion about proper methods and appropriate goals, so too rewilders confront sceptics who feel that they are just not getting it right. Aside from animal rights activists who see rewilders as maltreating semi-domestic grazers and analogue species (Van Maanen and Convery, 2016), aside from cost–benefit analysts who consider rewilding as representing a low priority in conservation pursuits (Caro, 2007), and aside from philosophers who point out the difficulty in identifying the right kinds of 'wild' to promote and unleash (Scotney, 2014), there remains the more general criticism about the hubris of rewilding. Because rewilders rely much less on historic conditions and former baselines to guide their projects, critics maintain that rewilding risks sallying into utterly new territory, willy-nilly creating places, objects, or organisms that have never existed before, with possibly disastrous consequences.

In short, critics feel that rewilders are playing god with natural systems, and that humans are not up to the task.

This critical view emphasises that ecological systems are dynamic and that human knowledge is inadequate. 'Scientific support for the main ecological assumptions behind rewilding, such as top-down control of ecosystems, is limited,' caution Nogués-Bravo and colleagues (2016); rewilding may produce unexpected and undesirable consequences, such as accelerated erosion or invasion by non-native species. Corlett (2016) adds that rewilding by analogue species risks producing deleterious side effects similar to those stemming from species introduced for biological control. A site must therefore be thoroughly studied to ascertain, for example, how far natural grazers can promote biodiversity and improve ecosystem functions. Indeed, for New Zealanders, 'wilding conifers' are invasive species running slipshod over the landscape (Howell, 2016). Rewilding infused with hubris can produce environmental disasters.

The best rejoinder to such critiques may be to suggest that rewilding is really curating. Curators cannot help but make choices about managing and displaying a special work, and sometimes such choices have undesirable consequences. Yet should curators do nothing, then entropic and spontaneous processes may usher in even less-desirable results. Rewilding critics should also understand that all ecological management – hands-on or hands-off – is susceptible to problematic and unpredictable results. Even well-planned and seemingly successful conservation projects, such as the eradication of feral goats on the Galapagos, as lauded by Nogués-Bravo and colleagues (2016), may portend negative side effects. A different team of ecologists point out, for example, that even if the eradication of invasive aliens offers undisputed benefits to native flora and fauna, 'empirical observation shows that these benefits can vary dramatically and unpredictably, and there may even be unexpected adverse consequences' (Courchamp et al., 2011). The moment environmental managers carry out any activity in the natural world, by reintroducing one species or eradicating another, it seems they are playing god with that environment. Exterminating a creature in the first place, say some, was itself a god-like act (Cohen, 2014). Our Anthropocene is the composite of all our human activities that modify the world, and rewilding is but one of these many activities – one like all the others that must be undertaken with care and foresight. Rewilders may remind their critics of the adage that 'We are as gods and might as well get good doing it' (Brand, 1968). Deep planning and responsible decisions are required of all ecological interventions.

Automated rewilding

An intriguing thought experiment proposed by Cantrell and colleagues (2017) suggests that the day will come when mass computing shall be better than human judgement for providing answers about how one can manage for wildness. Reseeding is already being carried out by drones that drop seeds in random fashion, for example, and eradicating marine invasives is already done by mini-submersibles prowling the ocean currents. Artificial intelligence could reach such sophistication as to make management decisions autonomous of people. A truly wild place might therefore be created if grazing, seeding, eroding, reproducing, and evolving could be managed according to criteria that are not linked to human values and bias. As if realising William Jordan's dream to restore ecosystems 'with a studied indifference to human interests' (Jordan, 2003, p. 87), artificial machine intelligence may one day provide these unbiased decisions. In effect, rewilders may one day be able to release a 'wildness creator' to go out and produce the wild (Cantrell et al., 2017). Surely in this scenario, rewilding could never be viewed as curating because humans will have finally been painted out of the picture.

Yet one rightly wonders if a true 'wildness creator' can ever be invented. Most obviously, sceptics of automated rewilding feel that every human machine will inevitably leave traces of the human who designed it. Intelligent systems 'are dependent on the assumptions inserted into the algorithms and the data that provide the foundation for their training and operation' (Galaz and Mouazen, 2017). Hal, the robot computer of *2001: Space Odyssey* who conspired to take over civilisation, failed to do so because a human was able to understand and outsmart it: any wildness created by a computer will be tainted by those who programmed it. Such scepticism, however, may be countered by the fact that artificial intelligence (AI) systems have already arrived at the level of surpassing some human abilities to perform logical thought processes. Proponents point out that AI is a hotly developing field that by transcending human bias will 'increase the accuracy' of managing river basins, for example, or offer 'methodological advantages' for modelling ecosystem services (Bagstad et al., 2017; Valizadeh et al., 2017). There is little question that AI will continue to be perfected, even to the point of being able to create wildness that is more wild than that produced by such methods as substituting analogue species for extinct species, or even by passive rewilding, whereby natural systems are simply abandoned to spontaneous processes, as such systems are already tainted by human modifications. It seems reasonable to believe that AI will someday become more proficient at editing out human imprints than can be done by relying on mere humans armed with mere human insights.

But in the last analysis, AI will probably never be utilised to create *wilder* wildness, just as it will never be utilised to make decisions about curating Spiral Jetty. Humans may indeed develop black boxes that can manage systems autonomously, but experience shows that we humans choose to remain in control, still insisting to be in charge of curating the art and the wild. Rewilders (and wilderness enthusiasts) cherish the wild because they see part of themselves within it. The wild cannot exist separate from people because we are the ones who define, enjoy, and create it. Humans decide which elements of the wild to promote, be it megafauna, high biodiversity, or landscape beauty – and such decisions may change according to our latest insights. Witness that certain organisms are being ushered to extinction by humans, such as smallpox and rinderpest pathogens, which are not placed on endangered species lists because we judge them as undesirable kinds of wild nature. Witness Josh Donlan's North American rewilding scheme that by employing dangerous predators is sometimes dismissed as junk science, whereas Franz Vera's Dutch counterpart, by utilising charismatic herbivores, is largely embraced by an enthusiastic public (Hall, 2014a). Witness wilderness areas, that once legally designated, attract many more tourists because this wilderness imprimatur is the mark of certified wild, resulting in a place that is less wild than before. Finally, witness doctors who when presented with diagnoses handed down by AI systems, and probably more reliable than those created through traditional medical science, will still not trust these diagnoses because they do not understand the underlying logic: 'human doctors still have to make the decisions', notes a journalist, '– and they won't trust an A.I. unless it can explain itself' (Kuang, 2017). Humans demand to stay in ultimate control of the wild. As a recent study concludes, 'rewilding, it would seem, is about who we think we are and how we co-constitute our sense of self' (Wynne-Jones et al., 2018).

The wild is *out there*, by definition beyond the control of people, but humans also demand to be the force behind the curtain that pushes the buttons and tweaks the dials. Nature-in-the-raw can only be viewed through the lens that we hold up to see it, but we always choose the lens to hold up. If a tree falls in the wilderness, humans do not care whether it makes a sound unless they are there to hear it. If rewilders finally invent a wildness creator, they will then dismantle it so as to stay within the wild and continue to control it.

Conclusions: curating art and rewilding nature

There have been attempts to classify rewilding into categories that depend on the human role, which span from passive to active, from no intervention to aggressive intervention (Lorimer et al., 2015; Corlett, 2016; Nogués-Bravo et al., 2016). But any purported rewilding that pursues former baselines, from pre-Columbian to pre-Holocene to pre-Pleistocene, is better viewed as *restoring* – the recreation of a former state that can never happen. True *rewilding* is forward-looking, the project of curating the kind of wild that one hopes to create. A rewilder can aim for approximations of nature-free-from-culture, but can never completely extract humanity from that nature. The Oostvaardersplassen is high art, the product of environmental scientists using semi-domestic grazing animals as paintbrushes to mix and streak hues across the landscape. But theirs is not a random process, for they are guided by such natural frameworks as food webs, evolutionary trends, dominance hierarchies, and seasonality, along with such human artefacts as semi-domestic grazers, invasive alien species, limited terrain, and climate change. Nearby Broken Circle/Spiral Hill is also a deeply humanised place, created by an artist sculpting with a crane, and also guided by wild nature and human values. Analogies between restoring art and restoring nature are easily over-extended, because their pasts and their creators are so different. Yet analogies between curating art and rewilding nature are so close as to not require an analogy: the wild is art. Art in the wild results from a 'wildness creator' that will always be human-controlled.

Although former wilderness is that place that one can never restore, there is a bright future for rewilders. 'The strength of rewilding is its flexibility, which derives from its focus on "upgrading" ecosystems processes, using the past as a source of insight and inspiration rather than a template for restoration, and willingness to mix nature, society, and a nature-based economy' (Schepers and Jepson, 2016). However, rewilding's conceptual flexibility also requires that rewilders proceed with extreme caution. To fend off hubris, rewilders must heed the wise advice for curating Spiral Jetty: 'We will just have to be patient and make no rash moves' (Anderson, 2004). Intervention in natural systems is necessary and desirable – and inevitable. Intervening for wildness requires deep study and careful planning so as to make as few blunders as possible. Rewilders are gods who are engineering ecosystems, and we must make sure that the kinds of wildness they create are the ones we want, and that the values they hold are the values we hold. We must therefore be careful about selecting our rewilders, because they are the ones deciding on when, where, and how to create the newly wild.

References

Alagona, P., Sandlos, J., and Wiersma, Y. (2012). Past imperfect: using historical ecology and baseline data for conservation and restoration projects in North America. *Environmental Philosophy*, **9**, 49–70.

Anderson, W. (2004). Restoring Spiral Jetty: what if? Artwatch International, April. www.artwatchinternational.org/restoring-spiral-jetty-what-if/ (accessed 18 November 2017).

Bagstad, K., Semmens, D., Ancona, Z., and Sherrouse, B. (2017). Evaluating alternative methods for biophysical and cultural ecosystem services hotspot mapping in natural resource planning. *Landscape Ecology*, **32**, 77–97.

Balaguer, L., Escudero, A., Martín-Duque, J., Mola, I., and Aronson, J. (2014). The historical reference in restoration ecology: re-defining a cornerstone concept. *Biological Conservation*, **176**, 12–20.

Beller, E., McClenachan, L., Trant, A., et al. (2017). Toward principles of historical ecology. *American Journal of Botany*, **104**, 645–648.

Boettger, S. (2006). Earthworks' contingencies. In King, E. and Levin, G. (Eds.), *Ethics and the visual arts* (pp. 217-233). New York, NY: Allworth Press.

Bradshaw, A. (1996). Underlying principles of restoration. *Canadian Journal of Fisheries and Aquatic Sciences*, **53**, 3-9.

Brand, S. (1968). *The Whole Earth Catalogue* (p. 2).

Cantrell, B., Martin, L., and Ellis, E. (2017). Designing autonomy: opportunities for new wildness in the Anthropocene. *Trends in Ecology & Evolution*, **32**, 156-166.

Caro, T. (2007). The Pleistocene re-wilding gambit. *Trends in Ecology & Evolution*, **22**, 281-283.

Chianese, R. (2013). Spiral Jetty. *American Scientist*, **101**, 20-21.

Clewell, A., and Aronson, J. (2013). *Ecological restoration: principles, values, and structure of an emerging profession*, 2nd edition. Washington, DC: Island Press.

Cohen, S. (2014). The ethics of de-extinction. *Nanoethics*, **8**, 165-178.

Corlett, R. (2016). Restoration, reintroduction, and rewilding in a changing world. *Trends in Ecology & Evolution*, **31**, 454-462.

Courchamp, F., Caut, S., Bonnaud, E., Bourgeois, K., Angulo, E., and Watari, Y. (2011). Eradication of alien invasive species: surprise effects and conservation successes. In Veitch, C., Clout, M., and Towns, D. (Eds.), *Island Invasives: Eradication and Management. Proceedings of the Conference on Island Invasives* (pp. 285-289). Gland, Switzerland: IUCN.

Cronon, W. (1995). The trouble with wilderness; or, getting back to the wrong nature. In *Uncommon ground: rethinking the human place in nature* (pp. 69-90). New York, NY: Norton.

Donlan, J., Berger, J., Bock, C., et al. (2006). Pleistocene rewilding: an optimistic agenda for twenty-first century conservation. *The American Naturalist*, **168**, 660-681.

Egan, D., and Howell, E. (2001). *The historical ecology handbook: a restorationist's giude to reference ecosystems*. Covelo, CA: Island Press.

Elliot, R. (1982). Faking nature. *Inquiry*, **25**, 81-93.

Eppich, R., Esmay, F., Learner, T., and Tang, A. (2011). Monitoring *Spiral Jetty*: Aerial Balloon Photography, at Heritage Portal. http://m.heritageportal.eu/Browse-Topics/PREVENTIVE-CONSERVATION/Spiral-Jetty-Paper-Rand-Eppich.pdf (accessed 19 November 2017).

Galaz, V., and Mouazen, A. (2017). New wilderness requires algorithmic transparency: a response to Cantrell et al. *Trends in Ecology & Evolution*, **32**, 628-629.

Hall, M. (2005). *Earth repair: a transatlantic history of environmental restoration*. Charlottesville, VA: University of Virginia Press.

Hall, M. (2014a). Extracting culture or injecting nature? Rewilding in transatlantic perspective. In Keulartz, J. and Drenthen, M. (Eds.), *Old World and New World perspectives on environmental philosophy* (pp. 17-35). New York, NY: Springer.

Hall, M. (2014b). Restoration and the search for counter narratives. In Isenberg, A. (Ed.), *The Oxford handbook of environmental history* (pp. 309-331). New York, NY: Oxford University Press.

Hancock, M., Summers, R., Amphlett, A., Willi, J., Servant, G., and Hamilton, A. (2010). Using cattle for conservation objectives in a Scots pine *Pinus sylvestris* forest: result of two trials. *European Journal of Forest Research*, **129**, 299-312.

Higgs, E., Falk, D., Guerrini, A., et al. (2014). The changing role of history in restoration ecology. *Frontiers in Ecology and Evolution*, **12**, 499-506.

Hobbs, R., Hallett, L., Ehrlich, P., and Mooney, H. (2011). Intervention ecology: applying ecological science in the twenty-first century. *BioScience*, **61**, 442-450.

Hodder, K., and Bullock, J. (2010). Nature without nurture? In Hall, M. (Ed.), *Restoration and history: the search for a usable environmental past* (pp. 223-235). New York, NY: Routledge.

Howell, C. (2016). Recreating the invasion of exotic conifers in New Zealand. In Randall, R., Lloyd, S., and Borger, C. (Eds.), *Twentieth Australasian Weeds Conference* (pp. 258-262). South Perth: Weeds Society of Western Australia.

Isola, M. (1996). Monumental art, but the wind and rain care not. *New York Times*, 24 November.

Jepson, P. (2016). A rewilding agenda for Europe: creating a network of experimental reserves. *Ecography*, **39**, 117-124.

Jordan, W. (2003). *The sunflower forest: ecological restoration and the new communion with nature*. Berkeley, CA: University of California Press.

Julavits, H. (2017). The art at the end of the world. *New York Times Magazine*, 7 July, MM44.

Katz, E. (1992). The big lie: human restoration of nature. *Research in Philosophy and Technology*, **12**, 231-244.

Kuang, C. (2017). Can A.I. be taught to explain itself? *The New York Times Magazine*, 21 November.

Lorimer, J., and Driessen, C. (2016). From 'Nazi cows' to cosmopolitan 'ecological engineers': specifying rewilding through a history of Heck cattle. *Annals of the American Association of Geographers*, **196**, 631-652.

Lorimer, J., Sandom, C., Jepson, P., Doughty, C., Barua, M., and Kirby, K. (2015). Rewilding: science, practice, and politics. *Annual Review of Environmental Resources*, **40**, 39-62.

Losin, P. (1988). The Sistine Chapel debate: Peter Losin replies. *Restoration & Management Notes*, **6**.

Lowenthal, D. (2015). *The past is a foreign country revisited* (pp. 464-494). Cambridge: Cambridge University Press.

Marijnissen, H. (2011). Na veertig jaar is de Emmer' cirkel gesloten. *De Verdieping Trouw*, 25 October. www.trouw.nl/home/na-veertig-jaar-is-de-emmer-cirkel-gesloten~ae4e23d9/ (accessed 22 November 2017).

Marris, E. (2011). *The rambunctious garden: saving nature in a post-wild world*. New York, NY: Bloomsbury.

Metera, E., Sakowski, T., Słoniewski K., and Romanowicz, B. (2010). Grazing as a tool to maintain biodiversity of grassland - a review. *Animal Science Papers and Reports*, **28**, 315-334.

Nogués-Bravo, D., Simberloff, D., Rahbek, C., and Sanders, N. (2016). Rewilding is the new Pandora's box in conservation. *Current Biology*, **26**, R87-R91.

Novick, P. (1988). *That noble dream: the 'objectivity question' and the American historical profession*. Cambridge: Cambridge University Press.

Pzabó, P. (2015). Historical ecology: past, present and future. *Biological Reviews*, **90**, 997-1014.

Schepers, F., and Jepson, P. (2016). Rewilding in a European context. *International Journal of Wilderness*, **22**, 25-30.

Scotney, R. (2014). Wilderness recognized: environments free from human control. In Keulartz, J. and Drenthen, M. (Eds.), *Old World and New World perspectives on environmental philosophy* (pp. 73-90). New York, NY: Springer.

SER (Society for Ecological Restoration). (2004). *The SER primer on ecological restoration*. Washington, DC: Society for Ecological Restoration International.

Tuchman, P. (2017). How do you sell a work of art built into the Earth? *New York Times*, 27 January.

Valizadeh, N., Mirzaei, M., Allawi, M., et al. (2017). Artificial intelligence and geo-statistical models for stream-flow forecasting in ungauged stations: state of the art. *Natural Hazards*, **86**, 1377-1392.

Van Maanen, E., and Convery, I. (2016). Rewilding: the realization and reality of a new challenge for nature in the twenty-first century. In Convery, I. and Davis, P. (Eds.), *Changing perceptions of nature* (pp. 303-319). Martlesham: Boydell Press.

Van Wieren, S., and Bruinderink, G. (2010). Ungulates and their management in the Netherlands. In Apollonio, M., Anderson, R., and Putman, R. (Eds.),

European ungulates and their management in the 21st century (pp. 165–183). Cambridge: Cambridge University Press.

Vermeulen, R. (2015). *Natural grazing: practices in the rewilding of cattle and horses*. Nijmegen, Netherlands: Rewilding Europe.

Wynne-Jones, S., Strouts, G., and Holmes, G. (2018). Abandoning or reimagining a cultural heartland? Understanding and responding to rewilding conflicts in Wales – the case of the Cambrian Wildwood. *Environmental Values*, **27**(4), 377–403.

CHAPTER TWELVE

Rewilding a country: Britain as a study case

CHRISTOPHER J. SANDOM
University of Sussex and Wild Business Ltd
SOPHIE WYNNE-JONES
Bangor University

Rewilding in Britain has become synonymous with George Monbiot and his book *Feral* (Monbiot, 2013), which is both celebrated and condemned for opening up debate and raising awareness, while also polarising and antagonising stakeholders. Rewilding in Britain did not start with Monbiot, with longer-standing discussion and experimentation advanced by members of the British Association of Nature Conservation, Wildland Network, Wildland Research Institute, and others (Taylor, 2005, 2011; Ward et al., 2006; Hodder et al., 2009; Sandom et al., 2013a). Monbiot has, however, transformed rewilding's potential. The explosion of interest following the publication of *Feral* led to the initiation of Rewilding Britain[1] to champion the cause, raise funds to support practical implementation and tackle barriers at a policy and legislative level. Supporters and trustees emphasise the 'positive' and 'optimistic' vision that rewilding poses. This is more than a 'Monbiot effect'; rather, rewilding seems to tap into something critical at this time of ecological and political anguish, namely a hope that things can be improved (Rewilding Britain interview #19 May 2017).[2] In this chapter, we explore the opportunity and risks that rewilding in Britain presents. We also provide an overview of rewilding-associated projects underway here, assessing their objectives, approach, and outcomes. Finally, we consider what the future might hold for rewilding in Britain.

The British context
Rewilding is a diverse concept. A key factor in this diversity is the context in which rewilding is applied. Internationally we see a division between North American approaches, which tend to advocate recreating wilderness

---

[1] www.rewildingbritain.org.uk/
[2] Quotes are derived from research interviews undertaken by Sophie Wynne-Jones, Graham Strouts, and Callum O'Neil.

(Foreman, 2004; Donlan et al., 2005)[3] and European rewilding which has been more accepting of anthropogenic legacies (Jørgensen, 2015; Jepson and Schepers, 2016). Rewilding in Britain can be aligned with a 'European' approach, but there are some notable points of distinction that make Britain an interesting case in its own right.

In contrast with some European countries, Britain is notable for its high population density in many parts of the country and accompanying intense land-use pressure. National Parks in Britain meet IUCN Category V (rather than Category II) requirements, acknowledging interaction between people and nature,[4] although classifications of 'wildness' are now being applied to land (e.g. Carver and Müller, 2014). The level of woodland cover is also among the lowest in Europe (12 per cent compared to 37 per cent European average in 2011[5]). In ecological terms, we lack many keystone species, particularly large carnivores due to historical extirpation with no opportunity for natural recolonisation. The impetus to reintroduce lost species here is, therefore, of particular interest and offers an arguably unique perspective given our island geography.

In political–economic and cultural terms, Britain is again noteworthy given the recent vote to leave the European Union, which is likely to be accompanied by considerable changes to agricultural and environmental policy. This, along with the publication of the 25-Year Environment Plan,[6] has energised discussion about the future of Britain's countryside, with rewilding being discussed as part of a post-Common Agricultural Policy (CAP) transition[7] (Wentworth and Alison, 2016). One issue that has been particularly contentious is whether the cultural value of farmed landscapes should be privileged (Convery and Dutson, 2008), whether traditional (i.e. low-intensity) farming delivers environmental benefits (Henle et al., 2008), and the degree to which ideas associated with rewilding such as species reintroduction threaten farmland and farming operations.[8] This links in with wider discussions on natural capital and ecosystem service delivery and the need to ensure better targeting of public monies to support these outputs (Helm, 2016). It also connects to debates on whether

---

[3] See http:/rewilding.org/rewildit/ and https://wildlandsnetwork.org/wildways/
[4] www.iucn.org/theme/protected-areas/about/protected-areas-categories/category-v-protected-landscapeseascape
[5] www.forestry.gov.uk/website/forstats2011.nsf/lucontents/4b2add432342111280257361003d32c5
[6] www.gov.uk/government/uploads/system/uploads/attachment_data/file/673203/25-year-environment-plan.pdf
[7] https://publications.parliament.uk/pa/cm201617/cmselect/cmenvaud/599/599.pdf; www.parliament.uk/business/committees/committees-a-z/commons-select/environmental-audit-committee/news-parliament-2015/future-of-natural-environment-after-the-eu-referendum-launch-16-17/
[8] www.nfuonline.com/assets/41019

sustainable food production is best achieved by a land-sharing or land-sparing approach (Garnett et al., 2013; Merckx and Pereira, 2015).

Rewilding has been the focus of several new 'position statements' from environmental NGOs (John Muir Trust, 2015b; RSPB, 2017; Woodland Trust, 2017) seeking to explore how it fits with their core objectives. This demonstrates an interesting expansion in the approach of conservationists in Britain, developing off the back of the Lawton report (Lawton et al., 2010) for 'more, bigger, better and joined-up' sites for conservation. While there is some contention within the sector as to whether rewilding supports the protection of valued sites and species, particularly those with dependencies on anthropogenic activities and continued intervention (e.g. heathland and hay meadows), the level of current engagement is marked when contrasted with earlier assessments (Taylor, 2005). However, it is noted that the preservationist approaches that rewilding reacts against are situated within wider European (and international) frameworks, and any departure from current structures and mechanisms will be challenging (Jepson and Schepers, 2016). There is also a concern that rewilding will create problems, both expected and unanticipated, in terms of novel ecological and socioecological interactions and emergences, and lead to a 'loosening' of hard-fought controls and protections (Nogues-Bravo et al., 2016). As such, rewilding poses both a risk and an opportunity to environmental policy-making.

A final point of interest is the regional-scale geographies of British rewilding, with some distinctions emerging between projects in the north and south, which can broadly be seen to reflect their respective upland and lowland geographies. Ecologically we see a preference for greater afforestation in upland environments that have been denuded and degraded from various forms of land clearance and heavy grazing, while in lowland areas there is an emphasis upon increasing herbivory to disturb tree encroachment in conservation areas. Despite taking opposing actions, in both cases, decisions are being made in an effort to restore what are perceived to be more natural interactions between vegetation succession and disturbance; practitioners are coping with differing starting points and ecological conditions. The specific ecological aspirations of different rewilding projects then connect with particular socioeconomic and cultural issues, depending on the levels of trade-off between stakeholders' preferred land uses. However, rewilding is not all about trees, despite a good deal of debate centring on this (Hodder et al., 2009; Wynne-Jones, 2012; Kirby and Watkins, 2015; Rotherham, 2017); neither is it about striving towards a particular habitat type, but rather creating diverse ecological systems that are dynamic and variable. Rewilding in Britain reflects this diversity, including mixed mosaic lowland habitats, marine and wetland environments, and often aspiring to catchment-scale approaches which explicitly aim to connect ecosystem processes through the landscape from upland to coast.

Defining rewilding in Britain

Given the plasticity of the term rewilding (Jørgensen, 2015; Gammon, 2017) it is important to be clear what we mean in the British context. In the following section, we discuss defining characteristics before moving to consider examples. Beginning with a broad interpretation, if rewilding means returning nature to a wilder state, we must determine what 'wild' means. Fisher and colleagues (2010) made a detailed assessment of what wild means in Scotland in relation to wild land across Europe (referred to as wild land, wildness, and wilderness). Their assessment culminated in mapping wilderness quality across Europe based on seven attributes: human population density, road density, distance from nearest road, rail density, distance from nearest railway line, naturalness of land cover, and terrain ruggedness. These seven attributes were applied so that human population density, infrastructure, and impact decrease wilderness quality, and natural vegetation communities and rugged terrain increase wilderness quality reflecting perceptions of wilderness.

Using such metrics is not straightforward, as Fisher and colleagues (2010) discuss; people's perceptions of wilderness and wildness vary considerably. Questions over desired landscape form are driven by aesthetic and cultural values. While unpopulated 'wild' landscapes have (a not unproblematic) appeal in a British context (particularly in the uplands, see e.g. MacDonald, 1998; Deary, 2016; Deary and Warren, 2017), a pastoral idyll is also dominant in the UK, which celebrates more agrarian visions (Daniels, 1993), demonstrating a range of pre- and post-Romantic influences on landscape aesthetics (Daniels, 1993; MacFarlane, 2009). Tensions here echo wider debates across Europe on the balance of preference for wild versus managed landscapes, an issue which remains closely wedded to place attachment and sense of identity (Bauer et al., 2009; Drenthen, 2009; Agnoletti, 2014; van Zanten et al., 2014; Navarro and Pereira, 2015b). This is reflected in concerns voiced by stakeholders:

> There are no wilderness areas in Wales and no potential for such ... We have a glorious landscape, which has been shaped by generations ... It is our cultural landscape and its values should be celebrated and not diminished through comparison with elsewhere ...
> (Pori Natur a Threftadaeth, 2008)

> ... [farmers] feel that they provide something of immense value ... in terms of a managed landscape ... we are part of a culture of shepherding, we are a pastoral society ... they are very much the guardians of that tradition.
> (Farming Representative Interview #18 2016)

Efforts to create 'wilderness' are also acknowledged to be problematic given the implicit erasure of 'peopled' histories and present-day exclusions (Cronon, 1996; Neumann, 2001; Adams and Mulligan, 2003; Pickerill, 2009; Jørgensen, 2015). While initially discussed in an American context, these difficulties are relevant to British landscapes (Toogood, 2003; Deary, 2016; Wynne-Jones et al., 2018). For example, following Fisher and colleagues (2010), increasing 'wilderness quality' equates with a reduction or removal of human populations, transport infrastructure, or human impact on land cover (and the livelihoods derived). This has justice implications, both in terms of impacts on communities affected, but also with regard to the decision-making processes surrounding such changes. Many remote rural communities are already suffering from out-migration and economic decline, and proposals for rewilding have been seen by some to compound the threats these areas face. This is particularly acute given uncertainties over farming in the face of subsidy decline and changing market conditions associated with Brexit (AHDB, 2017).

Responding to this issue, Prior and Brady (2017) have argued that aspirations for *wildness* are markedly different from a desire for *wilderness* and see rewilding as more aligned with ambitions for the former, given an emphasis on nature flourishing rather than an absence of people (see also Prior and Ward, 2016). This is similarly echoed by statements from Rewilding Britain and Rewilding Europe (Rewilding Britain, 2017a; Rewilding Europe, 2017). Both organisations (and several individual projects) now stress an explicit aim to support local communities to develop alternate economic activities directly linked to rewilding. This would involve a move away from reliance on extractive and consumptive uses, and transitioning towards eco-tourism and new 'nature-based economies' (Ayres and Wynne-Jones, 2014; Rewilding Europe, 2017; Wynne-Jones et al., 2018). Nonetheless, there are still substantial changes for communities (Navarro and Pereira, 2015a) and proposals for nature-based economies are not without their own difficulties. Moreover, it is apparent that some ontological tensions continue to persist in rewilding discourse around the degree of human separation from 'wild nature' (Arts et al., 2016). Some argue this provides a fundamental challenge to realising rewilding as a more progressive form of environmentalism which can overcome binary oppositions of nature and culture, and remains distant from 'fortress conservation' strategies that have a long and troubled history (Lorimer, 2015).

The next question to address is what 'nature flourishing' means for Britain. The wildness of nature is often equated to autonomy; i.e. nature having the freedom to determine its own form, function, and fate (Fisher and Parfitt, 2016; Prior and Ward, 2016; Wentworth and Alison, 2016; Rewilding Britain, 2017a; Woodland Trust, 2017). Nature determines its own fate through the interactions between organisms and between organisms and their environment (ecological processes) and natural selection (evolutionary processes), therein supporting biodiversity and ecosystem health. The potential for rewilding to support biodiversity has been central to the idea from the beginning; in their original introduction of rewilding, Soulé and Noss (1998) described it as complementary to traditional biodiversity conservation. Work since has echoed this sentiment, and often added the delivery of ecosystem services into the mix (Sandom et al., 2013c; Lorimer et al., 2015; Fisher and Parfitt, 2016; Svenning et al., 2016).

Unleashing ecological and evolutionary processes can simply be achieved through a passive approach, ceasing, or at least minimising, human impacts on other organisms' interactions. A concern associated with 'Passive Rewilding' is that the legacy of historical human impacts will still have a considerable bearing on how rewilded ecosystems develop today, with potentially damaging implications for biodiversity and society if applied the same way everywhere (Sandom et al., in press). This is particularly pertinent to the British context of highly degraded landscapes. In response, many rewilding proponents and practitioners propose Active Rewilding to restore ecosystems that will sustainably support native biodiversity as a precursor to Passive Rewilding. Active Rewilding focuses on restoring ecological processes, often through reintroducing species (Trophic Rewilding; Svenning et al., 2016), restoring physical structures or ecological conditions (Navarro and Pereira, 2015b). While there seems to be broad consensus and excitement for a nature-led approach to conservation among rewilding proponents, the different approaches to Active Rewilding have generated considerable discontent (Fisher, 2015), and are probably the reason why 'rewilding means different things to different people' is such a prevalent statement in the written submissions to the EAC inquiry[9] (Strouts, 2016; O'Neil, 2017).

What type of ecosystem will support biodiversity in Britain and what should be done to restore these ecosystems are the critical questions. The most heated exchanges are focused on whether vegetation succession (bottom-up processes) or herbivory and fire (top-down processes) should dominate naturally functioning ecosystems in Britain (Vera, 2000; Svenning, 2002; Mitchell, 2005). Britain is typically placed within the temperate closed-canopy forest biome (bottom-up dominance), but an alternative interpretation is that Britain, along with most of Europe, occurs within a region of ecosystem uncertainty that could be either closed-canopy woodland or open savanna-like wood–pasture (or a mixture) depending on the levels of herbivory at any particular time (bottom-up, top-down competition) (Bond, 2005; Bond et al., 2005). This debate has been fuelled by different interpretations of past environment reconstructions to determine what is natural, particularly in the early Holocene and the last interglacial (Vera, 2000; Whitehouse, 2006; Sandom et al., 2014). Different views have led to the varying approaches to Active

---

[9] www.parliament.uk/business/committees/committees-a-z/commons-select/environmental-audit-committee/inquiries/parliament-2015/future-of-the-natural-environment-after-the-eu-referendum-16-17/publications/

Rewilding, particularly regarding priorities and approaches to restoring the plant or herbivore communities.

What seems more certain is that diverse mosaics of habitat types are important for supporting Britain's biodiversity today (Hambler et al., 2010; Webb and Drewitt, 2010). This indicates that a mixed mosaic of habitats was the evolutionary context under which Britain's biodiversity evolved and thrived, and that sufficient remained in the early Holocene for biodiversity to be maintained. A potentially positive outcome of practitioners having different views and approaches to rewilding is that it may create the mixed mosaic of habitats needed to support biodiversity and provide ecologists an opportunity to better understand how nature works.

In terms of Active Rewilding, most attention has been given to the possibility of returning the big carnivores (following Soulé and Noss, 1998). Wolves and lynx still garner considerable attention in Britain (Sandom et al., 2012; White, 2015), but perhaps more importantly in the current British context is the return of beavers and their dam-building behaviour (Sandom and Macdonald, 2015; Puttock et al., 2017), wild boar and their rooting behaviour (Sims, 2005; Sandom et al., 2013a, 2013b), large herbivores (Kirby, 2004; Hodder et al., 2009), and smaller predators such as pine marten (Sheehy and Lawton, 2014; Vincent Wildlife Trust, 2017) and white-tailed eagle (Marquiss et al., 2004; RSPB, 2011). Restoring key elements of plant communities, important for restoring bottom-up successional processes, is also gaining greater attention. This tends to refer to returning seed sources to denuded vegetation communities through pioneer or enrichment planting (Benayas and Bullock, 2015). Much of what is possible and needed in Active Rewilding is determined by the scale and connectivity of the project. Rewilding emphasises the need for as large and well-connected projects as possible to give the greatest opportunity for ecological processes to function (Soulé and Noss, 1998; Rewilding Britain, 2017a). However, compromising on scale and connectivity, as is often required for practical reasons, increases the levels of management needed (Sandom et al., 2013c).

In summary, there seems to be consensus that rewilding means unleashing nature to determine its own fate, but views on how this should be done vary. Each form has its advantages and challenges. We see, broadly, that rewilding is not being equated with aspirations to recreate wilderness in light of the ecological and social tensions outlined. Instead, focus has been placed on increasing *wildness*. While some perceive rewilding as a threat to rural communities, others see it as an opportunity. Finally, while we present different schools of thought and various forms of rewilding (Passive, Active, Trophic), in practice, applying rewilding ideas is not this neat.

Rewilding in practice: what is happening on the ground?

In the first instance it is important to note that the term 'rewilding' is not always used to describe the approaches and projects that we review here as 'rewilding'. The projects we describe have different rationales behind their relative embrace or avoidance of the term rewilding, often connected to the levels of celebrity and infamy achieved by George Monbiot (O'Neil, 2017). In many cases this term has only become relevant in more recent years, given its popularisation, although the aspiration, and its scientific, aesthetic, and philosophic underpinnings, have deep roots which connect to a rich history of environmentalisms and ecological premises (Taylor, 2005; Lorimer et al., 2015).

We offer an overview of projects in Britain that we are aware of, which broadly fit criteria for different forms of rewilding. This includes projects listed by Rewilding Britain and Trees for Life as exemplars,[10,11] and projects that we have become aware of through various rewilding research and knowledge exchange events over the last 10 years, along with some reintroductions (both accidental and designed) that we think are relevant. Our overview is not exhaustive, and other authors might not include some examples that we have for the reasons discussed earlier.[12]

The geography of rewilding in Britain

Starting with Rewilding Britain's 13 examples, we can see that rewilding is occurring across Britain (Figure 12.1). There is, however, a bias towards the Scottish Highlands where wild(er)ness quality is already high (Fisher et al., 2010) and the structure of land ownership allows for large-scale projects (Warren, 2009). If the gold standard for rewilding is the creation of geographically large, connected, and ecologically intact ecosystems, under minimal human influence (Soulé and Noss, 1998), then rewilding in the Scottish Highlands offers the quickest wins. The Cairngorms and surrounding area in particular have seen a proliferation of rewilding relevant activity, with five of Rewilding Britain's 13 in this region. These projects are in areas characterised by open heather moorland, large deer populations, and remnant woodland. In the surrounding Highlands, the Community of Arran Seabed Trust (COAST) are pioneering rewilding in a marine context,[13] while Li and Coire Dhorrcail (in Knoydart) is again open moorland, deforested and grazed landscape (John

---

[10] See www.rewildingbritain.org.uk/rewilding/rewilding-projects/
[11] https://treesforlife.org.uk/blogs/article/10-exciting-rewilding-projects-happening-in-the-uk/
[12] Notably, we do not discuss 'unintentional' forms of rewilding that are occurring in both rural and peri-urban locations through land abandonment and a reduction of management (for example, in woodland regeneration in clear felled areas). We do, however, acknowledge that these offer exciting opportunities that would benefit from better documentation and monitoring.
[13] www.arrancoast.com/about-coast

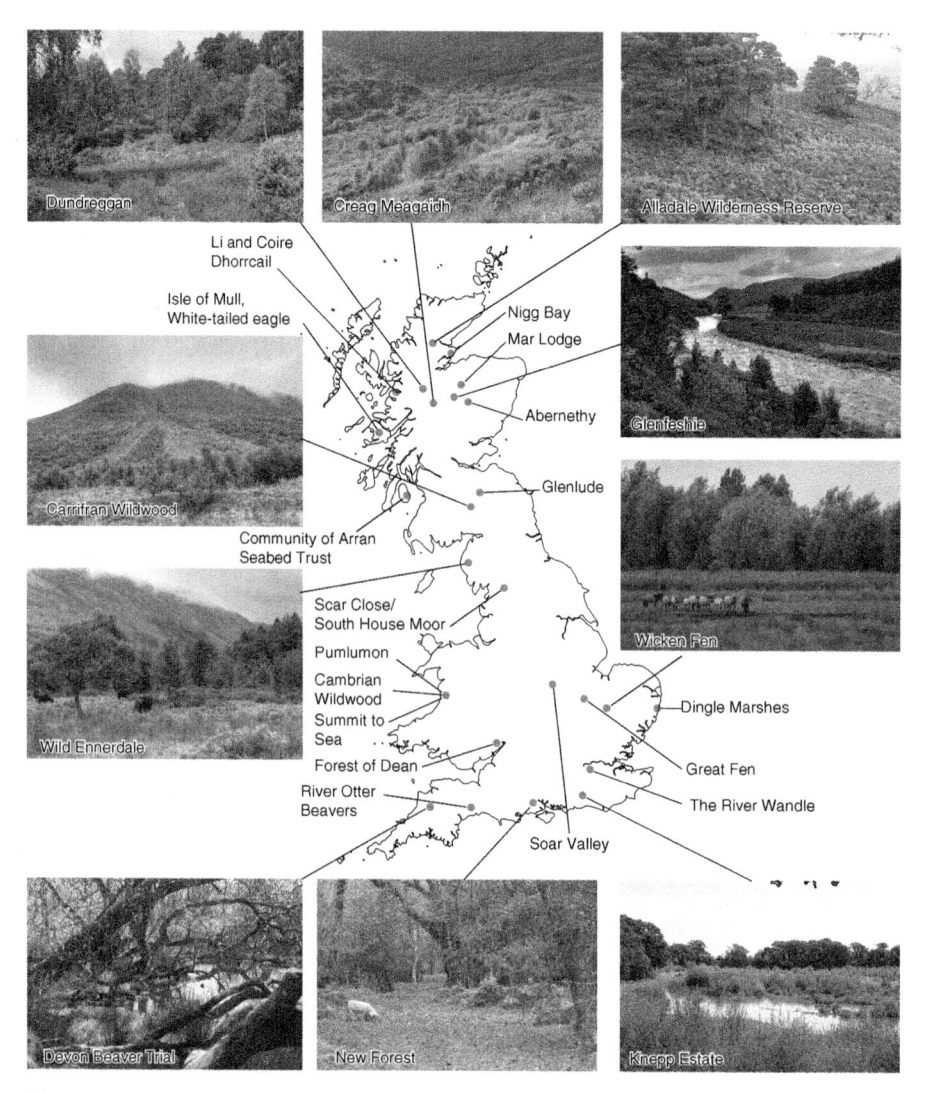

Figure 12.1. Rewilding-associated projects across Britain as identified by Rewilding Britain, Trees for Life, Mark Fisher, and the authors. Photos C. Sandom. (A black and white version of this figure will appear in some formats. For the colour version, please refer to the plate section.)

Muir Trust, 2015a). The Scottish borders and north of England also have long-standing projects: Carrifran Wildwood began as deforested and heavily grazed land (Newton and Ashmole, 2000), while Glenlude and Wild Ennerdale both have a history as commercial forestry (Wild Ennerdale, 2006; John Muir Trust, 2012). In Wales, the Cambrian Wildwood is aiming to restore lost woodland and increase ecological diversity in the uplands of Mid-Wales (WWLF, 2017). In the lowlands of England, Knepp Estate began as arable and pastoral land

(Knepp Estate, 2017), while the River Wandle project has seen the renaturalisation of one of the most heavily industrialised rivers in Britain (Pike et al., 2014).

Trees for Life have identified other projects that they associate with rewilding, including: Dingle Marshes, Suffolk; Great Fen, Cambridgeshire; Nigg Bay Managed realignment scheme, Cromarty Firth; Pumlumon Project, Welsh Uplands; and Soar Valley, Leicestershire.[14] This selection particularly highlights the relevance of restoring hydrological processes on marshland, fenland, and along the coast. Mark Fisher highlighted Scar Close and South House Moor within Ingleborough National Nature Reserve as regions where nature is being allowed to recover in the absence of livestock grazing pressure in his submission to the EAC inquiry. Other relevant sites include Wicken Fen in the south of England, which has a 100-year ambition to create a 5300-ha reserve of mixed mosaic wet habitats (The National Trust, 2009). North of the Cairngorms, Alladale Wilderness Reserve has ambitious, although controversial, aims to reintroduce wolves (*Canis lupus*), bear (*Ursus arctos*), lynx (*Lynx lynx*), European elk (*Cervus elaphus*), and wild boar (*Sus scrofa*) to a fenced reserve covering at least hundreds of square kilometres (Sandom et al., 2012, 2013a). The New Forest National Park is not a rewilding project as such, but roaming horses, pigs, cattle, and donkeys make for interesting ecological interactions that some would consider relevant (Hodder and Bullock, 2009).

Wild animals are also getting re-established elsewhere across Britain. The storm of 1987 and other escapes and releases freed farmed wild boar that are now established across much of the country, with a particularly large population in the Forest of Dean.[15] Beaver (*Castor fiber*) were officially reintroduced to Knapdale, Scotland, in 2009, but unofficially established themselves in Tayside in 2001 with a population now estimated to be over 100 individuals (Campbell et al., 2012). Across Scotland beavers are now officially accepted, protected and their range allowed to expand naturally,[16] although some tensions still persist.[17] Beavers have also been released without authorisation on the River Otter, Devon, England, and Natural England have now approved the managed re-release of beavers there (after testing) on a five-year trial basis.[18] Both Devon and Kent Wildlife Trusts are running trials with enclosed beaver. Associated research is highlighting the different ways beaver can impact landscapes. For example, extensive damming at the Devon beaver trial resulting in increased water storage, attenuated flow, and improved

[14] https://treesforlife.org.uk/blogs/article/10-exciting-rewilding-projects-happening-in-the-uk/
[15] www.forestry.gov.uk/pdf/FR_FeralWildBoarDeanCensus2016.pdf/$file/FR_FeralWildBoar DeanCensus2016.pdf
[16] www.nature.scot/professional-advice/safeguarding-protected-areas-and-species/protected-species/protected-species-z-guide/protected-species-beavers
[17] https://theferret.scot/scottish-beaver-cull-licensing/; see also Crowley et al., 2017.
[18] www.gov.uk/government/news/natural-england-approves-trial-release-of-beavers

water quality (Puttock et al., 2017), while the beavers at Knapdale have created over 13,000 m^2 of freshwater habitat (Jones and Campbell-Palmer, 2014).

The white-tailed eagle (*Haliaeetus albicilla*) reintroduction to Scotland returned an impressive predator, and the Isle of Mull has become a significant eco-tourist destination as a result. However, the threat of predation to livestock has also stoked human–wildlife conflict[19] (Simms et al., 2010). The red kite (*Milvus milvus*) survived in central Wales, and has now been reintroduced to England and Scotland. Pine marten (*Martes martes*) have been moved to Wales from Scotland to bolster the population there.[20] This is particularly interesting because evidence from Ireland suggests pine marten recovery can help native red squirrels coexist with grey squirrels (*Sciurus carolinensis*) (Sheehy and Lawton, 2014; Sheehy et al., 2018), due to their predation of the latter. Red squirrel (*S. vulgaris*) reintroductions are now being undertaken in the north of Wales, with further proposals as part of the Cambrian Wildwood (WWLF, 2017).[21]

Scale and connectivity

Rewilding is occurring in Britain at a wide range of scales. Rewilding Britain's exemplars include projects as small as 140 ha (Glenlude and Cambrian Wildwood, although the ambitions of the latter are greater) all the way up to 29,000 ha of Mar Lodge. Rewilding is being discussed at smaller scales still (the removal of livestock at Scar Close covers a 42-ha area), and even in urban contexts (Jorgensen and Keenan, 2012; Clacy, 2017; Rewilding Britain, 2017b), which has potential to engage people and promote innovation in conservation. However, most projects are trying to work at the largest scale they can. All of the Scottish Highland projects are measured in the thousands of hectares. South of the border, the largest projects are Wild Ennerdale, Wicken Fen, and the Knepp Estate, and also cover thousands of hectares. People's perceptions and ecological processes work across a huge range of spatial and temporal scales. For example, those living in the south-east of England are likely to identify and interpret wildness very differently to those living in the Scottish Highlands (Lutz et al., 1999; Bauer et al., 2009), while pollination delivered by a huge range of invertebrates operates at a different spatial scale to apex predators.

Even with rewilding projects distributed over the whole country, the prospect that corridors of wild land could ever connect them all is still a long way off. However, connecting smaller protected areas is the ambition of many rewilding projects. The Cambrian Wildwood project is now part of a wider 'Summit to Sea' vision covering 10,000 ha of the Cambrian Mountains. Trees for Life have a 200-year ambition of rewilding an enormous identified core

---

[19] See, for example, www.theguardian.com/environment/2015/may/25/isle-sea-eagles-mull-wildlife-tourists-sheep-farmers
[20] www.vwt.org.uk/projects/pine-marten-recovery-project/   [21] www.redsquirrels.info/

area.[22] The Great Fen Project wants to connect up protected areas of fenland in Cambridgeshire. Cairngorms Connect is seeking to connect projects across the Cairngorms. Larger protected areas that are better connected is the core message behind the Lawton *Making Space for Nature* report (2010), and it is an example of how rewilding and more traditional conservation can dovetail nicely.

Rewilding visions

Across the management plans of these rewilding projects is a consistent ambition to give nature greater freedom to 'take its course'. The owners of the Knepp Estate indicate that they believe their approach is 'radically different to conventional nature conservation in that it is not driven by specific goals or target species'.[23] Wicken Fen similarly identifies a contrast between working with natural processes and setting narrow species-driven goals (The National Trust, 2009). Mark Fisher describes the removal of livestock grazing at Scar Close as an opportunity to observe what will happen (EAC, submitted evidence). For some projects, explicit habitat targets have been set. At Carrifran Wildwood there is a stated ambition to recreate the past wooded landscape as far as possible (Newton and Ashmole, 2000), while the target for Dundregan is to restore 60 per cent of the site as a structurally diverse woodland (Trees for Life, 2017). However, in both of these examples there is a sense of restoring the ecosystems to a starting condition that will then be allowed to develop naturally through ecological processes (Trees for Life, personal communication, August 2017). Project management plans often highlight limits to how wild the sites will be allowed to go, with qualifying statements such as 'as far as possible', 'unless an estate asset is under risk', 'as much freedom as possible'. Qualifiers are likely to apply to most projects because of the diverse set of challenges rewilding projects face. The risks rewilding presents are not being ignored. For example, of the 110 written submissions to the EAC inquiry that referred to rewilding, 41 mentioned opportunities, while 33 highlighted risks.

None of the projects are taking a zero-management approach. All have undertaken some degree of restoration or have ongoing management. These actions include: removal of human infrastructure or physical impacts; (re)introducing missing or depleted fauna; planting missing flora; managing large herbivore population dynamics in the absence of large predators (e.g. culling or exclusion fencing); removal of non-native species; and creating keystone structures. The removal of human infrastructure and impacts is intended to restore hydrological processes, increase connectivity for wildlife, and increase the perception of wildness for visitors. Along the River Wandle,

---

[22] https://treesforlife.org.uk/work/core-area/   [23] https://knepp.co.uk

five weirs have been lowered or had fish passes added or improved.[24] At Knepp, four weirs have been removed, 1.5 miles of drainage canal filled in, and meanders restored to river Adur.[25] The application of wilder management plans is being implemented sensitively. For example, Carrifran's management plans report that the removal of buildings, walls, and other human structures will only occur if these structures are not of archaeological or cultural value.

Many projects are introducing species to their sites. Large herbivores have been introduced at most lowland sites to restore herbivory in order to restore plant–herbivore interactions. Fewer upland situated projects have introduced large herbivores, but there are exceptions in Wild Ennerdale and Alladale and other projects are considering it. Wild boar have been at Dundregan and Alladale, while Tamworth pigs have been introduced to the Knepp Estate. Knepp, Cambrian Wildwood, Glenlude, Li and Coire Dhorrcail, Abernethy, Dundreggan, and Alladale are all considering further species reintroductions.

Native tree planting is prevalent across upland sites. Two forms of planting are being used: 'pioneer planting' is establishing a seed source on open and treeless land, while 'enrichment planting' returns underrepresented species within a woodland community (Roberts, 2010). Tree planting and regeneration in the uplands is strongly associated with deer culling or exclusion fencing. At Creag Meagaidh deer density was brought down to < $5/km^2$, which has allowed natural tree regeneration. Many projects are seeking to remove non-native tree plantations, such as Sitka spruce. In areas with large homogeneous areas of non-natives, for example Wild Ennerdale and Glenlude, the proposal is to phase out plantations as a long-term strategy rather than clear felling.

Human dimensions

Rewilding is not just about the reconfiguration of natural systems – it also impacts on human livelihoods and well-being. To understand these human outcomes, we posed four key questions, which we address in turn.

(How) are the aims of rewilding projects connected to human benefits/needs? Rewilding has been acknowledged from the outset as a means to rework human–environment relationships (Taylor, 2005). 'Rewilding ourselves' is an explicit focus in the work of Trees for Life, Cambrian Wildwood, and in Monbiot's *Feral* (2013). For some practitioners, this is discussed in spiritual terms (Strouts, 2016). Others assert rewilding as a means to reconnect with nature and counteract 'nature deficit disorders' (Louv, 2008).[26] The majority of projects run specific engagement activities to achieve these ends, either supporting volunteers onsite to undertake

---

[24] www.wandletrust.org/tag/river-restoration/ [25] https://knepp.co.uk/river-restoration/
[26] See www.wildlifetrusts.org/rewilding

restoration activities, or enabling visitor access (through organised safaris/ wildlife watching, educational visits, or recreation activities such as camping, bushcraft, and walking). Attention is also being given to rewilding as a land management 'tool' to deliver ecosystem service benefits including flood mitigation,[27] carbon storage, and clean water. Coupled with this is an emphasis on income generation from payments for ecosystem services (PES) and ecotourism. This messaging is central to Rewilding Europe and has been taken up in earnest by Rewilding Britain over 2017 (Rewilding Britain, 2017a; Rewilding Europe, 2017). For individual projects, a mixed picture is evident. Many older projects have no focus on realising revenue from tourism or PES (Carrifran Wildwood), while more recent projects are lauding this as a key ambition (Cambrian Wildwood). Others are more ambivalent, acknowledging the tourist benefits that rewilding can bring and the importance of these for local economies, but not setting core targets around this (Wild Ennerdale). The differences in approach are largely reflected in the different organisational structures and business models that projects employ.

Who is doing rewilding? Our research demonstrates that a mix of organisations is involved in rewilding, but the NGO sector dominates. The John Muir Trust is the biggest land owner in Scottish rewilding-associated projects, while the National Trust has a notable presence in projects across Britain. Some public-sector organisations and statutory bodies, including Natural England, Forestry Commission, and Scottish Natural Heritage, are already engaged, and it is possible that further interest could be shown if revenue opportunities and clearer indicators of public benefit are forthcoming. New organisations and enterprises have also emerged, along with lots of partnerships, demonstrating the high levels of innovation and energy that rewilding has inspired. Bigger NGOs are working with community groups and small emerging charities. For example, the Woodland Trust has collaborated with Wales Wild Land Foundation on the Cambrian Wildwood, and John Muir Trust have partnered with the Borders Forest Trust at Carrifran and community members in Knoydart. Established organisations are also working together to realise larger-scale visions; for instance, at Wild Ennerdale and the Great Fen Project. Private actors are also involved, including United Utilities in Wild Ennerdale and large estates (Alladale, Glenfeshie in Scotland and Knepp in the south of England). The private estates have a strong emphasis upon developing tourist revenue streams,[28] but accessibility has been raised as a concern. For example, at Alladale, ambitions for large carnivore releases involve fencing to contain

---

[27] See, for example, www.rewildingbritain.org.uk/assets/uploads/files/publications/Rewilding%20and%20Flood%20Risk%20Management%20briefing.pdf
[28] Note: Glenfeshie are not branding themselves as a rewilding project (personal communication, July 2017).

the animals that comes into conflict with 'right to roam' legislation.[29] Moreover, the price tag attached to tourist experiences does not always suggest 'accessibility'. More broadly, there is a concern that rewilding could be construed as an elite venture, with wealthy 'philanthropists' buying up areas of land where traditional rural businesses are failing. This is particularly problematic in cultural terms as communities may feel themselves to be undermined (Wynne-Jones et al., 2018).

Who is paying for rewilding? Beyond the private estates noted, which were already owned by the individuals in question, the majority of ventures have been supported by charitable funding and public appeals. Their success suggests high levels of public buy-in, although it is noted that many appeals are couched in diverse terms and not simply targeting 'rewilding'. Large donors are important to charities and many philanthropists are excited by rewilding (Rewilding Britain interview #19 May 2017). Some nascent charities are even exploring options whereby a private buyer 'donates' land to a wider rewilding project while still maintaining legal ownership rights, although questions arise here over the levels of control afforded through resulting governance arrangements. Notably, not all the projects listed have sought land ownership, and Trees for Life in particular have pursued a strategy of partnership working to extend the Caledonian Forest. Ongoing costs vary depending on the approach taken. Many projects are pursuing an Active Rewilding approach with higher associated costs. Intentions to move towards lower intervention 'passive' approaches are hoped to reduce costs of management. The potential for cost savings has meant passive approaches are being lauded in policy circles, particularly in comparison with the costs of agri-environmental schemes and reserve upkeep (Interview NGO stakeholder #21 May 2017[30]). It is, however, too early to be clear whether these claims will be realised into the future.

(How) are local communities being involved? A critical question for all projects is the level of community support and processes of inclusion. 'People, communities and livelihoods' constitutes a core focus for Rewilding Britain in their work (Rewilding Britain, 2017a). Nonetheless, it is undeniable that earlier publicity surrounding *Feral* (Monbiot, 2013) has given rewilding a poor reputation among many rural land managers and community members (Sandom et al., in press). Equally, many of the projects listed have prompted

---

[29] See www.theguardian.com/environment/2014/sep/19/-sp-rewilding-large-species-britain-wolves-bears

[30] https://publications.parliament.uk/pa/cm201617/cmselect/cmenvaud/599/599.pdf; www.parliament.uk/business/committees/committees-a-z/commons-select/environmental-audit-committee/news-parliament-2015/future-of-natural-environment-after-the-eu-referendum-launch-16-17/

some degree of conflict over proposals for animal introductions, management visions, and landscape aesthetics. However, despite a difficult few years, efforts at amelioration are now being pursued, including open public meetings to explore rewilding in less-aggressive terms (Yorke, 2016) and evidence of considered engagement from individual projects.

This is not to suggest that remediation will be easy and many of the points raised earlier regarding the associations of rewilding with unpopulated wilderness continue to play out. These debates run deeper than conflicting cultural preferences, and connect to the troubled histories of Highland Clearances in eighteenth-century Scotland, where crofting communities were moved off the land to make way for sheep farming by the lairds (Toogood, 2003; Deary, 2016). Difficulties are also associated with rural–urban and insider–outsider divisions, with rewilding sometimes perceived as an agenda from metropolitan elites that is being imposed on rural communities.[31] Debates in Wales have even seen echoes of nationalistic conflict portraying rewilding as an English imposition and threat to Welsh identity (Wynne-Jones et al., 2018).

Projects have sought to overcome these challenges through a range of strategies. Acknowledging that large carnivores are not on the agenda in Wales has provided some assurance for farming interests (Wynne-Jones et al., 2018). Undertaking careful processes of stakeholder engagement for the reintroductions that have taken place (with assurances that problem animals will be removed) is an approach that has seen Vincent Wildlife Trust gain noted respect (Interview Land-Use Stakeholder 17 May 2016). In Scotland, SNH are offering compensation to farmers affected by sea eagle predation on livestock,[32] while supporting community buy-outs of estates has offered an important mechanism for the John Muir Trust to work with local people. In addition, acknowledgement that the Highlands were not always such an 'empty' wilderness has been critical to gaining acceptance of the rewilding vision (Deary, 2016; Deary and Warren, 2017). More broadly, burgeoning efforts to link rewilding with revitalised rural economies seems to be winning favour among community stakeholders. Some of the private estates are clearly seeing rewards, but we cannot tell whether their successes can be emulated as yet.

The need to be attentive to current pressures on rural communities is becoming increasingly imperative with Brexit, and sensitivities surrounding a sense of cultural loss have added weight to this. Acknowledging these difficulties, projects have sought to engage with local identities and the sense of place attached to rewilding areas; for example, through Wild

---

[31] See, for example, tensions over lynx reintroductions in Northumberland: www.hexham-courant.co.uk/news/Residents-question-lynx-proposal-73aaa858-9dab-4b32-ac83-f431da27be05-ds

[32] www.snh.scot/professional-advice/land-and-sea-management/managing-wildlife/sea-eagle-management-scheme

Ennerdale's 'Spirit of Place' activities in 2016–17 and connections with family legacies in the area.[33] While in Wales, efforts to connect with Welsh cultural heritage and linguistic traditions have been key for the Cambrian Wildwood (Wynne-Jones et al., 2018).

Overall, we have seen marked efforts to move beyond dividing lines and controversies, with rewilding advocates now actively seeking to identify areas of common ground amongst different parties. There are undeniably entrenched difficulties to work through, and the processes of diplomacy required offer no quick fixes, but rewilding is becoming a more distinctly peopled vision.

Outcomes

The discussion thus far has outlined how rewilding has been associated with a wide variety of opportunities and risks. Here we outline whether these have been realised at British sites. While it is not possible to provide a comprehensive review in the space of this chapter, the following section offers a brief summary of key trends and highlights future research priorities.

The recorded biodiversity benefits at rewilding sites across Britain are a clear highlight for many stakeholders. At Creag Meagaidh, black grouse went from sporadic recordings to five well-established leks and 30 birds (SNH, 2008). In 2012, eight species were recorded at Dundregan that have not been recorded in Britain before.[34] At the Knepp Estate, nightingale territories have doubled from nine in 1999 to 18 in 2005, male turtle doves have increased from three in 1999 to 16 in 2017, and purple emperor butterflies have gone from no records to reportedly the largest population in the UK.[35] However, benefits need to be considered against declines in other species. The development of woodland habitat at Carrifran has seen woodland bird species richness increase from two species in 1998 to 14 species in 2015 (Savory, 2016), with an aligned increase in abundance of woodland birds. However, abundance of meadow pipit has declined from ~280 birds to ~180 over the same period (Savory, 2016). At Wild Ennerdale bird richness in the grazed region on the valley bottom increased from 13 to 16 species, and the number of territories increased from 43 to 70 between 2008 and 2015 (Ullrich, 2015). However, numbers of green woodpecker, great spotted woodpecker, and song thrush have declined in the area. Nonetheless, these examples suggest that, at least for birds, more nature-led approaches to management may be delivering a net benefit to species richness and abundance so far. More comprehensive monitoring, with suitable control sites is needed to understand net biodiversity impacts across different rewilding approaches, in different circumstances.

---

[33] www.wildennerdale.co.uk/people/tom-rawling/
[34] https://treesforlife.org.uk/work/dundreggan/   [35] https://knepp.co.uk/the-results/

Considering the impact of rewilding on ecosystem services, research on the enclosed beavers in Devon found that mean peak discharge was reduced by 34 per cent ± 9 per cent in storm events. At the same site, sediment, nitrogen, and phosphorus concentrations were reduced after water had flowed through the 13 beaver dams that have been constructed (Puttock et al., 2017). Rewilding Britain and Friends of the Earth have compiled arguments for a more nature-based approach to flood mitigation (Rewilding Britain and Friends of the Earth, 2016).

So far, relatively few data appear to be available on greenhouse gas sequestration and emissions, soil restoration, or fire dynamics from our rewilding projects. However, various sites do identify opportunities in these areas. For example, Hodder and colleagues (2010) worked with four projects we have associated with rewilding to visualise and estimate the socioecological implications of their new management of the land in 2060. This analysis indicates that Knepp is anticipated to increase carbon storage by more than 50 per cent. The Knepp estate has also highlighted the potential for rewilding to restore degraded soils, calling for long-term set-aside schemes (> 20 years) to allow for soil recovery, although rewilding can conflict with food security by taking land out of production. In their EAC submission, The England and Wales Wild Fire Forum describe rewilding as a 'recipe for disaster', although this is associated with scrub encroachment from reduced herbivory (which is not the case for all sites). Hodder and colleagues (2010) also looked at how the new approaches to conservation influenced the cost of management. All four projects experienced increased costs of management, contrary to expectations for rewilding projects. This relates to the limited use of passive rewilding to date.

In terms of socioeconomic outcomes, a lot of attention has been paid to tourism. A report by Birnie and Barnard (2016) for Rewilding Britain highlighted a variety of potential benefits. Tourism currently generates £18.6 billion for Britain's rural economy, providing 340,000 full-time jobs. Importantly, over 65 per cent of the total spend on nature-based adventure tourism is thought to remain in local economies, ensuring a link between nature and local economic benefits. Nature-based tourism opportunities are present from summit to coast. Tourists going to see dolphins in Scotland's Moray Firth reportedly spend £9 million annually. Ospreys are worth an estimated £3.5 million to Scotland's economy, while a single pair in the RSPB Cors Dyfi reserve in Wales generate £350,000 per year. Glamping, camping, and safaris at the Knepp estate has turned over £230,000 in its fourth year (personal communication, 2018). Walkers holidaying in Scotland spent £174 million in 2012. These initial figures seem promising, but ongoing monitoring is required to see how markets develop or whether they reach saturation. Equally, it is important to determine *who* is gaining benefits/disbenefits, and how projects are responding to impacts arising. Initial research has begun to address these questions, as noted earlier, but social evaluation (and planning) needs to be an integrated part of projects going forward.

Overall, standardised monitoring of net effects on a variety of key biological indicator groups, a diverse suite of ecosystem services, and the economy, along with qualitative social and cultural monitoring would greatly aid opportunities to assess the outcomes of rewilding in different contexts. This would be particularly beneficial for determining when, where, and how rewilding can be useful. Monitoring is often expensive, time-consuming and requires specialist skills, so is often either implemented opportunistically or jettisoned.

The future of rewilding in Britain

Will we see more or less rewilding in the future? There are barriers and opportunities ahead. Our reporting demonstrates a marked rise in the levels of rewilding-relevant activity over the last 10 years. After years of professional debate, there is a wider embrace across the sector of a need for a new conservation paradigm. Rewilding also appears to hold great public appeal, given levels of engagement with popular publications and social media outlets, although less formal research has been conducted to confirm this. However, not everyone is on board, and those seeking to defend the interests of specific species and habitats with anthropogenic dependencies have raised noted concerns. Equally, those perceiving rewilding as a threat to their livelihoods and valued human landscapes have been understandably vocal. While some mitigations may be possible, these concerns prompt us to note that rewilding should not take place everywhere, although we acknowledge that trying to find anywhere without diverging opinion on the matter is unlikely.

Brexit has been noted as an opportunity for rewilding, enabling the retargeting of rural development payments and monies associated with CAP. High levels of vulnerability within upland farming suggest that land-use transition could be imminent, but rewilding advocates need to proceed cautiously so as to avoid wholesale cultural loss and animosity among rural communities. In terms of environmental legislation, it is likely that the legal repeal process will transfer key laws and obligations, and international obligations will remain. National and transferred laws and obligations can pose barriers to rewilding. Maintaining protected areas in good ecological condition is often associated with a need to manage. Personal communication with various reserve and land managers has highlighted the difficulties of taking more nature-led approaches when the land managed is designated as a Special Site of Scientific Interest (SSSI), for example. As a result, future and more ambitious rewilding may be associated with undesignated land.

Confusion still abounds about what rewilding is, and greater clarity will be needed if it is to become an important conservation approach in the future.

Based on our review and wider research, we propose that rewilding is still fundamentally defined as an approach that allows nature to decide, but suggest a framework for initiating and overseeing rewilding projects. This framework is constructed based on a number of assumptions that not all stakeholders will share, but we believe it incorporates enough flexibility to accommodate a diversity of visions and approaches depending on project priorities and circumstances.

1. Overarching goals of rewilding are to:
    a. create self-willed ecosystems, which are allowed the freedom to evolve and change as dictated by ecological processes;
    b. halt and ultimately reverse the decline of biodiversity; and
    c. increase the delivery of a diverse suit of ecosystem services to support society.
2. Guiding principles for implementing rewilding projects are:
    a. rewilding is an agreement among owners/managers, policy-makers, and all engaged stakeholders to accept the outcomes of allowing nature to manage itself as far as possible;
    b. rewilding can include actions to kick-start ecological processes to get a target ecosystem into a more ecologically represented and intact starting condition, after which nature is allowed to manage itself; and
    c. management can be used in rewilding projects to avoid undesirable biodiversity and socioeconomic outcomes, but the more management that is used the less wild the project is considered.
3. Rewilding actions to kick-start ecological processes include:
    a. species reintroduction and translocation;
    b. species removal and management;
    c. human infrastructure removal or remediation; and
    d. altering human activity to avoid humans determining the outcomes of ecosystems, by lessening human domination of ecological processes.
4. Under these guidelines, rewilding can be:
    a. carried out at any scale, but bigger and more connected is better;
    b. applied to any area of land or water;
    c. a diverse set of projects that include many different visions for wild land; and
    d. determined a success or failure depending on whether the rewilding generates a net positive impact on biodiversity and ecosystem services, and reduces the amount of long-term management required to achieve these positive impacts.

5. Under these guidelines, rewilding cannot be used to manage for a particular target species, habitat, or ecosystem outcome.
6. Key assumptions include:
    a. rewilding is an idea that sits within and is complementary to existing ecological restoration and biodiversity conservation approaches;
    b. rewilding is not appropriate everywhere;
    c. rewilding must be implemented by willing and supportive communities; and
    d. people are part of nature, but communities and visitors in rewilding areas agree to take steps to give precedence to ecological processes and systems.

There are still plenty of difficulties when applying this framework, especially when trying to take a wilder approach at ecologically smaller scales with an impoverished ecosystem, a context we have seen many British projects trying to grapple with. Yet we also contend that many of the lessons currently being learned in the British context are central to developing a more viable conception of rewilding in broader international terms. These include points around levels of human involvement, the extent and rationale of management, and working with incumbent legislative/regulatory frameworks and cultural value systems. We believe rewilding in Britain offers interesting insights about how to give nature more freedom in human-dominated landscapes. The north–south, upland–lowland divide usefully highlights the diverse visions that more nature-led approaches will deliver in response to varying circumstances. Finally, as an island, Britain provides an interesting ongoing case study in exploring how a country might choose to actively rewild itself.

Acknowledgements

CJS acknowledges support from the NERC Knowledge Exchange Fund (NE/P005926/1). SWJ acknowledges support from the Sêr Cymru National Research Network for Low Carbon, Energy & Environment Research Development Fund. We would also like to thank all the practitioners and policy-makers who have discussed rewilding in Britain with us; Thomas Dando, Betsy Brown, and Georgina Pashler for support assessing the literature; Graham Strouts and Callum O'Neil for conducting research interviews which have informed this chapter.

References

Adams, W.M., and Mulligan, M. (2003). *Decolonizing nature: strategies for conservation in a post-colonial era*. London: Earthscan.

Agnoletti, M. (2014). Rural landscape, nature conservation and culture: some notes on research trends and management approaches from a (southern) European perspective. *Landscape and Urban Planning*, **126**, 66–73.

AHDB. (2017). Brexit Scenarios and Impact Assessment. Horizon Market Intelligence.

Arts, K., Fischer, A., and van der Wal, R. (2016). Boundaries of the wolf and the wild: a conceptual examination of the relationship between rewilding and animal reintroduction. *Restoration Ecology*, **24**, 27–34.

Ayres, S., and Wynne-Jones, S. (2014). Cambrian Wildwood – new ventures in a wilder landscape. *ECOS*, **3**, 23.

Bauer, N., Wallner, A., and Hunziker, M. (2009). The change of European landscapes: human-nature relationships, public attitudes towards rewilding, and the implications for landscape management in Switzerland. *Journal of Environmental Management*, **90**, 2910–2920.

Benayas, J.M.R., and Bullock, J.M. (2015). Vegetation restoration and other actions to enhance wildlife in European agricultural landscaps. In Pereira, M.P. and Navarro, L.M. (Eds.), *Rewilding European landscapes* (pp. 127–142). Cham: SpringerOpen.

Birnie, N., and Barnard, F. (2016). Socio-economic benefits of rewilding in the Highlands of Scotland. Report for Rewilding Britain by Conservation Capital.

Bond, W.J. (2005). Large parts of the world are brown or black: a different view on the 'Green World' hypothesis. *Journal of Vegetation Science*, **16**, 261–266.

Bond, W.J., Woodward, F.I., and Midgley, G.F. (2005). The global distribution of ecosystems in a world without fire. *New Phytologist*, **165**, 525–537.

Campbell, R., Harrington, A., Ross, A., and Harrington, L. (2012). *Distribution, population assessment and activities of beavers in Tayside*. Scottish Natural Heritage Commissioned Report No. 540. Inverness: SNH.

Carver, S., and Müller, C. (2014). *Wildness study in Wales*. Leeds: Wildland Research Institute, University of Leeds.

Clacy, C. (2017). Placing the wild: conserving 'wild-life' in post-industrial urban Britain. In *Decolonising Wild-life*. London: Royal Geographical Society-Institute of British Geographers Annual Conference.

Convery, I., and Dutson, T. (2008). Rural communities and landscape change: a case study of wild Ennerdale. *Journal of Rural and Community Development*, **3**, 104–118.

Cronon, W. (1996). The trouble with wilderness or, getting back to the wrong nature. *Environmental History*, **1**, 7–28.

Crowley, S.L., Hinchliffe, S., and McDonald, R.A. (2017). Nonhuman citizens on trial: the ecological politics of a beaver reintroduction. *Environment and Planning A*, **49**, 1846–1866.

Daniels, S. (1993). *Fields of vision: landscape imagery and national identity in England and the United States*. Cambridge: Polity Press.

Deary, H. (2016). Restoring wildness to the Scottish Highlands: a landscape of legacies. In Hourdequin, M. and Havlick, D.G. (Eds.), *Restoring layered landscapes* (pp. 95–111). Oxford: Oxford University Press.

Deary, H., and Warren, C.R. (2017). Divergent visions of wildness and naturalness in a storied landscape: practices and discourses of rewilding in Scotland's wild places. *Journal of Rural Studies*, **54**, 211–222.

Donlan, J. (2005). Re-wilding North America. *Nature*, **436**, 913–914.

Drenthen, M. (2009). Ecological restoration and place attachment: emplacing non-places? *Environmental Values*, **18**, 285–312.

Fisher, M. (2015). A challenge to Rewilding Britain. www.self-willed-land.org.uk/articles/challenge_RB.htm

Fisher, M., and Parfitt, A. (2016). The challenge of wild nature conserving itself. *ECOS*, **37**, 27–34.

Fisher, M., Carver, S., Kun, Z., McMorran, R., Arrel, K., and Mitchell, G. (2010). *Review of status and conservation of wild land in Europe*. Project commissioned by the Scottish Government. Leeds: Wildland Research Institute, University of Leeds.

Foreman, D. (2004). *Rewilding North America: a vision for conservation in the 21st century*. Washington, DC: Island Press.

Gammon, A.R. (2017). Rewilding – a process or a paradigm? *ECOS*, **38**.

Garnett, T., Appleby, M.C., Balmford, A., et al. (2013). Sustainable intensification in agriculture: premises and policies. *Science*, **341**, 33-34.

Hambler, C., Henderson, P.A., and Speight, M.R. (2010). Extinction rates, extinction-prone habitat, and indicator groups in Britain and at larger scales. *Biological Conservation*, **144**, 713-721.

Helm, D. (2016). British agricultural policy after BREXIT. Natural Capital Network – Paper 5. www.dieterhelm.co.uk/natural-capital/environment/agricultural-policy-after-brexit/

Henle, K., Alard, D., Clitherow, J., et al. (2008). Identifying and managing the conflicts between agriculture and biodiversity conservation in Europe – a review. *Agriculture Ecosystems and Environment*, **124**, 60-71.

Hodder, K.H., and Bullock, D. (2009). Really wild? Naturalistic grazing in modern landscapes. *British Wildlife*, **20**, 37-43.

Hodder, K.H., Buckland, P.C., Kirby, K.K., and Bullock, J.M. (2009). Can the mid-Holocene provide suitable models for rewilding the landscape in Britain? *British Wildlife*, **20**, 4-15.

Hodder, K., Douglas, S., Newton, A., et al. (2010). *Analysis of the costs and benefits of alternative solutions for restoring biodiversity*. Final report, Defrea project WC0758/CR0444. Bournemouth: Bournemouth University.

Jepson, P., and Schepers, F. (2016). Making space for rewilding: creating an enabling policy environment. www.google.co.uk/url?sa=t&rct=j&q=&esrc=s&source=web&cd=1&cad=rja&uact=8&ved=0ahUKEwjnsbW-g9_ZAhVJB8AKHetVCS0QFggsMAA&url=https%3A%2F%2Fwww.rewildingeurope.com%2Fwp-content%2Fuploads%2F2016%2F05%2FMaking-Space-for-Rewilding-Policy-Brief1.pdf&usg=AOvVaw16NmFiTMG0AsXIyvqJyuFz

John Muir Trust. (2012). Management Plan Glenlude Estate.

John Muir Trust. (2015a). Li and Coire Dhorrcail Management Plan.

John Muir Trust. (2015b). Rewilding: restoring ecosystem for nature and people. www.johnmuirtrust.org/assets/000/000/397/rewilding_policy_agreed0315_published_original.pdf?1434628289

Jones, S., and Campbell-Palmer, R. (2014). *The Scottish Beaver Trial: the story of Britain's first licensed release into the wild*. Final report.

Jorgensen, A., and Keenan, R. (2012). *Urban wildscapes*. Abingdon: Routledge.

Jørgensen, D. (2015). Rethinking rewilding. *Geoforum*, **65**, 482-488.

Kirby, K.J. (2004). A model of a natural wooded landscape in Britain as influenced by large herbivore activity. *Forestry*, **77**, 405-420.

Kirby, K., and Watkins, C. (2015). *Europe's changing woods and forests: from wildwood to managed landscapes*. Wallingford: CABI.

Knepp Estate. (2017). Knepp Wildland. https://knepp.co.uk

Lawton, J.H., Brotherton, P.N.M., Brown, V.K., et al. (2010). *Making space for nature: a review of England's wildlife sites and ecological network*. London: Defra.

Lorimer, J. (2015). *Wildlife in the Anthropocene – conservation after nature*. Minneapolis, MN: University of Minnesota Press.

Lorimer, J., Sandom, C., Jepson, P., Doughty, C., Barua, M., and Kirby, K.J. (2015). Rewilding: science, practice, and politics. *Annual Review of Environment and Resources*, **40**, 39-62.

Louv, R. (2008). *Last child in the woods: saving our children from nature-deficit disorder*. Chapel Hill, NC: Algonquin Books.

Lutz, A.R., Simpson-Housley, P., and De Man, A.F. (1999). Wilderness – rural and urban attitudes and perceptions. *Environment and Behavior*, **31**, 259-266.

MacDonald, F. (1998). Viewing Highland Scotland: ideology, representation and the 'natural heritage'. *Area*, **30**, 237-244.

Marquiss, M., Madders, M., Irvine, J., et al. (2004). The impact of white-tailed eagles on

sheep farming on Mull: final report. www
.gov.scot/Resource/Doc/47060/0014566.pdf
Merckx, T., and Pereira, H.M. (2015). Reshaping
agri-environmental subsidies: from
marginal farming to large-scale rewilding.
Basic and Applied Ecology, **16**, 95–103.
Mitchell, F.J.G. (2005). How open were
European primeval forests? Hypothesis
testing using palaeoecological data. *Journal
of Ecology*, **93**, 168–177.
Monbiot, G. (2013). *Feral: searching for
enchantment on the Frontiers of rewilding*.
London: Penguin.
Navarro, L.M., and Pereira, H.M. (2015a).
Rewilding abandoned landscapes in
Europe. In *Rewilding European landscapes* (pp.
3–23). New York, NY: Springer.
Navarro, L.M., and Pereira, H.M. (2015b).
Rewilding European landscapes. New York,
NY: Springer.
Neumann, R.P. (2001). *Imposing wilderness:
struggles over livelihood and nature preservation
in Africa*. Berkeley, CA: University of
California Press.
Newton, A., and Ashmole, P. (2000). Carrifran
Wildwood Project – native woodland
restoration in the Southern Uplands of
Scotland; management plan.
Nogues-Bravo, D., Simberloff, D., Rahbek, C.,
and Sanders, N.J. (2016). Rewilding is the
new Pandora's box in conservation. *Current
Biology*, **26**, R87–R91.
O'Neil, C. (2017). What rewilding means in
Britain. MSc, Bangor University.
Pickerill, J. (2009). Finding common ground?
Spaces of dialogue and the negotiation of
Indigenous interests in environmental
campaigns in Australia. *Geoforum*, **40**,
66–79.
Pike, T., Bedford, C., Davies, B., and Brown, D.
(2014). A catchment plan for the River
Wandle. www.wandletrust.org/wp-content
/uploads/2014/12/
Wandle_Catchment_Plan_-_Sept_2014_-
_full_document.pdf
Pori Natur a Threftadaeth. (2008). Statement on
rewilding and wilderness in Wales.

Prior, J., and Brady, E. (2017). Environmental
aesthetics and rewilding. *Environmental
Values*, **26**, 31–51.
Prior, J., and Ward, K.J. (2016). Rethinking
rewilding: a response to Jorgensen.
Geoforum, **69**, 132–135.
Puttock, A., Graham, H.A., Cunliffe, A.M.,
Elliott, M., and Brazier, R.E. (2017).
Eurasian beaver activity increases water
storage, attenuates flow and mitigates
diffuse pollution from
intensively-managed grasslands. *Science of
the Total Environment*, **576**, 430–443.
Rewilding Britain. (2017a). Rewilding
principles. www.rewildingbritain.org.uk
/rewilding/rewilding-principles
Rewilding Britain. (2017b). Urban rewilding.
www.rewildingbritain.org.uk/blog/urban-
rewilding
Rewilding Britain and Friends of the Earth.
(2016). Rewilding and flood risk
management. www.rewildingbritain.org
.uk/assets/uploads/files/publications/
Rewilding%20and%20Flood%20Risk%
20Management%20briefing.pdf
Rewilding Europe. (2017). Annual review 2016.
www.rewildingeurope.com/wp-content
/uploads/publications/rewilding-europe-
annual-review-2016/index.html
Roberts, J. (2010). RSPB Abernethy National
Nature Reserve: environmental statement
for forest expansion proposals. https://scot
land.forestry.gov.uk/images/corporate/pdf/
Abernethy-Forest-Expansion-ES-2011.pdf
Rotherham, I.D. (2017). *Shadow woods – a search
for lost landscapes*. Lulu.com.
RSPB. (2011). *Wildlife at work: the economic impact
of white-tailed eagles on the Isle of Mull*. Sandy:
RSPB.
RSPB. (2017). Policy note: rewilding.
Sandom, C.J., and Macdonald, D.W. (2015).
What next? Rewilding as a radical future
for the British countryside. In MacDonald,
D.W. and Feber, R.E. (Eds.), *Wildlife
conservation on farmland, Vol. 1: Managing for
nature on lowland farms* (pp. 291–316).
Oxford: Oxford University Press.

Sandom, C., Bull, J., Canney, S., and Macdonald, D.W. (2012). Exploring the value of wolves (*Canis lupus*) in landscape-scale fenced reserves for ecological restoration in the Scottish Highlands. In Somers, M.J.J. and Hayward, M.W. (Eds.), *Fencing for conservation* (pp. 245-276). New York, NY: Springer.

Sandom, C.J., Hughes, J., and Macdonald, D.W. (2013a). Rewilding the Scottish Highlands: do wild boar, *Sus scrofa*, use a suitable foraging strategy to be effective ecosystem engineers? *Restoration Ecology*, **21**, 336-343.

Sandom, C.J., Hughes, J., and Macdonald, D.W. (2013b). Rooting for rewilding: quantifying wild boar's *Sus scrofa* rooting rate in the Scottish Highlands. *Restoration Ecology*, **21**, 329-335.

Sandom, C., Donlan, C.J., Svenning, J.C., and Hansen, D. (2013c). Rewilding. In MacDonald, D.W. and Willis, K.J. (Eds.), *Key topics in conservation biology 2* (pp. 430-451). Chichester: Wiley.

Sandom, C.J., Ejrnaes, R., Hansen, M.D.D., and Svenning, J.C. (2014). High herbivore density associated with vegetation diversity in interglacial ecosystems. *Proceedings of the National Academy of Sciences of the United States of America*, **111**, 4162-4167.

Sandom, C.J., Dempsey, B., Bullock, D., et al. (in press). Rewilding in the English uplands: policy and practice. *Journal of Applied Ecology*.

Savory, C.J. (2016). Colonisation by woodland birds at Carrifran Wildwood: the story so far. *Scottish Birds*, **32**, 135-149.

Sheehy, E., and Lawton, C. (2014). Population crash in an invasive species following the recovery of a native predator: the case of the American grey squirrel and the European pine marten in Ireland. *Biodiversity and Conservation*, **23**, 753-774.

Sheehy, E., Sutherland, C., O'Reilly, C., and Lambin, X. (2018). The enemy of my enemy is my friend: native pine marten recovery reverses the decline of the red squirrel by suppressing grey squirrel populations. *Proceedings of the Royal Society of London B: Biological Sciences*, **285**, 20172603.

Simms, I.C., Ormston, C.M., Somerwill, K.E., et al. (2010). *A pilot study into sea eagle predation on lambs in the Gairloch area – final report*. Scottish Natural Heritage Commissioned Report 370. Inverness: SNH.

Sims, N. (2005). The ecological impacts of wild boar rooting in East Sussex. DPhil, University of Sussex.

SNH. (2008). The story of Creag Meagaidh National Nature Reserve. www.snh.scot/story-creag-meagaidh-national-nature-reserve

Soulé, M., and Noss, R. (1998). Rewilding and biodiversity: complementary goals for continental conservation. *Wild Earth*, **8**, 19-28.

Strouts, G. (2016). Rewilding discourses: evaluating different discourses of rewilding amongst land-use stakeholders in the UK. MSc, Bangor University.

Svenning, J.C. (2002). A review of natural vegetation openness in north-western Europe. *Biological Conservation*, **104**, 133-148.

Svenning, J.C., Pedersen, P.B.M., Donlan, C.J., et al. (2016). Science for a wilder Anthropocene: synthesis and future directions for trophic rewilding research. *Proceedings of the National Academy of Sciences of the United States of America*, **113**, 898-906.

Taylor, P. (2005). *Beyond conservation*. Abingdon: Routledge.

Taylor, P. (2011). *Rewilding*. ECOS.

The National Trust. (2009). Wicken Fen Vision. www.nationaltrust.org.uk/wicken-fen-nature-reserve/features/wicken-fen-vision

Toogood, M. (2003). Decolonizing highland conservation. In Adams, W. and Mulligan, M. (Eds.), *Decolonizing nature* (pp. 152-171). Abingdon: Earthscan.

Trees for Life. (2017). Dundreggan Vision. https://treesforlife.org.uk/work/dundreggan/vision/

Ullrich, P. (2015). Bird Monitoring 2015: Valley Bottom Grazing Survey.

van Zanten, B.T., Verburg, P.H., Koetse, M.J., and van Beukering, P.J. (2014). Preferences for European agrarian landscapes: a meta-analysis of case studies. *Landscape and Urban Planning*, **132**, 89–101.

Vera, F. (2000). *Grazing ecology and forest history*. Wallingford: CABI Publishing.

Vincent Wildlife Trust. (2017). Pine Marten Recovery Project. www.vwt.org.uk/projects/pine-marten-recovery-project/

Ward, V., Fisher, M., and Carver, S. (2006). Re-wilding projects in the UK – the database. *ECOS – British Association af Nature Conservationists*, **27**, 5.

Warren, C. (2009). *Managing Scotland's environment*, 2nd edition. Edinburgh: Edinburgh University Press.

Webb, J., and Drewitt, A. (2010). Managing for species: integrating the needs of England's priority species into habitat management. *Natural England Research Reports*, **24**, 1–129.

Wentworth, J., and Alison, J. (2016). Rewilding and ecosystem services. Parlimentary Office of Science and Technology. http://researchbriefings.parliament.uk/ResearchBriefing/Summary/POST-PN-0537#fullreport

White, C. (2015). *Analysis for the reintroduction of lynx to the UK*. Los Angeles, CA: AECOM.

Whitehouse, N.J. (2006). The Holocene British and Irish ancient forest fossil beetle fauna: implications for forest history, biodiversity and faunal colonisation. *Quaternary Science Reviews*, **25**, 1755–1789.

Wild Ennerdale. (2006). Wild Ennerdale Stewardship Plan. www.wildennerdale.co.uk/wordpress/wp-content/uploads/2013/02/Stewardship-Plan-Text.pdf

Woodland Trust. (2017). Rewilding: working with nature. www.google.co.uk/url?sa=t&rct=j&q=&esrc=s&source=web&cd=2&cad=rja&uact=8&ved=0ahUKEwiv44-hgd_ZAhWsAsAKHcpjAWwQFggvMAE&url=https%3A%2F%2Fwww.woodlandtrust.org.uk%2Fmediafile%2F100819260%2Fps-wt-170717-rewilding.pdf%3Fcb%3D805ac92ad2fe421dbd6bfff7de044614&usg=AOvVaw1sJY339NweRjk61jcmKrtQ

WWLF (Wales Wildland Foundation). (2017). Cambrian Wildwood project description. www.cambrianwildwood.org

Wynne-Jones, S. (2012). Heartlands and wildwoods. *ECOS*, **33**.

Wynne-Jones, S., Strouts, G., and Holmes, G. (2018). Rewilding in Wales: reimagining or abandoning a cultural heartland? *Environmental Values*, **27**, 377–403.

Yorke, R. (2016). Rewilding in the UK – hidden meanings, real emotions. *ECOS*, **37**, 53–59.

CHAPTER THIRTEEN

Bringing back large carnivores to rewild landscapes

JOHN D.C. LINNELL and CRAIG R. JACKSON
Norwegian Institute for Nature Research

Although there are multiple constructions of the concept of rewilding (Jørgensen, 2015) one aspect that they all have in common is a focus on the restoration of ecological processes. Such processes are diverse, including both abiotic (e.g. fire, flood, wind, avalanche) and biotic (e.g. competition, trampling, herbivory, predation, disease) factors. The removal of these processes was often a result of deliberate anthropogenic agency due to some conflicts with human interests. Therefore, restoring these processes can be a technically and sociopolitically challenging task as the original causes of conflict are generally still present, and human activity has fundamentally changed the structure and functioning of the landscape. Among rewilding advocates there is a frequent reference for the need to restore all the strongly interactive species at all trophic levels (Soulé et al., 2005), among which large carnivores are frequently regarded as having strong top-down structuring effects on ecosystems (Soule et al., 2003; Chapter 16). While the restoration of each process has its own set of challenges, those associated with large carnivores are currently very much in focus among conservation practitioners and the public alike. This chapter seeks to identify key challenges and potential ways to address them, before rounding off with a discussion of the extent to which we can expect large carnivores to resume their former ecological functions in the rewilded landscapes of the Anthropocene. We limit our analysis to cases where large carnivores have returned to landscapes from which they have been absent for periods of decades or longer, thus excluding cases where populations have increased within a given site where they have been continuously present. This separation is made to focus on the unique challenges of trying to restore species and the associated ecosystem processes that have been lost, which is the specific subject of rewilding.

Active and passive ways to restore large carnivores

Rewilding is a relatively new concept, so few cases of large carnivore recovery have been conducted within the explicit frames of a rewilding project. However, there are many cases where large carnivores have returned to landscapes from which they had been exterminated during recent centuries (Table 13.1) under the auspices of species, or ecosystem, restoration programmes. Restoration has generally occurred through either natural expansion from a remnant source population, or through reintroduction, where individuals have been translocated from either a wild or captive source. There are also some examples of combined approaches where extra individuals have been added to an existing remnant population (Table 13.1). Most examples for which there is any degree of documentation come from the temperate areas of the Northern hemisphere, except for some special cases from fenced reserves in southern Africa.

Table 13.1. *Some examples of relatively well-documented large carnivore recovery by natural expansion and reintroduction.*

Natural expansion

Grey wolves *Canis lupus*	Recolonisation of northern Montana from Canada	Boyd et al. (1995)
Grey wolves	Recolonisation of large parts of Minnesota, Michigan, and Wisconsin from a refuge in northern Minnesota	Wydeven et al. (2009)
Grey wolves	Recolonisation of Fennoscandia from Russia	Wabakken et al. (2001); Jansson et al. (2012)
Grey wolves	Recolonisation of large parts of the Alps from central Italy	Marucco (2016)
Grey wolves	Recolonisation of large parts of western Poland, Germany and even Denmark from eastern Poland	Nowak et al. (2017)
Brown bears *Ursus arctos*	Gradual expansion through parts of the Cantabrian mountains of Spain	Gonzalez et al. (2016)
Brown bears	Recolonisation of large parts of Sweden and border areas of Norway from refuges in central Sweden	Swenson et al. (1995)
Brown bears	Recolonisation of large parts of the Greater Yellowstone Ecosystem from Yellowstone National Park	Schwartz et al. (2006)
Eurasian lynx *Lynx lynx*	Recolonisation of large parts of Scandinavia from refuges in central Sweden	Linnell et al. (2009)

Table 13.1. (cont.)

Wolverines *Gulo gulo*	Recolonisation of southern Norway from a refuge in central Norway and neighbouring areas of central Sweden	Landa et al. (2000)
Mountain lions *Puma concolor*	Recolonisation of large areas to both the west and east from multiple refuges in the Rocky Mountains and south-western states	Hornocker and Negri (2010)
Asiatic lions *Panthera leo*	Expansion from a protected area refuge to wider human-dominated landscapes around Gir reserve in India	Singh (2017); Singh and Gibson (2011)
African wild dogs *Lycaon pictus*	Laikipia, Kenya	Woodroffe (2011)
African wild dogs	Savé Valley Conservancy, Zimbabwe	Lindsey et al. (2009)
Lions *Panthera leo*	Savé Valley Conservancy, Zimbabwe	Lindsey et al. (2009)
Reintroduction		
Grey wolves	Reintroduction from wild Canadian source into central Idaho and Yellowstone National Park, with subsequent expansion across large areas of the central Rocky Mountains and adjacent states	Fritts et al. (2001)
Mexican wolves *Canis lupus baileyi*	Reintroduction from captivity into parts of New Mexico and Arizona	Brown and Parsons (2001)
Red wolves *Canis rufus*	Reintroduced from captivity to North Carolina	Phillips et al. (2003)
Brown bears	Reintroduction from a wild source in Slovenia into the Italian Alps	De Barba et al. (2010)
Brown bears	Reintroduction from a wild source in Slovenia to the French part of the Pyrenees	Chapron et al. (2009)
Black bears *Ursus americanus*	Reintroduction from a wild source in Minnesota to Arkansas	Clark et al. (2002)
Eurasian lynx	Reintroduction of wild source lynx into multiple sites in the Swiss Alps and Jura and Vosges mountains in France	Linnell et al. (2009)
Eurasian lynx	Reintroduction of captive-bred lynx in the Harz mountains of central Germany	Linnell et al. (2009)

Table 13.1. (cont.)

African wild dogs	Reintroductions into several fenced protected areas in South Africa; for example, Balule Nature Reserve, Hluhluwe-iMfolozi Park, Karongwe Game Reserve, Madikwe Game Reserve, Marakele National Park, Pilanesberg National Park, uMkhuze Game Reserve, Venetia Limpopo Nature Reserve	Gusset et al. (2008)
African wild dogs	Reintroduced into Serengeti National Park from the neighbouring Loliondo Game Controlled Area, Tanzania	Masenga (2017)
African wild dogs	Reintroduction from a wild source in South Africa to the Northern Tuli Game Reserve, Botswana	Jackson et al. (2012)
Cheetahs *Acinonyx jubatus*	Reintroduction into several fenced protected areas in South Africa; for example, Amakhala Game Reserve, Blaauwbosch Game Reserve, Kwandwe Game Reserve, Lalibela Game Reserve, Samara Private Game Reserve, Shamwari Game Reserve	Hayward et al. (2007)
Cheetahs	Captured on livestock ranches and reintroduced to Matusadona National Park, Zimbabwe	Purchase and Vhurumuku (2005)
Cheetahs	Cheetahs from various reserves in South Africa reintroduced into Liwonde National Park, Malawi	Mkoka (2017)
Lions	Reintroduction into numerous (> 40) fenced protected areas in South Africa; for example, Addo Elephant National Park, Kariega Game Reserve, Kwandwe Game Reserve, Lalibela Game Reserve, Pumba Private Game Reserve, Scotia Game Reserve, Shamwari Game Reserve, Phinda Game Reserve, Pilansberg National Park, Madikwe Game Reserve, Tembe Elephant Park, Marakele National Park	Hayward et al. (2007), Hunter et al. (2007), Slotow and Hunter (2009)

Table 13.1. *(cont.)*

Spotted hyaenas *Crocuta crocuta*	Reintroductions into fenced reserves in South Africa; for example, Addo Elephant National Park and Pumba Game Reserve	Hayward et al. (2007)
Leopards *Panthera pardus*	Reintroductions into fenced reserves in South Africa; for example, Addo Elephant National Park, Kwandwe Game Reserve, Shamwari Game Reserve and Pumba Game Reserve	Hayward et al. (2007)
Tiger *Panthera tigris*	Reintroduction into Panna Tiger Reserve, India	Sarkar et al. (2016)
Recovery and reintroduction		
Florida panthers	Expansion of remnant Florida population following introduction of additional animals from Texas	Weeks et al. (2017)

Large carnivore reintroductions have been reviewed multiple times in the literature (e.g. Reading and Clark, 1996; Breitenmoser et al., 2001) and have been subject to books on the subject (Hayward and Somers, 2009). There is therefore a good collective understanding of the topic among practitioners. The existence of a Reintroduction Specialist Group within the IUCN's Species Survival Commission has been instrumental in exchanging experiences to foster the development of detailed best practices formalised within a set of guidelines (IUCN/SSC, 2013). In contrast, the experience with natural recovery is much more fragmented. Various synthesis works have gathered together some of the experience from grizzly bears (*Ursus arctos*), wolves (*Canis lupus*), and cougars (*Puma concolor*) in the American West (Clark et al., 2005; Clark and Rutherford, 2014) and for wolves in the Great Lakes area (Nie, 2003; Wydeven et al., 2009), for example, but this has not led to either holistic or structured best-practice guidelines. The European experience is largely unpublished, although another IUCN Specialist Group, the Large Carnivore Initiative for Europe, serves to network practitioners and other experts and acts as an organic repository for collective experience. In the rest of this chapter we will focus on the broad similarities and differences between the two approaches of natural recovery and reintroduction.

Special aspects of large carnivores

There are a number of special features of large carnivores that are highly relevant for their recovery and explain why there are both great opportunities

Box 13.1. Rewilding: the special case of carnivore reintroductions in South Africa

Reintroductions form a vital component of carnivore rewilding. To this end, South Africa has advanced the field of reintroduction biology pertaining to the re-establishment of African carnivore populations. Large carnivore populations were extirpated from much of the country, but are now widely distributed following numerous reintroductions. Restored populations are almost without exception located within fenced conservation areas, as electrified predator-proof fencing is a legal prerequisite for large carnivore reintroduction in South Africa. Furthermore, the average size of such 'rewilded' sites is small, typically less than 500 km^2, but some are as small as 15 km^2 (Power, 2002). Consequently, most reintroduced carnivore populations are geographically isolated and frequently surrounded by a matrix of incompatible land-use types, where conservation efforts are entirely reliant on fences to separate wildlife populations from the surrounding human-dominated landscapes. While fences may, to a large degree, afford carnivores respite from threatening processes and reduce most of the social–political issues that are so dominant in other situations (Packer et al., 2013, Snyman et al., 2015), their presence creates numerous ecological and management complications (Slotow and Hunter, 2009; Creel et al., 2013; Miller and Funston, 2014). The small fenced-off ecosystems restrict dispersal, gene flow and seasonal movement patterns that are particularly important for many of the prey species (Hayward and Kerley, 2009; Woodroffe et al., 2014; Durant et al., 2015). In addition to ecological problems, maintaining fencing infrastructure is extremely costly and labour-intensive (Lindsey et al., 2012; Creel et al., 2013). The artificial nature of these small-scale, isolated ecosystems necessitates continual intensive management of both predators and prey that carries excessive long-term economic costs (Davis-Mostert et al., 2009; Creel et al., 2013; Durant et al., 2015). For example, carnivores such as lions rapidly attain high densities and deplete prey populations, requiring invasive management interventions (Hunter, 1998; Power, 2002; Tambling and du Toit, 2005; Snyman et al., 2015). While the definition of rewilding may vary somewhat, it is generally considered to involve a largely self-sustaining ecosystem (post-reintroduction) where human intervention is minimised. Although the South African reintroduction efforts have advanced the science of reintroduction biology and contributed to the conservation of numerous species, the underlying model consequently deviates appreciably from the rewilding ideology.

and many challenges. On the plus side, these are mainly biologically robust species (high reproductive rates and high adult survival) with very flexible habitat needs (most species have very wide natural distributions and have shown a widespread ability to adapt to human-modified landscapes). Together, these characteristics imply that there is substantial opportunity for their recovery. More challenging is their predatory nature, which is the source of many of the conflicts with humans (see below) when they predate on livestock, companion animals, shared wild quarry, and even on people (Quigley and Herrero, 2005). Another consequence of their position at the top of the trophic pyramid is that they need very large amounts of space, with individual home ranges in the order of hundreds or even thousands of square kilometres (Duncan et al., 2015). When this is linked to their solitary and often territorial nature, it implies that large carnivore populations typically occupy tens or hundreds of thousands of square kilometres of landscapes. Because of the near-universal loss of wilderness areas, this implies that all populations inevitably share space with humans and their activities (see Box 13.1 for exceptions). While the large carnivores have generally shown a surprising ability to adapt to this shared space (e.g. Gehrt et al., 2010; Chapron et al., 2014), their presence has inspired a great deal of controversy among some sectors of the public and stakeholders. These controversies focus on both the material and economic conflicts caused by carnivore predation (e.g. Kaczensky, 1999) and on a wide range of social and political conflicts that focus more on the symbolism of large carnivore conservation than on the actual physical animal (Nie, 2003; Dickman, 2010; Redpath et al., 2013). Elements of these conflicts can be found for many other species groups, but large carnivores hold a special position in human consciousness and tradition (Kruuk, 2002), for better and for worse. In the following sections we expand on all of these aspects.

The special aspects of reintroductions

The active reintroduction of large carnivores involves many special challenges that are not present in natural recovery situations. Reintroductions are highly challenging and complex operations that require substantial capacity, training, funding, and commitment. The IUCN guidelines and many other references discuss reintroductions in detail, and the particulars are likely to vary with species and location, so we do not describe specific approaches here. Rather, we just outline the range of issues that need to be considered and explain why they are important.

First, repopulating a landscape necessitates the identification of carnivore populations from which suitable individuals can be sourced. Of importance is whether reintroduced carnivores originate from a wild population or a captive environment. In contrast to captive animals, free-ranging carnivores typically have better-developed hunting skills and a greater fear of humans, factors that have a profound effect on post-release success rates (Jule et al., 2008). Source individuals should also be familiar with other large carnivore species present at the release site. For example, cheetahs (*Acinonyx jubatus*) and African wild dogs (*Lycaon pictus*) unfamiliar with lions (*Panthera leo*) may suffer high rates of post-release mortality (Scheepers and Venzke, 1995). Wild-caught animals are thus the preferred option when considering founders for carnivore reintroductions. Additionally, source animals should ideally be genetically similar to the region's historic population (Slotow and Hunter, 2009).

If the source animals are drawn from wild populations one of the most crucial elements lies with the capture and handling of the animals. The capture of wild animals for any purpose (be it research marking or reintroduction) is always fraught with risk for the animals. Capture methods for large carnivores are diverse, and include leg-hold traps, snares, box and culvert traps, the use of hounds to tree animals, and darting from helicopters or other vehicles. While no methods are free from stress or without risk of injury or mortality (e.g. Arnemo et al., 2006) there is a great deal of experience from researchers working with most species groups so that there is little excuse for failing to learn from others' experience to use the best technique for the right species in each setting. However, it is crucial that funders, stakeholders, and the public are prepared for the risks that are unavoidable aspects of such complex and invasive operations. Almost all capture of large carnivores requires chemical immobilisation of the animals. This is a specialised area of expertise within veterinary science, and again there is considerable experience so that for most species there is an up-to-date understanding of the best drug combinations (e.g. Kreeger and Arnemo, 2012). In addition to the practical aspects of choosing the best methods are a range of complex legal issues. Many countries have very stringent legislation (animal welfare, veterinary, conservation, security) that may limit the choice of methods in many areas, so that there will often be difficult trade-offs to make regarding regulations, efficiency, and humaneness. In contrast, some developing countries may have poorly developed legislation and approval procedures when applying for permits, which often create delays and much uncertainty.

Once individuals have been caught they need to be transported. Transports may move multiple animals at once, but capture is usually a process that takes single animals at a time and often requires a great deal of waiting. As a result, animals may need to be kept alive in captivity for various periods until enough individuals have been gathered to justify a transport. This period may also be associated with a veterinary quarantine period. The issue of possible disease transfer during animal translocation is important and must be formally considered (Griffith and Scott, 1993). Holding wild animals in captivity is a very complex procedure, as all their needs must be met by humans. Wild animals are also often very stressed in captivity. At this stage zoo veterinarians or zoo husbandry staff are often the most qualified and experienced. Transporting is also a specialised activity, where a great deal of attention to detail needs to be spent on designing transport cases, and the use of sedatives that may be needed to reduce stress.

The manner in which translocated individuals are released influences their post-release movements, survival, and reproduction. Although logistically and financially more demanding, 'soft-release' protocols are typically more successful (Gusset et al., 2008). Soft-release strategies involve the confinement of translocated animals at the release site for a period of days or months prior to release. After release, reintroduced animals need to establish themselves in an unfamiliar landscape, and soft releases tend to reduce post-release movements, which can be particularly extensive for highly mobile large carnivores. A soft-release period may also permit the formation of social groups that may enhance post-release survival and reproduction. For some species with complex social organisation, like African wild dogs, this period is essential (Gusset et al., 2008). During a soft-release holding period, artificial food provision may lead to carnivores associating people and/or vehicles with food, and thereby unwanted and potentially dangerous effects following release. Adopting simple strategies may minimise such risks; for example, delivering carcasses into the carnivore enclosure via a remote pulley system that eliminates the need for direct human presence.

A substantial period of post-release monitoring is necessary to enable timely management interventions, reaction to any conflicts, and documentation of factors affecting the reintroduction's success. This long-term commitment is time-consuming, labour-intensive, and costly. The nature and intensity of management and monitoring may change over time. Initially, frequent monitoring of the animals' health and movements should be prioritised. Monitoring health is crucial to ensure that there are no residual negative effects of capture and transportation. Movement is crucial to determine whether animals stay within the release site and to analyse their behaviour. Depending on movement there may be a need to redirect public information and conflict mitigation/reaction activity. Longer-term activities may shift to address other challenges. For example, in order to maintain a population's genetic integrity, periodic additional translocations may be required, particularly when founder populations are small. This requires genetical analysis at regular intervals. In favourable conditions, large carnivore populations may increase rapidly. When this occurs in isolated ecosystems where dispersal or emigration are restricted, high population densities may result. Growing carnivore populations may drive unsustainable levels of predation and severe declines in the populations of main prey species, necessitating monitoring of prey and potentially intensive management interventions (Tambling and du Toit, 2005; Miller and Funston, 2014). This has emerged as a recurrent problem following lion reintroduction into small, fenced conservation areas in South Africa and, in some instances, has led to lions being euthanised or the use of other interventions such as contraception and translocation (Slotow and Hunter, 2009; Miller and Funston, 2014).

Dealing with this complex range of highly specialised tasks requires a well-coordinated team. Previous studies have repeatedly underlined the need for a large and well-trained staff and teams that are capable of learning from previous experience and adapting their procedures as they learn (Clark et al., 1994; Maehr et al., 2001). Because reintroductions typically require the interactions of multiple agencies or authorities (often with very different working styles, backgrounds, and philosophies) there is a crucial need to coordinate their activities. Reintroductions are almost always a subject of great public attention, and when it concerns large carnivores a lot of this attention may well be controversial. This can place great pressure on any team as all their activities will be subject to public scrutiny. As such, it is essential to include dedicated communication staff in any team, and to invest considerable resources in both proactive information and reaction to whatever direction the public debate takes. Demonstrating that decisions are evidence-based and following established best practices may also assist in legitimising a reintroduction project to a potentially sceptical public.

Key message: Reintroductions are highly complex, expensive, and long-term projects that require secure long-term funding and well-trained teams with solid institutional support.

The special aspects of natural recoveries

Natural recovery involves far less in the way of *a priori* technical competence. The animals are wild and need no special handling. The only requirement is proximity to a source population and the availability of connected habitat through which individuals can disperse (Woodroffe, 2003). Most large carnivores are capable of very long-distance natal dispersal movements and have a broad tolerance of dispersal habitat. Wolves and African wild dogs especially have been recorded making very long dispersal movements (Ciucci et al., 2009; Davies-Mostert et al., 2012; Razen et al., 2016), even exceeding 1000 km (Wabakken et al., 2007; Masenga et al., 2016), through human-modified landscapes. Eurasian lynx (*Lynx lynx*), tigers (*Panthera tigris*), cougars, and brown bears have also shown a remarkable ability to disperse hundreds of kilometres (Sweanor et al., 2000; Thompson and Jenks, 2005; Samelius et al., 2012; Athreya et al., 2014). This exceptional dispersal ability allows reproducing wolf packs to appear in areas many hundreds of kilometres from any known source. The major species difference lies in the sex bias in dispersal. Statistically, most carnivore species show a male bias in dispersal frequency, and sometimes in distance, but some females in most species do demonstrate substantial dispersal movements (e.g. Masenga et al., 2016), which greatly facilitates colonisation. One exception is with brown bears, where females show extreme philopatry (Swenson et al., 1998b; Kojola and Heikkinen, 2006; Jerina and Adamic, 2008). The result is that the area of distribution of the reproductive portion of bear populations expands very slowly in contrast to the other species, although young male bears can also appear hundreds of kilometres from known areas of reproduction (Rosen and Bath, 2009).

This dispersal ability of most large carnivore species allows them to rapidly recolonise areas within a radius of hundreds of kilometres from known source populations. While it is most likely that recolonisation will occur adjacent to known source populations, it can also occur at some distance, which makes it a rather unpredictable process. Therefore, although there may be some benefits in laying proactive ground work for recolonisation in areas close to sources, it is probably more important to have the measures in place to rapidly react to recolonisation when and where it does occur. The only measure that can potentially enhance recolonisation is to ensure that there are continuous areas of habitat. During dispersal, most species show a remarkable ability to cross unfavourable habitat and even linear barriers (Blanco et al., 2005), although highways and railroads can be major sources of mortality (Kaczensky et al., 2003). Therefore, providing safe crossing structures can assist movements of carnivores and thereby help recolonisation (Duke et al., 2001). Spatial analysis of habitat connectivity using geographic information systems has been widely used to help anticipate area of likely expansion and to target the location of connectivity enhancing structures or measures (e.g. Carroll et al., 2001; Kramer-Schadt et al., 2004; Elliot et al., 2014). Response to human-modified landscapes will vary between species, but may also vary between different demographic groups of the same species. For example, adult female lions are far more risk-averse than dispersing males (Elliot et al., 2014), which may result in an increased sex bias in dispersal ability.

Key message: Most large carnivores can disperse long distances and have a good capacity to recolonise areas after prolonged periods of absence.

Essential general aspects of large carnivore recovery

Apart from the different processes associated with how the carnivores initially recolonise an area, the follow-up to reintroduction and natural recovery are broadly similar. Several decades of experience have repeatedly underlined that the greatest challenges are associated with the social and political aspects, rather than the ecological (Reading and Clark, 1996; Breitenmoser et al., 2001). As a result, there is a need to focus on the public outreach, institutional and governance arrangements that are needed to make decisions about the management of the carnivores in their recolonised landscapes (Clark et al., 2005; Clark and Rutherford, 2014).

Movements and scales

The overriding challenge associated with large carnivores is that they need so much space. As a result, carnivore populations rarely re-establish themselves within a specific site or protected area (with the potential exception of a few of the very largest protected areas or the special case of South Africa's fenced reserves). Rather, they re-establish themselves in regions that stretch across thousands of square kilometres that inevitably cross many jurisdictional borders (Linnell and Boitani, 2012). These include the borders between different protected areas, between protected areas and multi-use landscapes, and between different municipalities, counties and even between different countries. For example, the naturally recolonising wolf population in the Alps stretches across six countries, many of which have also delegated wildlife management to subnational entities, and the reintroduced wolf population centred on Yellowstone National Park spans six states in a range managed by a bewildering array of federal and state agencies and tribal and private land owners, all with different priorities, skills, mandates, interests, and functions. The wide dispersal movements that we mentioned above, which are so useful in promoting natural recolonisation, create an additional challenge as an individual carnivore can occasionally turn up anywhere within a radius of many hundreds of kilometres around the areas where they are predictably resident. The sudden and unexpected arrival of such individuals can trigger intense controversy and conflict (e.g. Rosen and Bath, 2009). These huge area requirements impose many organisational challenges as there is a need for massive human resources to cover huge areas, for multiple institutions and agencies to cooperate, and for governance structures that can manage the almost impossible task of giving voice to a multitude of local interests while ensuring large-scale coordination of effort (Clark and Rutherford, 2014; Linnell, 2015).

As a result, large carnivore population recovery cannot be undertaken alone by any single private body or land manager. Such actors can be instrumental in managing individual sites that contribute to a wider effort and may be especially important at initial stages of recovery by providing refuges or stepping stones. However, in the medium to long term, large carnivore population recovery requires the involvement of multiple government agencies, at multiple administrative scales, often including a key role for international bodies to coordinate activity across national borders.

Key message: Large carnivore population recovery occurs at vast spatial scales that require buy-in and coordination of a wide range of land owners, land managers, and administrative jurisdictions. In many cases these scales will cross international borders.

Suitable habitat

With the exception of some very few carnivore species that are habitat specialists (e.g. the giant panda, *Ailuropoda melanoleuca*), most of the extant large carnivores are generalist species that can survive across a wide range of habitat types. Although many species show a statistical preference for areas with low levels of human activity (Mladenoff et al., 2009), most species have also shown a far greater tolerance (Bouyer et al., 2015) for human habitat modification (both farmland and urbanisation) than is often realised (e.g. Zimmermann et al., 2010; Woodroffe, 2011; Athreya et al., 2013; Yirga et al., 2013; Singh, 2017). The ability of some large carnivores like cougars and leopards (*Panthera pardus*) to survive on the edges of heavily urbanised landscapes, including the edges of some of the largest megacities on Earth, demonstrates the behavioural flexibility of these species (see examples in Gehrt et al., 2010). The implication is that large carnivores can potentially recover in a wide range of areas, both natural and human-modified, provided enough prey is available. Therefore, one of the major necessary steps to facilitate successful large carnivore recovery is to ensure that a diversity of wild prey species are available, and that their populations are managed in a way that ensures they are at a high enough density to support predation. This has been recognised, for example, in recent studies exploring the potential for tiger reintroduction in central and southeast Asia (Chestin et al., 2017; Gray et al., 2017), and the recent recovery of large carnivores across Europe (Chapron et al., 2014) was greatly facilitated by the recovery of their prey several decades earlier (Linnell and Zachos, 2011). Prey populations can be enhanced through strategies as diverse, and complementary, as limiting competition with livestock through sustainable rangeland management, regulating their harvest by humans and, if needed, via reintroduction.

It is also important to note that some of the more adaptable large carnivores, like wolves and leopards in parts of Iberia and India, respectively, or spotted hyaenas (*Crocuta crocuta*) in Ethiopia, can recover to a considerable extent even in landscapes that lack suitable wild prey (Vos, 2000; Blanco and Cortes, 2009; Yirga et al., 2012; Athreya et al., 2013; Odden et al., 2014) by living almost entirely on livestock and domestic animals or garbage (Butler et al., 2014). This is clearly not a desirable situation within a rewilding context, but demonstrates the ability of many large carnivores to adapt to human-modified landscapes (Oriol-Cotterill et al., 2015). Clearly there are limits to what some species can tolerate in terms of intensive agriculture, transport infrastructure density, and urbanisation, so there is a need to engage in landscape planning at very large scales. Probably the key issue is to ensure that connectivity is maintained in the landscape, which implies that linear transport, or border security (Linnell et al., 2016) infrastructure is constructed in a way that allows permeability for large mammals. Experience with natural recovery has shown that there is considerable space for large carnivore recovery (Chapron et al., 2014; Milanesi et al., 2017), even in a world with a widespread human footprint across all ecosystems. However, just because habitat is available does not guarantee that large carnivores will be allowed to avail of it because of social and political constraints.

Key message: Most large carnivores are very tolerant of habitat modification by humans, so wilderness is not a requirement for them to fulfil their ecological needs.

Conflicts – economic and social

The many reviews that have been written about large carnivore reintroduction in recent decades have consistently underlined the finding that managing the social and political aspects of large carnivore recovery are far more important, and challenging, than the ecological aspects (Clarke et al., 1994; Reading and Clarke, 1996; Breitenmoser et al., 2001; Sharpe et al., 2001; Hayward and Sommers, 2009; Clark and Rutherford, 2014). This is largely because large carnivores are associated with a wide range of conflicts with human interests. These conflicts were often the drivers of the original declines, and most conflict dimensions are just as present today as in the past. The challenge for today is to adopt different approaches to these conflicts that do not require the extermination of large carnivores. A first and crucial step in this direction is to recognise the complexity of these conflicts in the modern-day context.

The most obvious conflicts stem from the predatory nature of large carnivores. They rarely distinguish between domestic and wild prey, and wherever large carnivores and unprotected livestock share space there is likely to be a degree of depredation on livestock. Depredation rates vary widely in space and time and with species, but in general appear to be highest in areas where traditional forms of animal husbandry have been abandoned following prolonged periods without large carnivores, and where social and economic factors make the relative costs of labour (required to guard livestock) very high compared to the value of the livestock. This is especially the case in central and western Europe and North America. There are a range of livestock husbandry measures, both traditional and modern, that can be put into place to greatly reduce carnivore depredation (Breitenmoser et al., 2005). Electric fencing of pastures and the use of livestock guarding dogs with shepherds and night-time enclosures are examples that are effective. However, any changes to husbandry systems are inevitably going to be expensive to implement on the large scales that are necessary, and often meet with opposition from producers. As a consequence, their introduction requires technical assistance (preferably embedded within agricultural outreach programmes) as well as social, cultural, and economic incentives. Some large carnivore species may also kill domestic dogs, which depending on local cultural circumstances may also be a major source of conflict (i.e. where they are companion animals or valuable working dogs) (Butler et al., 2014).

Paying compensation for livestock lost to predators is a commonly used measure to redress some of the costs of asking livestock producers to tolerate carnivores (Nyhus et al., 2005). Despite being widespread, there is little evidence that compensation actually achieves its desired goals of buying tolerance for carnivores (Naughton-Treves et al., 2003) and it may actually hinder the uptake of effective livestock protection measures (Bulte and Rondeau, 2006). In extreme cases compensation may actually destabilise the sustainability of pastoral systems (Naess et al., 2011). Therefore, current thinking is very much more in favour of paying incentives rather than compensation (Schwerdtner and Gruberb, 2007; Zabel and Horn-Müller, 2008). Where compensation is paid, it is now recommended to make payments dependent on the adoption of measures to prevent livestock depredation.

The second major conflict associated with predation is when recreational hunters perceive large carnivores as competitors for shared prey. This is a major source of conflict in Europe, North America, and southern Africa, where hunting is an institutionalised and carefully regulated activity with strong social, cultural, and economic components. The actual impact of recovering large carnivores on prey populations is highly variable and hard to predict (Pitman et al., 2017; Chapter 16), and in cases where there is an impact the only potential response is to adjust hunting quotas of shared prey to accommodate the increased mortality because it is not possible (or desirable in a rewilding context) to stop wild large carnivores from killing wild prey. In some of the northern and eastern European systems large carnivores are also recreationally hunted in order to maintain carnivore populations at a density that manages their impacts on shared prey in order to minimise conflicts with hunters.

The final predation conflict is linked to the exceptional cases when large carnivores pose a risk to human safety. A range of species have been shown to kill humans on occasion (Löe and Røskaft, 2004; Packer et al., 2005; Quigley and Herrero, 2005), although the risk posed by some species is often hotly contested (e.g. for wolves, see Linnell and Alleau, 2015). A key consequence of this risk is that recovery planning needs to explicitly recognise that fear may be widespread among stakeholders in carnivore recovery areas (Røskaft et al., 2003) and may not be in any way related to the objective level of risk posed by a given species (Lescureux et al., 2011). Responses should include providing information to prevent attacks and planning reactive responses to cases of carnivore attacks on humans, but must also go beyond to try and understand some of the root causes of fear.

In addition to these conflicts that have a clear physical and economic manifestation come a whole suite of underlying social, cultural, and political conflicts. Many human-dimensions studies have focused on the variation in attitudes that individual people have when sharing landscapes with large carnivores. Attitudes typically differ widely between people, ranging from extremely positive to negative, although there is a fairly predictable set of factors such as gender, age, occupation, and education that explain a lot of the variation (Kansky and Knight, 2014; Kansky et al., 2014). Further studies have tried to link attitudes with the risk of individuals engaging in behaviours that negatively impact carnivore recovery (Marchini and Macdonald, 2012; Hazzah et al., 2017). However, the most serious conflicts are not associated with individual attitudes, but with group actions that can undermine the entire carnivore conservation agenda. These are typically expressed as social conflicts (see Redpath et al., 2013) between different groups of people with different values and interests which clash over the way large carnivores are managed (e.g. when and where to allow lethal control and hunting, if at all, or if carnivores should be allowed to recolonise certain landscapes or if reintroductions should go ahead). While these conflicts can have their origins in the economic conflicts described above, they are typically magnified by far deeper value-based divisions such as the rural–urban, modern–traditional, or doministic–mutualistic gradients or wider issues of power and trust between rural people and governments (Manfredo, 2008; Manfredo et al., 2017a, 2017b; Skogen et al., 2017). In other words, the underlying conflicts may often be about what the carnivores represent symbolically, rather than about the carnivores themselves. While the symbolic carnivore in traditional societies in many developing countries may be associated with positive views (e.g. Saunders, 1998; Ghosal, 2013; Li et al., 2014), large carnivores have traditionally had very negative images in Western culture in recent centuries (Boitani, 1995). In the context of modern conflicts, large carnivores (especially wolves) have taken on a whole new symbolism linked to political struggle over values (Nie, 2003; Skogen et al., 2006). Both research and experience have revealed that many rural stakeholders are opposed in principle to large carnivore recovery, and that this opposition may be greater towards reintroductions than natural recoveries (Morzillo et al., 2010) because reintroductions are deliberate actions by other groups. The consequence of these conflicts are serious efforts by opposition groups to undermine the conservation agenda by seeking legislative or policy change to weaken conservation goals and promoting a culture where illegal killing of carnivores may be more acceptable (Manfredo et al., 2017a). In some Western and Asian settings this has gone so far that the goal of carnivore conservation has been portrayed as an environmental justice issue, with the conservationists as the aggressors and rural stakeholders as the victims (Noam, 2007; Jacobsen and Linnell, 2016).

Fortunately, this is not universal, as studies have failed to find parallel discourses in other settings, for example in the case of jaguars in Brazil (Bredin et al., 2018).

These conflicts seem to be especially linked to areas where carnivores have been absent for generations, and where people have lost the practical and psychological adaptations required to live in the presence of these potentially dangerous species. At best, it will take a long time for people to recover these adaptations, implying that a large carnivore recovery programme in an area where carnivores have long been absent will require decades of constant efffort to address concerns and conflicts, both real and perceived. As the recovery front expands, this will require bringing in new areas and new people that are being asked to share their landscapes. At worst, some of the conflicts may become so symbolic, political, and institutionalised that it may never be possible to de-escalate them (Nie, 2003). However, there is extensive variation in human tolerance for the proximity of large carnivores. There are many cases like leopards and lions in India, brown bears in south-eastern Europe or spotted hyaenas in Ethiopia where people appear to be willing to accept these species as very close neighbours. Unfortunately, our understanding of the full range of cultural, psychological, and social factors promoting coexistence lags far behind our understanding of conflict (Carter and Linnell, 2016). There are also very clear differences in human tolerance for different carnivore species.

This discussion of widespread conflicts shows how important it is to firmly anchor any carnivore recovery strategy or reintroduction programme within the full set of governmental and non-governmental institutions that are relevant for a given site. Recovery is never finished. Once the initial recolonisation phase has passed there will be a constant and never-ending need for management of people and carnivores. In effect, carnivore related issues need to be mainstreamed into all sectors (such as agriculture, forestry, transport, energy, security). Management actions will inevitably be controversial with some stakeholders, but there is evidence that the decision-making process can be as important as the actual decisions made, provided communities are fully engaged and decisions are viewed as legitimate. Accordingly, there is currently a great deal of effort ongoing into trying to understand how to craft effective decision-making institutions that facilitate broad stakeholder engagement with administrations and scientists (Clark and Rutherford, 2014; Redpath et al., 2017). There is still much work to do before agreement on context-dependent best practices emerge, and there may well be some aspects of carnivore management that will remain unsolvable. One of the most challenging problems again results from the scale issues we have mentioned above. It is very hard to build institutions that can have the desired level of local involvement while simultaneously engaging in large-scale coordination (Linnell, 2015).

Key message: Large carnivore recovery can be associated with a multitude of conflicts including a range of economic, psychological, social, cultural, and political elements. Addressing these requires a broad institutionalisation of the recovery progress within governmental and non-governmental organisations and constant interaction with a range of stakeholders.

Limiting mortality

The aforementioned ability of carnivores to tolerate high degrees of human habitat modification is totally dependent on minimising human-caused mortality. As large-bodied mammals, the life histories of large carnivores are especially sensitive to adult mortality, which is primarily due to anthropogenic causes. There has been widespread misinterpretation of correlational studies (e.g. Mladenoff et al., 2009) about carnivore distribution and human impacts (e.g. Woodroffe, 2000; Linnell et al., 2001). It is not often the habitat modifications per se which limit carnivore distribution, but the way in which these modifications either facilitate human-caused mortality directly and/or enhance conflicts which indirectly leads to retaliatory mortality (Basille et al., 2013). Humans kill large carnivores for a wide variety of reasons, including recreation, self-defence, hatred, economic return, retaliation for conflicts, and ritual. The extent to which these causes can be regulated is highly variable and subject to the socioeconomic environment and institutional structures. Law enforcement is one obvious element, but it will never be possible to police entire landscapes when considering the massive ranges of large carnivores. The persistence of large carnivores in human-dominated landscapes thus depends on a widespread social willingness to voluntarily obey laws and accept regulations. However, illegal killing persists even in some of the most socially, economically, and institutionally developed countries (Kaczensky et al., 2011; Kaltenborn and Brainerd, 2016), implying that it will never be possible to entirely prevent because it is driven by many different motivations (Carter et al., 2017) and may be both deliberate or accidental. We also shouldn't forget that even protected areas are subject to widespread illegal killing (Woodroffe and Ginsberg, 1998). Management activities should therefore seek to keep illegal killing within acceptable levels, and should adopt multiple strategies, including but not limited to law enforcement. Within any given context it is crucial to understand the local motivations for illegal killing (e.g. Marchini and Macdonald, 2012) to find the most appropriate, and practical, approach to reduce it. However, effort should be scaled proportionally to the expected impact on population performance.

Virtually all large carnivore recovery programmes include a certain degree of lethal control of individual carnivores. This is usually in response to various types of conflict, such as killing livestock or humans, the development of problematic behaviour such as showing too bold behaviour or raiding garbage, or dispersal into areas of highly unsuitable habitat. It may also be used to slow the growth rate of rapidly increasing carnivore populations following successful recovery in order to facilitate human adaptation to carnivore presence (Linnell et al., 2010; Miller and Funston, 2014). For example, the highly acclaimed wolf reintroduction project in Yellowstone and Idaho has routinely used state agents to kill wolves that depredate livestock. In the period from 1987 when wolves recolonised northern Montana (the reintroductions to Yellowstone and Idaho were in 1995 and 1996) until 2005, a total of 396 wolves were lethally controlled in response to livestock depredation (Bangs et al., 2006). The utility of lethal control to address conflict is highly controversial, with different analyses producing different results (Bradley et al., 2015). At least part of this controversy concerns the criteria used to assess conflict (i.e. economic conflicts versus social conflicts) and the scales of analysis. However, given the potential negative side effects of non-lethal measures like translocation (Athreya et al., 2011) it seems likely that lethal control will always be needed in some circumstances (Huber et al., 2008), although there is a growing awareness of the greater utility of investment in proactive conflict prevention measures (e.g. Breitenmoser et al., 2005). At the landscape scale it is necessary to plan the distribution of human activities – for example, through land-use zoning, spatially differentiated regulations, incentives or subsidies – to try and minimise conflict causing activities in areas where large carnivores are recovering and promote the modification of such activities to make them less vulnerable to carnivores.

Accidental anthropogenic mortality may also be significant in the early stages of population recovery. One common issue is that associated with collisions between carnivores and vehicles along transport infrastructure (e.g. Taylor et al., 2002). This is not just a developed world problem, as the developing world is currently entering a phase of large-scale construction of transport infrastructure. The science of road ecology (see van der Ree et al., 2015) is accumulating a huge body of experience about infrastructure mitigation strategies. For example, the complementary use of fencing and wildlife crossing structures (green bridges or underpasses) can both limit mortality and facilitate connectivity for large carnivores (e.g. Duke et al., 2001), but is usually only practical for major highways or high-speed railways. The use of habitat and movement analyses can help place such structures in the optimal places.

Key message: Regulating human-caused mortality of large carnivores is essential to facilitate population recovery. However, large carnivores can tolerate a controlled offtake, which opens the potential for limited offtake as lethal control or sports hunting (see next section) without hindering recovery.

Extracting consumptive and non-consumptive benefits from large carnivores

Many approaches have been tried to generate revenue from large carnivores in an effort to offset their costs and try to improve public opinion (Dickman et al., 2011). Large carnivores are iconic and charismatic. As such, they have the potential to generate benefits to local stakeholders through consumptive and/or non-consumptive means. Carnivores often act as a major attraction for eco-tourism operations which generate considerable income in various parts of the world. Tourists venture to the Pantanal (Brazil) in search of jaguars (*Panthera onca*) or hope to see tigers in Ranthambhore (India), wolves in Yellowstone (USA), cheetahs in the Serengeti (Tanzania) or lions in Tsavo (Kenya) (e.g. Duffield et al., 2008; Lyngdoh et al., 2017; Tortato and Izzo, 2017). However, the feasibility thereof may differ regionally due to ecological, behavioural, and sociopolitical factors. For example, eco-tourism is viable in the Maasai Mara (Kenya), where carnivores are easily observed in the open landscape and habituated to the presence of tourist vehicles. Yellowstone National Park and some locations in Alaska also offer some locations for viewing wolves and brown bears, and snow leopard (*Panthera uncia*) tourism is increasing in parts of the Himalayas. In contrast, the mountainous and densely forested habitats that are home to many Eurasian carnivores do not open for the same possibilities. Observing comparatively low-density and shy species in such closed habitats is extremely challenging and alternative non-consumptive benefit-generating approaches are required. For example, brown bears have become a focal species for eco-tourism ventures in the northern Dinaric mountains of Croatia/Slovenia as well as in Finland and Estonia. However, this requires the use of feeding stations, which can inadvertently have effects on bear ecology. In addition, visitors are able to look for carnivore tracks and signs, which provides an alternative and indirect 'carnivore experience' that is not dependent on direct encounters or observations. However, few of these temperate-zone cases offer the same opening for low-threshold large-scale tourism as the African savanna parks.

Planning recovery must also consider how a population is to be managed when it has recovered, and a key debate that frequently turns up concerns hunting. Although controversial (Treves, 2009; Macdonald et al., 2016; Lute et al., 2018), sports and trophy hunting is widely used to generate substantial economic benefits (Funston et al., 2013). The trophy hunting sector tends to be more robust than traditional eco-tourism, being viable in many countries and regions that attract low numbers of conventional tourists, and is able to generate revenues under a wider range of scenarios (Lindsey, 2008). Consequently, trophy hunting can secure financial benefits in places where eco-tourism in not viable. Many developing countries are dependent on foreign visitors to realise economic wildlife-derived benefits. Political instability, terrorism or fear of disease tend to have a greater impact on the foreign eco-tourism industry compared to the foreign hunting fraternity (Lindsey, 2008). Sports hunting has not been clearly linked to any recent population extinction among large carnivores, and in fact many species like brown bears and lynx in Scandinavia, brown bears in the western Balkans, or cougars in most western US states have increased over a period of many decades while being subject to recreational harvest. As well as bringing some economic benefits (Knott et al., 2014), it is also often claimed that hunting helps build local tolerance and in many cases it has been viewed as a central element in promoting tolerance for carnivore recovery (Swenson et al., 1998a; Huber et al., 2008; Hornocker and Negri, 2010; Linnell et al., 2010; Majic et al., 2011). Several reintroductions have been initiated by hunters (Linnell et al., 2009). However, sustainable trophy/sports hunting of large carnivores is technically a challenging activity, and requires a good scientific basis, effective monitoring, hunter compliance, good governance, and strong institutions. These institutional prerequisites are certainly not universally in place, so its compatibility with conservation will have to be evaluated on a case-by-case basis. When developed correctly it is also possible for hunters to contribute substantially to the scientific basis for management (e.g. Nilsen et al., 2012) and permits the development of a sustainable system through adaptive management. It must be recognised that hunting of any form may well have unexpected effects on species life histories (e.g. Bischoff et al., 2017), but these must be seen in the context of the almost universal impact of all other human activities on species ecology and habitat. The key point, however, is to recognise that allowing sports hunting can potentially serve many purposes (e.g. Linnell et al., 2017) and is not automatically incompatible with conservation (Loveridge et al., 2007), and can certainly bring some benefits, to some stakeholders, in some contexts. Therefore, both consumptive and non-consumptive approaches to gaining revenue and building tolerance for recovering or reintroduced carnivores exist, although the extent to which the different approaches are viable or acceptable will vary greatly from location to location (Funston et al., 2013).

Key message: Both non-consumptive (e.g. eco-tourism) and consumptive (e.g. sports hunting) approaches exist to generate benefits and reduce conflicts from large carnivores. Both can play a role in fostering support for recovery and reintroduction. However, the relative utility of each will vary dramatically with social, institutional, and ecological circumstances.

Large carnivore recovery and rewilding

Experience from all continents has shown that large carnivore recovery is possible on wide spatial scales and across a diversity of landscapes if the right social, political, and institutional frameworks are in place to regulate human behaviour. This is very good news for large carnivore conservation as it opens up possibilities for the conservation of large and interconnected populations. However, what does this mean for rewilding? Obviously having top predators back in an ecosystem is a prerequisite for it to resume its full set of ecological functions. Unfortunately, just because predators return does not mean that all the other human threats and influences on the rest of the ecosystem disappear. Neither does the return of predators to an area imply that either their dynamics or their predator–prey relationships will necessarily be restored to something 'natural' (e.g. with minimal human influence). This is because the large spatial scale at which carnivore (and often herbivore) populations exist implies that they will automatically be spread across areas where human influences are strong (Boitani and Linnell, 2015; Kuijper et al., 2016). We believe that large carnivores reveal a fundamental scale mismatch that is often overlooked in the rewilding debate. Rewilding is often viewed as a site-specific activity where the goal is to restore the full suite of ecological processes and where direct human activity is minimised. In terms of issues like vegetation dynamics, some disturbance dynamics, and a great deal of herbivory, this can be achieved at relatively small scales (e.g. tens to hundreds of square kilometres). Predation can be easily added to such areas, but on those small scales we are more often talking about predator individuals spending part of their time on a site rather than populations residing on the site (with the exception of a handful of the planet's largest protected areas). The spatiotemporal dynamics between predators and prey that are obviously essential to modulating any eventual trophic cascades (which are so often touted as being essential to rewilding) will be driven by factors over much larger scales and which will therefore inevitability be strongly influenced by a range of off-site anthropogenic factors. The conflicts caused by large carnivores will inevitably lead to some forms of management intervention focused on the carnivores.

The 'large carnivore scale' and the controversies around the ecological process of predation force consideration of what to many may be an unpleasant reality; we simply don't have sufficient space left (the Anthropocene has gone too far) for purely 'natural processes' at large scales (like predator–prey dynamics). The question then becomes what type of compromises (i.e. how much human influence) are rewilders willing to accept? There is enormous scope to restore all trophic interactions to novel ecosystems (*sensu* Hobbs et al., 2013), albeit in highly modified ways, to create a wild Anthropocene. When working towards this goal, the inclusion of large carnivores into the mix of restored processes is very much possible, but also very challenging, and will provide the ultimate litmus test of humanity's commitment to coexistence with wildlife (Boitani and Linnell, 2015; Carter and Linnell, 2016).

Conclusion

There are three main take-home messages that we want to communicate. (1) There is enormous scope for large carnivore recovery and reintroduction, even in shared and human-modified landscapes. (2) Large carnivore population recovery is not just about bringing some individuals back to a single site (although this may be a necessary first step). Recovery is about finding ways of managing the entire landscape, and all human activities and interests within this landscape, in a manner that minimises the full diversity of social and economic conflicts posed by large carnivores while accommodating the needs of large carnivores and their prey. (3) The return of large carnivores will restore many ecosystem processes, but issues of scale imply that top-down predator–prey–habitat dynamics will inevitably be modified by humans.

References

Arnemo, J.M., Ahlqvist, P., Andersen, R., et al. (2006). Risk of capture-related mortality in large free-ranging mammals: experiences from Scandinavia. *Wildlife Biology*, **12**, 109–113.

Athreya, V., Odden, M., Linnell, J.D.C., and Karanth, K.U. (2011). Translocation as a tool for mitigating conflict with leopards in human-dominated landscapes of India. *Conservation Biology*, **25**, 133–141.

Athreya, V., Odden, M., Linnell, J.D.C., Krishnaswamy, J., and Karanth, K.K. (2013). Big cats in our backyards: persistence of large carnivores in a human dominated landscape in India. *PLoS ONE*, **8**, e57872.

Athreya, V., Navya, R., Punjabi, G.A., et al. (2014). Movement and activity pattern of a collared tigress in a human-dominated landscape in central India. *Tropical Conservation Science*, **7**, 75–86.

Bangs, E.E., Jimenez, M.D., Niemeyer, C., et al. (2006). Non-lethal and lethal tools to manage wolf–livestock conflict in the northwestern United States. *Proceedings of the Vertebrate Pest Conference*, **22**, 7–16.

Basille, M., Van Moorter, B., Herfindal, I., et al. (2013). Selecting habitat to survive: the impact of road density on survival in a large carnivore. *PLoS ONE*, **8**(7), e65493.

Bischof, R., Bonenfant, C., Rivrud, I.M., et al. (2017). Regulated hunting re-shapes the life-history of brown bears. *Nature Ecology and Evolution*. doi.org/10.1038/s41559-017-0400-7

Blanco, J.C., and Cortes, Y. (2009). Ecological and social constraints of wolf recovery in Spain. In Musiani, M., Boitani, L., and Paquet, P.C. (Eds.), *A new era for wolves and people: wolf recovery, human attitudes, and policy* (p. 41). Calgary: University of Calgary Press.

Blanco, J.C., Cortés, Y., and Virgós, E. (2005). Wolf response to two kinds of barriers in an agricultural habitat in Spain. *Canadian Journal of Zoology*, **83**, 312–323.

Boitani, L. (1995). Ecological and cultural diversities in the evolution of wolf human relationships. In Carbyn, L.N., Fritts, S.H., and Seip, D.R. (Eds.), *Ecology and conservation of wolves in a changing world* (pp. 3–12). Alberta, Canada: Canadian Circumpolar Institute.

Boitani, L., and Linnell, J.D.C. (2015). Bringing large mammals back: large carnivores in Europe. In Pereira, H.M. and Navarro, L.M. (Eds.), *Rewilding European landscapes* (pp. 67–84). Berlin: Springer.

Bouyer, Y., Gervasi, V., Poncin, P., Beudels-Jamar, R., Odden, J., and Linnell, J.D.C. (2015). Tolerance to anthropogenic disturbance by a large carnivore: the case of Eurasian lynx in south-eastern Norway. *Animal Conservation*, **18**, 271–278.

Boyd, D.K., Paquet, P.C., Donelon, S., Ream, R.R., Pletscher, D.H., and White, C.C. (1995). Transboundary movements of a recolonizing wolf population in the Rocky Mountains. In Carbyn, L.N., Fritts, S.H., and Seip, D.R. (Eds.), *Ecology and conservation of wolves in a changing world* (pp. 135–140). Alberta, Canada: Canadian Circumpolar Institute.

Bradley, E.H., Robinson, H.S., Bangs, E.E., et al. (2015). Effects of wolf removal on livestock depredation recurrence and wolf recovery in Montana, Idaho, and Wyoming. *Journal of Wildlife Management*, **79**, 1337–1346.

Bredin, Y.K., Lescureux, N., and Linnell, J.D.C. (2018). Local perceptions of jaguar conservation and environmental justice in Goiás, Matto Grosso and Roraima states (Brazil). *Global Ecology and Conservation*, **13**, e00369.

Breitenmoser, U., Breitenmoser-Würsten, C., Carbyn, L.N., and Funk, S.M. (2001). Assessment of carnivore reintroductions. In Gittleman, J.L., Funk, S.M., Macdonald, D.W., and Wayne, R.K. (Eds.), *Carnivore conservation* (pp. 241–281). Cambridge: Cambridge University Press.

Breitenmoser, U., Angst, C., Landry, J.M., Breitenmoser-Würsten, C., Linnell, J.D.C., and Weber, J.M. (2005). Non-lethal techniques for reducing depredation. In Woodroffe, R., Thirgood, S., and Rabinowitz, A. (Eds.), *People and wildlife: conflict or coexistence?* (pp. 49–71). Cambridge: Cambridge University Press.

Brown, W.M., and Parsons, D.R. (2001). Restoring the Mexican gray wolf to the desert southwest. In Maehr, D.S., Noss, R.F., and Larkin, J.L. (Eds.), *Large mammal restoration: ecological and sociological challenges in the 21st century* (pp. 169–186). London: Island Press.

Bulte, E.H., and Rondeau, D. (2006). Why compensating wildlife damages may be bad for conservation. *Journal of Wildlife Management*, **69**, 14–19.

Butler, J.R.A., Linnell, J.D.C., Morrant, D., Athreya, V., Lescureux, N., and McKeown, A. (2014). Dog eat dog, cat eat dog: social–ecological dimensions and implications of dog predation by wild carnivores. In Gompper, M. (Ed.), *Free-ranging dogs and wildlife conservation* (pp. 117–143). Oxford: Oxford University Press.

Carroll, C., Noss, R.F., and Paquet, P.C. (2001). Carnivores as focal species for conservation planning in the Rocky Mountain region. *Ecological Applications*, **11**, 961–980.

Carter, N.H., and Linnell, J.D.C. (2016). Co-adaptation is key to coexisting with large carnivores. *Trends in Ecology and Evolution*, **31**, 575–578.

Carter, N.H., Lopez-Bao, J.V., Bruskotter, J.T., et al. (2017). A conceptual framework for understanding illegal killing of large carnivores. *Ambio*, **46**, 251–264.

Chapron, G., Wielgus, R., Wuenette, P.-Y., and Camarra, J.-J. (2009). Diagnosing mechanisms of decline and planning for recovery of an endangered brown bear (*Ursus arctos*) population. *PLoS ONE*, **4**, e7568.

Chapron, G., Kaczensky, P., Linnell, J.D.C., et al. (2014). Recovery of large carnivores in Europe's modern human-dominated landscapes. *Science*, **346**, 1517–1519.

Chestin, I.E., Paltsyn, M.Y., Pereladova, O.B., Legorova, L.V., and Gibbs, J.P. (2017). Tiger re-establishment potential to former Caspian tiger (*Panthera tigris virgata*) range in Central Asia. *Biological Conservation*, **205**, 42–51.

Ciucci, P., Reggioni, W., Maiorano, L., and Boitani, L. (2009). Long-distance dispersal of a rescued wolf from the northern Apennines to the western Alps. *Journal of Wildlife Management*, **73**, 1300–1306.

Clark, J.D., Huber, D., and Servheen, C. (2002). Bear reintroductions: lessons and challenges. *Ursus*, **13**, 335–346.

Clark, S.G., and Rutherford, M.B. (2014). *Large carnivore conservation: integrating science and policy in the North American west*. Chicago, IL: University of Chicago Press.

Clark, T.W., Reading, R.P., and Clarke, A.L. (1994). *Endangered species recovery: finding the lessons, improving the process*. Washington, DC: Island Press.

Clark, T.W., Rutherford, M.B., and Casey, D. (2005). *Coexisting with large carnivores: lessons from Greater Yellowstone*. Washington, DC: Island Press.

Creel, S., Becker, M.S., Durant, S.M., et al. (2013). Conserving large populations of lions – the argument for fences has holes. *Ecology Letters*, **16**, 1413-1413.

Davies-Mostert, H.T., Mills, M.G.L., and Macdonald, D.W. (2009). A critical assessment of South Africa's managed metapopulation recovery strategy for African wild dogs. In Hayward, M.W. and Somers, M.J. (Eds.), *Reintroduction of top-order predators* (pp. 10-42). London: Wiley-Blackwell.

Davies-Mostert, H.T., Kamler, J.F., Mills, M.G., et al. (2012). Long-distance transboundary dispersal of African wild dogs among protected areas in southern Africa. *African Journal of Ecology*, **50**, 500-506.

De Barba, M., Waits, L.P., Garton, E.O., et al. (2010). The power of genetic monitoring for studying demography, ecology and genetics of a reintroduced brown bear population. *Molecular Ecology*, **19**, 3938-3951.

Dickman, A.J. (2010). Complexities of conflict: the importance of considering social factors for effectively resolving human–wildlife conflict. *Animal Conservation*, **13**, 458-466.

Dickman, A.J., Macdonald, E.A., and Macdonald, D.W. (2011). A review of financial instruments to pay for predator conservation and encourage human–carnivore coexistence. *Proceedings of the National Academy of Sciences of the United States of America*, **108**, 13937-13944.

Duffield, J.W., Neher, C.J., and Patterson, D.A. (2008). Wolf recovery in Yellowstone: Park visitor attitudes, expenditures, and economic impacts. *George Wright Forum*, **25**, 13-19.

Duke, D.L., Hebblewhite, M., Paquet, P.C., Callaghan, C., and Percy, M. (2001). Restoring a large carnivore corridor in Banff National Park. In Maehr, D.S., Noss, R.F., and Larkin, J.L. (Eds.), *Large mammal restoration: ecological and sociological challenges in the 21st century* (pp. 261-292). London: Island Press.

Duncan, C., Nilsen, E.B., Linnell, J.D.C., and Pettorelli, N. (2015). Life history attributes and resource dynamics determine intraspecific home range sizes in Carnivora. *Remote Sensing in Ecology and Conservation*, **1**, 39-50.

Durant, S.M., Becker, M.S., Creel, S., et al. (2015). Developing fencing policies for dryland ecosystems. *Journal of Applied Ecology*, **52**, 544-551.

Elliot, N.B., Cushman, S.A., Macdonald, D.W., and Loveridge, A.J. (2014). The devil is in the dispersers: predictions of landscape connectivity change with demography. *Journal of Applied Ecology*, **51**, 1169-1178.

Fritts, S.H., Mack, C.M., Smith, D.W., et al. (2001). Outcomes of hard and soft releases of reintroduced wolves in central Idaho and the Greater Yellowstone Area. In Maehr, D.S., Noss, R.F., and Larkin, J.L. (Eds.), *Large mammal restoration: ecological and sociological challenges in the 21st century* (pp. 125-148). London: Island Press.

Funston, P.J., Groom, R.J., and Lindsey, P.A. (2013). Insights into the management of large carnivores for profitable wildlife-based land uses in African savannas. *PLoS ONE*, **8**(3), e59044.

Gehrt, S.D., Riley, S.P.D., and Cypher, B.L. (2010). *Urban carnivores: ecology, conflict, and conservation*. Baltimore, MD: Johns Hopkins University Press.

Ghosal, S. (2013). Intimate beasts: exploring relationships between humans and large carnivores in western India. PhD thesis, Norwegian University of Life Sciences, Ås, Norway.

Gonzalez, E.G., Blanco, J.C., Ballesteros, F., Alcaraz, L., Palomero, G., and Doadrio, I.

(2016). Genetic and demographic recovery of an isolated population of brown bear *Ursus arctos* L., 1758. *PeerJ*, **4**, e1928.

Gray, T.N.E., Crouthers, R., Ramesh, K., et al. (2017). A framework for assessing readiness for tiger *Panthera tigris* reintroduction: a case study from eastern Cambodia. *Biodiversity and Conservation*, **26**, 2383–2399.

Griffith, B., and Scott, J.M. (1993). Animal translocation and potential disease transmission. *Journal of Zoo and Wildlife Medicine*, **24**, 2231–2236.

Gusset, M., Ryan, S.J., Hofmeyr, M., et al. (2008). Efforts going to the dogs? Evaluating attempts to re-introduce endangered wild dogs in South Africa. *Journal of Applied Ecology*, **45**, 100–108.

Hayward, M.W., and Kerley, G.I.H. (2009). Fencing for conservation: restriction of evolutionary potential or a riposte to threatening processes? *Biological Conservation*, **142**, 1–13.

Hayward, M.W., and Somers, M.J. (2009). *Reintroduction of top-order predators*. London: Wiley-Blackwell.

Hayward, M.W., Adendorff, J., O'Brien, J., et al. (2007). Practical considerations for the reintroduction of large, terrestrial, mammalian predators based on reintroductions to South Africa's Eastern Cape Province. *The Open Conservation Biology Journal*, **1**, 1–11.

Hazzah, L., Bath, A., Dolrenry, S., Dickman, A., and Frank, L. (2017). From attitudes to actions: predictors of lion killing by Maasai warriors. *PLoS ONE*, **12**(1), e0170796.

Hobbs, R.J., Higgs, E.S., and Hall, C.M. (2013). *Novel ecosystems: intervening in the new ecological world order*. Oxford: Wiley-Blackwell.

Hornocker, M., and Negri, S. (2010). *Cougar: ecology and conservation*. Chicago, IL: University of Chicago Press.

Huber, D., Kusak, J., Majic-Skrbinsek, A., Majnaric, D., and Sindicic, M. (2008). A multidimensional approach to managing the European brown bear in Croatia. *Ursus*, **19**, 22–32.

Hunter, L.T. (1998). The behavioural ecology of reintroduced lions and cheetahs in the Phinda Resource Reserve, Kwazulu-Natal, South Africa. PhD doctoral dissertation, University of Pretoria.

Hunter, L.T.B., Pretorius, K., Carlisle, L.C., et al. (2007). Restoring lions *Panthera leo* to northern KwaZulu-Natal, South Africa: short-term biological and technical success but equivocal long-term conservation. *Oryx*, **41**, 196–204.

IUCN/SSC. (2013). *Guidelines for reintroductions and other conservation translocations*. Gland, Switzerland: IUCN Species Survival Commission.

Jackson, C.R., McNutt, J.W., and Apps, P.J. (2012). Managing the ranging behaviour of African wild dogs (*Lycaon pictus*) using translocated scent marks. *Wildlife Research*, **39**, 31–34.

Jacobsen, K.S., and Linnell, J.D.C. (2016). Perceptions of environmental justice and the conflict surrounding large carnivore management in Norway – implications for conflict management. *Biological Conservation*, **203**, 197–206.

Jansson, E., Ruokonen, M., Kojola, I., and Aspi, J. (2012). Rise and fall of a wolf population: genetic diversity and structure during recovery, rapid expansion and drastic decline. *Molecular Ecology*, **21**, 5178–5193.

Jerina, K., and Adamic, M. (2008). Fifty years of brown bear population expansion: effects of sex-biased dispersal on rate of expansion and population structure. *Journal of Mammalogy*, **89**, 1491–1501.

Jørgensen, D. (2015). Rethinking rewilding. *Geoforum*, **65**, 482–488.

Jule, K.R., Leaver, L.A., and Lea, S.E.G. (2008). The effects of captive experience on reintroduction survival in carnivores: a review and analysis. *Biological Conservation*, **141**, 355–363.

Kaczensky, P. (1999). Large carnivore depredation on livestock in Europe. *Ursus*, **11**, 59–72.

Kaczensky, P., Knauer, F., Krze, B., Jonozovic, M., Adamic, M., and Grossow, H.

(2003). The impact of high speed, high volume traffic axes on brown bears in Slovenia. *Biological Conservation*, **111**, 191-204.

Kaczensky, P., Jerina, K., Jonozovic, M., et al. (2011). Illegal killings may hamper brown bear recovery in the Eastern Alps. *Ursus*, **22**, 37-46.

Kaltenborn, B., and Brainerd, S.M. (2016). Can poaching inadvertently contribute to increased public acceptance of wolves in Scandinavia? *European Journal of Wildlife Research*, **62**, 179-188.

Kansky, R., and Knight, A.T. (2014). Key factors driving attitudes towards large mammals in conflict with humans. *Biological Conservation*, **179**, 93-105.

Kansky, R., Kidd, M., and Knight, A.T. (2014). Meta-analysis of attitudes toward damage-causing mammalian wildlife. *Conservation Biology*, **28**, 924-938.

Knott, E.J., Bunnefeld, N., Huber, D., Reljic, S., Kerezi, V., and Milner-Gulland, E.J. (2014). The potential impacts of changes in bear hunting policy for hunting organisations in Croatia. *European Journal of Wildlife Research*, **60**, 85-97.

Kojola, I., and Heikkinen, S. (2006). The structure of the expanded brown bear population at the edge of the Finnish range. *Annales Zoologica Fennici*, **43**, 258-262.

Kramer-Schadt, S., Revilla, E., Wiegand, T., and Breitenmoser, U. (2004). Fragmented landscapes, road mortality and patch connectivity: modelling influences on the dispersal of Eurasian lynx. *Journal of Applied Ecology*, **41**, 711-723.

Kreeger, T.J., and Arnemo, J.M. (2012). *Handbook of wildlife chemical immobilization*. Author.

Kruuk, H. (2002). *Hunter and hunted: relationships between carnivores and people*. Cambridge: Cambridge University Press.

Kuijper, D.P.J., Sahlen, E., Elmhagen, B., et al. (2016). Paws without claws? Ecological effects of large carnivores in anthropogenic landscapes. *Proceedings of the Royal Society of London B: Biological Sciences*, **283**, 20161625.

Landa, A., Linnell, J.D.C., Swenson, J.E., Røskaft, E., and Moskness, I. (2000). Conservation of Scandinavian wolverines in ecological and political landscapes. In Griffiths, H.I. (Ed.), *Mustelids in a modern world: conservation aspects of small carnivore-human interactions* (pp. 1-20). Leiden: Backhuys Publishers.

Lescureux, N., Linnell, J.D.C., Mustafa, S., et al. (2011). Fear of the unknown: local knowledge and perceptions of the Eurasian lynx Lynx lynx in western Macedonia. *Oryx*, **45**, 600-607.

Li, J., Wang, D., Yin, H., et al. (2014). Role of Tibetan Buddhist monasteries in snow leopard conservation. *Conservation Biology*, **28**, 87-94.

Lindsey, P.A. (2008). Trophy hunting in Sub Saharan Africa, economic scale and conservation significance. *Best Practices in Sustainable Hunting*, **1**, 41-47.

Lindsey, P., du Toit, R., Pole, A., and Romañach, S. (2009). Savé Valley Conservancy: a large scale African experiment in cooperative wildlife management. In Suich, H., Child, B., and Spenceley, A. (Eds.), *Evolution and innovation in wildlife conservation: parks and game ranches to transfrontier conservation areas* (pp. 163-184). Abingdon: Earthscan.

Lindsey, P.A., Masterson, C.L., Beck, A.L., and Romañach, S. (2012). Ecological, social and financial issues related to fencing as a conservation tool in Africa. In Somers, M.J.J. and Hayward, M.W. (Eds.), *Fencing for conservation* (pp. 215-234). New York, NY: Springer.

Linnell, J.D.C. (2015). Defining scales for managing biodiversity and natural resources in the face of conflicts. In Redpath, S.M., Guitiérrez, R.J., Wood, K.A., and Young, J.C. (Eds.), *Conflicts in conservation: navigating towards solutions* (pp. 208-218). Cambridge: Cambridge University Press.

Linnell, J.D.C., and Alleau, J. (2015). Predators that kill humans: myth, reality, context and the politics of wolf attacks on people. In Angelici, F.M. (Ed.), *Problematic wildlife – a cross-disciplinary approach* (pp. 357–372). Berlin: Springer.

Linnell, J.D.C., and Boitani, L. (2012). Building biological realism into wolf management policy: the development of the population approach in Europe. *Hystrix – Italian Journal of Mammalogy*, **23**, 80–91.

Linnell, J.D.C., and Zachos, F.E. (2011). Status and distribution patterns of European ungulates: genetics, population history and conservation. In Putman, R., Apollonio, M., and Andersen, R. (Eds.), *Ungulate management in Europe: problems and practices* (pp. 12–53). Cambridge: Cambridge University Press.

Linnell, J.D.C., Swenson, J.E., and Andersen, R. (2001). Predators and people: conservation of large carnivores is possible at high human densities if management policy is favourable. *Animal Conservation*, **4**, 345–349.

Linnell, J.D.C., Breitenmoser, U., Breitenmoser-Würsten, C., Odden, J., and von Arx, M. (2009). Recovery of Eurasian lynx in Europe: what part has reintroduction played? In Hayward, M.W. and Somers, M.J. (Eds.), *Reintroduction of top-order predators* (pp. 72–91). Oxford: Wiley-Blackwell.

Linnell, J.D.C., Broseth, H., Odden, J., and Nilsen, E.B. (2010). Sustainably harvesting a large carnivore? Development of Eurasian lynx populations in Norway during 160 years of shifting policy. *Environmental Management*, **45**, 1142–1154.

Linnell, J.D.C., Trouwborst, A., Boitani, L., et al. (2016). Border security fencing and wildlife: the end of the transboundary paradigm in Eurasia? *PLoS Biology*, **14**(6), e1002483.

Linnell, J.D.C., Trouwborst, A., and Fleurke, F.M. (2017). When is it acceptable to kill a strictly protected carnivore? Exploring the legal constraints on wildlife management within Europe's Bern Convention. *Nature Conservation*, **2**, 129–157.

Löe, J., and Røskaft, E. (2004). Large carnivores and human safety: a review. *Ambio*, **33**, 283–288.

Loveridge, A.J., Searle, A.W., Murindagomo, F., and Macdonald, D.W. (2007). The impact of sport hunting on the population dynamics of an African lion population in a protected area. *Biological Conservation*, **134**, 548–558.

Lute, M.L., Carter, N.H., López-Bao, J.V., and Linnell, J.D.C. (2018). Conservation professionals agree on challenges to coexisting with large carnivores but not on solutions. *Biological Conservation*, **218**, 223–232.

Lyngdoh, S., Mathur, V.B., and Sinha, B.C. (2017). Tigers, tourists and wildlife: visitor demographics and experience in three Indian tiger reserves. *Biodiversity and Conservation*, **26**, 2187–2204.

Macdonald, D.W., Jacobsen, K.S., Burnham, D., Johnson, P.J., and Loveridge, A. (2016). Cecil: a moment of a movement? Analysis of media coverage of the death of a lion, *Panthera leo*. *Animals*, **6**, 26.

Maehr, D.S., Noss, R.F., and Larkin, J.L. (2001). *Large mammal restoration: ecological and sociological challenges in the 21st century*. Washington, DC: Island Press.

Majic, A., de Bodonia, A.M.T., Huber, D., and Bunnefeld, N. (2011). Dynamics of public attitudes toward bears and the role of bear hunting in Croatia. *Biological Conservation*, **144**, 3018–3027.

Manfred, M.J. (2008). *Who cares about wildlife? Social science concepts for exploring human–wildlife relationships and conservation issues*. New York, NY: Springer.

Manfredo, M.J., Bruskotter, J.T., Teel, T.L., et al. (2017a). Why social values cannot be changed for the sake of conservation. *Conservation Biology*, **31**, 772–780.

Manfredo, M.J., Teel, T.L., Sullivan, L., and Dietsch, A.M. (2017b). Values, trust, and cultural backlash in conservation governance: the case of wildlife management in the United States. *Biological Conservation*, **214**, 303–311.

Marchini, S., and Macdonald, D.W. (2012). Predicting ranchers' intention to kill jaguars: case studies in Amazonia and Pantanal. *Biological Conservation*, **147**, 213-221.

Marucco, F. (2016). *Proceedings II Conference LIFE WolfAlps – the wolf population in the Alps: status and management*, Cuneo, 22 July 2016. Project LIFE 12 NAT/IT/00080 WOLFALPS. Italy.

Masenga, E.H. (2017). Behavioural ecology of free-ranging and reintroduced African wild dog (*Lycaon pictus*) packs in the Serengeti ecosystem, Tanzania. Doctoral thesis, Norwegian University of Science and Technology.

Masenga, E.H., Jackson, C.R., Mjingo, E.E., et al. (2016). Insights into long-distance dispersal by African wild dogs in East Africa. *African Journal of Ecology*, **54**, 95-98.

Milanesi, P., Breiner, F.T., Puopolo, F., and Holderegger, R. (2017). European human-dominated landscapes provide ample space for the recolonization of large carnivore populations under future land change scenarios. *Ecography*, **40**, 1359-1368.

Miller, S.M., and Funston, P.J. (2014). Rapid growth rates of lion (*Panthera leo*) populations in small, fenced reserves in South Africa: a management dilemma. *South African Journal of Wildlife Research*, **44**, 43-55.

Mkoka, C. (2017). Liwonde National Park gets five cheetahs. *The Times*, Malawi.

Mladenoff, D.J., Clayton, M.K., Pratt, S.D., Sickley, T.A., and Wydeven, A.P. (2009). Change in occupied wolf habitat in the northern Great Lakes Region. In Wydeven, A.P., Van Deelen, T., and Heske, E.J. (Eds.), *Recovery of gray wolves in the Great Lakes region of the United States: an endangered species success story* (pp. 119-138). New York, NY: Springer.

Morzillo, A.T., Mertig, A.G., Hollister, J.W., Garner, N., and Liu, J. (2010). Socioeconomic factors affecting local support for black bear recovery strategies. *Environmental Management*, **45**, 1299-1311.

Naess, M.W., Bardsen, B.J., Pedersen, E., and Tveraa, T. (2011). Pastoral herding strategies and governmental management objectives: predation compensation as a risk buffering strategy in the Saami reindeer husbandry. *Human Ecology*, **39**, 489-508.

Naughton-Treves, L., Grossberg, R., and Treves, A. (2003). Paying for tolerance: rural citizens' attitudes toward wolf depredation and compensation. *Conservation Biology*, **17**, 1500-1511.

Nie, M.A. (2003). *Beyond wolves: the politics of wolf recovery and management*. London: University of Minnesota Press.

Nilsen, E.B., Broseth, H., Odden, J., and Linnell, J.D.C. (2012). Quota hunting of Eurasian lynx in Norway: patterns of hunter selection, hunter efficiency and monitoring accuracy. *European Journal of Wildlife Research*, **58**, 325-333.

Noam, Z. (2007). Eco-authoritarian conservation and ethnic conflict in Burma. *Policy Matters*, **15**, 272-287.

Nowak, S., Myslajek, R.W., Szewczyk, M., Tomczak, P., Borowik, T., and Jedrzejewska, B. (2017). Sedentary but not dispersing wolves *Canis lupus* recolonizing western Poland (2001–2016) conform to the predictions of a habitat suitability model. *Diversity and Distributions*, **23**, 1353-1364.

Nyhus, P.J., Osofsky, S.A., Ferraro, P., Madden, F., and Fischer, H. (2005). Bearing the costs of human–wildlife conflict: the challenges of compensation schemes. In Woodroffe, R., Thirgood, S., and Rabinowitz, A. (Eds.), *People & wildlife: conflict or co-existence* (pp. 107-121). Cambridge: Cambridge University Press.

Odden, M., Athreya, V., Rattan, S., and Linnell, J.D.C. (2014). Adaptable neighbours: movement patterns of GPS-collared leopards in human dominated landscapes in India. *PLoS ONE*, **9**, e112044.

Oriol-Cotterill, A., Valeix, M., Frank, L.G., Riginos, C., and Macdonald, D.W. (2015).

Landscapes of coexistence for terrestrial carnivores: the ecological consequences of being downgraded from ultimate to penultimate predator by humans. *Oikos*, **124**, 1263-1273.

Packer, C., Ikanda, D., Kissui, B., and Kushnir, H. (2005). Conservation biology: lion attacks on humans in Tanzania. *Nature*, **436**, 927-928.

Packer, C., Loveridge, A., Canney, S., et al. (2013). Conserving large carnivores: dollars and fence. *Ecology Letters*, **16**, 635-641.

Phillips, M.K., Henry, V.G., and Kelly, B.T. (2003). Restoration of the red wolf. In Mech, L.D. and Boitani, L. (Eds.), *Wolves: behavior, ecology, and conservation* (pp. 272-288). Chicago, IL: University of Chicago Press.

Pitman, R.T., Fattebert, J., Williams, S.T., et al. (2017). The conservation costs of game ranching. *Conservation Letters*, **10**, 403-413.

Power, R. (2002). Prey selection of lions *Panthera leo* in a small, enclosed reserve. *Koedoe*, **45**, 67-75.

Purchase, G., and Vhurumuku, G. (2005). *Evaluation of a wild-wild translocation of cheetah (Acinonyx jubatus) from private land to Matusadona National Park, Zimbabwe (1994-2005)*. Bulawayo: Zambezi Society.

Quigley, H., and Herrero, S. (2005). Characteristics and prevention of attacks on humans. In Woodroffe, R., Thirgood, S., and Rabinowitz, A. (Eds.), *People and wildlife: conflict or coexistence?* (pp. 27-48). Cambridge: Cambridge University Press.

Razen, N., Brugnoli, A., Castagna, C., et al. (2016). Long-distance dispersal connects Dinaric-Balkan and Alpine grey wolf (*Canis lupus*) populations. *European Journal of Wildlife Research*, **62**, 137-142.

Reading, R.P., and Clark, T.W. (1996). Carnivore reintroductions: an interdisciplinary examination. In Gittleman, J.L. (Ed.), *Carnivore behavior, ecology and evolution* (pp. 296-336). London: Cornell University Press.

Redpath, S.M., Young, J., Evely, A., et al. (2013). Understanding and managing conservation conflicts. *Trends in Ecology and Evolution*, **28**, 100-109.

Redpath, S.M., Linnell, J.D.C., Festa-Bianchet, M., et al. (2017). Don't forget to look down - collaborative approaches to predator conservation. *Biological Reviews*, **92**, 2157-2163.

Rosen, T., and Bath, A. (2009). Transboundary management of large carnivores in Europe: from incident to opportunity. *Conservation Letters*, **2**, 109-114.

Røskaft, E., Bjerke, T., Kaltenborn, B.P., and Linnell, J.D.C. (2003). Patterns of self reported fear towards large carnivores among the Norwegian public. *Evolution and Human Behaviour*, **24**, 184-198.

Samelius, G., Andren, H., Liberg, O., et al. (2012). Spatial and temporal variation in natal dispersal by Eurasian lynx in Scandinavia. *Journal of Zoology*, **286**, 120-130.

Sarkar, M.S., Ramesh, K., Johnson, J.A., et al. (2016). Movement and home range characteristics of reintroduced tiger (*Panthera tigris*) population in Panna Tiger Reserve, central India. *European Journal of Wildlife Research*, **62**, 537-547.

Saunders, N.J. (1998). *Icons of power: feline symbolism in the Americas*. London: Routledge.

Scheepers, J.L., and Venzke, K.A.E. (1995). Attempts to reintroduce African wild dogs *Lycaon pictus* into Etosha National Park, Namibia. *South African Journal of Wildlife Research*, **25**, 138-140.

Schwartz, C.C., Haroldson, M.A., White, G.C., et al. (2006). Temporal, spatial, and environmental influences on the demographics of grizzly bears in the Greater Yellowstone Ecosystem. *Wildlife Monographs*, **161**(1), 1-68.

Schwerdtner, K., and Gruberb, B. (2007). A conceptual framework for damage compensation schemes. *Biological Conservation*, **134**, 354-360.

Sharpe, V., Norton, B., and Donnelley, S. (2001). *Wolves and human communities: biology, politics, and ethics.* Washington, DC: Island Press.

Singh, H.S. (2017). Dispersion of the Asiatic lion *Panthera leo persica* and its survival in human-dominated landscape outside the Gir forest, Gujarat, India. *Current Science*, **112**, 933-940.

Singh, H.S., and Gibson, L. (2011). A conservation success story in the otherwise dire megafauna extinction crisis: the Asiatic lion (*Panthera leo persica*) of Gir forest. *Biological Conservation*, **144**, 1753-1757.

Skogen, K., Mauz, I., and Krange, O. (2006). Wolves and eco-power. A French-Norwegian analysis of the narratives of the return of large carnivores. *Journal of Alpine Research*, **94**, 78-87.

Skogen, K., Krange, O., and Figari, H. (2017). *Wolf conflicts: a sociological study*. Oxford: Berghahn Books.

Slotow, R., and Hunter, L.T.B. (2009). Reintroduction decisions taken at the incorrect social scale devalue their conservation contribution. In Hayward, M.W. and Somers, M.J. (Eds.), *Reintroduction of top-order predators* (pp. 43-71). London: Wiley-Blackwell.

Snyman, A., Jackson, C.R., and Funston, P.J. (2015). The effect of alternative forms of hunting on the social organization of two small populations of lions *Panthera leo* in southern Africa. *Oryx*, **49**, 604-610.

Soulé, M.E., Estes, J.A., Berger, J., and Martinez del Rios, C. (2003). Ecological effectiveness: conservation goals for interactive species. *Conservation Biology*, **17**, 1238-1250.

Soulé, M., Estes, J.A., Miller, B., and Honnold, D.L. (2005). Strongly interacting species: conservation policy, management, and ethics. *BioScience*, **55**, 168-176.

Sweanor, L.L., Logan, K.A., and Hornocker, M.G. (2000). Cougar dispersal patterns, metapopulation dynamics, and conservation. *Conservation Biology*, **14**, 798-808.

Swenson, J.E., Wabakken, P., Sandegren, F., Bjärvall, A., Franzén, R., and Söderberg, A. (1995). The near extinction and recovery of brown bears in Scandinavia in relation to the bear management policies of Norway and Sweden. *Wildlife Biology*, **1**, 11-25.

Swenson, J.E., Sandegren, F., and Söderberg, A. (1998a). Geographic expansion of an increasing brown bear population: evidence for presaturation dispersal. *Journal of Animal Ecology*, **67**, 819-826.

Swenson, J.E., Sandegren, F., Bjärvall, A., Söderberg, A., and Wabakken, P. (1998b). Status and management of the brown bear in Sweden. In Servheen, C., Herrero, S., and Peyton, B. (Eds.), *Bears, status survey and conservation action plan* (pp. 111-113). Gland, Switzerland: IUCN/SSC Bear and Polar Bear Specialist Groups.

Tambling, C.J., and Du Toit, J.T. (2005). Modelling wildebeest population dynamics: implications of predation and harvesting in a closed system. *Journal of Applied Ecology*, **42**, 431-441.

Taylor, S.K., Buergelt, C.D., Roelke-Parker, M.E., Homer, B.L., and Rostein, D.S. (2002). Causes of mortality of free-ranging Florida panthers. *Journal of Wildlife Diseases*, **38**, 107-114.

Thompson, D.J., and Jenks, J.A. (2005). Long-distance dispersal by a subadult male cougar from the Black Hills, South Dakota. *Journal of Wildlife Management*, **69**, 818-820.

Tortato, F.R., and Izzo, T.J. (2017). Advances and barriers to the development of jaguar tourism in the Brazilian Pantanal. *Perspectives in Ecology and Conservation*, **15**, 61-63.

Treves, A. (2009). Hunting for large carnivore conservation. *Journal of Applied Ecology*, **46**, 1350-1356.

van der Ree, R., Smith, D.J., and Grilo, C. (2015). *Handbook of road ecology*. Oxford: Wiley Blackwell.

Vos, J. (2000). Food habits and livestock depredation of two Iberian wolf packs (*Canis lupus signatus*) in the north of

Portugal. *Journal of Zoology, London*, **251**, 457–462.

Wabakken, P., Sand, H., Liberg, O., and Bjärvall, A. (2001). The recovery, distribution, and population dynamics of wolves on the Scandinavian peninsula, 1978–1998. *Canadian Journal of Zoology*, **79**, 710–725.

Wabakken, P., Sand, H., Kojola, I., et al. (2007). Multistage, long-range natal dispersal by a Global Positioning System-collared Scandinavian wolf. *Journal of Wildlife Management*, **71**, 1631–1634.

Weeks, A.R., Heinze, D., Perrin, L., et al. (2017). Genetic rescue increases fitness and aids rapid recovery of an endangered marsupial population. *Nature Communications*, **8**(1).

Woodroffe, R. (2000). Predators and people: using human densities to interpret declines of large carnivores. *Animal Conservation*, **3**, 165–173.

Woodroffe, R. (2003). Dispersal and conservation: a behavioral perspective on metapopulation persistence. In Festa-Bianchet, M. and Apollonio, M. (Eds.), *Animal behavior and wildlife conservation* (pp. 33–48). Washington, DC: Island Press.

Woodroffe, R. (2011). Demography of a recovering African wild dog (*Lycaon pictus*) population. *Journal of Mammalogy*, **92**, 305–315.

Woodroffe, R., and Ginsberg, J.R. (1998). Edge effects and the extinction of populations inside protected areas. *Science*, **280**, 2126–2128.

Woodroffe, R., Hedges, S., and Durant, S.M. (2014). To fence or not to fence. *Science*, **344**, 46.

Wydeven, A.P., Van Deelen, T., and Heske, E.J. (2009). *Recovery of gray wolves in the Great Lakes region of the United States: an endangered species success story*. New York, NY: Springer.

Yirga, G., De Iongh, H.H., Leirs, H., Gebrihiwot, K., Deckers, J., and Bauer, H. (2012). Adaptability of large carnivores to changing anthropogenic food sources: diet change of spotted hyena (*Crocuta crocuta*) during Christian fasting period in northern Ethiopia. *Journal of Animal Ecology*, **81**, 1052–1055.

Yirga, G., Ersino, W., De Iongh, H.H., et al. (2013). Spotted hyena (*Crocuta crocuta*) coexisting at high density with people in Wukro district, northern Ethiopia. *Mammalian Biology*, **78**, 193–197.

Zabel, A., and Holm-Müller, K. (2008). Conservation performance payments for carnivore conservation in Sweden. *Conservation Biology*, **22**, 247–251.

Zimmermann, A., Baker, N., Inskip, C., et al. (2010). Contemporary views of human-carnivore conflicts on wild rangelands. In du Toit, J., Kock, R., and Deutsch, J.C. (Eds.), *Wild rangelands: conserving wildlife while maintaining livestock in semi-arid ecosystems* (pp. 129–151). London: Wiley-Blackwell.

CHAPTER FOURTEEN

Rewilding cities

MARCUS OWENS and JENNIFER WOLCH
University of California, Berkeley

As landscape architects and urban planners, we view urban rewilding within a broader process described by the late twentieth-century urban theorist Henri Lefebvre as 'the production of space', where urban space serves as a 'concrete abstraction' of social structures. Lefebvre builds on a longer tradition of urban theory that views urbanisation in relation to specific modes of production, or ways of transforming nature and natural resources into wealth that is then fixed in the buildings and infrastructures that give form to urban space. Within this tradition, rooted in the experience of 'the West' realising itself as such in the Renaissance and the Enlightenment, the concept of landscape and the discipline of landscape architecture, respectively, concerned the modes of representing nature – specifically the surface earth, and ways of reproducing pleasing or otherwise desirable aspects of it in urban space (Cosgrove, 1998).

As these modes of production, representation, and reproduction have evolved with new technological practices, so too have ideas about nature and urbanism. Since Lefebvre, many urban scholars have gone on to show how analysing urban space as a 'concrete abstraction' can serve as a means of critiquing interrelated concepts of race, gender, sexuality, colonialism and the very idea of nature as a social construct (Dear, 1991). Concurrently, others turned their attention to science and technology to show how the generation of scientific knowledge and sociotechnical practices relate to one another, as well as the broader spatial process of urbanisation (Latour et al., 1998).

A first goal of this chapter is to show how thinking through the relationship between urbanism and technological change can help us understand future ecologies, specifically the concept of rewilding as it relates to historical concepts of nature and wilderness. After presenting a theoretical argument for understanding rewilding through contemporary processes of urbanisation, a second portion of the chapter builds on approaches that consider scientific and technological practices in relation to urban space by examining modes of representation that characterise urban rewilding in cities of North America and Europe, with specific examples for California. This begins with an overview of the evolution of the camera trap and the ways it is currently used to survey large urban animals. It then focuses on the ways mobile computing and social images help define how we understand urban rewilding. Finally, with advances in computational power and the proliferation of cameras and images, it considers the ways urban wildlife is surveyed by non-human algorithms. The case of automated collision avoidance systems in autonomous vehicles is especially pertinent given the ecological importance of roadkill and habitat connectivity with regard to rewilding schemes. Considering the array of ecological forecasting stemming from the potential of autonomous vehicles to reconfigure urban space in cities around the world, a final portion of this chapter expounds on these speculations to envision dynamics of urban rewilding in the future.

Wilderness and urban transformation

Our study of the role of technology in rewilding urban environments must first be situated within a broader genealogy of wilderness, urbanisation, and the planning and design disciplines. Basic aspects of the Judeo-Christian idea of the wilderness are reflected in the foundational texts of planning and architecture, namely Alberti's Renaissance interpretations of ancient Vitruvian texts. Therein, the city is a geometrically self-contained object, outside of which exists the uncivilised natural world beholden to the laws of the polis. As the basis for Western architecture and city planning, Alberti's treatises were enshrined in legal code as the Law of the Indies as the organising logic for the colonisation of the new world (Cosgrove, 1998). In this regard, Alberti's architectural philosophy neatly coalesced with a Baconian approach to science and technology, where the aim of technological progress is the transformation of nature for the improvement of the human condition, leading to a mechanistic view of nature and animals later espoused by Descarte (Pickles, 2004; Elden, 2010; Merchant, 2013). Connecting the relationship between the spatial logic of Renaissance city planning and mercantilist expansion with Alberti's linear perspectival method for mathematically rendering figure–ground relationships in realistic drawings and oil paintings, landscape theorist W.J.T. Mitchell notes how oil paintings in particular ruptured the muralistic attachment of visual representation to architecture, permitting landscape paintings to be bought and sold, just as the emergent mercantilist bourgeoisie aimed to do with land itself (Mitchell, 2002). Particularly insightful for our study on the role of digital images in rewilding urban environments, Alberti's mechanically reproducible 'objectivity' rendered in distanciated realist framings of the world provides scholars of media and software a way to understand the concept of the virtual, and how screens, windows, cameras, and images affect understanding of the environment (Kittler, 2001; Friedberg, 2006).

The idea of nature as seen through a screen or window corresponds closely to the development of the modern romantic idea of wilderness, where an aristocratic gaze upon the fallow, enclosed estate gave form to aesthetic principles of landscape such as the picturesque, the sublime, and the pastoral by the end of the Enlightenment. This move away from a Hobbesian view of nature and the idealisation of the Rousseauian noble savage manifested in the fashion of 'wilderness gardens' on English estates and the proliferation of 'Rousseau islands' as the 'wild' style of English landscape gardening spread across Europe (Di Palma, 2014). As patterns of urbanisation in North America and Europe transformed with the industrial revolution, an emergent class of professional city planners and landscape architects such as Pierre Joseph Lenne in Prussia, Joseph Paxton in the United Kingdom and later Frederick Law Olmsted in the USA adapted these aesthetic concepts and strategies to design new types of public open spaces. Their aim was to improve urban health among residents subjected to the polluted and crowded environments of the early industrial city and to socialise an emerging proletariat. For example, urban park development projects such as Boston Fenns, designed by Olmsted, incorporated proto-ecological principles that addressed public health in tangible ways by improving water quality as well as recreational opportunities (Whiston Spirn, 1995). Moving beyond discrete public spaces, by the end of the nineteenth century the city planning professionals continued to develop strategies for improving the environmental quality of the city. Prime examples are the City Beautiful movement led by architects such as Daniel Burnham in Chicago and Ebenezer Howard's Garden City movement in the UK.

During this time, the concept of wilderness also began to take a more definitive form, especially aided by developments in photographic representation. The sublime grandeur of Yosemite documented with pioneer photographer Carleton Watkins' glass plate images circulated through the cities of the US East Coast, combined with the writing of Frederick Law Olmsted to give birth to what would become the first National Park. The theorist Peter Grusin describes this process of media circulation within an assemblage of railroad networks, telegraph lines, urban galleries and a state apparatus capable of regulating a remote territory as the workings of a 'nature machine' that functions to produce a specific public consciousness of a wilderness space such as Yosemite (Grusin, 2008). A similar dynamic can be identified in the establishment of the first Federal Wildlife Refuge by Theodore Roosevelt with the support of Frank Chapman of the Museum of Natural History's photographic studies of the birds on Pelican Island in Florida (Chapman, 1900). Nineteenth-century conservation ideology and scientific photographic technology also led to archival documentation of species approaching extinction. One prominent example was the London Zoo's quagga, the first extinct species to be photographed (Guggisberg, 1977), although whether the quagga was actually a distinct species remains open to question. In these cases, wilderness scenes and endangered wild animals are represented as fragile or pristine specimens requiring careful conservation through segregation from urban development and human interference, as indicated by their presentation behind glass plate photographs (Brower, 2011).

The rapid development of the US National Park system from the turn of the twentieth century onwards, along with the post-World War Two construction of the US interstate highway system, allowed massive numbers of people to experience wilderness areas directly. Soon, advances in ecology and environmental science led to growing concerns about species endangerment associated with wilderness visitation. The National Park Service's Mission 66 programme built a wide array of visitation facilities to contain visitors to limited places for wilderness viewing and education, while the 1963 Wilderness Act served to protect specific tracts of land from habitation or motorised vehicles. Propelled by Cold War technologies repurposed for environmental monitoring and wildlife tracking (Benson, 2010), National Parks were recast as 'nature laboratories' according to the pervasive logic of climax ecology, adding another argument for restriction of public access to select wilderness areas (Kupper, 2016).

To remedy the conflict between demands for restricted access with a growing and ever more mobile public, the National Park Service turned to leaders in the fields of media and design, including Walt Disney and Eero Saarinen, to mediate a detached view of these wilderness spaces to a mass audience (Allaback, 2000; Kim, 2004; Carr, 2007). The cinematic conception of a 'pristine wilderness' advanced to North American and western European mass audiences in the middle of the twentieth century occurred alongside the fear of a nuclear winter, a 'population bomb', and space exploration, helping to forge a more explicitly ecological second-wave of environmentalism. In terms of visual representation, the circulation of views of the Earth from outer space bolstered ideas such as Marshall MacLuhan's widely popularised notion of Earth as a 'global village' or Buckminster Fuller's 'spaceship earth', helping to promote a shift in consciousness away from specifically nationalist themes of earlier environmental movements, towards an awareness of the risks facing the larger planet arising from resource-intensive consumer lifestyles that then became the target for change (Kirk, 2007; Masco, 2010; Castells, 2011). This new lifestyle environmentalism emerged against the backdrop of the financial and energy crises of the early 1970s, corresponding to a shift of industrial production to the 'third world', ambiguously termed the 'Global South' after the end of the Cold War.

Environmental protection legislation in Europe and North America and the rise of 'ecological' urban planning and design in the 1970s occurred in tandem with radically changing metropolitan form leading to polycentric, economically polarised and socially segregated cities with far-flung exurbs destroyed and fragmented animal habitats. As globalisation reshaped industrial geography the dense cores of declining industrial cities such as Detroit were hollowed out, opening up large swaths of land for wildlife. Contemporary urban form, with its wide range of densities and interspersion with fingers of wildland, had contradictory impacts on the ability of non-human animals to survive in large conurbations. On the one hand, flexible species are able to penetrate even rather dense urban areas that enjoy some connectivity with larger bioreserves, bringing animals into frequent contact with people. Perhaps the best example of this pattern in North America is the spread of coyote populations throughout most metropolitan areas. On the other hand, the desire for 'indoor–outdoor' living, the ideal of living in 'harmony' with nature, and demand for easy proximity to the outdoors has led to a pattern of suburban development that abuts wildland areas, often leading to human-wildlife conflict and culling campaigns. Concerns about habitat fragmentation and loss and species endangerment or local extirpation have stimulated interventions such as metropolitan wildlife corridors and bioreserves designed to allow wildlife living on urban-wildland interfaces to move among habitat patches and larger wildland reserve zones.

Urban rewilding and twenty-first-century urbanism

Contemporary discourse about urban wilderness and rewilding is situated within this larger narrative of urbanisation and changing urban form. Scholars such as Graham and Marvin (2001) built on studies of uneven urban development to focus on the role of infrastructures, especially information communications technologies, resulting in an increasingly stratified urban experience underpinned by divestment in common infrastructure and privatisation. Similarly, the urban theorist Manuel Castells' (2011) idea of a 'network society' suggests that contemporary cities are best understood not as a set of physical places, but rather as systems of networked spaces animated by global as well as local flows of people, products, images, and data. A feedback loop of resource exploitation, privatisation, and technological change driving urban spatial restructuring at the end of the twentieth century has also transformed ways of understanding wilderness and more generally the environment. In addition, the rise of environmental justice as both an analytical framework (Bullard, 1994) and political movement highlighted how urban ecosystem services not only reflect but often reinforce social, especially racial, difference, and how sustainability planning and ecological restoration projects could trigger housing value increases and displacement – so-called 'green gentrification' (Dooling, 2009; Bryson, 2013; Wolch et al., 2014). Others have noted how ecological sustainability functions as the 'narrative twin' to bandwidth, in which the optimisation of resource consumption through data consumption functions to justify future investment in research and growth (Halpern et al., 2013). Along these lines theorists have articulated a 'meta-infrastructure of planetary-scale computation' combined with the emergence of 'information as a historical agent of economic and geographic command' that destabilises the nation-state as a geopolitical organising logic (Bratton, 2016).

The implications of planetary-scale computation for the concept of wilderness are laid out in a recent thought experiment-provocation by landscape ecologists and designers Cantrell et al. (2017). They ask, could autonomous ecological restoration technologies using artificial intelligence offset human impact on an ecosystem, creating a new type of wildness for the Anthropocene? How might the interaction between autonomous machines and non-human animals sit with our inherited concepts of wilderness and urbanisation? And by extension, how might increasingly wet cities or the rise of autonomous mobility alter urban form and the possibilities for face-to-face human–animal interaction?

Of course, we are a long way from a singular, autonomous ecological restoration machine that is the subject of Cantrell et al.'s provocation. And while the Anthropocene may be upon us, New York City will be wetter, but not underwater or dominated by (for example) autonomous vehicles for some time. More practically, however, we can point to concrete ways in which urban land-use responses to climate change – in both developed and developing worlds – can be expected to radically reorder urban–wilderness relationships, and how non-human animal relationships with increasingly autonomous machines may reconfigure visceral human relationships with wildlife and the environment, arguably a defining characteristic of urban rewilding. It is already clear from a variety of empirical research on environmental and urban sensing (Cuff et al., 2008; Gabrys, 2014) and illustrated in the cases presented below that the proliferation of images – environmental or otherwise – have superseded the singular human gaze that once formed the basis for linear perspective and the idea of landscape. To this end, in a play on Panofsky, the media theorist Lev Manovich suggests that the database has replaced perspective as the symbolic form of the contemporary era (Manovich, 1999).

This would suggest that urban rewilding is produced through images-as-data, proliferating beyond the capacity of human cognition, and that increasingly automated analytical processes may shape what it means to 'be wild' into the twenty-first century. At the same time, mass consumption, technology, and geopolitics drive anthropogenic climate change and automation technologies are on the verge of profoundly reshaping urban form in global cities, albeit in different ways in different locations. In some cases, this may allow room for wildlife requiring new ways of managing and navigating encounters with humans.

Rewilding and changing ideas about urban wilderness

By the last decades of the twentieth century, modern ideas of wilderness as articulated by North American and European scholars were widely contested, despite their longevity and cultural power. Perhaps most relevant to the concept of urban rewilding, William Cronon urged people to 'look in the wilderness in their own backyard' in his famous 1996 essay 'The Trouble with Wilderness' (Cronon, 1996). The concept of rewilding similarly emerged during this period, a conceptual branding tool mobilised by activist Dave Foreman of the NGO Earth First!, built upon scientific principles borrowed from fields such as conservation biology and restoration ecology (Foreman, 2004). This parallels the aesthetic concept of urban wilderness, reconceiving areas where economic restructuring left abandoned industrial zones overgrown by vegetation leading to a return of wildlife (Mathey and Rink, 2010).

'Rewilding' was thus understood as an urban process that occurred in three ways: first, through deindustrialisation and the resultant liberation of urban space that could be rewilded; second, through spontaneous, autonomous occupation of such space by non-human animals; and third, through purposive implementation of an ecological programme, often initiated through representations of nature as an agent acting to reclaim urban space. In many cases, with the Naturpark Süd Gelände in Berlin as one prominent example, the institutionalisation of these new urban wilderness spaces resulted from decades-long battles by coalitions of neighbourhood residents and ecologists (Lachmund and Jasanoff, 2004; Gandy, 2013) against developers and the state interested in economic redevelopment. This politicisation of ecological data and destabilisation of power and expertise at work in the development of urban wilderness continues to underpin many of the contemporary urban rewilding examples discussed in this chapter.

The move away from a singular conception of nature was theorised by Haraway with the figure of the 'cyborg' – part human, part animal, part machine – as refutation of Western nature/culture dualism and teleological models of technological progress inherited from Bacon and Descartes (Haraway, 2013). Urban geographers used Haraway's notion of the cyborg to describe patterns of late twentieth-century urbanisation (Gandy, 2005), with the sprawling post-war Southern California landscape often serving as a prototype for these developments. Southern California was the site and oft-evoked reference point in the pathbreaking *Uncommon Ground* conference and book, and as ground zero for urban conservation biology as developed through analysis of urban coastal areas in San Diego County (Soulé, 1985). In terms of urban form, the Los Angeles School of urban theory also emerged during this period to make sense of the sprawling polycentric landscape of Southern California, elevating Los Angeles as the archetypical postmodern city in contradistinction to long-standing master narratives of urban form based on industrial metropoles such as Chicago, Berlin, Paris, and Manchester in the nineteenth and early twentieth centuries (Dear, 2002).

Within this milieu, Jennifer Wolch developed the concept of 'zoöpolis', to specifically describe 'the emergence of ethics, practices and politics of caring for animals and nature' in cities (Wolch and Emel, 1998). The idea of zoöpolis reflects Cronon's call to find the wilderness close by, examining the contours of human–wildlife encounters in everyday urban spaces. Along these lines, in his book *Biophillic Cities*, Timothy Beatley builds on Wolch's concept to describe the relevance of new modes of mediating positive encounters with nature. This includes webcams that show sea lions at Pier 39 in San Francisco and falcons in downtown Richmond, Virginia, as important efforts to 'help expand the base of political and popular support [showing that] conservation is not to be viewed as simply something that trained scientists and professional resources do in faraway places. Rather, it has very local relevance and ... clear connections to where people live and work' (Beatley, 2011).

Beatley's observation about the epistemic qualities of wildlife images corresponds to the tradition of civic environmentalism that emerged in the USA in the wake of the environmental movements of the 1970s and 1980s mentioned earlier. This movement was complemented by a wide array of literature developed since the 1980s exploring the splintering of society into a vast array of interest and advocacy groups and the devolution of public services traditionally provided by the state to a complex array of non-profit institutions that might enhance civic and/or client participation but which were arguably outside of traditional democratic controls (Wolch, 1990; Rose, 1996). Along these lines, we are reminded that the 'rewilding' term comes from the practices of non-profit organisations, in this case Earth First!, competing to create brands and build constituencies and resources for their causes. In this way, the advent of civic environmentalism (Rothman, 2004; Karvonen and Yocom, 2011), environmental stewardship (Svendsen, 2010; Fisher et al., 2012), citizen or 'street' science (Foth et al., 2011; Catlin-Groves, 2012), and ecological gentrification (Dooling, 2009; Birge-Liberman, 2010) are particularly relevant to questions of urban rewilding.

Digital images and urban wildlife

Operating from the traditions of landscape and human–animal studies, we examine the visual artefacts of urban rewilding that emerge from methods of sensing and communicating the presence of wildlife in the urban landscape. This section introduces developments in information communications technologies which manifest convergences between camera traps and motion pictures, allowing us to better understand reconfigurations of the virtual and the visceral aspects of urban rewilding in the age of automation. Stemming from a need to reduce time-consuming processes of image analysis, web-based cameras that automatically identify the presence of wildlife in the frame are increasingly deployed in monitoring projects (Yasuda and Kawakami, 2002). However, beyond efficiency and data-processing optimisation, the voluntary labour involved in processing large data sets of wildlife images also serves to produce a lively constituency for conservation in the model of the 'nature machine' described above. Along these lines, we go on to consider the role of local social networks and web-map visualisations of urban wildlife monitored with GIS, RFID or other tracking technologies in mediating human encounters in rewilded future ecologies. Spurred by the proliferation of images with social media, we discuss the implications of massive data sets of wildlife imagery, specifically automated sensing and the ability of artificial intelligence to identify and react to wildlife in the landscape. Finally, we conclude with remarks about the collapsing of the visceral and the virtual characteristics of urban rewilding, and speculations on implications for future ecologies.

Camera traps

Following Muybridge's *Horse in Motion*, camera trapping for wildlife became widely associated with George Shiras, whose 'flashlight' photographs were featured in *National Geographic Magazine* and awarded a gold medal at the Paris World Exhibition in 1900. Shiras documented a variety of wildlife around the world, developing an array of strategies for particular species, including baited triggers, stringing trip wire across likely migration routes, and even tying trip wires to dislodged sticks from a beaver dam (Nesbit, 1926; Guggisberg, 1977; Kucera and Barrett, 2011). Into the twentieth century camera trapping developed alongside sport hunting and colonial land management that characterised the world of conservation during this period. This is reflected in one of the earliest detailed guides on outdoor photography, William Nesbit's *How to Hunt with the Camera*, where he describes 'flashlight trap photography' as 'a most fascinating sport and is deservedly becoming more and more popular' (Nesbit, 1926). In terms of scientific analysis and the production of wildlife images as data points, Frank M. Chapman of the American Museum of Natural History used camera traps to carry out a census of the Barro Colorado research island in Panama in 1928. In addition to capturing stealthy and/or rare carnivores, more common animals such as squirrels and raccoons often served as subjects for the development of camera trap technologies.

Driven by broader development of photographic and information communications technologies later in the twentieth century, initial advances in camera traps centred on hardware: the automation of components, miniaturisation, portable batteries, electric lights, digitalisation, infrared beams as triggering mechanisms, etc. Decreasing costs led to the diversification of applications of camera traps, and by the 1990s, camera trap methods had expanded from documenting the existence of rare animals to encompass a variety of techniques for evaluating wildlife population dynamics that increasingly hinge on software for their analysis (Nichols et al., 2011). Particularly *urban* applications include the use of camera traps to design rabies vaccine distribution mechanisms for foxes in Zurich (Hegglin et al., 2004), as well as determining the design of highway underpasses for population connectivity outside Los Angeles (Ng et al., 2004). At the same time as analysis of camera trap photographs has become increasingly sophisticated, revealing ever more nuanced understanding of wildlife populations, camera trap technology has also entered the consumer market for homeowners and hunters, merging into the domain of social images commonly associated with the web and social media (Kays and Slauson, 2008). Much debate, exploration, and experimentation continues among wildlife ecologists in regards to sampling methods and analysis techniques of camera trap visual data (O'Connell et al., 2010) and this diversity of methods is reflected in ecological studies of California's urban wildlife.

Regardless of method, camera trap projects in the Bay Area and Los Angeles tend to be operated by governmental land management agencies, such as the East Bay Regional Parks District or the National Park Service. Such projects also tend to involve partnerships with conservation NGOs and equipment manufacturers. For example, the East Bay Regional Parks District has partnered mountain lion conservation groups such as Panthera, Felidae, and the Bay Area Puma Project to assist in efforts to monitor movements of mountain lions in the East Bay hills. In addition to NGOs, many camera trap projects also rely on individual volunteers to perform the labour-intensive tasks of collecting data from remote camera traps and processing large photo data sets.

For example, in Los Angeles, the Springs Fire Wildlife Project, part of the larger Mediterranean Coast Inventory and Monitoring Network, relies on volunteer trail runners to cover up to 16 miles a day to rotate 10 cameras between 90 randomly selected points in the Santa Monica Mountains. Characteristic of urban rewilding images, of the 1047 photos triggered in a US Geological survey of mesocarnivores in San Francisco's Presidio, 90 featured the target carnivores, while some 819 featured people and/or companion animals (Boydston, 2005). Also in the Bay Area, the One Tam Wildlife Picture Index Project uses volunteers to gather data from trail cameras, as well as a larger cohort of crowd-sourced volunteer 'click workers' to sift through the accumulated images and identify animals gathered from the index. The One Tam Project is unique in that it is a picture index over a large territory, therefore producing an even greater number of images (of wildlife, and false positives including domestic animals, humans, and 'empty' landscapes) than point-specific camera traps such as those deployed on highway overpasses. The volunteer image analysts function not only to provide data entry service for the ecological monitoring projects, but also provide a way for

Figure 14.1. A camera-trapped coyote in San Francisco's Presidio, outfitted with a tracking collar, and identifying ear marks.

the public to get involved with park agencies in ways alluded to by Beatley. Many volunteers may not be able to climb the hills north of San Francisco, but processing images on a computer in their own home creates an accessible interface for the park and a new constituency for non-profit park partner organisations. Real-time camera-trapped wildlife imagery therefore provides a means to raise awareness about a given species or habitat concern in the broader population (Locke et al., 2005) and ultimately raise resources.

Webcams and live feeds

While the reliance on volunteers may serve a social purpose in regards to garnering support or developing awareness for conservation initiatives, there are difficulties in gathering data for scientific analysis from such camera networks. These challenges account for a turn towards web-based transmission and automatic sensing deployed in a camera trap project described in Yasuda and Kawakami (2002) and a convergence with webcam imagery. Similarly, the use of aerial drones for wildlife surveys also increasingly relies on automation to process the massive amounts of visual data collected (Chabot and Bird, 2015). While drone technology was not designed explicitly for urban wildlife monitoring, we can think of drone footage in relation to webcam images and live feeds that emerged as with the advent of live television as well as the needs of the space race, globalised markets, and 24-hour communications cycles. Here, the 1936 Olympics are purported to be the first 'live' television event, while early closed circuit television (CCTV) systems were designed by Siemens AG to monitor V2 rocket launches (Zielinski, 1999). In terms of environmental communication, the live landscape image came into its own in the post-World War Two period, most famously with the broadcast of the Moon landing. The use of CCTV in the 1980s as in-house information television for large Japanese firms to keep abreast of the latest stock market and trade reports coincides with the development of 24-hour satellite television instituted around the same time and the economic processes of globalisation. One unique aspect of the webcam image is its accessibility via a web browser, permitting global broadcast across the internet beginning in 1993 and proliferating with commercial webcams entering the mass market in 1994.

However, the use of webcams extends beyond passive voyeurism, a common critique of landscape painting and photography. Like the Japanese corporations' utilitarian adoption of CCTV, in 1995 the Huntington Beach firm Surfline positioned webcams up and down the California coast and beyond to help surfers decide when and where to find the best waves. But perhaps the most famous environmental webcam of the past decade is the 'spillcam' installed by British Petroleum and hosted on the website of the US House of Representatives Subcommittee on Energy Independence and Global Warming after the government forced BP to make the view public (Goldenberg and correspondent, 2010).

Judging by the many thousands of wildlife-themed live feeds on the internet, it is clear that webcams have become an important but undertheorised component of contemporary ecological aesthetics and conservation practice. Many zoos and aquaria include live feeds, as well as rural conservation areas across the world. In California urban areas, live feeds of landscapes at the Golden Gate National Recreation Area and the Santa Monica Mountains National Recreation Area are offered on National Park Service websites. The webcam image's durational media type makes it suited to observe animal habitats, such as nests, dens, and watering holes. These generate intimate images that would be impossible to observe otherwise. For example, the Midwest Raptor Resource Center, a non-profit organisation founded in 1998 to 'create, improve and directly maintain' nest sites, claims to have put up the first nest cam the same year it was founded. Similar organisations operate webcams in the Bay Area, such as the SF Bay Osprey cam operated by the Golden Gate Audubon Society with support from public and private sources including the City and Port of Richmond, PG&E, Richmond Museum Association, HD On Tap, Rosie the Riveter National Historical Park, S.S. Red Oak Victory Ship & Museum, SF Bay Osprey Coalition, Craig Newmark of Craig's List, and other private individuals.

Most of the San Francisco Bay Area's nest cams are operated by commercial entities whose buildings feature a nest. One exception is also among the first nest cams in the Bay Area, placed on San Jose City Hall in 2007 in collaboration with the UC Santa Cruz Predatory Bird Research Group; the San Jose Peregrine

Falcon Alliance raised $25,000 for upgrades and a falcon nest box with notable corporate support coming from local tech companies including Adobe, Discover Video, Sony, Data Display Audio Visual Company, and Extron Electronics. However, more common are nest cams, hosted by companies that produce technologies supporting wildlife camera systems, located near coastal wetlands that are part of the Pacific Flyway. For example, the microprocessor and defence contractor PMC mounted a HawkCam featuring a nest on its headquarters that soon became popular among its software engineers, as well as its neighbour Yahoo! also located near the bayshore in Sunnyvale. The cam prompted the company to take conservation measures and adapt its building maintenance procedures (although in Silicon Valley fashion, PMC was purchased by Microsemi in 2016 and the webcam has been shut off). Similarly, the utilities giant PG&E has maintained a webcam atop its tower in the Financial District of San Francisco since 2008 in collaboration with the UC Santa Cruz Predatory Bird Research Group.

A second kind of urban wildlife webcam in California are those trained on seals in La Jolla and sea lions in San Francisco's Pier 39. In both cases, these cams present mysteriously resurgent populations of large marine mammals in

Figure 14.2. A coyote spotting on the local social network Nextdoor.com, in San Francisco.

busy, affluent urban waterfronts. However, despite a similar subject, each of these webcams reflects substantially different contexts and results. In the case of San Francisco, the cam is operated by Tandem Partnerships, a subsidiary of the operator of the privately owned shopping centre that was developed in what had been a historic working waterfront until the 1960s. More explicitly than the nest cams, the Pier 39 Sea Lion cam appears more directly linked to a commercial endeavour, as it is widely advertised as a tourist destination rather than a broader conservation project. In the case of La Jolla, seal and sea lion populations arrived in the 1980s and grew over the ensuing decades. The presence of these marine mammals was vigorously opposed by beach access groups who resisted ceding their beach to the seals and sea lions; today, the creatures are protected by a Marine Mammal Reserve and are a major tourist attraction – although still a source of controversy among locals. The short-lived Seal Cam was established by Western Alliance for Nature with the support of EarthCam, a service provider for a global network of webcams. The cameras were approved by the mayor in 2013, but deactivated shortly thereafter.

Urban rewilding on the social web

In addition to camera traps and webcams, social transformations ushered in with the 'web 2.0' have impacted human relationships with urban wildlife in a number of ways. Defined by user-generated content, 'web. 2.0' urban wildlife platforms were spurred by the proliferation of social networks, and particularly by smart phones which permitted computing to move outside. Urban wildlife media have the potential to go viral around the world, and can dovetail with entrepreneurial urban strategies, as reflected in the case of the branding and marketing of the South Congress Bridge or 'Bat Bridge' in Austin, Texas. In terms of this chapter's emphasis on the dynamics of urban rewilding and the relationship between the visceral and the virtual, we are particularly interested in 'local social networks' and the way they can be used as a means of mediating everyday human–wildlife encounters. To this end, web 2.0 urban wildlife platforms also relate to a larger domain of citizen science, and an increasingly politicisation (and monetisation) of spatial data at the intersection of 'citizen science' and 'civic tech' (Foth et al., 2011; Catlin-Groves, 2012). Often this overlaps with branding, marketing and user research efforts by non-profit agencies and consultancies. Unsurprisingly given its proximity to Silicon Valley, California has been a leader in this domain. For instance, San Francisco's Golden Gate National Recreation Area claims status as the first park to develop a website (for Alcatraz) in the mid-1990s, as well as a twitter feed in the next decade.

Two such platforms are iNaturalist and Coyote Cacher, both at the forefront of rewilding efforts in California. The more 'beta' of the two platforms, Coyote

Cacher, was developed by the University of California Division of Agriculture and Natural Resources South Coast Research and Extension Center in Irvine in 2016. The portal allows residents to report and receive email message alerts about coyote sightings in their area. Sightings within the last 30 days are also visualised on a web-map, and in 2017 the project team developed a mobile app. In addition to providing residents with information on navigating encounters with coyotes, Coyote Cacher also provides the project's research scientists baseline information regarding the success of community hazing efforts by analysis of coyote reporting metadata (Kan-Rice, 2017). The app is part of a broader programme to evaluate the effectiveness of hazing efforts, which includes collaring and tracking coyotes subject to citizen- and state-organised hazing.

The second platform, iNaturalist, is a mobile app that allows users to catalogue and geo-tag flora and fauna. Developed by Ken-Ichi Ueda at the UC Berkeley School of Information in 2008, iNaturalist is an open-source software relying on a number of public APIs and open-source software tools to help users identify and map flora and fauna they find in their surroundings. These tools include Catalogue of Life, uBio, Ruby on Rails, jQuery, Google Maps, and Mark James' Silk Icons. Launched as a business in 2011, this crowd-sourced approach to documenting urban nature has resulted in the discovery of new species of frog and the first photograph of a snail documented by Captain Cook. iNaturalist was acquired by the California Academy of Sciences in San Francisco in 2014, and serves a means for the institution to engage in public outreach and the urban environment, with particular uses for urban wildlife management. Working in conjunction with the California Academy of Science, the Golden Gate Parks Conservancy, a non-profit fundraising and volunteer management organisation supporting ecological restoration projects in the Golden Gate National Recreation Area, has used iNaturalist to organise 'Bioblitzes' in the former military lands that make up this urban National Park. iNaturalist is also promoted by the Presidio Trust as a venue to report photos and observations of coyote sightings, building on a longer tradition of civic environmentalism, or the cultivation of environmentalist constituencies for which the Golden Gate National Recreation Area has become known (Rothman, 2004).

Like many social apps, iNaturalist deploys design strategies derived from casino slot machines, aiming to spur motivation in the user. iNaturalist therefore goes beyond the novelty of viral animal images on the web or the data aggregation of traditional citizen science projects through its use of gamification to spur users to gather information about the environment (Bowser et al., 2014; Nechwatal, 2016). In this regard, the platform seems well poised to transition into applying augmented reality to gamify the process of documenting urban flora and fauna. This would be a significant development in light of the breakout success of PokemonGo in 2016, as well as the increasing popularity of Snapchat in paving the way for broader social acceptance of augmented reality in everyday life, something at which Google Glasses and Google Goggles have thus far largely failed. Moreover, like the metadata analytics of Coyote Cacher, the wealth of data produced by social wildlife image platforms like iNaturalist provides scientists with additional layers through which to understand urban rewilding.

Automation and future ecologies of rewilding

As automated analysis of wildlife images helps conservation biologists better plan for connective corridors, similar imaging technologies are at work in a variety of robots and machines designed to act autonomously in the environment. In particular, we focus on autonomous vehicles – in Silicon Valley pioneered by Google and Tesla – given their anticipated transformative effects on the built environment more broadly. However, a variety of other speculative and developing applications of autonomous rewilding technologies bring forth the ethical and philosophical issues of artificial intelligence implied by Cantrel et al. in their thought experiment of the autonomous ecological restoration system. Nevertheless, an overview of the encounter between an autonomous vehicle and wildlife provides a general template for machine–animal interactions in cities.

Autonomous vehicles rely on a number of devices to orient themselves in an environment, including radar, LIDAR, image and ultrasonic sensors, as well as data aggregated from the cloud such as traffic or weather conditions. Conceivably, data on large endangered animals warranting intensive monitoring, such as panthers, could be integrated into this cloud data structure as in 'real-time virtual fences' being developed for managing human–elephant conflict in Africa (Jachowski et al., 2014). Environmental sensors could also be embedded into road infrastructure to identify wildlife in the vicinity and communicate this information to the autonomous vehicle, as is already done with conventional signage and human drivers (Huiser and McGowen, 2012; Bland, 2015). However, in most cases, especially for more common animals of minimal value to humans, wildlife is likely to be sensed by the automobile itself, through a combination of LIDAR and image sensors. In this way, we may think of an animal's relationship to autonomous vehicles as somewhat similar to a camera trap network with integrated automated image analysis: the animal's body meets an infrared beam, which triggers the production and analysis of an image. However, rather than a single image or even a photo burst, in the case of autonomous vehicles, it is likely that these two processes will be simultaneous and continuous as the vehicle and animal both move across the environment, with the beam measuring the distance of the animal from the vehicle, and the image serving to identify the animal and take appropriate action.

In addition to swerving and braking of an autonomous automobile, we can imagine a range of other potential interactions between machines and wildlife, given emerging applications of security robots to gather data and dissuade loitering in San Francisco (McCormick, 2017) – for example, micro-drone swarms that identify, track, and strategically haze carnivores encroaching on sensitive areas. In practice, predicting animal movement remains a major problem, recalling the role of animal locomotion at the genesis of motion pictures. For example, Volvo's 'Large Animals Detection System', designed to identify caribou and moose, was completely thrown by the locomotion of kangaroos (Zhou, 2017). Ostensibly, with additional environmental and experiential data, machine learning pattern recognition will improve beyond human abilities to predict a given wild animal's behaviour.

However, even with 'perfected' technical systems where all wildlife can be sensed, identified, and accurately predicted, autonomous vehicles present major ethical and practical challenges. These can be roughly summed up in the moral philosophy thought experiment known as the 'trolley problem' in which an actor must make a decision between two negative outcomes, in the archetypical case allowing a runaway trolley to collide into a group of five people, or throwing a switch and killing only one. In the case of autonomous vehicles, this is further complicated by the fact that consumer demands for manufacturer passenger safety guarantees will be made for autonomous vehicles whose decision-making algorithms will be 'black-boxed' and apt to be adaptive and generated from unsupervised deep machine learning processes. While it can be safely assumed that human lives will continue to take precedence over non-human animals, algorithms at work in autonomous vehicles will still need to construct a hierarchy of animal life upon which to base decisions about collision avoidance manoeuvres.

New ecologies are apt to emerge with the introduction of autonomous vehicles. Forecasts are based on two promises: a reduction in accidents because of superior sensing and navigating abilities of autonomous vehicles, and changes in urban form that lend themselves to wildlife habitat. However, it is difficult to anticipate the ways in which technologies will be adapted and integrated into a highly fragmented and polarised world (Bijker, 1997) that is also confronting the impacts of climate change, many as yet poorly understood.

Nevertheless, the potential of autonomous vehicles to ameliorate road ecology and thus contribute to the goal of urban rewilding is immense. Roadkill is a major ecological as well as traffic safety problem, with billions of animals killed yearly. Some estimates suggest that 1 million animals a day are killed on US roads alone (Beckmann et al., 2012), amounting to some $4 billion in damage and roughly 200 human deaths (Bland, 2015). Reduced collisions with wildlife could transform the dynamics of road ecology, particularly with regard to large predators who require vast territories to maintain a minimum viable population (MVP). In particular, despite efforts at creating an underpass system and habitat connectivity, highways currently restrict the Florida panther to the everglades in Southern Florida, an ecological sink with limited carrying capacity. A similar situation exists for the mountain lions of Los Angeles County, although it is less dire from the perspectives of genetic diversity and/or drift.

In addition to the ethical concerns surrounding autonomous vehicles and collision avoidance, which extend beyond roadkill, there has been much speculation on the ecological effects and potential for a broader urban spatial restructuring these technologies could usher in. Indeed, Richard Foreman, a pioneer in the field of road ecology, has long promoted a reconfiguration of the transportation system to a utopia of autonomous vehicles running on elevated throughways that connected dense, self-contained settlements. With more recent advances in autonomous vehicles, manufacturers have contracted landscape architecture and urban design academics and professionals to envision the spatial impacts of the introduction of these technologies (Berger et al., 2017), and many others working in the broader field of 'smart cities' or 'city science' are considering the implications of autonomous vehicles.

There is considerable convergence of opinion that autonomous vehicles will create massive changes in urban form, generally as a result of some combination of two main dynamics. The first, like Richard Foreman's utopia, revolves around the potential densification of cities due to a freeing up of space that is currently dedicated to human-driven cars, a vision often linked to changing ownership models from private to public or 'shared', or a product of service. As noted above, autonomous vehicles could reduce street widths (or increase throughput of existing infrastructure) and reconfigure the distribution of parking and related uses in the urban landscape. Both types of land-use change could permit reconfigurations of the streetscape and urban ground plane, a densification of urban fabric and the possibility for new open space networks that provide wildlife habitat (both stepping stone and connected corridors/ networks). The second dynamic that some forecasters note highlights the potential decentralisation occasioned by autonomous vehicles. By increasing throughput on existing overloaded infrastructure and freeing up humans from driving, thereby extending feasible commuting times, autonomous vehicles could help produce an 'infinite suburbia' (Berger et al., 2017). In this case, autonomous vehicles could impact wildlife habitats even with reduced human–wildlife collisions, with increasingly far-flung development causing habitat fragmentation from fencing and other changes in landscape ecology resulting from increased human settlement, no matter how 'sustainable'. For example, conflicts stemming from encounters with domestic animals already lead to demands from residents in urban–wildland interface zones to exterminate predators and pests.

Exactly how these dynamics play out is heavily dependent on responses from local governments, especially urban land use and transportation planners whose actions are themselves shaped and constrained by politicians, real estate developers, business interests, environmental and community groups, and a variety of other actors at the state and national scale. Regardless of the incremental, business-as-usual approach to urban land-use decisions that characterises most cities, the threat of climate change is apt to play an ever-growing role in decisions about how to think about, and redeploy, urban land freed up by widespread adoption of autonomous vehicles. Such decisions will increasingly be shaped by the imperative to protect low-lying districts, coastal neighbourhoods, areas adjacent to urban rivers and deltas likely to be inundated, either through sea walls, barrages and levees, and/or green infrastructure and related adaptation strategies to make cities more permeable and thus less at risk of flooding. If adopted at the regional scale, the combination of climate change adaptation strategies with land-use resources opened up for reuse by the advent of autonomous vehicles could revolutionise the extent and distribution of urban green space and thus wildlife habitat. Terrestrial animals might have movement corridors from large bioreserves at the metropolitan edge, saltwater intrusion into urban groundwater reserves and waterways could bring invasive aquatic wildlife to riparian zones, and inundation may push wildlife trying to escape rising waters into seeking higher and more urbanised ground. Such dynamics may lead to more resident–wildlife encounters, and thus a more visceral urbanism.

Conclusions

In this chapter we have attempted to highlight the role of land-use change and images in the production of urban wilderness, spaces for wild animals in the city, the types of animals that may take up urban residence – and their corollaries in the imagination of urban populations. We have shown how climate change and adaptations of urban infrastructure to cope with climate change as well as the appearance of autonomous vehicles may result in profound changes in urban form and the interdigitating of built and permeable green space throughout many cities. Such changes signal the potential for far higher rates of urban rewilding, human–wildlife encounters and the introduction of unfamiliar 'climate refugee' species migrating from wildland habitat rendered inhospitable by climate change, challenging local knowledge of animal behaviour based on historical bioregional wildlife assemblages. These dynamics are likely to multiply familiar, place-specific controversies about urban rewilding: whether suburban deer herds should be culled, badgers exterminated, squirrels poisoned, or fox, coyotes, bears, cougars, or yet more exotic species tolerated or violently vanquished from urban space and city life.

We have also highlighted the ways in which the aesthetics of urban wilderness at the turn of the twentieth century, and thus our understanding of urban rewilding, are increasingly shaped by virtual imagery rather than direct experience with urban wildlife. Imagery is produced by scientists as well as citizen scientists and constituted by global circulations of digital images on social media. Never before has it been easier to watch virtual urban wildlife – whether hawk nestlings growing up in Manhattan, cougars strolling through Silicon Valley backyards, or even leopards walking across Mumbai parking lots. Digital formats, in turn, open the door to second-order analysis of metadata, subject to deep-learning machine pattern recognition. These technologies may benefit wild animals. Deployed in the analysis of camera trap or webcam imagery, digital imagery can help urban ecologists and wilderness managers understand the natural history of urban wildlife and contend with the challenges they face either from humans or changing environmental conditions, as well as help urban designers create green infrastructure that allows animal movement across metropolitan space. Used as collision avoidance mechanisms in autonomous vehicles, such imagery may also create future landscape ecologies reshaped by alleviation of roadway ecological sinks. There may, however, be a wide variety of less-anticipated dangers for urban wildlife, including the use of digital imagery to support urban extermination campaigns, or the creation through autonomous vehicle technology of 'infinite suburbs' that could dramatically increase the scale of habitat fragmentation and destruction. The emergence of cities that at once promise a more visceral urbanism and a variety of virtual ways to experience urban wildlife is destined to reshape human–animal relations. As a result, our ideas about what constitutes wilderness and wildness, and who has rights to the city, are apt to radically change over the course of the twenty-first century and beyond.

References

Allaback, S. (2000). Mission 66 visitor centers: the history of a building type. http://hdl.handle.net/2027/umn.31951d02714653.

Beatley, T. (2011). *Biophilic cities: integrating nature into urban design and planning*. Washington, DC: Island Press.

Beckmann, J.P., Clevenger, A.P., Huijser, M., and Hilty, J.A. (2012). *Safe passages: highways, wildlife, and habitat connectivity*. Washington, DC: Island Press.

Benson, E. (2010). *Wired wilderness: technologies of tracking and the making of modern wildlife*. Baltimore, MD: Johns Hopkins University Press.

Berger, A., Kotkin, J., and Balderas-Guzmán, C. (2017). *Infinite suburbia*. Princeton, NJ: Princeton Architectural Press.

Bijker, W.E. (1997). *Of bicycles, bakelites, and bulbs: toward a theory of sociotechnical change*. Cambridge, MA: MIT Press.

Birge-Liberman, P. (2010). (Re)Greening the city: urban park restoration as a spatial fix. *Geography Compass*, **4**, 1392–1407.

Bland, A. (2015). Will driverless cars mean less roadkill? Smithsonian. 2 November 2015. www.smithsonianmag.com/innovation/will-driverless-cars-mean-less-roadkill-180957103/.

Boydston, E.E. (2005). *Behavior, ecology, and detection surveys of mammalian carnivores in the Presidio*. Henderson, NV: US Geological Survey.

Bowser, A., Wiggins, A., Shanley, L., Preece, J., and Henderson, S. (2014). Sharing data while protecting privacy in citizen science. *Interactions*, **21**(1), 70–73.

Bratton, B.H. (2016). *The stack: on software and sovereignty*, 1st edition. Cambridge, MA: MIT Press.

Brower, M. (2011). *Developing animals: wildlife and early American photography*. Minneapolis, MN: University of Minnesota Press. https://muse.jhu.edu/book/24655.

Bryson, J. (2013). The nature of gentrification. *Geography Compass*, **7**, 578–587.

Bullard, R.D. (1994). *Unequal protection: environmental justice and communities of color*. San Francisco, CA: Sierra Club Books.

Cantrell, B., Martin, L.J., and Ellis, E.C. (2017). Designing autonomy: opportunities for new wildness in the Anthropocene. *Trends in Ecology & Evolution*, **32**, 156–166.

Carr, E. (2007). *Mission 66: modernism and the National Park dilemma*. Amherst, MA: University of Massachusetts Press in association with Library of American Landscape History.

Castells, M. (2011). *The rise of the network society: the information age: economy, society, and culture* (Vol. 1). Oxford: John Wiley & Sons.

Catlin-Groves, C.L. (2012). The citizen science landscape: from volunteers to citizen sensors and beyond. *International Journal of Zoology*, **2012**, 349630.

Chabot, D., and Bird, D.M. (2015). Wildlife research and management methods in the 21st century: where do unmanned aircraft fit in? *Journal of Unmanned Vehicle Systems Virtual Issue*, **1**, 137–155.

Chapman, F.M. (1900). *Bird studies with a camera; with introductory chapters on the outfit and methods of the bird photographer*. New York.

Cosgrove, D.E. (1998). *Social formation and symbolic landscape*. New York, NY: Wiley.

Cronon, W. (1996). The trouble with wilderness: or, getting back to the wrong nature. *Environmental History*, **1**, 7–28.

Cuff, D., Hansen, M., and Kang, J. (2008). Urban sensing: out of the woods. *Communications of the ACM*, **51**, 24–33.

Dear, M. (1991). *The postmodern urban condition*, 1st edition. Oxford: Wiley-Blackwell.

Dear, M. (2002). Los Angeles and the Chicago School: invitation to a debate. *City & Community*, **1**, 5–32.

Di Palma, V. (2014). *Wasteland: a history*. New Haven, CT: Yale University Press.

Dooling, S. (2009). Ecological gentrification: a research agenda exploring justice in the city. *International Journal of Urban and Regional Research*, **33**, 621–639.

Elden, S. (2010). Land, terrain, territory. *Progress in Human Geography*, **34**, 799–817.

Fisher, D.R., Campbell, L.K., and Svendsen, E.S. (2012). The organisational structure of urban environmental stewardship. *Environmental Politics*, **21**, 26–48.

Foreman, D. (2004). *Rewilding North America: a vision for conservation in the 21st century*. Washington, DC: Island Press.

Foth, M., Forlano, L., Satchell, C., Gibbs, M., and Donath, J. (2011). *From social butterfly to engaged citizen: urban informatics, social media, ubiquitous computing, and mobile technology to support citizen engagement*. Cambridge, MA: MIT Press.

Friedberg, A. (2006). *The virtual window: from Alberti to Microsoft*. Cambridge, MA: MIT Press.

Gabrys, J. (2014). Programming environments: environmentality and citizen sensing in the smart city. *Environment and Planning D: Society and Space*, **32**(1), 30–48.

Gandy, M. (2005). Cyborg urbanization: complexity and monstrosity in the contemporary city. *International Journal of Urban and Regional Research*, **29**(1), 26–49.

Gandy, M. (2013). Marginalia: aesthetics, ecology, and urban wastelands. *Annals of the Association of American Geographers*, **103**(6), 1301–1316.

Goldenberg, S., and US Environment Correspondent. (2010). BP switches on live video from oil leak. *Guardian*, 21 May 2010, sec. Environment. www.theguardian.com/environment/2010/may/20/deepwater-horizon-oil-spill-live-web-footage.

Graham, S., and Marvin, S. (2001). *Splintering urbanism: networked infrastructures, technological mobilities and the urban condition*. London: Routledge.

Grusin, R. (2008). *Culture, technology, and the creation of America's National Parks*, 1st edition. Cambridge: Cambridge University Press.

Guggisberg, C.A.W. (1977). *Early wildlife photographers*. New York, NY: Taplinger Publishing Company.

Halpern, O., LeCavalier, J., Calvillo, N., and Pietsch, W. (2013). Test-bed urbanism. *Public Culture*, **25**, 272–306.

Haraway, D.J. (2013). *Simians, cyborgs, and women: the reinvention of nature*. Abingdon: Routledge.

Hegglin, D., Bontadina, F., Gloor, S., et al. (2004). Baiting red foxes in an urban area: a camera trap study. *Journal of Wildlife Management*, **68**(4), 1010–1017.

Huiser, M., and McGowen, P.T. (2012). Reducing wildlife–vehicle collisions. In Beckmann, J.P., Clevenger, A.P., Huijser, M., and Hilty, J.A. (Eds.), *Safe passages: highways, wildlife, and habitat connectivity*. Washington, DC: Island Press.

Jachowski, D.S., Slotow, R., and Millspaugh, J.J. (2014). Good virtual fences make good neighbors: opportunities for conservation. *Animal Conservation*, **17**(3), 187–196.

Kan-Rice, P. (2017). New mobile app to track close encounters with coyotes. *ANR Blogs* (blog). 13 February 2017. http://ucanr.edu/blogs/blogcore/postdetail.cfm?postnum=23229.

Karvonen, A., and Yocom, K. (2011). The civics of urban nature: enacting hybrid landscapes. *Environment and Planning – Part A*, **43**(6), 1305.

Kays, R.W., and Slauson, K.M. (2008). Remote cameras. In Long, R.A., McKay, P., Zielinski, W.J., and Ray, J.C. (Eds.), *Noninvasive survey methods for carnivores* (pp. 110–140). Washington, DC: Island Press.

Kim, J. (2004). Mission 66. In Colomina, B., Brennan, A., and Kim, J. (Eds.), *Cold War hothouses: inventing postwar culture, from cockpit to playboy*. New York, NY: Princeton Architectural Press.

Kirk, A.G. (2007). *Counterculture green: the whole Earth catalog and American environmentalism*. CultureAmerica. Lawrence, KS: University Press of Kansas.

Kittler, F.A. (2001). Perspective and the book. *Grey Room*, no. 05, 38–53.

Kucera, T.E., and Barrett, R.H. (2011). A history of camera trapping. In O'Connell, A.F., Nichols, J.D., and Karanth, K.U. (Eds.), *Camera traps in animal ecology* (pp. 9–26). New York, NY: Springer.

Kupper, P. (2016). Nature's laboratories: exploring the intersection between science and National Parks. In Howkins, A., Orsi, J., and Fiege, M. (Eds.), *National Parks beyond the nation: global perspectives on 'America's best idea'* (Vol. 1). Norman, OK: University of Oklahoma Press.

Lachmund, J., and Jasanoff, S. (2004). Knowing the urban wasteland: ecological expertise as local process. In Jasanoff, S. and Long-Martello, M. (Eds.), *Earthly politics: local and global environmental governance* (pp. 241–261). Cambridge, MA: MIT Press.

Latour, B., Hermant, E., and Shannon, S. (1998). *Paris Ville Invisible*. La Découverte Paris. www.bruno-latour.fr/sites/default/files/downloads/A-PARIS%20TOTAL.pdf.

Locke, S.L., Cline, M.D., Wetzel, D.L., Pittman, M.T., Brewer, C.E., and Harveson, L.A. (2005). A web-based digital camera for monitoring remote wildlife. *Wildlife Society Bulletin*, **33**(2), 761–765.

Manovich, L. (1999). Database as symbolic form. *Convergence: The International Journal of Research into New Media Technologies*, **5**(2), 80–99.

Masco, J. (2010). Bad weather: on planetary crisis. *Social Studies of Science*, **40**, 7–40.

Mathey, J., and Rink, D. (2010). Urban wastelands – a chance for biodiversity in cities? Ecological aspects, social perceptions and acceptance of wilderness by residents. In Müller, N., Werner, R., and Kelcey, J.G. (Eds.), *Urban biodiversity and design* (pp. 406–424). Oxford: Blackwell-Wiley.

McCormick, E. (2017). Big Brother on wheels? Fired security robot divides local homeless people. *Guardian*, 17 December 2017, sec. US news. www.theguardian.com/us-news/2017/dec/16/san-francisco-homeless-robot.

Merchant, C. (2013). *Reinventing Eden: the fate of nature in western culture*. Abingdon: Routledge.

Mitchell, W.J.T. (2002). *Landscape and power*. Chicago, IL: University of Chicago Press.

Nechwatal, S. (2016). Using gamification to connect classrooms to communities. *Interaction*, **44**(3), 38.

Nesbit, W. (1926). *How to hunt with the camera: a complete guide to all forms of outdoor photography*, 1st edition. Boston, MA: E.P. Dutton & Co.

Ng, S.J., Dole, J.W., Sauvajot, R.M., Riley, S.P.D., and Valone, T.J. (2004). Use of highway undercrossings by wildlife in southern California. *Biological Conservation*, **115**(3), 499–507.

Nichols, J.D., O'Connell, A.F., and Karanth, K.U. (2011). Camera traps in animal ecology and conservation: what's next? In O'Connell, A.F., Nichols, J.D., and Karanth, K.U. (Eds.), *Camera traps in animal ecology: methods and analyses* (pp. 253–263). New York, NY: Springer Science & Business Media.

O'Connell, A.F., Nichols, J.D., and Karanth, K.U. (Eds.) (2010). *Camera traps in animal ecology: methods and analyses*. New York, NY: Springer Science & Business Media.

Pickles, J. (2004). *A history of spaces: cartographic reason, mapping, and the geo-coded world*. London: Routledge.

Rose, N. (1996). The death of the social? Re-figuring the territory of government. *International Journal of Human Resource Management*, **25**(3), 327–356.

Rothman, H. (2004). *The new urban park: Golden Gate National Recreation Area and civic environmentalism*. Lawrence, KS: University Press of Kansas.

Soulé, M.E. (1985). What is conservation biology? *BioScience*, **35**(11), 727–734.

Svendsen, E.S. (2010). Civic environmental stewardship as a form of governance in New York City. http://originwww.nrs.fs.fed.us/nyc/local-resources/downloads/Svendsen_dissertation.pdf.

Whiston Spirn, A. (1995). Constructing nature: the legacy of Frederick Law Olmstead. In Cronon, W. (Ed.), *Uncommon ground: toward reinventing nature* (pp. 91–113). London: W.W. Norton.

Wolch, J.R. (1990). *The shadow state: transformations in the voluntary sector*. New York, NY: Foundation Center.

Wolch, J.R., and Emel, J. (1998). *Animal geographies: place, politics, and identity in the nature–culture borderlands*. New York, NY: Verso.

Wolch, J.R., Byrne, J., and Newell, J.P. (2014). Urban green space, public health, and environmental justice: the challenge of making cities 'just green enough'. *Landscape and Urban Planning*, **125**, 234–244.

Yasuda, M., and Kawakami, K. (2002). New method of monitoring remote wildlife via the internet. *Ecological Research*, **17**, 119–124.

Zhou, N. (2017). Volvo admits its self-driving cars are confused by kangaroos. *Guardian*, 1 July 2017, sec. Technology. www.theguardian.com/technology/2017/jul/01/volvo-admits-its-self-driving-cars-are-confused-by-kangaroos.

Zielinski, S. (1999). *Audiovisions: cinema and television as tntr'actes in history*. Amsterdam: Amsterdam University Press.

CHAPTER FIFTEEN

The role of translocation in rewilding

PHILIP J. SEDDON
University of Otago
DOUG P. ARMSTRONG
Massey University

Where the goals of a rewilding project cannot be met through the management of species living within the project area, due to local, regional or even global extinctions, then one option is to establish or re-establish populations of selected species through translocation. Such translocations might aim to restore ecosystem functions, such as grazing and seed dispersal, or to re-establish populations of *keystone species* such as top predators, or *ecosystem engineers* that can modify habitats for other species. If the goal of such translocations is enhancement of native biodiversity, and/or ecosystem resilience and stability, then under the IUCN definition (IUCN, 2013), these would be called *conservation translocations* (see Box 15.1 for definitions of italicised terms). There is a distinction between translocations where the primary objective is to improve the status of the focal species and translocation for rewilding where objectives relate to ecosystem functions (Sandom et al., 2013; Corlett, 2016). Translocation for rewilding might entail *reintroduction*, where releases occur within the *indigenous range* of the species with the primary aim of restoring some ecological function, or involve a *conservation introduction* where releases take place outside the indigenous range to provide an *ecological replacement* for an extinct species (Seddon et al., 2014). Suitable ecological replacements could be closely related taxa, such as subspecies (Seddon et al., 2005), or phylogenetically different but functionally equivalent species (Donlan et al., 2005).

Trophic rewilding seeks to restore top-down interactions to promote biodiverse ecosystems (Svenning et al., 2016). The relevant forms of conservation translocation for trophic rewilding would include both reintroduction of indigenous species and the ecological replacement of extinct forms. Ecological rewilding aims to allow restoration of natural processes (Corlett, 2016). Pleistocene rewilding has a more specific agenda – to restore the ecosystem functions lost through Pleistocene extinctions of mega-vertebrates (Donlan et al., 2005). For Pleistocene rewilding suitable functional substitutes are required to take the place of globally extinct species, necessitating ecological replacement translocations.

Box 15.1. Definitions

Conservation translocation	the intentional movement and release of a living organism where the primary objective is some conservation benefit: this will usually comprise improving the conservation status of the focal species locally or globally, and/or restoring natural ecosystem functions or processes (IUCN, 2013).
De-extinction	the creation of functional proxies of extinct species using selective breeding, cloning or genomic engineering technology (IUCN, 2016).
Ecological replacement	intentional movement and release of an organism outside its indigenous range to perform a specific ecological function (IUCN, 2013).
Ecosystem engineer	a species that can affect other organisms by creating, modifying, maintaining or destroying habitats (Byers et al., 2006).
Indigenous range	known or inferred distribution generated from historical (written or verbal) records or physical evidence of the species occurrence (IUCN, 2013).
Keystone species	species that are so important in determining the ecological functioning of a community that they warrant special conservation efforts, and whose loss would precipitate further extinctions; for restoration, keystone species are necessary to help re-establish and sustain ecosystem structure and stability (Mills et al., 1993).
Reinforcement	intentional movement and release of an organism into an existing population of conspecifics (IUCN, 2013).
Reintroduction	intentional movement and release of an organism inside its indigenous range from which it has disappeared (IUCN, 2013).
Structured decision-making	an organised approach to identifying and evaluating options and making choices in complex decision situations characterised by uncertain science, diverse stakeholders, and difficult trade-offs, designed to assist decision-makers to derive solutions that are rigorous, inclusive, defensible, and transparent (Gregory et al., 2012).
Translocation	human-mediated movement of organisms from one site for release in another (IUCN, 2013).

Review of conservation translocations for rewilding: how prevalent, what type, where, why?

To broadly assess the prevalence of rewilding as a driver of conservation translocations we reviewed all 242 terrestrial faunal translocation projects summarised in the IUCN Species Survival Commission (SSC) Reintroduction Specialist Group's (RSG) 'Global Re-introduction Perspectives' series (Soorae, 2008, 2010, 2011, 2013, 2016). We sought to identify those projects that highlighted the restoration of ecosystem functions or processes as at least one of the project goals or justifications. To count as a rewilding-related translocation, a project needed to do more than improve the conservation status of only the target species. We also excluded projects that had as incentives only to restore biotic or abiotic conditions for the target species, or that aimed to use the target species as an indicator of ecosystem health.

Only 15 of the 242 projects (6 per cent) had ecosystem restoration goals stated explicitly; there was no trend over time, and too few rewilding projects to draw conclusions about taxonomic representation, although large mammal translocations were the most numerous (Figure 15.1). The 15 projects were spread over most of the IUCN global regions, with over half taking place in Oceania, Europe, and North America, combined – reflecting the general

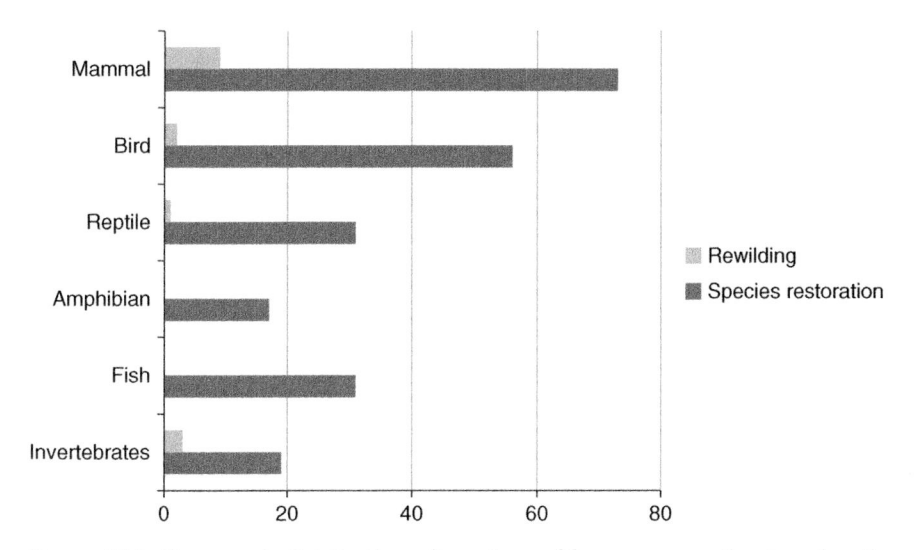

Figure 15.1. Taxonomic distribution of numbers of fauna conservation translocation projects that have a primary aim of improving the conservation status of the target species (species restoration) versus restoration of ecosystem functions and processes (rewilding). Proportions are based on a review of 242 projects summarised in the IUCN/SSC Reintroduction Specialist Group publications (Soorae, 2008, 2010, 2011, 2013, 2016).

Table 15.1. *Summary of 50 European rewilding projects involving conservation translocations, derived from the* Rewilding Europe *project-partner summaries (www.rewildingeurope.com/european-rewilding-network/; accessed 10 August 2017).*

Country	Translocation type	Reptiles	Birds	Mammals
Austria	Reintroduction		Northern bald ibis	
Belarus	Reintroduction			European bison
Bulgaria	Reinforcement			Fallow and red deer
	Reintroduction	Hermann's tortoise	Saker falcon, Griffon vulture	European bison
	Ecological replacement			Horses, cattle
Denmark	Reintroduction			Beaver
Germany	Reintroduction			European bison, Eurasian lynx
Italy	Reintroduction		Lesser kestrel, white stork, little bustard, raven, griffon vulture	Red deer, roe deer, alpine ibex, chamois
Latvia	Ecological replacement			Horses, cattle
Netherlands	Reintroduction			Red deer, beaver, European bison
	Ecological replacement			Cattle
Poland	Reintroduction			Eurasian lynx
Portugal	Ecological replacement			Sheep, donkeys, horses
Romania	Reintroduction			European bison
Spain	Reintroduction		Black (cinereous) vulture	Iberian ibex, wild rabbit,
Slovenia	Reintroduction			Eurasian lynx
Sweden	Reintroduction			European bison, Pere David's deer
Switzerland	Reintroduction		Bearded vulture	Alpine ibex
Ukraine	Ecological replacement			Cattle, Przewalski's horse, domestic horses
United Kingdom	Reintroduction			Red deer, fallow deer
	Ecological replacement			Cattle, horses, pigs
Project totals		1	10	38

conservation translocation activity in these regions (65 per cent of global translocations; Seddon and Armstrong, 2016). Clearly, many rewilding projects are not represented in the published RSG summaries, and possibly rewilding projects using ecological replacements are less likely to submit summaries because the RSG publication is called a 're-introduction' perspective. Nevertheless, this brief review does indicate that by far the majority of conservation translocations taking place globally to date are motivated by species restoration goals, and not by the restoration of ecosystem functions. This could change as there develops a greater focus on ecosystem-level restorations and the translocation of ecosystem engineers, keystone species, and species in higher trophic levels, and with increasing attention given to rewilding in all its guises.

To consider the range of translocation types and species that are the specific focus of rewilding projects, we explored the web links of Rewilding Europe (2017) and identified 50 translocation projects involving 26 species across 17 European countries (Table 15.1). Thirteen of the 14 ecological replacements involved the release of domesticated descendant versions of ancestral forms of horses, cattle, sheep, and pigs. Most (76 per cent) translocations involved mammals, and the most common reintroductions were those of European bison, Eurasian lynx (*Lynx lynx*), and Alpine and Iberian ibex (*Capra* sp.). The most common of the nine species of birds being reintroduced were raptors, including saker falcon (*Falco cherrug*), lesser kestrel (*Falco naumanni*), and three vulture species. Similar summaries could be compiled for other regions, but the organisation of project summaries elsewhere did not facilitate extraction of information on rewilding projects relying on translocations. We assume that the same general patterns, of the prevalence of reintroductions over ecological replacements and the taxonomic bias towards larger mammals and birds, would apply to rewilding programmes in regions such as in North America (Rewilding Institute, 2017).

Conservation translocation success

It might seem strange for a conservation tool that has been applied in one form or another for over 100 years (Seddon and Armstrong, 2016) – and for one with its own associated scientific discipline, reintroduction biology (Ewen et al., 2012) – that we still have no universal, agreed-upon set of criteria with which to evaluate conservation translocation success. We know when a project fails, e.g. when all the released animals die, but it is harder to say when it has succeeded. Early attempts to frame reintroduction success criteria referred to the establishment of a self-sustaining population (Griffith et al., 1989), but placed no metrics or time frames to this, thus opening the possibility whereby once successful reintroductions subsequently fail (Seddon, 1999). Other suggested criteria used rules of thumb that lack cross-taxa applicability, such as having '500 free-living individuals' (Beck et al., 1994). It is perhaps an indication of the maturation of the discipline of reintroduction biology that we have moved away from a simplistic, one-size-fits-all view of translocation success.

One set of issues arises due to the special characteristics of any reintroduced population, i.e. that it is derived from a subset of some relatively small group of founders that survives and breeds and produces a growing population. For long-lived, slow-growing species this will be a lengthy process, so at what point, if ever, does a reintroduction practitioner declare success? A newly establishing population might initially grow rapidly, and spectacular early population growth might be mistaken for a high probability of long-term population persistence. A demographic framework for assessing reintroduction outcomes recognises three population stages: establishment, growth, and regulation (Sarrazin, 2007), suggesting that any evaluation of ultimate success or failure needs to focus on the regulation phase, when the population fluctuates around some carrying capacity.

More recently there has been an attempt to consider translocation success in terms of well-established frameworks. Robert and colleagues (2015) used the IUCN Red List categorisation process that assigns an extinction risk to given species to assess whether metrics such as population size, which are used to characterise remnant populations, might be useful to assess the conservation status of reintroduced populations at the regulation phase. They determined that the IUCN Red List criterion of current population size is a useful proxy for risk of extinction, providing some confirmation of the need for reintroduction practitioners to evaluate ultimate success only after the new population has reached some carrying capacity; the challenge then being determination of when a re-established population has reached capacity, as for many reintroduced species there is a lack of information on vital rates and their variability. It is clear that any assessment of translocation success must be time-bound and species-specific, and that self-sustainability becomes more aspirational than achievable, particularly with the recognition that the persistence of many free-ranging populations is dependent on some level of ongoing management intervention (Mallon and Stanley Price, 2013).

The RSG summaries have neatly avoided the minefield of project success evaluation by allowing project managers to define their own success indicators, which might relate to demographic, ecological, social or management goals, and then to classify the overall project as being highly successful, successful, partially successful or a failure. Such qualification of assessment suitably reflects the complexity of most projects, but perhaps does not facilitate any overall assessment of conservation translocation success rates. Such project evaluations suggest 58 per cent of > 200 recent reintroduction projects across all taxa were considered fully successful by all project-specific criteria, and only 5 per cent were classified as complete failures (Soorae, 2008, 2010,

2011, 2013, 2016). However, it is hard to know how to treat these data, because the criteria for success differ between projects and some metrics have little to do with the demographics of the released population. Also there might exist a bias, whereby successful projects are more likely to be self-selected to provide summaries, and because over time success can go to failure, but probably never vice versa. It therefore seems unlikely that conservation translocations are much more successful than we think, and nor should we expect improved rates of success over time because more challenging translocations will be attempted as knowledge and capacity grows. Chauvenet and colleagues (2016) and Converse and Armstrong (2016) advocate for the use of *structured decision-making* to define clear objectives as a basis for quantitative assessment of outcomes. This provides a framework for transparent translocation planning that can reduce uncertainty and incorporate the varying risk profiles of different human stakeholders, and will be of most use in choosing among management alternatives. However, the process does not necessarily provide a clear pathway to cross-project and -taxa summary evaluations of ultimate success rates, as formalised demographic objectives and time frames will be species- and even project-specific, and dependent to some extent on the robustness of underlying population models (Converse and Armstrong, 2016). Nevertheless, greater application of decision-making tools in translocation planning and monitoring should facilitate future assessments of outcomes.

We are left then with past reviews of translocation outcomes suggesting success rates as low as 26–38 per cent, but that ignore outcome shifts over time and which, necessarily, apply a one-size-fits-all set of success criteria (Griffith et al., 1989, Wolf et al., 1998, Fischer and Lindenmayer, 2000). These have given rise to a perception that most reintroductions end in failure. This is perhaps misleading, but not a bad thing in that it emphasises that any translocation is not a trivial exercise, but rather a complex undertaking that requires careful planning, execution, and follow-up. This means there is indeed an issue to address: given some (undefined) low probability of success, what are the implications for rewilding projects that are dependent on conservation translocations, and how might the probability of success vary according to the taxon, the ecosystem, and the type of conservation translocation being attempted?

Critical risk factors influencing translocation outcomes

Results of an analysis of 181 bird and mammal translocation outcomes highlighted the importance of three variables: the perceived quality of habitat in the release area, the location of the release site relative to the indigenous range of the translocated species, and the number of animals released; there was also evidence that translocations of omnivorous species tend to be more successful than either herbivores or carnivores (Wolf et al.,

1998). Subsequent work has emphasised the importance of release site selection and the risks of translocation failure due to habitat mismatch (Armstrong et al., 2002, Cook et al., 2010, White et al., 2012), prompting the IUCN to emphasise matching site suitability to the needs of the candidate species by stating that sites should: 'meet the candidate species' total biotic and abiotic needs through space and time and for all life stages' (IUCN, 2013).

Habitat and indigenous range

Prior assessment of habitat quality is essential. Correlative and mechanistic species distribution models can guide decision-making (Kearney et al., 2010; Osborne and Seddon, 2012) and, when possible, a trial release with intensive post-release monitoring can provide valuable information on a species habitat requirements (Kemp et al., 2015). It is particularly important to assess habitat quality for proposed conservation introductions such as ecological replacements, where animals are to be released outside their indigenous range. This appropriately places the emphasis on whether any proposed release area provides for the needs of the candidate species, because records of prior occupancy are necessarily absent.

Some conservation introductions for rewilding present an interesting variation on ecological replacement. Many rewilding projects in Europe (Table 15.1) release domesticated livestock to restore the ecosystem functions lost through effective extinction of the ancestral form, for example, the release of breeds of cattle as ecological replacements for the extinct aurochs. It could be argued that such descendent forms might inherit the indigenous range of their ancestors, but the long process of domestication, livestock breeding, and husbandry could be viewed as having significantly changed the ecology of free-ranging livestock. As a consequence, although the risks might be reduced in using livestock instead of wild species as ecological replacements, the need to evaluate the suitability of any release site remains. The release area must meet all the needs of the animals being released, both for animal welfare and for other project objectives. Conservation introductions, releasing animals outside their indigenous range, are intuitively more likely than reintroductions to run the risk of there being a habitat mismatch (although climate change and other effects could improve the suitability of sites outside the indigenous range), and this risk will probably be greatest for ecological replacements using unrelated species as functional substitutes. The IUCN Guidelines (2013) also highlight some other critical risk factors, including the presence of new threats, the possibility of unexpected and deleterious outcomes, and the need for social acceptance.

New threats in release areas

One of the key requirements before any reintroduction is to identify and remove or reduce the causes of the previous extinction (IUCN, 2013). However, attention also needs to be paid to the possibility of new threats to the target species, arising within the release area. For example, hihi (*Notiomystis cincta*) is a New Zealand endemic bird whose North Island-wide distribution was reduced to only a single remnant island population due to habitat loss and the impacts of introduced mammalian predators (Taylor et al., 2005). Reintroductions of hihi have taken place on predator-free offshore islands and fenced sanctuaries, and in the absence of this threat, and with use of supplementary feeding, hihi populations can do well (Chauvenet et al., 2012). However, modelling indicates that hihi population dynamics are climate-driven, and that under predicted climate change the region of hihi habitat will shift, isolating some of the largest re-established hihi populations in increasingly unsuitable sites (Chauvenet et al., 2013), with no opportunity for unassisted dispersal to suitable areas across intervening predator-dense landscapes. Anticipation of new threats will be facilitated by having good basic biological knowledge of the target species, including understanding habitat requirements and the degree of vulnerability to future climate change.

Unanticipated ecological effects

There is a risk that any translocated species could have major unintended impacts on other species or on ecosystem functions within the release area. These might be challenging to predict because the behaviour or population demographics of the target species might not be the same as at the origin of the founder animals, possibly due to interactions with different species assemblages, variant habitat conditions or release from predation or from parasite pressures (Selbach et al., 2018). The risks of unanticipated ecological impacts will be greater for rewilding involving ecological replacement, where a species is moved outside its indigenous range and into a new set of environmental conditions (see case studies 5 and 7).

Social acceptance of translocations

Socioeconomic risks include the risk of direct, harmful impacts on people and their livelihoods from released organisms and more indirect ecological impacts that negatively affect ecosystem services (IUCN, 2013). Translocations of keystone species or ecosystem engineers have a high likelihood of facing public opposition because of real or perceived threats to livelihoods and landscapes (see case studies 1 and 2). Adverse public attitudes will be greatest for reintroductions of such species when these have been absent from an area for many years and also for ecological replacement releases outside the target species indigenous range, because local communities will have no relevant experience of the species with which to evaluate the probability of adverse effects.

Implications for rewilding of critical factors affecting translocation outcomes

Given these critical factors affecting the likelihood of translocation success, impact, and social acceptability, what are the implications for rewilding? Risk, uncertainty, potential for deleterious effects and public attitudes will vary depending on the type of translocation being proposed, and on the particular species to be released, and will thus influence overall project success.

Reintroduction for rewilding

For reintroductions, risk and uncertainty tend to increase with time since the original extirpation and with the functional role of target species, so that even though the focal species might be indigenous to a release area, changes in land use and other social factors might make their restoration both risky and, at least initially, socially unacceptable. This has been the case for the reintroduction of both top predators and ecosystem engineers.

Case study 1: Wolf reintroductions

Top predators promote species richness in some ecosystems (Sergio et al., 2008) and the loss of large predators has led to increased abundance of herbivore prey, in turn leading to increased grazing and associated changes to community structure and composition (Nilsen et al., 2007). The grey wolf (*Canis lupus*) is a top predator that has been historically extirpated from much of its former range due to human persecution, but is now the focus of a number of restoration efforts throughout Europe and North America, including wolf recovery programmes in Montana, Idaho, Arizona, and New Mexico over the last ~15 years (Ripple and Beschta, 2004). However, public attitudes to wolf reintroduction can be negative, with rural communities expressing concern over the possibility of livestock losses and potential risks to humans (Pate et al., 1996). In general people with the most positive attitudes towards wolves tend to be those with least experience of them, whereas the people who interact directly with wolves tend to have more negative attitudes (Williams et al., 2002; Ericsson and Heberlein, 2003). Wolves were eradicated from the Scottish Highlands by 1769, but their reintroduction has been suggested as a means to control dense deer populations that threaten reforestation efforts, reduce bird densities, and compete with livestock (Nilsen et al., 2007). But although the wider Scottish public is supportive, farmers have expressed concerns (Wilson, 2004). Local community support and tangible benefits from the reintroduction of potentially disruptive species are critical to reduce opposition or even attempts to disrupt or sabotage a project (Nilsen et al.,

2007). The restoration of any large mammalian predator will inevitably face opposition from human communities for which such predators are no longer part of the landscape, and wolves seem to attract special concern. It is possible that the restoration of wolves as a new normal part of the environment could reset public attitudes.

Case study 2: Beaver reintroduction in Scotland
In 2009-2010 16 Eurasian beavers (*Castor fiber*) were translocated from Norway and released into Knapdale Forest in Mid Argyll, as part of the Scottish Beaver Trial (Gaywood, 2015), marking the return of beavers to Scotland after an absence of some 500 years (O'Connell et al., 2008). Beavers are considered ecosystem engineers, able to change the geomorphology and hydrology of a landscape and thereby increase both habitat and species diversity (Rosell et al., 2005). Since the second half of the twentieth century, reintroductions of beavers across Europe have resulted in a total population of > 300,000 animals, but the reintroduction of beavers to the UK has encountered concerns about their impacts on riparian ecosystems that have reached new functional and economic equilibria in the centuries since beaver extirpation (South and Macdonald, 2000). Beaver activities were predicted to restore wetlands, reduce flooding, increase water storage, improve water quality, and reduce siltation (Gurnell, 1998; Rosell et al., 2005; Stringer and Gaywood, 2016). This habitat restoration was predicted to have a positive effect on otters (*Lutra lutra*), great crested newts (*Triturus cristatus*), and water voles (*Arvicola amphibious*) (Stringer and Gaywood, 2016), but was also predicted to have a negative impact on Atlantic hazelwood trees and aspen (*Populus tremula*) (Stringer and Gaywood, 2016). Other anticipated negative impacts of beavers included loss of forestry land due to flooding, prevention of fish migration, and a reduction in fish spawning habitat (Sheridan, 2014). In Wyoming, USA, 89 per cent of land owners with beavers on their property had damage to their crops, pasture, and irrigation systems (McKinstry and Anderson, 1999). In Poland, beavers also cause damage to farmland and crops and are culled annually (Dzięciołowski and Gozdziewski, 1999). Beavers reintroduced to Germany in the 1960s caused damage to crops, forestry and irrigation channels, and flooding, and as a result beavers are now culled annually (Pillai and Heptinstall, 2013). Nevertheless, although in 2007 over half of Knapdale residents were against the reintroduction of beavers (Carrell, 2007), following the 2010 releases most residents supported beavers living in Knapdale, and many believed that they had improved tourism and the economy (The Scottish Beaver Trial Team, 2014). The principal risks of beaver reintroduction relate to impacts on human-dominated production or recreational landscapes. Where stakeholder perceptions can be managed and community support gained, beaver reintroductions appear to have a good

probability of success, to the extent that burgeoning beaver populations might need active culling.

Rewilding using ecological replacements

Ecological replacements might appear to be inherently risky because any replacement species is, by definition, not indigenous to the release area. However, the level of risk and the degree of public acceptance will vary greatly depending on the type of replacement selected. We can think of four categories of ecological replacements: close relatives to the extirpated species, e.g. races or subspecies; the use of domesticated descendent versions, e.g. cattle for aurochs; taxonomically very different forms, e.g. different Orders or even different Classes; directed back-breeding or genomic manipulations to produce functional proxies, i.e. de-extinction.

Case study 3: One kokako for another
New Zealand has been home to a number of endemic species of the Callaeidae family, including huia (*Heteralocha acutirostris*), saddleback (*Philesurmus* sp.), and kokako (*Callaeas* sp.). The huia is extinct as a result of overexploitation and predation by introduced mammals, whereas saddleback and kokako persist on offshore island or mainland areas where introduced mammals are subjected to intensive lethal control. Kokako were once classified as subspecies, but are now considered to be represented as two species, one in the North Island (NI) (*C. wilsoni*) and one in the South Island (SI) (*C. cinerea*). The SI kokako is believed to have gone extinct within the last 10 years, although there are ongoing efforts to locate possible relict populations (Milne and Stocker, 2014). In contrast, NI kokako have benefitted from serial reintroductions into predator-controlled areas (Basse et al., 2003) and currently their increasing population is estimated at 2800 birds (Birdlife International, 2017). In 2009, 10 NI kokako were translocated to a release site on the west coast of the SI in an attempt to establish a population as part of effort to restore kokako to SI forests. Unfortunately the project failed when all the released animals died, presumably due to predation by stoats due to inadequate predator control in the release area. The ecological replacement of one congeneric for another might be viewed as premature by those who believe that the SI kokako is still extant and consider there to be a chance of interbreeding and loss of the unique SI genetic lineage. Nevertheless, the ecology of NI kokako is well understood and the techniques for successful translocation have been developed, and given adequate management of the key threats the probability of success would hinge primarily on the performance of the substitute species in the different environments of the South Island.

Case study 4: Giant tortoises as ecological replacements
In island ecosystems the loss of large island frugivores can result in significant cascading effects (Hansen et al., 2009). Consequently, one of the most significant applications of ecological replacement has been the restoration of herbivory and seed dispersal functions in island ecosystems. There is evidence of the ecosystem-engineering role of giant tortoises, both as important dispersers of large-seeded plants, and through their grazing and trampling which creates and maintains native vegetation communities (Gibbs et al., 2008). To restore grazing functions and the seed dispersal of large-seeded plants, Aldrabra giant tortoises *Aldrabrachelys gigantea* have been introduced to Mauritian offshore islands to replace the extinct Mauritian *Cylindraspis* species (Hansen and Galetti, 2009; Griffiths et al., 2011). As a result, seed dispersal has resumed, and there are indications that passage through the tortoise gut improves seed germination success (Griffiths et al., 2011). There are also plans to use ecologically similar species of giant tortoise to reinstate lost processes following giant tortoise extinction in Madagascar, the Galapagos, the Mascarenes, the Seychelles, and the Caribbean (Hansen et al., 2010). Although not taxonomically very close to the extinct species, i.e. not subspecies, functionally it makes sense to substitute one giant tortoise for another. The ecological roles and performance of the substitutes are adequately known and the predicted benefits of any release can be well monitored in spatially restricted areas.

Case study 5: Tortoises as unrelated ecological replacements for birds
The greatest level of risk might derive where an ecological replacement is taxonomically, if not functionally, far removed from the extirpated species. Where no suitable near relatives or taxonomically close replacements are available, radical substitutions could be considered. Possibly the most extreme current example is the use of tortoises as replacements for moa-nalo, a group of extinct birds in Hawaii. Confiscated African spurred tortoises (*Centrochelys sulcata*) have been released into an experimental enclosure in the limestone system of the Makauwahi Cave Reserve in an attempt to restore the forest understorey grazing functions lost with the extinction of endemic large flightless duck-like geese – the moa nalo, Hawaiian for 'lost fowl' (Burney et al., 2012). Over 100 native plant species have been reintroduced to the reserve, and controlled trials found that within six months exotic plants dominated plots lacking tortoises (Burney et al., 2012).

The least risk and likely the greatest potential for public support might be through the use of domesticated descendants of the extirpated species, as these are well known to the general public and their biology is understood, although some uncertainties might arise through the ecological impacts of feral domestic animals such as pigs and cattle. In parts of the world both feral cattle and particularly feral pigs are known to cause significant ecological change following their introduction for food or for farming with subsequent accidental escape (Bengsen et al., 2014).

A last category of ecological replacement for rewilding arises through *de-extinction*, a catch-all term to describe any of three pathways to the creation of at least a functional proxy of an extinct species (Shapiro, 2017). The first pathway, back-breeding, involves the selection of domestic animals to produce a wild-type phenotype, under the assumption that descendant forms carry the potential to express the phenotype of an extinct ancestor. The end result will be a phenotypic proxy of an extinct form, although genomic information could be used to better guide the process. The second pathway is via interspecies cloning. For mammals, this requires an appropriate surrogate host to carry an embryo to term, and unless there are cryopreserved gametes (eggs and sperm), a host cell is required in which to insert the genetic material from a preserved somatic (body) cell from the extinct animal. Without suitably preserved cells cloning is not an option, thus restricting cloning as a de-extinction pathway to only very recent extinctions. In addition, the use of a surrogate means there will be genetic components inherited from the host, there will be epigenetic effects whereby the host environment might turn on or off the activity of some genes, and there will be post-natal differences due to learning, the rearing environment, diet and the resulting microbiome (Shapiro, 2017). The third pathway is where there are no cryopreserved cells, meaning the genome of the extinct form must be deciphered from any available, and inevitably deteriorated, tissue. There will be gaps, but these gaps could be filled with the best approximation of the extinct sequences, likely from a nearest living relative, which can be used to create modified cell lines by replacing DNA sequences of the extant species with synthesised DNA in the extinct species sequence. Nuclei from such cells could then be used in cloning. The result would be the creation of some hybrid form with some expression of hybrid traits. There are two high-profile de-extinction projects underway, following the back-breeding and the genomic engineering pathways, respectively.

Case study 6: Back-breeding aurochs as ecological replacements
The aurochs (*Bos primigenius*) was the ancestor of all modern breeds of cattle and a keystone species that once grazed, fertilised, and trampled much of Europe and Asia, from Atlantic to Pacific coasts, and from Scandinavia to India and North Africa, but was hunted to extinction by 1627 (Ajmone-Marsan et al., 2010). The loss of grazing pressure from free-ranging herds of aurochs and other large herbivores, along with the more recent declines in livestock herding across Europe (McKnight, 2014), has resulted in an increase in woody forest vegetation and a decline in more open and mosaic vegetated landscapes and a decline in species richness (van Wieren, 1995). The Netherlands-based Tauros

Programme, started in 2008, aims to use selective breeding, guided by genetic analyses, of some carefully chosen domestic breeds of cattle to produce a Tauros, a functional equivalent to the aurochs (Stokstad, 2015). While some original genetic sequences might be preserved and expressed, generations of hybridisation and multiple selection pressures will limit the degree of genetic similarity between the final form and its aurochs ancestor. Nevertheless, the intent is that tauroses will be used to repopulate large parts of a rewilded Europe to restore herbivore-dominated landscapes. At one level, tauroses could be viewed as a variant of the use of feral descendent domesticated forms, i.e. another, albeit phenotypically more primitive, breed of cattle. As such, the risk of a rewilding project requiring translocation of tauros will be equivalent as there might be little risk of unexpected ecological impacts, and local communities might be reasonably accepting of free-ranging cattle, providing these did not pose threats to health or livelihoods.

Case study 7: Woolly mammoth/Asian elephant hybrids as ecological replacements

For hundreds of thousands of years the woolly mammoth (*Mammuthus primigenius*) roamed over vast areas of Eurasia and North America, until disappearing by about 3700 years BP due to hunting pressure exacerbated by climate-induced habitat collapse (Nogues-Bravo et al., 2008). George Church's lab at Harvard University's Wyss Institute is working to create a genetically modified Asian elephant (*Elephas maximus*) with mammoth-derived adaptations to cold climates relating to ear size, subcutaneous fat, hair, and haemoglobin (Shapiro, 2015a). The rationale for attempting this is the restoration of the mammoth steppe biome, the vast areas of species-rich semi-arid grasslands that were lost with the decline of keystone megaherbivores. Small-scale experiments in Siberia indicate that the reintroduction of extant large herbivores, such as bison and muskox (*Ovibos moschatus*), can reinstate nutrient cycling, seed dispersal, and soil turnover processes, and cause a shift in dominance from wet moss tundra to grass steppe (Zimov et al., 1995). It is hypothesised that restoration of populations of herbivores, including perhaps an Asian elephant/woolly mammoth hybrid (Shapiro, 2015b), could enhance grassland restoration over huge areas, and perhaps even result in a slowing of the release of carbon from thawing Arctic soils (Zimov et al., 2012). Apart from the considerable technical and ethical challenges involved in genetically modifying Asian elephants, the free release of a genetically modified organism (GMO) as an ecological replacement for a species that went extinct thousands of years ago into a release site that has undergone ecological change since the original extinction is possibly the most risky and uncertain rewilding project being considered. Nothing is known of the ecology of the proposed GMO, there are uncertainties around the ability of founder animals to survive and reproduce, and the predicted ecosystem-level changes would require growth and persistence of large populations of the genetically engineered elephants. In addition, global climate change might make release areas unsuitable or render the anticipated carbon-sequestration benefits moot.

Summary: relative risks and uncertainties

The relative probability of success of any conservation translocation sits along a spectrum of outcomes that depend on risk factors and uncertainties (Figure 15.2) – no project is a guaranteed winner, but also no proposed release is doomed just because it is risky. The greater the uncertainly, the broader will be the prior distribution of outcomes and hence the lower the mean

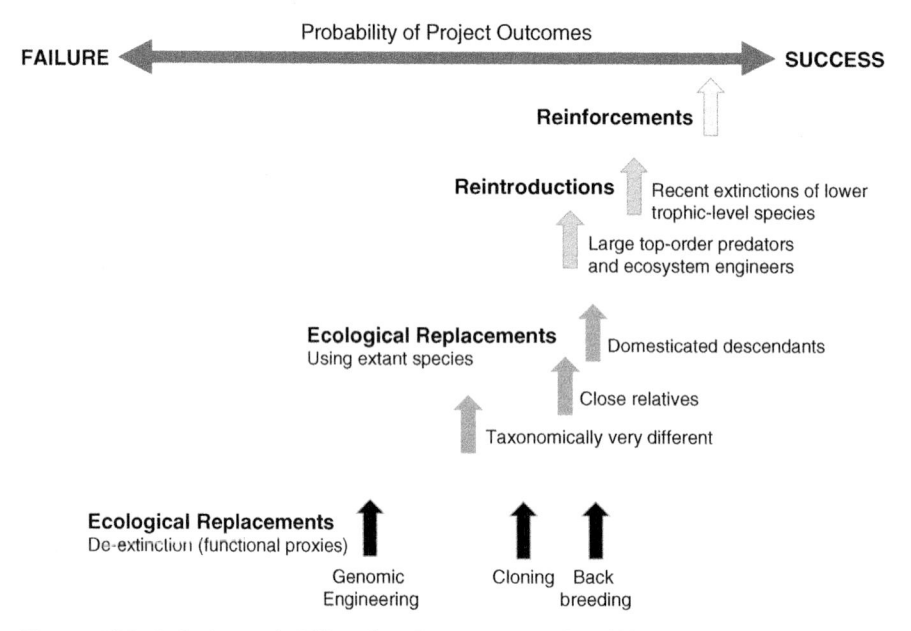

Figure 15.2. Relative probability of project outcomes for different types of conservation translocation that could be undertaken for rewilding: reinforcements, reintroductions, and ecological replacements. Reintroductions are separated into recent extinctions of lower-trophic-level species and reintroductions of large top-order predators or ecosystem engineers. Ecological replacement projects can use either extant species, or apply genetic manipulations to create functional proxies of an extinct species; for the latter, the risks and uncertainties in the technological creation of founder groups of proxy species are not taken into account. Outcome probability takes into account risk and uncertainty, including perceived probability of negative impacts on ecosystems, and negative social attitudes. The placement of arrows within translocation type indicates approximate relative probability of outcomes, all other things being equal. Vertical alignment of arrows between translocation types is less reliably indicative of relative probabilities of outcomes due to likely marked variation in risk factors and uncertainties between the different types of projects. (A black and white version of this figure will appear in some formats. For the colour version, please refer to the plate section.)

probability of success. Thus, high uncertainty might mean that one project is not pursued over another due to a perceived low probability of success. Because of the many project options, and the many factors involved in each, it is hard to generalise about which type of translocations for rewilding would have the greatest chance of success, but we can suggest some very general patterns.

- Projects using individuals of extant species whose biology is well known, and which are to be released into a population of conspecifics, i.e. reinforcements, will have less uncertainty and a higher probability of success than reintroductions, because for reinforcements the adequacy of the release area is demonstrable by the performance of the recipient population.
- Reintroductions into areas from which the target species has been missing for a short time will have less uncertainty than for areas where the species has been absent for a long time, due to there being less potential for radical ecosystem change over shorter time frames.
- Reintroductions of lower-trophic-level species will attract less public opposition than the reintroduction of larger top predators or ecosystem engineers.

Ecological replacements might seem to carry greater risk and uncertainty than reintroductions, but this will depend strongly on the type of replacement being proposed; domesticated descendant forms and very close relatives would seem not to be inherently more risky or uncertain than reintroductions of indigenous species that have been absent from the proposed release area for a long time, and might even attract more public support than the reintroduction of species with potential to cause deleterious effects in human-dominated landscapes.

Similarly, ecological replacements derived through so-called de-extinction pathways might seem to pose greater risks and uncertainties than the use of extant species as substitutes, but this will depend on the pathway used. Thus, back-breeding is qualitatively equivalent to using domesticated descendants, and cloning (technical lab challenges aside) might not carry any greater risks or uncertainties than some ambitious reintroductions or the use of unrelated substitutes. Genomic engineering does, however, involve significantly greater uncertainties because nothing is known of the free-ranging ecology of the GMO founder animals, and as a result might carry a lower probability of success than cloning or back-breeding. The vertical alignment of the de-extinction options relative to other ecological replacements, or to reintroductions (Figure 15.2), is less clear, and might depend strongly on other factors, including trophic level, release area suitability, the presence of new threats and public attitudes to the genetic modification of wild species.

Conclusions

For many rewilding efforts the passive recolonisation of key species is not an option. If natural recolonisation is unlikely to occur (Stewart et al., 2017), some form of conservation translocation might need to be applied in order to meet project goals. Any translocation carries some level of risk and uncertainty, and the successful establishment of a viable population is never assured. Much of the focus of the discipline of Reintroduction Biology over the last three decades has been to improve the practice of conservation translocations by applying the learning from past projects that have been designed and implemented to address *a priori* hypotheses; to apply decision-making tools to choose among multiple management alternatives; and that consider all aspects of the translocation process, from welfare of individual animals (Harrington et al., 2013), to population establishment and ecosystem effects (Taylor et al., 2017). While much translocation activity to date has been motivated primarily by the conservation of the target species, this body of learning is now available to be applied to rewilding where the objectives relate more to the performance of the target species within the recipient system.

Acknowledgements

This chapter was improved by the insightful comments of Arian Wallach and Nathalie Pettorelli.

References

Ajmone-Marsan, P., Fernado Garcia, J., and Lenstra, J.A. (2010). On the origin of cattle: how aurochs became cattle and colonized the world. *Evolutionary Anthropology*, **19**, 148–157.

Armstrong, D.P., Davidson, R.S., Dimond, W.J., et al. (2002). Population dynamics of reintroduced forest birds on New Zealand islands. *Journal of Biogeography*, **29**, 609–621.

Basse, B., Flux, I., and Innes, J. (2003). Recovery and maintenance of North Island kokako *Callaeas cinerea wilsoni* populations through pulsed pest control. *Biological Conservation*, **109**, 259–270.

Beck, B.B., Rapaport, L.G., Stanley Price, M., and Wilson, A.C. (1994). Reintroduction of captive-born animals. In Olney, P.J.S., Mace, G.M., and Feistner, A.T.C. (Eds.), *Creative conservation: interactive management of wild and captive animals* (pp. 265–286). London: Chapman & Hall.

Bengsen, A.J., Gentle, M.N., Mitchell, J.L., Pearson, H.E., and Saunders, G.R. (2014). Impacts and management of wild pigs *Sus scrofa* in Australia. *Mammal Review*, **44**, 135–147.

BirdLife International. (2017). Species factsheet: *Callaeas wilsoni*. www.birdlife.org (accessed 17 August 2017). Recommended citation for factsheets for more than one species: BirdLife International (2017) IUCN Red List for birds. www.birdlife.org (accessed 17 August 2017).

Burney, D.A., Juvik, J.O., Pigot Burney, L., and Diagne, T. (2012). Can unwanted suburban tortoises rescue native Hawaiian plants? *The Tortoise*, **2012**, 104–115.

Byers, J.E., Cuddington, K., Jones, C.G., et al. (2006). Using ecosystem engineers to restore ecological systems. *Trends in Ecology & Evolution*, **21**, 493–500.

Carrell, S. (2007). Scotland, home to free prescriptions, council houses and the wild beaver. www.theguardian.com/politics/2007/dec/24/uk.scotland

Chauvenet, A.L.M., Ewen, J.G., Armstrong, D.P., et al. (2012). Does supplemental feeding affect the viability of translocated populations? The example of the hihi. *Animal Conservation*, **15**, 337–350.

Chauvenet, A.L.M., Ewen, J.G., Armstrong, D.P., and Pettorelli, N. (2013). Saving the hihi under climate change: a case for assisted colonization. *Journal of Applied Ecology*, **50**, 1330–1340.

Chauvenet, A.L.M., Canessa, S., and Ewen, J.G. (2016). Setting objectives and defining the success of reintroductions. In Jachowski, D.S., Slowtow, R., Angermeier, P.L., and Millspaugh, J.J. (Eds.), *Reintroduction of fish and wildlife populations* (pp. 105–122). Oakland, CA: University of California Press.

Converse, S.J., and Armstrong, D.P. (2016). Demographic modelling for reintroduction decision-making. In Jachowski, D.S., Slowtow, R., Angermeier, P.L., and Millspaugh, J.J. (Eds.), *Reintroduction of fish and wildlife populations* (pp. 123–146). Oakland, CA: University of California Press.

Cook, C.N., Morgan, D.G., and Marshall, D.J. (2010). Reevaluating suitable habitat for reintroductions: lessons learnt from the eastern barred bandicoot recovery program. *Animal Conservation*, **13**, 184–195.

Corlett, R.T. (2013). The shifted baseline: prehistoric defaunation in the tropics and its consequences for biodiversity conservation. *Biological Conservation*, **163**, 13–21.

Corlett, R.T. (2016). Restoration, reintroduction, and rewilding in a changing world. *Trends in Ecology & Evolution*, **31**, 453–462.

Donlan, J., Greene, H.W., Berger, J., et al. (2005). Re-wilding North America. *Nature*, **436**, 913–914.

Dzięciołowski, R., and Gozdziewski, J. (1999). The reintroduction of European beaver, *Castor fiber*, in Poland. In Busher, P.E. and Gozdziewski, J. (Eds.), *Beaver protection, management, and utilization in Europe and North America* (pp. 31–35). New York, NY: Springer.

Ericsson, G., and Heberlein, T.A. (2003). Attitudes of hunters, locals, and the general public in Sweden now that the wolves are back. *Biological Conservation*, **111**, 149–159.

Ewen, J.G., Armstrong, D.P., Parker, K.A., and Seddon, P.J. (Eds.) (2012). *Reintroduction biology: integrating science and management*. Conservation Science and Practice no. 9. Chichester: Wiley-Blackwell.

Fischer, J., and Lindenmayer, D.B. (2000). An assessment of the published results of animal relocations. *Biological Conservation*, **96**, 1–11.

Gaywood, M.J. (2015). *Beavers in Scotland: a report to the Scottish Government*. Inverness, UK: Scottish National Heritage.

Gibbs, J.P., Marquez, C., and Sterling, E.J. (2008). The role of endangered species reintroduction in ecosystem restoration: tortoise–cactus interactions on Espanola Island, Galapagos. *Restoration Ecology*, **16**, 88093.

Gregory, R., Failing, L., Harstone, M., Long, G., McDaniels, T., and Ohlson, D. (2012). *Structured decision making: a practical guide to environmental management choices*. Chichester: Wiley-Blackwell.

Griffith, B., Scott, J.M., Carpenter, J.W., and Reed, C. (1989). Translocation as a species conservation tool: status and strategy. *Science*, **245**, 477–480.

Griffiths, C.J., Hansen, D.M., Jones, C.G., Zuel, N., and Harris, S. (2011). Resurrecting extinct interactions with extant substitutes. *Current Biology*, **21**, 762–765.

Gurnell, A.M. (1998). The hydrogeomorphological effects of beaver dam-building activity. *Progress in Physical Geography*, **22**, 167–189.

Hansen, D.M., and Galetti, M. (2009). The forgotten megafauna. *Science*, **324**, 42–43.

Hansen, D.M., Dinlan, J.C., Griffiths, C.J., and Campbell, K.J. (2010). Ecological history and latent conservation potential: large and giant tortoises as a model or taxon substitions. *Ecography*, **33**, 272-284.

Harrington, L.A., Moehrenschlager, A., Gelling, M., Atkinson, R.P.D., Hughes, J., and McDonald, D.W. (2013). Conflicting and complementary ethics in animal welfare considerations in reintroductions. *Conservation Biology*, **27**, 486-500.

IUCN/SSC. (2013). *Guidelines for reintroductions and other conservation translocations*. IUCN/SSC Re-introduction Specialist Group, IUCN, Gland, Switzerland and Cambridge, United Kingdom. www.iucnsscrsg.org/

IUCN/SSC. (2016). *IUCN SSC guiding principles on creating proxies of extinct species for conservation benefit*. Version 1.0. Gland, Switzerland: IUCN Species Survival Commission.

Kearney, M.R., Wintle, B.A., and Porter, W.P. (2010). Correlative and mechanistic models of species distribution provide congruent forecasts under climate change. *Conservation Letters*, **3**, 203-213.

Kemp, L., Norbury, G., Groenewegen, R., and Comer, S. (2015). The roles of trials and experiments in fauna reintroduction programs. In Armstrong, D.P., Hayward, M.W., Moro, D., and Seddon, P.J. (Eds.), *Advances in reintroduction biology of Australian and New Zealand fauna* (pp. 73-90). Victoria, Australia: CSIRO Publishing.

Mallon, D.P., and Stanley Price, M.R. (2013). The fall of the wild. *Oryx*, **47**, 467-468.

McKinstry, M.C., and Anderson, S.H. (1999). Attitudes of private and public land managers in Wyoming, USA, toward beaver. *Environmental Management*, **23**, 95-101.

McKnight, R. (2014). Rewilding the European landscape: an unconventional approach to land management. City Wild. Paper 8. http://digitalcommons.esf.edu/citywild/8

Mills, L.S., Soulé, M.E., and Doak, D.F. (1993). The keystone species concept in ecology and conservation. *BioScience*, **43**, 219-224.

Milne, A., and Stocker, R. (2014). Evidence for the continued existence of the South Island kokako *Callaeas cinerea* drawn from reports collected between January 1990 and June 2012. *Notornis*, **61**, 137-143

Nilsen, E.B., Milner-Gulland, E.J., Schofield, L., Mysterud, A., Stenseth, N.C., and Coulsen, T. (2007). Wolf reintroduction to Scotland: public attitudes and consequences for red deer management. *Proceedings of the Royal Society of London B: Biological Sciences*, **274**, 995-1002.

Nogues-Bravo, D., Rodriguez, J., Hortal, J., Batra, P., and Araujo, M.B. (2008). Climate change, humans, and the extinction of the woolly mammoth. *PLoS Biology*. doi: 10.1371/journal.pbio.0060079

O'Connell, M.J., Atkinson, S.R., Gamez, K., Pickering, S.P., and Dutton, J.S. (2008). Forage preferences of the European beaver *Castor fiber*: implications for re-introduction. *Conservation and Society*, **62**, 190-194.

Osborne, P.E., and Seddon, P.J. (2012). Selecting suitable habitats for reintroductions: variation, change and the role of species distribution modelling. In Ewen, J.G., Armstrong, D.P., Parker, K.A., and Seddon, P.J. (Eds.), *Reintroduction biology: integrating science and management*. Conservation Science and Practice no. 9 (pp. 73-104). Chichester: Wiley-Blackwell.

Pate, J., Manfredo, M.J., Bright, A.D., and Tischbein, G. (1996). Coloradans' attitudes towards reintroducing the grey wolf into Colorado. *Wildlife Society Bulletin*, **24**, 421-428.

Pillai, A., and Heptinstall, D. (2013). Twenty years of the Habitats Directive: a case study on species reintroduction, protection and management. *Environmental Law Review*, **15**, 27-46.

Rewilding Europe. (2017). www.rewildingeurope.com (accessed 10 August 2017).

Rewilding Institute. (2017). https://rewilding.org/rewildit/our-programs/ (accessed 10 August 2017).

Ripple, W.J., and Beschta, R.L. (2004). Wolves and the ecology of fear: can predation risk structure ecosystems? *BioScience*, **54**, 755-766.

Robert, A., Colas, B., Guigon, I., et al. (2015). Defining reintroduction success using IUCN criteria for threatened species: a demographic assessment. *Animal Conservation*, **18**, 397–406.

Rosell, F., Bozser, O., Collen, P., and Parker, H. (2005). Ecological impact of beavers *Castor fiber* and *Castor canadensis* and their ability to modify ecosystems. *Mammal Review*, **35**, 248–276.

Sandom, C., Donlan, C.J., Svenning, J.-C., and Hansen, D. (2013) Rewilding. In MacDonald, D.W. and Willis, K.J. (Eds.), *Key topics in conservation biology 2* (pp. 430–451). Chichester: John Wiley & Sons.

Sarrazin, F. (2007). Introductory remarks – a demographic frame for reintroductions. *EcoScience*, **14**, iv-v.

Seddon, P.J. (1999). Persistence without intervention: assessing success in wildlife reintroductions. *Trends in Ecology & Evolution*, **14**, 503.

Seddon, P.J., and Armstrong, D.P. (2016). Reintroduction and other conservation translocations: history and future developments. In Jachowski, D.S., Slowtow, R., Angermeier, P.L., and Millspaugh, J.J. (Eds.), *Reintroduction of fish and wildlife populations* (pp. 7–28). Oakland, CA: University of California Press.

Seddon, P.J., Soorae, P.S., and Launay, F. (2005). Taxonomic bias in reintroduction projects. *Animal Conservation*, **8**, 51–58.

Seddon, P.J., Griffiths, C.J., Soorae, P.S., and Armstrong, D.P. (2014). Reversing defaunation: restoring species in a changing world. *Science*, **345**, 406–412.

Selbach, C., Seddon, P.J., and Poulin, R. (2018). Parasites lost: neglecting a crucial element in de-extinction. *Trends in Parasitology*, **34**, 9–11.

Sergio, F., Caro, T., Brown, D., et al. (2008). Top predators as conservation tools: ecological rationale, assumptions, and efficacy. *Annual Review of Ecology Evolution and Systematics*, **39**, 1–19.

Shapiro, B. (2015a). Mammoth 2.0: will genome engineering resurrect extinct species? *Genome Biology*, **16**, 228.

Shapiro, B. (2015b). *How to clone a mammoth: the science of de-extinction*. Princeton, NJ: Princeton University Press.

Shapiro, B. (2017). Pathways to de-extinction: how close can we get to resurrections of an extinct species? *Functional Ecology*, **31**, 996–1002.

Sheridan, D. (2014). Return of beavers would damage salmon, say anglers. www.the times.co.uk/article/return-of-beavers-would-damage-salmon-say-anglers-9mn8xmcwdm9

Soorae, P.S. (2008). *Global reintroduction perspectives: re-introduction case studies from around the globe*. Abu Dhabi, UAE: IUCN/SSC Re-introduction Specialist Group.

Soorae, P.S. (2010). *Global reintroduction perspectives: additional case-studies from around the globe*. Abu Dhabi, UAE: IUCN/SSC Re-introduction Specialist Group.

Soorae, P.S. (2011). *Global reintroduction perspectives: more case-studies from around the globe*. Abu Dhabi, UAE: IUCN/SSC Re-introduction Specialist Group and Environment Agency.

Soorae, P.S. (2013). *Global re-introduction perspectives: 2013. Case-studies from around the globe*. Gland, Switzerland: IUCN/ SSC Re-introduction Specialist Group and Abu Dhabi, UAE: Environment Agency.

Soorae, P.S. (2016). *Global re-introduction perspectives: 2016. further case studies from around the globe*. Gland, Switzerland: IUCN/ SSC Re-introduction Specialist Group and Abu Dhabi, UAE: Environment Agency.

South, A.S.R., and Macdonald, D. (2000). Simulating the proposed reintroduction of the European beaver (*Castor fiber*) to Scotland. *Biological Conservation*, **93**, 103–116.

Stewart, F.E.C., Volpe, J.P., Taylor, J.S., et al. (2017). Distinguishing reintroduction from recolonization with genetic testing. *Biological Conservation*, **214**, 242–249.

Stokstad, E. (2015). Bringing back the aurochs. *Science*, **350**, 1144-1147.

Stringer, A.P., and Gaywood, M.J. (2016). The impacts of beavers *Castor* spp. on biodiversity and the ecological basis for their reintroduction to Scotland, UK. *Mammal Review*, **46**, 270-283.

Svenning, J.-C., Pedersen, P.B.M., Donlan, C.J., et al. (2016). Science for a wilder Anthropocene: synthesis and future directions for trophic rewilding research. *Proceedings of the National Academy of Sciences of the United States of America*, **113**, 898-906.

Taylor, G., Canessa, S., Armstrong, D.P., Seddon, P.J., and Ewen, J.G. (2017). Is reintroduction biology an effective applied science? *Trends in Ecology & Evolution*, **32**, 873-880.

Taylor, S., Castro, I., and Griffiths, R. (2005). *Hihi/Stitchbird (Notiomystis cincta) Recovery Plan 2004-09*. Wellington, New Zealand: Department of Conservation.

The Scottish Beaver Trial Team. (2014). *Scottish Beaver Trial: Project Update No. 22* (p. 7). www.scottishbeavers.org.uk/about-the-trial/trial-progress-reports/

van Wieren, S.E. (1995). The potential role of large herbivores in nature conservation and extensive land use in Europe. *Biological Journal of the Linnaean Society*, **56**, 11-23.

White Jr, T.H., Collar, N.J., Moorhouse, R.J., Sanz, V., Stolen, E.D., and Brightsmith, D.J. (2012). Psittacine reintroductions: common denominators of success. *Biological Conservation*, **148**, 106-115.

Williams, C.K., Ericsson, G., and Heberlein, T.A. (2002). A quantitative summary of attitudes toward wolves and their reintroduction. *Wildlife Society Bulletin*, **30**, 575-584.

Wilson, C.J. (2004). Could we live with reintroduced large carnivores in the UK? *Mammal Review*, **34**, 211-232.

Wolf, C.M., Garland Jr, T., and Griffith, B. (1998). Predictors of avian and mammalian translocation success: reanalysis with phylogenetically independent contrasts. *Biological Conservation*, **86**, 243-255.

Zimov, S.A., Chuprynin, V.I., Oreshko, A.P., Chapin III, F.S., Reynolds, J.F., and Chapin, M.C. (1995). Steppe-Tundra transition: a herbivore-driven biome shift at the end of the Pleistocene. *The American Naturalist*, 146, 765-794.

Zimov, S.A., Zimov, N.S., and Chapin III, F.S. (2012). The past and future of the mammoth steppe ecosystem. In Louys, J. (Ed.), *Paleontology in ecology and conservation* (pp. 193-225). Berlin: Springer-Verlag.

CHAPTER SIXTEEN

Top-down control of ecosystems and the case for rewilding: does it all add up?

MATT W. HAYWARD
University of Newcastle
SARAH EDWARDS
University of Pretoria
BRONWYN A. FANCOURT
University of Tasmania
JOHN D.C. LINNELL and ERLEND B. NILSEN
Norwegian Institute for Nature Research

In its simplest form, 'top-down' control refers to directional regulation within an ecosystem, where species occupying higher trophic levels exert controlling influences on species at the next lower trophic level (Terborgh et al., 1999). Thus, top-down control can describe top predators controlling smaller predators or prey, or herbivores exerting a controlling influence on plant biomass. By contrast, 'bottom-up' control refers to abiotic resources and species at the lowest trophic level (producers) regulating the abundance of species at the next highest trophic level (herbivores), which in turn can influence species at higher trophic levels (predators). The top-down control hypothesis was first proposed by Camerano (1880) but refined by Hairston and colleagues (1960), who argued that herbivores, under usual conditions, cannot be limited by either weather or food, and must therefore be limited by predation (i.e. top-down control). This somewhat simplistic view triggered much debate (e.g. Murdoch, 1966; Ehrlich and Birch, 1967; Slobodkin et al., 1967), highlighting many exceptions to, and logical gaps in, the original hypothesis. To this day, the relative contributions of both top-down and bottom-up forces in shaping terrestrial, marine, and freshwater aquatic ecosystems are hotly debated (Sih et al., 1985; Linnell and Strand, 2000; Elmhagen and Rushton, 2007; Laundré et al., 2014), illustrating that causal relationships between species and their limiting factors are far more complex than the simplistic construct first suggested.

Ecosystems are complex, dynamic structures regulated by a plethora of factors that exert control over the distribution, abundance, and density of species (e.g. Brown, 1984; Woodward et al., 1990; Pounds et al., 2006).

The strength and direction of controlling influences can vary in both space and time, and differ between species, life history or successional stages, and with varying habitat complexity, human disturbance, social group size, community composition, resource availability, dietary breadth, productivity, and environmental conditions (Sih et al., 1985; Polis et al., 1989; Elmhagen and Rushton, 2007). For example, while crabs exert top-down control as one of the major predators of sea turtle eggs and hatchlings (Marco et al., 2015; Morgan et al., 2017), the direction of control reverses during later life-history stages, when adult turtles prey on crabs (Tomas et al., 2001). Predators are often food-limited (i.e. > 50 per cent of predator density is explained by prey density; Fuller and Sievert, 2001; Karanth et al., 2004; Hayward et al., 2007), but the degree of top-down limitation by predators on prey has been less frequently studied and the two estimates we know of that estimated the decline in prey abundance following predator recolonisation at individual sites range from 25 per cent to 33 per cent (Jędrzejewska and Jędrzejewski, 1998; Georgiadis et al., 2007). An assumption of top-down control of ecosystems forms the basis upon which many rewilding proposals and trophic cascade studies are built, aiming to restore ecosystem function predominantly in systems where predators have been removed or reduced. Below, we examine the history of the theory, the potential mechanisms and effects of top-down control of ecosystems, review the evidence for top-down control and associated trophic cascades, and critically evaluate the premise that rewilding can restore ecosystem function through top-down control of ecosystems.

Modern predator–prey theory is often thought to have arisen with the seminal work of Holling (1959), who used Solomon (1949) as a basis to define the 'components' of predation as functional and numerical responses, respectively. The functional response defines the number of prey items (or biomass) taken by an individual predator in a given time window in response to changing prey density, and the numerical response is the increase of the predator population as a response to this prey consumption. A key component in most contemporary predator–prey theory is the dependence of predator population dynamics on prey population state variables, such as abundance or biomass (Owen-Smith, 2011), and the opposite dependence of prey dynamics on predator population state variables and behaviour (Fuller and Sievert, 2001; Karanth et al., 2004; Hayward et al., 2007). Even in simple one prey–one predator systems, early theory and laboratory experiments suggested that population cycles could arise (reviewed by May, 1976), and that landscape heterogeneity is crucial for prey persistence (Gause et al., 1936; Huffaker, 1958). In more complex systems with several prey and predator species, patterns of prey species selection (Sinclair et al., 2003) and intraguild competition and predation (Palomares and Caro, 1999) become part of the story. Contemporary research thus highlights that the ecological context in which the predator–prey interactions take place will strongly modulate the outcome of the interaction (Pasanen-Mortensen et al., 2017). Furthermore, it is important to bear in mind that whereas general research on population dynamics over the last decades has highlighted the contributions of environmental and demographic stochasticity to the observed fluctuations (Lande et al., 2003), much of the underlying predator–prey theory is still formulated as deterministic processes. The influence of stochastic factors, and interactions between stochastic and deterministic factors (Wilmers et al., 2006, 2007) in shaping predator–prey dynamics, is currently not well understood and significantly affects our ability to predict the outcome of predator–prey interactions. In the rest of this chapter we address these issues with the objective of exploring the ecological realism of one of the main assumptions of rewilding; namely, that top-down control is a near-universal feature of ecosystems (Soulé, 2010).

Functional responses and trophic interactions

Functional responses describe the rate at which a predator consumes prey at different prey and/or predator densities (Holling, 1959; Abrams and Ginzburg, 2000). Differently shaped functional response curves might give rise to different dynamics, and are an important component in determining the degree of top-down limitation (Abrams, 2000; Abrams and Ginzburg, 2000; Vucetich et al., 2002). One classical dimension of this question relates to the distinction between Holling's Type I, II, and III functional responses (Holling, 1959). Type I responses are almost exclusive to filter feeders (Jeschke et al., 2004), whereas several studies have tried to distinguish between Type II and Type III responses in carnivore–ungulate systems. These have largely concluded that sufficient field data are rarely available (Marshal and Boutin, 1999). Other studies have assessed how the parameters in the functional response equations are modulated by different variables, such as predator age, sex or season (Nilsen et al., 2009b) and there is increasing evidence that distinct functional responses may be less clear under natural conditions (Chan et al., 2017).

A more profound and conceptual question is to what extent kill rates are determined by prey density alone or if they are also modulated by the density of predators (Arditi and Ginzburg, 1989; Abrams, 1994; Akcakaya et al., 1995; Abrams and Ginzburg, 2000). While 'classical' functional responses assume that predator kill rates increase following prey density increases (Holling, 1959; Skalski and Gilliam, 2001), predator-dependent theory predicts that predator kill rates will also be affected by predator density itself. Ratio-dependence is a particular form of predator-dependence where the response depends on the ratio of prey to predators (Abrams and Ginzburg, 2000). While relatively few studies have tried to distinguish between classical functional responses or predator-dependence for carnivore–ungulate systems (but see Skalski and Gilliam, 2001; Vucetich et al., 2002), the distinction is far from trivial and has far-reaching implications for ecosystem dynamics and the potential for top-down limitation. In a very simplified view, predation will be less important in determining prey dynamics and abundance under a predator- or ratio-dependent response (Abrams and Ginzburg, 2000; Vucetich et al., 2002). More precisely, the different responses give rise to very different assumptions about the effects of primary productivity on prey and predator abundance, and the possibilities for equilibrium points that are both low and stable (Abrams and Ginzburg, 2000). In general, while ratio-dependent theory predicts a positive relationship between prey and predator abundance, no such relationship is expected under prey-dependent theory. Predator- (or ratio-) dependent functional responses can emerge due to prey and predator behaviour such as predator avoidance, group hunting, and interference competition among predators (Abrams and Ginzburg, 2000; Fryxell et al., 2007). Understanding how spatial heterogeneity and resource distribution affect predator–prey dynamics is fundamental to our ability to predict the strength of top-down forces. Based on the meta-analysis by Hebblewhite (2013), evidence from recent studies seems to favour ratio-dependence in that there is a positive relationship between prey and predator abundance.

Age- and size-specific predation: patterns and effects

Large carnivores rarely kill prey individuals at random, and often certain species, sex and age categories are at higher risk than others (Hayward and Kerley, 2005; Hayward et al., 2017). From a broad perspective, solitary predators generally prefer to kill prey of a similar size to themselves (Hayward et al., 2016), whereas group hunters take much larger prey (Hayward et al., 2014). When categories of individual prey species are considered (i.e. sex and age class), these patterns become more complex. At the most fundamental level, the ability of a predator to subdue adult individuals from the prey population is contingent on the size relationship between them (Radloff and du Toit, 2004; Gervasi et al., 2015). Ungulates, like most mammals, produce offspring that are small compared to their adult body mass (on average only 10 per cent; Blueweiss et al., 1978). For this reason, adults and juveniles experience very different body-mass relationships with their predators, and many mammalian predators that cannot prey upon adults of a given prey species may still be able to subdue juveniles (Gervasi et al., 2012, 2015). We still do not have a complete understanding of the factors that determine the propensity of a carnivore to kill adult versus juvenile prey. However, a recent review of the topic (Gervasi et al., 2015) found that both prey and predator body mass contributed (as expected from prey preference research predictions). However, other factors related to both prey and predators also contributed to the proportion of juveniles among the killed prey individuals. Furthermore, the proportion of juveniles among the prey individuals killed was lower for felids than for canids; male predators killed fewer juveniles; and the proportion of juveniles killed was higher for herd-living ungulates compared to solitary species (Gervasi et al., 2015). It is worth noting, however, that although this review included a total of 159 predator–prey dyads (i.e. different predator–prey combinations), it still only included data from 12 carnivore species (seven felid species, four canid species, and one hyaena species) and 37 ungulate species. Consequently, we still lack considerable knowledge about the sex and age composition of killed individuals from most predator–prey systems, and the generalisations that have been drawn so far are based on a rather limited set of species.

A major reason why understanding the age and sex composition among killed individuals is so important is the fact that among long-lived, iteroparous species with long generation times, such as ungulates, adult survival has a strong functional relationship with population growth rate (Gaillard et al., 2000, 2005; Gaillard and Yoccoz, 2003). Because of the high elasticity of population growth rate (lambda) to adult survival, small perturbations in adult survival have a far greater effect than similar perturbations to juvenile survival or recruitment. When predators in a predator–prey dyad also kill adult prey individuals, the effect of predation will often therefore be profound (Gervasi et al., 2012). In ungulates and other long-lived species, adult survival is generally high and stable over time (Gaillard and Yoccoz, 2003). Therefore, it is often reported that variation in recruitment rather than adult survival contributes more to observed variation in population growth (Gaillard et al., 1998). However, the majority of the studies that underlie these generalisations are from predator-free areas, whereas variation in adult survival might be more influential in areas where large mammalian predators (and/or hunting) are important components of the system. In a comparative analysis of Norwegian and French roe deer (*Capreolus capreolus*) populations, Nilsen and colleagues (2009a) showed how predators changed these relationships such that roe deer in sympatry with Eurasian lynx (*Lynx lynx*) were subject to considerable predation on adults that drove variation in population growth rates. Conversely, Owen-Smith and Mason (2005) reported from a suite of large ungulates in South Africa's Kruger National Park that survival rates of adults varied more over multi-year periods compared with mainly annual fluctuations in juvenile survival, and concluded that elevated predation risk might have been the main reason for the change in adult survival. Therefore, while variation in juvenile survival and recruitment might have been responsible for short-term fluctuations in prey abundance, changes in adult survival due to predation might have been responsible for the changes between time periods. Because the proportion of juveniles among killed individuals depends on properties of both the prey and the predators, the demographic impact of predation might vary within systems as well. This is well documented from

Scandinavian ecosystems, where the effects of predation by Eurasian lynx and red fox (*Vulpes vulpes*) on roe deer was compared with predation from wolves (*Canis lupus*) and brown bear (*Ursus arctos*) on moose (*Alces alces*). In short, the impact of lynx predation on roe deer was far higher than wolf predation on moose, which was again far higher than red fox predation on roe deer and bear predation on moose (Gervasi et al., 2012). These patterns, to a limited extent, related to the differences in kill or predation rates, but were a result of different propensities to kill adult prey individuals. In conclusion, the age distribution among prey individuals is strongly related to the demographic effect of predation among carnivore–ungulate systems, implying that there are very different potentials for top-down effects in different predator–prey systems.

Effects of system productivity on top-down limitation

As noted in the introduction, the impact of predation on prey populations is dependent on ecological context. System productivity has been listed among the more influential context-dependent factors that determine the level of top-down limitation on prey populations (Oksanen and Oksanen, 2000; Melis et al., 2009). If predators exhibit Type II functional responses (but not Type III), predation rates increase when prey abundance decreases (Hebblewhite, 2013; Melis et al., 2013) without a proportionate decrease in predator numbers, which is when inversely density-dependent predation rates should be expected.

In a pan-European analysis of predation impact on roe deer population dynamics, Melis and colleagues (2009) used data on roe deer population density from 72 localities to investigate how environmental factors such as productivity and predator occurrence determined abundance levels. Their analysis indicated that predators had a greater suppressive role in less productive areas than areas with higher productivity, and this supports other studies (Melis et al., 2010; Letnic and Ripple, 2017). All else being equal, in a rewilding setting, one could expect greater proportional impacts of a carnivore reintroduction programme in less productive areas with already low prey densities than in more productive areas with higher prey densities.

Mechanisms of top-down control

There are several direct and indirect mechanisms by which species might exert top-down control within an ecosystem.

Killing. This direct form of control occurs where a predator kills and eats the victim (consumptive) or kills the victim without eating it. Consumptive killing commonly occurs where a predator kills a prey individual to satisfy its nutritional or energy requirements (Lourenço et al., 2014). It can also include intraguild predation, which is the killing and eating of other predators that use similar, often limiting, resources and are thus potential competitors (Polis et al., 1989). Killing without eating can occur when prey are killed surplus to the predator's dietary needs (Kruuk, 1972; Short et al., 2002). Additionally, some predators kill (without consuming) intraguild members in an extreme form of interference competition, theoretically to remove a potential competitor (Palomares and Caro, 1999).

Competition. This indirect form of control can occur through either exploitation or interference (Birch, 1957). Exploitation competition occurs when one species uses a limited resource, such as food or shelter sites, thereby depriving others from using it. The most common form of exploitation competition occurs where two or more predator species have a high degree of dietary overlap (Glen and Dickman, 2008) and the competitively dominant species consumes most of the prey. Exploitation competition can also occur over resources such as optimal den or shelter sites. For example, red foxes in Norway occupy lower-altitude dens that were previously occupied by arctic foxes (*Vulpes lagopus*; which are now restricted to higher altitudes), suggesting that red foxes are depriving arctic foxes from using those den sites (Hersteinsson et al., 1989; Linnell et al., 1999). Interference competition occurs where a species is restricted from accessing a resource, typically through antagonistic interactions with more dominant competing species or predators (Linnell and Strand, 2000). A chronic risk of antagonistic interactions might prevent the inferior competitor from accessing resources at times or in areas where the dominant competitor is active (Durant, 1998; Hayward and Slotow, 2009).

Effects of top-down control

Top-down control can result in several different lethal and non-lethal outcomes, including changes in abundance, demographics, activity, and fitness.

(i) *Abundance.* Where a top predator kills many individuals of a smaller predator or prey species, in some cases the abundance of the 'controlled' species might be reduced to an alternative stable state, or possibly even local extinction (Polis et al., 1989; Palomares and Caro, 1999).

(ii) *Demographics.* Selective predation of a particular demographic class of prey may alter key population demographics, such as sex ratios or age structure (O'Kane and MacDonald, 2016). For example, in sexually dimorphic species where females are smaller than males, smaller predators may remove females more often than males as their smaller size could make them easier targets (Gervasi et al., 2015). Over time, this may skew the prey species sex ratio in favour of males, thereby reducing the population's reproductive capacity.

(iii) *Activity and habitat use.* Species at high risk of predation are said to be living in a 'landscape of fear' (Laundré et al., 2001). To avoid antagonistic or fatal encounters with predators or dominant competitors, prey species might alter their spatial and/or temporal activity to forage in areas or at times when predators are less active or absent (Hayward and Slotow, 2009; Broekhuis et al., 2013; Fancourt et al., 2015; Winnie and Creel, 2017).

(iv) *Fitness.* Where the costs of predator avoidance or increased vigilance are high, prey species may suffer a loss of fitness. By avoiding antagonistic interactions, some species may be forced to forage in suboptimal areas or at suboptimal times where prey or food is less available or of lower quality, potentially reducing foraging success or efficiencies (Barnier et al., 2014). For dietary specialists or in areas where food resources are limited, reduced foraging success might lead to poorer body condition, increased susceptibility to parasites or disease, lower reproductive output and possibly death (Lima, 1998; Preisser et al., 2005; Creel et al., 2007).

When investigating the potential effects of top-down control in ecosystems, many studies focus on the lethal aspect of predation – that is, control manifests as the death and removal of the prey or competitor (Lima, 1998). However, a growing body of evidence suggests that the non-lethal effects of predation can be far more pervasive (Creel and Christianson, 2008), with trait-mediated effects generally as strong as, or stronger than, the effects of direct consumption (Preisser et al., 2005).

Much of the research on top-down control infers dynamics by simplifying food webs into a few linear trophic levels, and these thereby fail to acknowledge the dynamically important complexity of the multiple, reticulate connections within food webs (Polis and Strong, 1996). For example, grizzly bears (*Ursus arctos*) are omnivores (Mattson et al., 1991) and so may regulate both herbivores and plants (as occurred where *Orconectes rusticus* omnivory complicates trophic cascades in the littoral zone; Lodge et al., 1994), and do not fit neatly into discrete, homogeneous trophic levels. Similarly, the ontogenetic shift in diet as predators grow further obscures discrete trophic levels. Therefore, although Type II functional responses could be predicted to more strongly trigger trophic cascades, they often indicate apparent competition in which a predator is subsidised by a species of prey that is typically bottom-up controlled, while suppressing a rarer, more preferred second species of prey (Polis and Strong, 1996). This, and the diffusion of the direct effects of consumption and productivity by the high degree of connectance in food webs, suggest that top-down control and trophic cascades are relatively uncommon in nature (Polis and Strong, 1996).

Evidence for top-down control

There is a growing literature base reporting top-down control in ecosystems, with a particular emphasis on the ability of apex predators to control mesopredators (Sih et al., 1985; Polis et al., 1989; Terborgh et al., 1999; Ritchie and Johnson, 2009; Newsome et al., 2017). However, many studies have yielded highly variable and often contradictory results (including from the same study sites) regarding the effects of top-down control, both within and between terrestrial, marine, and freshwater ecosystems (Sih et al., 1985; Shurin et al., 2002; Laundré et al., 2014). Furthermore, reviews of studies underpinning the evidence base for top-down control have raised concern about the validity of many conclusions, citing methodological weaknesses, sampling biases, failure to account for detectability issues, misinterpreting weak inference and hypotheses for robust evidence, and failure to consider alternative explanations for observed patterns (Sih et al., 1985; Allen et al., 2011, 2013, 2017; Fleming et al., 2012; Hayward and Marlow, 2014; Lourenço et al., 2014; Ford and Goheen, 2015).

Evidence that predators kill both prey (Kruuk, 1972; Hornsby, 1982; Judge et al., 2012; Fancourt, 2015; McGregor et al., 2015) and smaller predators (Polis et al., 1989; Palomares and Caro, 1999; Linnell and Strand, 2000) is well documented. However, in many cases, the naïve assumption is that lethal removal of prey individuals by a predator directly affects population abundance of the prey. However, rarely do studies demonstrate or quantify the extent of any limitation or suppression of population abundance, often inferring such impacts from dietary analyses or indices derived from correlative observational studies. Dietary studies, while informative, often mask the complete picture of predation impact as they do not incorporate surplus kills that are not consumed (Kruuk, 1972) or the abundance of prey items. Dietary studies from gut samples can yield different results from studies using scat samples due to the differing digestibility of certain prey items and the absence of identifiable remnants in scats (Balestrieri et al., 2011). Additionally, distinguishing whether prey species were killed or simply scavenged as carrion is often difficult, thereby precluding the ability to robustly quantify any impact of predation from such studies.

Controlled experiments produce stronger inferences as to the effects of top-down control than do observations alone; however, experiments are often impractical and context dependent with large predators, particularly where the questions are broad in scope (Sih et al., 1985). For example, Moseby and colleagues (2012) demonstrated that dingoes (*Canis familiaris*) can kill red foxes and feral cats (*Felis catus*) as a form of interference competition in a large predator-proof fenced enclosure in arid Australia, but the applicability of such experiments to natural, unfenced landscapes is unknown. Fences would have prevented foxes and cats from escaping lethal interactions with dingoes and may have forced them to feed and shelter in areas from which they would normally disperse, rendering them easier targets than if they were outside the enclosure. Furthermore, canids routinely use fences and other linear features on the landscape to increase their predatory effectiveness (van Dyk and Slotow, 2003; Davies-Mostert et al., 2013; Dickie et al., 2017). The frequency and population impact of such killings is also unknown, but is unlikely to be limiting given that foxes and cats were able to establish and spread across Australia following their introduction, despite the long-established presence of dingoes (Hayward and Marlow, 2014; Allen et al., 2017).

Observational studies can address broad questions but are more open to alternative interpretations (Sih et al., 1985). Correlative observational studies comparing differences or changes in abundance between two sympatric predators (or predator and prey species) provide only weak inference as to the degree of any control, as often there will be a range of alternative and/or confounding factors that may act in conjunction with, or completely separately to, predation in exerting control over a species. Furthermore, such studies often infer relationships in abundance using unvalidated activity indices that ignore detectability issues and fail to differentiate between changes in activity and changes in abundance. Such indices typically assume a linear relationship between the index and the true abundance (Pollock et al., 2002; Stephens et al., 2005), but most activity indices are unreliable surrogates for abundance (Karanth et al., 2011). For example, Brook and colleagues (2012) used a negative correlation in camera trap detection rates (i.e. naïve occupancy estimates) as an activity index to infer that dingoes suppress feral cat abundance. However, this index was never validated and was based on a mean of 12 cat detections per site over several months. The observed negative correlation in photos may simply reflect fine-scale predator avoidance rather than suppression of abundance, as illustrated by Fancourt (2016).

While the non-lethal effects of top-down control are increasingly being considered in the literature (Lima, 1998; Winnie and Creel, 2017), the costs of the risk of predation are rarely quantified beyond food-risk, mesocosm-based approaches, and are instead an inferred construct of the effects of predation risk on prey (Gallagher et al., 2017). Recent studies have demonstrated the influence of predation risk on prey vigilance (Périquet et al., 2010), diet quality (Barnier et al., 2014), demographics (O'Kane and MacDonald, 2016), and temporal and spatial activity (Durant, 1998; Hayward and Slotow, 2009).

With top-down impacts being context-specific, they are unlikely to apply equally to different systems (Haswell et al., 2017; Morgan et al., 2017). For example, drawing on evidence from aquatic systems to validate weak inference in terrestrial systems is rarely appropriate, as reviews have shown the strength of top-down control is often far greater and of more importance in aquatic systems than on land (Sih et al., 1985; Polis et al., 1989; Shurin et al.,

2002). Likewise, structurally complex environments have more refuges and potentially reduce predator foraging efficiency, both of which may allow the coexistence of more species and dampen the strength of any top-down control (Menge and Sutherland, 1976).

The literature base is fraught with a range of biases that raise further questions about the importance and validity of any reported evidence of top-down control. As Sih and colleagues (1985) observed, 'That individual-level studies show a higher frequency of response [than community and population-level responses] suggests that investigators choose (consciously or not) to do detailed studies on prey that appear to be affected by predators. This bias is not surprising and not unwise, but it does perhaps yield an inflated impression of the importance of predation.' Publication biases further exacerbate the uncertainty, with many journals only publishing studies that show a significant effect whereas nil-effect studies might not be considered 'globally relevant' (Csada et al., 1996).

The evidence for and against top-down control of ecosystems following rewilding

Trophic cascades within Yellowstone National Park; wolves, elk, and aspen

Perhaps the best-known and most debated example of a terrestrial trophic cascade arising from rewilding centres around wolves, elk *Cervus canadensis*, and quaking aspen *Populus tremuloides* across the northern range of Yellowstone National Park (Yellowstone; Figure 16.1).

The reintroduction of wolves in 1995/1996 following a 70-year absence has presented a unique opportunity in which to study a trophic cascade through time (Ripple and Beschta, 2012). However, it is worth highlighting that the wolf reintroduction occurred alongside increases in grizzly bear populations, changes in fish populations (an important bear prey resource), increases in drought frequency and severity, increased human hunting pressure in elk winter migration areas to the north and east of Yellowstone, and massive fires in the late 1980s, which confound linking cause and effect without experimentation (Figure 16.2). Nonetheless, the lack of recruitment within aspen stands has been the subject of intense scientific and public debate. Although many factors have been studied that could have potentially contributed to the decline of aspen during the period of wolf absence, some authors attribute the herbivory of young aspen by elk as playing a critical role (Ripple et al., 2001). Specifically, it has been hypothesised that the historical aspen decline in Yellowstone was caused by the removal of a trophic cascade involving wolves, elk, and aspen. Therefore, it follows that the rewilding of the ecosystem via the reintroduction of wolves might lead to increases in aspen recruitment by reinstating these trophic interactions.

Figure 16.1. Key top-down regulating species discussed in the text: (A) wolves and their effect on Yellowstone's (B) elk population (note bark stripping on the aspen); and (C) dingoes in Australia. (A) Mark Kent (CC BY-SA2), (B) Kristal Kraft CC BY-SA2, and (C) Paul Balfe CC BY2. (A black and white version of this figure will appear in some formats. For the colour version, please refer to the plate section.)

Quaking aspen is the most widely distributed tree in North America and native to Yellowstone (Houston, 1982). The species usually occurs in clonal stands of genetically identical stems, originating as suckers growing from a common root system (Jones and DeByle, 1985). Although the regeneration of aspen seedlings is rare, established root systems can persist for centuries (Knight, 1994), through the recruitment of young suckers into the mature canopy, following the death of older, mature trees. For recruitment to occur, young saplings and seeds must grow above the browsing height of ungulates (~200 cm) (Beschta and Ripple, 2009). Aspen is a primary food of elk during

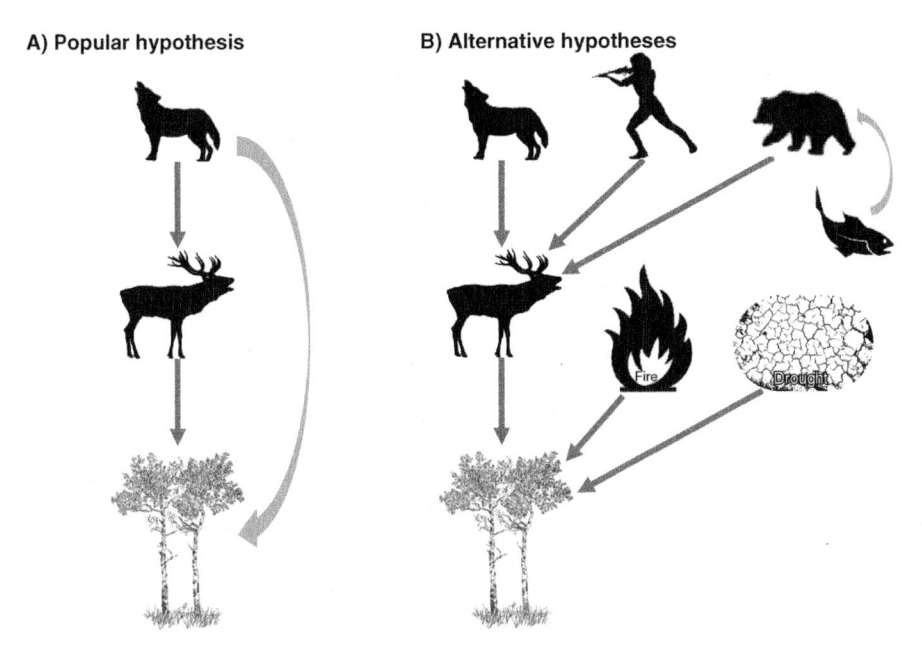

Figure 16.2. Excessive simplification of the Yellowstone system may have led to the belief that strong top-down limitation of elk by wolves has led to a trophic cascade (A), whereas the high degree of connectance with other key organisms within this ecosystem illustrates numerous other factors that may more realistically influence elk and aspen dynamics (B). Negative interactions are shown with dark grey, straight line arrows, positive interactions with lighter grey, curved arrows.

winter, when they browse the leaders off aspen suckers and prevent their growth into the overstorey (Romme et al., 1995). Aspen historically covered 4–6 per cent of Yellowstone's northern range (Houston, 1982), and although the species is considered a minor one for the area, it has local importance as the only deciduous forest type present, supporting a high abundance and diversity of breeding birds and other animals (DeByle, 1985). The percentage of aspen cover declined over the last century in Yellowstone (Romme et al., 1995) to just 1–2 per cent of the northern range (Larsen and Ripple, 2003). The decline has been a cause of concern, with the species now considered at risk of disappearing from many of the Rocky Mountains' National Parks (Fortin et al., 2005). A change in the character of stands, from having variable age classes to its current state of deteriorating stands dominated by large circumference, mature trees and declining older stems (National Research Council, 2002), has accompanied the decline, suggesting a halt in recruitment (Larsen and Ripple, 2003).

Evidence suggests that the majority of the aspen overstorey within the current northern range originated during the 1870s–1880s, which is viewed as the only major period of recruitment since 1800 (Romme et al., 1995). The years between 1871 and 1890 may represent unusually favourable conditions for aspen development, which included frequent fires, favourable climatic conditions, and low levels of ungulate browsing (Ripple and Larsen, 2000). Low levels of browsing were due to the period of market hunting that occurred from 1872 until 1886, when elk and other large herbivores were shot within the park for their hides (Schullery and Whittlesey, 1992). Since the 1920s, elk abundance has been limited by management removals, regulated hunting activities in areas bordering Yellowstone, and natural factors (Ripple et al., 2001). Between 1921 and 1999, no evidence of significant recruitment of new stems into the overstorey was found (Romme et al., 1995; Ripple et al., 2001). However, Ripple and Beschta (2012) found evidence of overstorey recruitment (i.e. aspen suckers attaining heights above ungulate browsing height) in 2006, and by 2010, the majority of the 97 surveyed stands showed releasing aspen.

The timeline of aspen recruitment and presence of wolves within Yellowstone suggests a cessation of aspen recruitment during the period wolves were absent, with a resumption of recruitment following wolf reintroduction, indicating a tri-trophic cascade interaction (Ripple et al., 2001). However, the specific mechanisms underlying the cascade cannot be identified from such a simple correlational timeline, and require further analysis. Two possible mechanisms involving wolf agency exist: (1) the direct effect of wolf regulation on elk abundance and an associated decline in the browsing of aspen, that is, the cessation in recruitment was caused by over-browsing by over-abundant elk (Kay, 1990); or (2) that wolves produce indirect effects on elk, which alter elk foraging behaviour and habitat selection.

Within Yellowstone, elk are a preferred prey of wolves, and have annually comprised 80 per cent of all observed kills since reintroduction (Phillips and Smith, 1997). Therefore, a negative association between prey abundance and aspen browsing following the reintroduction of wolves might be expected, due to the direct, lethal effects of wolf on elk. Indeed, the total elk population has been estimated at 18 per cent less than the average before wolf reintroduction (Farnes et al., 1999). In particular, elk numbers within the Northern Yellowstone Elk Management Unit have shown a significant decline since 1999, from 14,500 to 8300, the lowest count since 1971–1972. This population reduction is believed to be due to changes driven by wolf reintroduction, given that elk populations elsewhere in Montana have grown beyond management levels, despite high levels of harvest (Creel et al., 2005). Such facts may suggest the trophic cascade within Yellowstone has been mediated primarily by changes in elk population numbers following wolf reintroduction. However, between 1930 and 1968, during wolf absence, Yellowstone's elk population was artificially maintained (by hunting) at 3000–6000 individuals (Romme et al., 1995), a time during which no significant effects on aspen recruitment were detected (Ripple and Larsen, 2000). Hence, this direct trophic chain mechanism seems unlikely.

The second possible mechanism for the influence of wolves may also affect the primary productivity of the ecosystem by non-lethal effects on prey, which may ultimately have stronger influences on food webs than lethal effects (Beckerman et al., 1997). The non-lethal effects of wolves may influence elk via predation risk effects, for example, increased vigilance, and changes in foraging behaviour and habitat selection. When such a trait-mediated indirect interaction (Abrams, 1996) is strong enough to bring about changes in the structure of the ecosystem, it is referred to as a behaviourally mediated trophic cascade (Abrams, 1984) or a non-consumptive effect (Middleton et al., 2013a). It has been hypothesised that changes in the habitat selection and foraging behaviour of elk following wolf extirpation may have resulted in the cessation of aspen overstorey recruitment from the 1920s to 1999, coinciding with the time wolves were mainly absent from Yellowstone (Ripple and Larsen, 2000). Similar results have been found for European systems (Kuijper et al., 2013).

The potential behaviourally mediated trophic cascade between wolf, elk, and aspen has been the focus of many studies that have investigated the influence of wolves on elk movements and foraging behaviour, as well as those comparing the status of woody browse between areas representing differing levels of wolf use or predation risk. The results of some such studies suggest a link between browsing pressure and elk distribution. For example, using wolf VHF collar telemetry from three packs to define low and high wolf-use areas, Ripple and colleagues (2001) compared the number of elk pellet groups (as an index of elk activity), aspen sucker heights, and the percentage of browsed suckers in high and low wolf-use areas. They found high wolf-use areas to have significantly lower elk pellet group counts (a proxy for habitat use) than low wolf-use areas for all habitat types investigated, suggesting foraging behaviours may have been altered by predation risk. Furthermore, within riparian/wet meadow habitats, mean aspen sucker heights were significantly higher in the high wolf-use areas than the low wolf-use areas. Similar results were found by Hebblewhite and colleagues (2005) in the Bow Valley of Canada's Banff National Park. Here, high human activity around Banff excluded the presence of wolves from the central part of the valley, while the adjacent areas had been naturally recolonised since 1986. The indirect effect of wolves on aspen sapling density and browsing intensity followed the trophic cascade hypothesis. Sapling density was twice as high in the high wolf-use area than within the low wolf-use area, while browsing intensity was higher in the low wolf-use area. Furthermore, elk density, survival, calf recruitment, and pellet density were all higher within the low wolf-use area. The study concluded the high human activity, and the exclusion of wolves, strongly mediated cascade effects, supporting the wolf-caused trophic cascade hypothesis.

As an alternative approach to defining high and low wolf-use areas, Ripple and Beschta (2003) compared the status of cottonwood (*Populus* spp.) and other woody plants between sites defined as high and low predation risk areas for elk within Yellowstone's Soda Butte Creek. Following the foraging theory of Lima and Dill (1990), Ripple and Beschta (2003) argued that prey must balance the demands of food intake and safety. Accordingly, in low predation risk areas, elk should browse more heavily, and reduce the height or regeneration success of plants. Conversely, when foraging under high-risk conditions, elk must either move into safer habitats and/or increase vigilance levels. Therefore, the effects of browsing in high predation risk sites will be lower. High-risk areas were defined as those with low visibility, on the assumption that this would influence the ability to detect approaching wolves (a coursing predator), and/or the presence of escape barriers, such as high-cut banks, stream terraces and gullies, while low-risk predation sites were categorised as open habitats. As predicted, results showed high-risk sites to have lower percentages of browsed stems and taller woody plants than low-risk predation sites. Young cottonwoods were tallest within localised sites where predation risk was deemed relatively high, thus representing effective refuges from the effects of browsing. Ripple and Beschta (2003) argued that a simple decrease in browsing pressuring following the reduction of elk population size would not have resulted in the patchy plant recovery seen in the study area, and that the results were indicative of a behaviourally mediated trophic cascade.

While these studies suggest support for trophic cascade effects between wolf, elk, and aspen, they do not provide information regarding the specific underlying mechanisms of the trophic cascade. Fortin and colleagues (2005) hypothesised that due to the link between browsing pressure and elk distribution, the mechanisms should be related to factors controlling elk movements. In an attempt to shed light on such factors, Fortin et al. (2005) examined the movement of elk in the presence of high and low wolf use areas within Yellowstone. Data showed that as the risk of an encounter with wolves increased, elk decreased their preference for aspen stands. Furthermore, within high wolf-use areas, elk showed a strong selection for conifer forests, a finding consistent with other wolf–elk observation studies (Fortin et al., 2005). This shift away from aspen stands in areas of high wolf use was suggested as leading to localised reductions in aspen browsing, and that Yellowstone's trophic cascade is due to an underlying behavioural mechanism.

Creel and colleagues (2005) argued that although the conclusion of Fortin et al. (2005), and others, of wolves producing behavioural responses in elk causing cascading effects on the plant communities is plausible, it is weakened by the lack of data regarding spatial and temporal variation in predation risk, and the responses of elk to such variation. Creel et al. (2005) found elk responded to the presence of wolves within a day or less, and on a spatial scale of approximately 1 km or less. In agreement with Fortin et al. (2005), they found when wolves were present, elk reduced their preferred use of open grassland habitats and moved into the protective cover of wooded areas. However, Creel et al. (2005) concluded that although elk anti-predator behaviour could promote a trophic cascade, the hypothesis of declining elk numbers being responsible for the changes in aspen stands could not be ruled out, given that both elk behaviour and numbers have changed since the reintroduction of wolves into the park.

The evidence presented so far may have persuaded the reader of the existence of a behaviourally mediated trophic cascade between wolf, elk, and aspen within Yellowstone. However, the picture is not so clear-cut. A large body of research showing conflicting results to those already reviewed must also be considered. Middleton and colleagues (2013a) argued that the active hunting mode of wolves does not equate to a prediction causing anti-predator behaviours strong enough to impact prey demography. The hunting mode of top predators has been identified as a factor influencing the strength and occurrence of non-consumptive effects (Creel, 2011). Ambush predators are believed to produce localised point-source cues, which can be used predictably by prey to associate with certain habitat features. In contrast, active hunters, which are generally more wide-ranging, produce more diffuse cues, meaning prey are left with little information with which to justify costly anti-predator behaviour (Luttbeg and Schmitz, 2000). Using simultaneous GPS data from wolves and elk, Middleton et al. (2013a) found that although wolves do induce anti-predator behaviour, such as increased movement, vigilance and displacement rates, these responses were not linked to costly reductions in feeding or changes in habitat use. Furthermore, encounters between elk and wolves only occurred on average once every nine days for migratory elk, and were inconsequential in the context of the annual cycle of fat losses and gains or pregnancy. Middleton et al. (2013a) attribute the results to the hunting mode of wolves, and concluded it was unlikely wolves were creating a behaviourally mediated trophic cascade within the Yellowstone ecosystem.

Kauffman and colleagues (2010) further questioned the results of previous studies, concluding that evidence for a behaviourally mediated trophic cascade within Yellowstone was unconfirmed due to the reliance on predation risk gradients based on untested assumptions regarding elk and wolf behaviour. As an alternative approach to high and low wolf-use areas (or high and low predation-risk areas) based on habitat features, they created a landscape-level risk map based on the spatial distribution of 774 elk killed by wolves during the first 10 years of wolf recovery (Kauffman et al., 2010), arguing that kill distribution was a better indicator of predation risk in other elk-wolf systems (e.g. Creel and Winnie, 2005). The use of actual predation data showed that, as the number of wolf packs within the northern range increased, predation events became uncoupled from wolf territories. Using this map, they found the current variation in aspen stands to be directly linked to stand productivity, rather than predation-risk gradients, and that the average proportion of browsed roots was actually positively associated with predation risk. Evidence from elk exclosures, designed to experimentally protect aspen from elk browsing, across a range of predation risks, showed that differences in growth or survival between exclosures and control plots were undiminished in stands with greater predation risks. Furthermore, exclosures across the range of predation risk improved annual average survival and growth over three years, and aspen exposed to elk browsing failed to increase in height, suggesting a continued suppression by elk. Such evidence implies wolves confer no protective advantages to aspen. Kauffman et al. (2010) attribute their contrasting results to creating a predation risk map from actual predation events, whereas Ripple et al. (2001) had defined high predation-risk areas as the territory core and periphery areas.

Kauffman et al. (2010) also questioned the claims of the cessation of aspen recruitment from 1920 onwards, as stated by Ripple and Larsen (2000). Using tree rings, they did not observe abrupt changes in recruitment, and found that the historical recruitment failure showed a wide variation amongst sampled stand, with the last year of recruitment ranging from 1892 to 1956. Again, the contrasting results were attributed to methodological differences; here, increment scores per tree were taken from a random selection of 9–14 trees per stand, whereas Ripple and Larsen (2000) extracted increment scores from the single largest tree within each stand. Kaufmann et al. (2010) believed the methods used by Ripple and Larsen (2000) were more descriptive of stand origin, i.e. the age of the oldest tree, than about recruitment.

The evidence for a behaviourally mediated trophic cascade following the reintroduction of wolves into Yellowstone is therefore conflicting, with no clear agreement within the literature having been reached. However, the question remains, could rewilding Yellowstone with wolves have brought about changes in the ecosystem to restore its original condition, and thus provide support for the concept of rewilding with carnivores as a valid conservation management tool? Marshall and colleagues (2013) argued that the relative effects of removal and restoration of a top predator are unlikely to be symmetrical, if the removal of predators creates feedback loops that reinforce the influence of removal, and resist the effects of the subsequent restoration. The relationship between willow *Salix* spp., wolf, elk, and beaver *Castor canadensis*, within the riparian habitats of Yellowstone's northern range has been examined with regard to restoring the system to a pre-wolf removal status.

Within this system, willow provide building material and food for beaver; however, tall willow is needed for dam construction – the presence of which could be limited by elk browsing. Beaver dams raise local water tables and allow the deposition of fine-grained sediment, creating suitable conditions for willow. Once beaver dams drain, moist mineral sediments are exposed, providing an ideal habitat for willow establishment (Read, 1958). Historically, the northern range was dominated by riparian zones with an abundance of willow; however, today there is a significant reduction in willow abundance and stature (Kay, 1990).

Marshall and colleagues (2013) tested the hypothesis that moderating browsing alone allows willow to recover to their threshold height of 2 m needed for recovery against the hypothesis that recovery is dependent upon a combination of moderating browsing and restoring a disturbance by beaver. Using a 10-year experiment, with elk exclosures and artificial beaver dams, willow growth and biomass accumulation was measured in response to two levels of herbivory and two levels of water table depth at four sites originally dammed by beaver. The results showed that the decade of protection from browsing alone was not sufficient to allow willow to grow past the 2 m threshold; however, protection from browsing in combination with raising the water table with artificial beaver dams did. They concluded the top-down effects of wolf restoration alone were insufficient to restore Yellowstone's riparian habitats, and that the current landscape state may be resilient to the trophic cascade effects of wolves alone due to the absence of tall willow needed for the return of beaver and the absence of beaver needed for willow establishment. Such results demonstrate that some ecosystems may be resilient to predator restoration, and thus it is fundamentally important to prevent the loss of apex predators from food webs in the first place. Yet further studies have revealed that changes to freshwater fish species composition following the introduction of a non-native trout may have influenced grizzly bear diet such that they have increased their predation on elk calves (Middleton et al., 2013b), providing yet another potential mechanism by which eventual changes to elk populations can be explained by factors other than the agency of wolves.

The conflicting evidence from Yellowstone suggests some potential for a behaviourally mediated trophic cascade following the restoration of wolves, but the methods used to study such a system influence the conclusions reached. However, the success of reintroducing apex predators to achieve rewilding is likely to be hampered if ecosystem changes during their absence result in a new stable state that resists returning to 'original' conditions, with or without apex predators. Human influence on ecosystems rarely consists of just the removal of the apex predators. Typically humans restructure the entire ecosystem, from the soil to the vegetation and the fauna.

What happened to predator–prey theory in the Yellowstone wolf debate?

A very large part of the theory of predator–prey interactions has been driven by research on wolf–ungulate systems. Much empirical and theoretical research has been stimulated by decades-long controversies concerning predator management in Alaska and Canada (e.g. Messier, 1991, 1994, 1995a, 1995b; Orians et al., 1997) and opportunistic responses to natural experiments, like on Isle Royale (Vucetich et al., 2002). Wolf–prey systems are among the best-studied ecological interactions on Earth (Eberhardt et al., 2003; Mech and Peterson, 2003). However, this large body of knowledge has simply underlined just how complex multi-predator/multi-prey systems can be, and how they are influenced by a wide range of environmental factors. This makes it very difficult to predict the impact of any predator on its prey in the context of rewilding. It is surprising how little this body of theory and data has been used to inform the Yellowstone trophic cascades debate (Allen et al., 2017). At the very least, it should have been used to temper conclusions about the universality of Yellowstone-specific findings.

As outsiders looking in to the trophic cascade debate, we see a tension has developed between proponents of the Yellowstone trophic cascade and people with evidence that does not support this. We recognise that robust methods of replication, controls and testing multiple hypotheses are fundamental to the scientific method, but that this is not always possible in large-scale, natural, field-based studies. That said, our view is that the most robust science suggests trophic cascades are not evident in Yellowstone. Despite this, there seems to be a popular view among the broader public that trophic cascades are an indisputed fact. This belief will delay our ability to improve understanding of food webs and trophic interactions as the critical mass of the populace will need to have their views overturned despite believing apex predators will solve multiple problems.

Conclusions

Whether rewilding can restore top-down control of ecosystems remains equivocal. As we have shown, there are various studies promoting the concept that are countered by other, more robust studies rejecting it. Only weak inferences can be drawn from weak scientific methods – illustrated by the conjecture surrounding the trophic cascades reported in Yellowstone and Australia. The field has been blighted by methodological weaknesses, selective use of data to support arguments, and too much over-generalisation to date (Winnie and Creel, 2017), and until robust methods are employed, it will remain mired in debate. There has also been a lamentable lack of discussion of predator–prey theory and the importance of local context in the rewilding discussions. Too many studies have relied on relative abundance indices that have not been validated against actual abundance while ignoring the well-known effect of differential detectability on assessing interspecific interactions (Hayward et al., 2015). Proposed and actual experimental interventions offer further hope (Moseby et al., 2012; Newsome et al., 2015), although there are many logistical challenges when working on the necessary scales.

Trophic cascades are more commonly reported from the more simplistic trophic systems of the Americas and Australia that lost large predators in the late Pleistocene extinctions (Van Valkenburgh et al., 2016). This reiterates the long-held view that trophic cascades and top-down regulation are an artefact of simplifying complex food webs into 'idealised' trophic levels while ignoring the complexity of full food webs (Polis and Strong, 1996). Limited evidence in more complex systems in Africa, where a diverse guild of predators persists, suggests trophic cascades are not as common (Ford et al., 2015). Hence, intact ecosystems might be more resilient to the direct fluctuations arising from

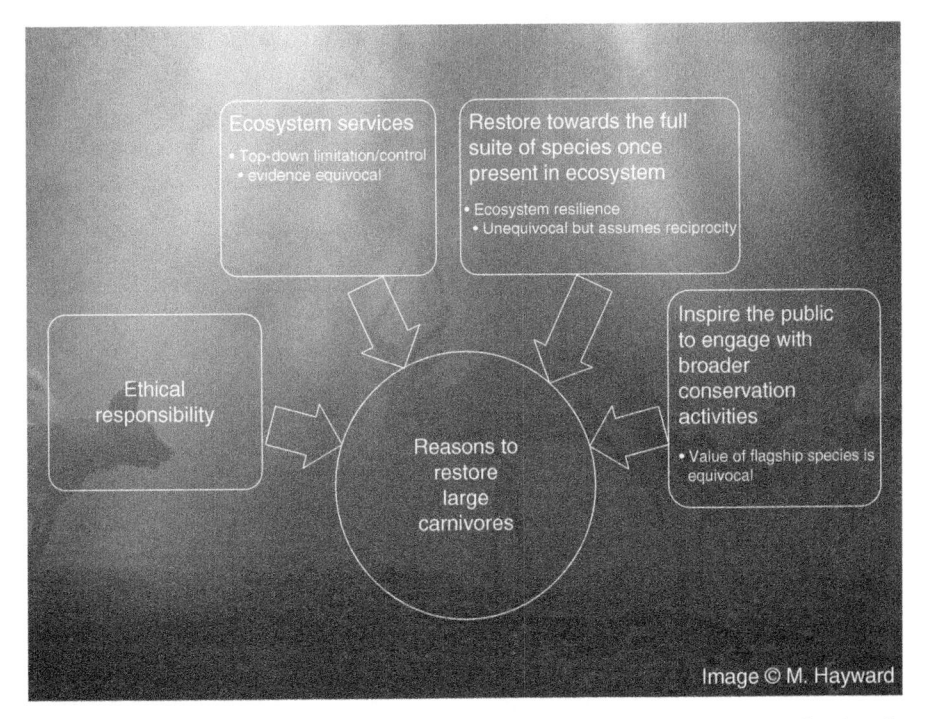

Figure 16.3. The rationale for rewilding with large carnivores may not need to invoke the ecosystem services carnivores can perform, but rather rely on other justifications. In essence, this belies the neoliberal 'new conservation' arguments (where biodiversity must be shown to benefit humanity) versus the traditional, ecocentric view of conservation (where proponents argue we should conserve biodiversity for its own sake) (Kopnina et al., 2018).

single predator–single prey interactions or the discovery of trophic cascades is less likely to be evident when the full complexity of food webs is considered. There is also a lack of clarity on what else has changed within the ecosystems that are proposed for rewilding. Are there irreversible changes (e.g. development, habitat alteration, climate change, and introduced species) that prevent or limit the return of some lost functionality (i.e. the assumption of reciprocity; Marshall et al., 2013; Ng'weno et al., 2017)? Could these changes cause rewilding to have new and perhaps negative impacts? Clearly, there are still questions to be answered before we can be completely confident that rewilding is a benign activity.

Nonetheless, 'Gloom has limited capacity to motivate, whereas hope is the elixir of action' (Morton, 2017, p. 122). Rewilding is a conservation activity that may drive action in repairing ecosystems and engaging the broader public and politicians in protecting our natural heritage. We are unaware of any conservation decisions that have been enacted primarily because of the ecological importance of large carnivores, and vice versa, and recognising that large carnivores may, at times, not produce big ecological effects in no way diminishes their conservation value. Whether rewilding leads to top-down control of ecosystems may not be as important as aiming for the full suite of biodiversity that could occur at specific locations (Figure 16.3). It is just that we cannot yet confidently invoke top-down control as a means towards that end.

References

Abrams, P.A. (1984). Foraging time optimization and interactions in food webs. *The American Naturalist*, **124**, 80–96.

Abrams, P.A. (1994). The fallacies of ratio-dependent predation. *Ecology*, **75**, 1842–1850.

Abrams, P.A. (1996). Dynamics and interactions in food webs with adaptive foragers. In Polis, G.A. and Winemiller, K.O. (Eds.), *Food webs* (pp. 113–121). Boston, MA: Springer.

Abrams, P.A. (2000). The evolution of predator–prey interactions: theory and evidence. *Annual Review of Ecology and Systematics*, **31**, 79–105.

Abrams, P.A., and Ginzburg, L.R. (2000). The nature of predation: prey dependent, ratio dependent or neither? *Trends in Ecology & Evolution*, **15**, 337–341.

Akcakaya, H.R., Arditi, R., and Ginzburg, L.R. (1995). Ratio-dependent predation – an abstraction that works. *Ecology*, **76**, 995–1004.

Allen, B.L., Engeman, R.M., and Allen, L.R. (2011). Wild dogma: an examination of recent 'evidence' for dingo regulation of invasive mesopredator release in Australia. *Current Zoology*, **57**, 568–583.

Allen, B.L., Fleming, P.J.S., Allen, L.R., Engeman, R.M., Ballard, G., and Leung, L.K.P. (2013). As clear as mud: a critical review of evidence for the ecological roles of Australian dingoes. *Biological Conservation*, **159**, 158–174.

Allen, B.L., Allen, L.R., Andrén, H., et al. (2017). Can we save large carnivores without losing large carnivore science? *Food Webs*, **12**, 64–75.

Arditi, R., and Ginzburg, L.R. (1989). Coupling in predator prey dynamics – ratio dependence. *Journal of Theoretical Biology*, **139**, 311–326.

Balestrieri, A., Remonti, L., and Prigioni, C. (2011). Assessing carnivore diet by faecal samples and stomach contents: a case study with Alpine red foxes. *Central European Journal of Biology*, **6**, 283-292.

Barnier, F., Valeix, M., Duncan, P., et al. (2014). Diet quality in a wild grazer declines under the threat of an ambush predator. *Proceedings of the Royal Society of London B: Biological Sciences*, **281**, 20140446.

Beckerman, A.P., Uriarte, M., and Schmitz, O.J. (1997). Experimental evidence for a behavior-mediated trophic cascade in a terrestrial food chain. *Proceedings of the National Academy of Sciences of the United States of America*, **94**, 10735-10738.

Beschta, R.L., and Ripple, W.J. (2009). Large predators and trophic cascades in terrestrial ecosystems of the western United States. *Biological Conservation*, **142**, 2401-2414.

Birch, L.C. (1957). The meanings of competition. *The American Naturalist*, **91**, 5-18.

Blueweiss, L., Fox, H., Kudzma, H., Nakashima, D., Peters, R., and Sams, S. (1978). Relationship between body size and some life-history parameters. *Oecologia*, **37**, 257-272.

Broekhuis, F., Cozzi, G., Valeix, M., McNutt, J.W., and Macdonald, D.W. (2013). Risk avoidance in sympatric large carnivores: reactive or predictive? *Journal of Animal Ecology*, **82**, 1098-1105.

Brook, L.A., Johnson, C.N., and Ritchie, E.G. (2012). Effects of predator control on behaviour of an apex predator and indirect consequences for mesopredator suppression. *Journal of Applied Ecology*, **49**, 1278-1286.

Brown, J.H. (1984). On the relationship between abundance and distribution of species. *The American Naturalist*, **124**, 255-279.

Camerano, L. (1880). Dell'equilibrio dei viventi merce la reciproca distruzione. *Accademia delle Scienze di Torino*, **15**, 393-414 (translated in the cited source by C.M. Jacobi and J.E. Cohen, 1994, into: On the equilibrium of living beings by means of reciprocal destruction). *Frontiers of theoretical biology* (ed. S.A. Levin), pp. 360-380. Dordrecht: Springer-Verlag.

Chan, K., Boutin, S., Hossie, T.J., Krebs, C.J., O'Donoghue, M., and Murray, D.L. (2017). Improving the assessment of predator functional responses by considering alternate prey and predator interactions. *Ecology*, **98**, 1787-1796.

Creel, S. (2011). Toward a predictive theory of risk effects: hypotheses for prey attributes and compensatory mortality. *Ecology*, **92**, 2190-2195.

Creel, S., and Christianson, D. (2008). Relationships between direct predation and risk effects. *Trends in Ecology & Evolution*, **23**, 194-201.

Creel, S., and Winnie, J.A. (2005). Responses of elk herd size to fine-scale spatial and temporal variation in the risk of predation by wolves. *Animal Behaviour*, **69**, 1181-1189.

Creel, S., Winnie, J.J., Maxwell, B., Hamlin, K., and Creel, M. (2005). Elk alter habitat selection as an antipredator response to wolves. *Ecology*, **86**, 3387-3397.

Creel, S., Christianson, D., Liley, S., and Winnie, J.A. (2007). Predation risk affects reproductive physiology and demography of elk. *Science*, **315**, 960-960.

Csada, R.D., James, P.C., and Richard, H.M.E. (1996). The 'file drawer problem' of non-significant results: does it apply to biological research? *Oikos*, **76**, 591-593.

Davies-Mostert, H.T., Mills, M.G.L., and Macdonald, D.W. (2013). Hard boundaries influence African wild dogs' diet and prey selection. *Journal of Applied Ecology*, **50**, 1358-1366.

DeByle, N.V. (1985). Wildlife. In DeByle, N.V., and Winokur, B.P. (Eds.), *Aspen: ecology and management in the western United States*. Fort Collins, CO: US Forest Service General Technical Report.

Dickie, M., Serrouya, R., McNay, R.S., and Boutin, S. (2017). Faster and farther: wolf movement on linear features and implications for hunting behaviour. *Journal of Applied Ecology*, **54**, 253-263.

Durant, S.M. (1998). Competition refuges and coexistence: an example from Serengeti carnivores. *Journal of Animal Ecology*, **67**, 370-386.

Eberhardt, L.L., Garrott, R.A., Smith, D.W., White, P.J., and Peterson, R.O. (2003). Assessing the impact of wolves on ungulate prey. *Ecological Applications*, **13**, 776-783.

Ehrlich, P.R., and Birch, L.C. (1967). The 'balance of nature' and 'population control'. *The American Naturalist*, **101**, 97-107.

Elmhagen, B., and Rushton, S.P. (2007). Trophic control of mesopredators in terrestrial ecosystems: top-down or bottom-up? *Ecology Letters*, **10**, 197-206.

Fancourt, B.A. (2015). Making a killing: photographic evidence of predation of a Tasmanian pademelon (*Thylogale billardierii*) by a feral cat (*Felis catus*). *Australian Mammalogy*, **37**, 120-124.

Fancourt, B.A. (2016). Avoiding the subject: the implications of avoidance behaviour for detecting predators. *Behavioral Ecology and Sociobiology*, **70**, 1535-1546.

Fancourt, B.A., Hawkins, C.E., Cameron, E.Z., Jones, M.E., and Nicol, S.C. (2015). Devil declines and catastrophic cascades: is mesopredator release of feral cats inhibiting recovery of the eastern quoll? *PLoS ONE*, **10**, e0119303.

Farnes, P., Heydon, C., and Hansen, K. (1999). Snowpack distribution across Yellowstone National Park. *Final Report CA*, **1268**, 1-9017.

Fleming, P.J.S., Allen, B.L., and Ballard, G.-A. (2012). Seven considerations about dingoes as biodiversity engineers: the socioecological niches of dogs in Australia. *Australian Mammalogy*, **34**, 119-131.

Ford, A.T., and Goheen, J.R. (2015). Trophic cascades by large carnivores: a case for strong inference and mechanism. *Trends in Ecology & Evolution*, **30**, 725-735.

Ford, A.T., Goheen, J.R., Augustine, D.J., et al. (2015). Recovery of African wild dogs suppresses prey but does not trigger a trophic cascade. *Ecology*, **96**, 2705-2714.

Fortin, D., Beyer, H.L., Boyce, M.S., Smith, D.W., Duchesne, T., and Mao, J.S. (2005). Wolves influence elk movements: behavior shapes a trophic cascade in Yellowstone National Park. *Ecology*, **86**, 1320-1330.

Fryxell, J.M., Mosser, A., Sinclair, A.R.E., and Packer, C. (2007). Group formation stabilizes predator-prey dynamics. *Nature*, **449**, U1041-U1044.

Fuller, T.K., and Sievert, P.R. (2001). Carnivore demography and the consequences of changes in prey availability. In Gittleman, J.L., Funk, S.M., Macdonald, D.W., and Wayne, R.K. (Eds.), *Carnivore conservation* (pp. 163-178). Cambridge: Cambridge University Press and the Zoological Society of London.

Gaillard, J.M., and Yoccoz, N.G. (2003). Temporal variation in survival of mammals: a case of environmental canalization? *Ecology*, **84**, 3294-3306.

Gaillard, J.M., Festa-Bianchet, M., and Yoccoz, N.G. (1998). Population dynamics of large herbivores: variable recruitment with constant adult survival. *Trends in Ecology & Evolution*, **13**, 58-63.

Gaillard, J.M., Festa-Bianchet, M., Yoccoz, N.G., Loison, A., and Toigo, C. (2000). Temporal variation in fitness components and population dynamics of large herbivores. *Annual Review of Ecology and Systematics*, **31**, 367-393.

Gaillard, J.M., Yoccoz, N.G., Lebreton, J.D., et al. (2005). Generation time: a reliable metric to measure life-history variation among mammalian populations. *The American Naturalist*, **166**, 119-123.

Gallagher, A.J., Creel, S., Wilson, R.P., and Cooke, S.J. (2017). Energy landscapes and the landscape of fear. *Trends in Ecology & Evolution*, **32**, 88-96.

Gause, G.F., Smaragdova, N.P., and Witt, A.A. (1936). Further studies of interaction between predators and prey. *Journal of Animal Ecology*, **5**, 1-18.

Georgiadis, N.J., Ihwagi, F., Olwero, J.G.N., and Romanach, S.S. (2007). Savanna herbivore dynamics in a livestock-dominated landscape. II. Ecological, conservation, and management implications of predator restoration. *Biological Conservation*, **137**, 473–483.

Gervasi, V., Nilsen, E.B., Sand, H., et al. (2012). Predicting the potential demographic impact of predators on their prey: a comparative analysis of two carnivore-ungulate systems in Scandinavia. *Journal of Animal Ecology*, **81**, 443–454.

Gervasi, V., Nilsen, E.B., and Linnell, J.D.C. (2015). Body mass relationships affect the age structure of predation across carnivore–ungulate systems: a review and synthesis. *Mammal Review*, **45**, 253–266.

Glen, A.S., and Dickman, C.R. (2008). Niche overlap between marsupial and eutherian carnivores: does competition threaten the endangered spotted-tailed quoll? *Journal of Applied Ecology*, **45**, 700–707.

Gorini, L., Linnell, J.D.C., May, R., et al. (2012). Habitat heterogeneity and mammalian predator–prey interactions. *Mammal Review*, **42**, 55–77.

Hairston, N.G., Smith, F.E., and Slobodkin, L.B. (1960). Community structure, population control, and competition. *The American Naturalist*, **94**, 421–425.

Haswell, P.M., Kusak, J., and Hayward, M.W. (2017). Large carnivore impacts are context-dependent. *Food Webs*, **12**, 3–13.

Hayward, M.W., and Kerley, G.I.H. (2005). Prey preferences of the lion (*Panthera leo*). *Journal of Zoology*, **267**, 309–322.

Hayward, M.W., and Marlow, N. (2014). Will dingoes really conserve wildlife and can our methods tell? *Journal of Applied Ecology*, **51**, 835–838.

Hayward, M.W., and Slotow, R. (2009). Temporal partitioning of activity in large African carnivores: tests of multiple hypotheses. *South African Journal of Wildlife Research*, **39**, 109–125.

Hayward, M.W., O'Brien, J., and Kerley, G.I.H. (2007). Carrying capacity of large African predators: predictions and tests. *Biological Conservation*, **139**, 219–229.

Hayward, M.W., Lyngdoh, S., and Habib, B. (2014). Diet and prey preferences of dholes (*Cuon alpinus*): dietary competition within Asia's apex predator guild. *Journal of Zoology*, **294**, 255–266.

Hayward, M.W., Boitani, L., Burrows, N.D., et al. (2015). Ecologists need to use robust survey design, sampling and analysis methods. *Journal of Applied Ecology*, **52**, 286–290.

Hayward, M.W., Kamler, J.F., Montgomery, R.A., et al. (2016). Prey preferences of the jaguar *Panthera onca* reflect the post-Pleistocene demise of large prey. *Frontiers in Ecology and Evolution*, **3**, e148. http://dx.doi.org/110.3389/fevo.2015.00148

Hayward, M.W., Porter, L., Lanszki, J., et al. (2017). Factors affecting the prey preferences of jackals (Canidae). *Mammalian Biology – Zeitschrift für Säugetierkunde*, **85**, 70–82.

Hebblewhite, M. (2013). Consequences of ratio-dependent predation by wolves for elk population dynamics. *Population Ecology*, **55**, 511–522.

Hebblewhite, M., White, C.A., Nietvelt, C.G., et al. (2005). Human activity mediates a trophic cascade caused by wolves. *Ecology*, **86**, 2135–2144.

Hersteinsson, P., Angerbjörn, A., Frafjord, K., and Kaikusalo, A. (1989). The arctic fox in Fennoscandia and Iceland: management problems. *Biological Conservation*, **49**, 67–81.

Holling, C.S. (1959). The components of predation as revealed by a study of small mammal predation of the European pine sawfly. *The Canadian Entomologist*, **91**, 293–320.

Hornsby, P. (1982). Predation of the Euro *Macropus robustus* (Marsupialia, Macropodidae) by the European fox *Vulpes vulpes* (Placentalia, Canidae). *Australian Mammalogy*, **5**, 225–227.

Houston, D. (1982). *The northern Yellowstone elk: ecology and management*. New York, NY: Macmillan Publishing.

Huffaker, C.B. (1958). Experimental studies on predation: dispersion factors and predator-prey oscillations. *Hilgardia*, **27**, 343-383.

Jędrzejewska, B., and Jędrzejewski, W. (1998). *Predation in vertebrate communities: the Białowieża primeval forest as a case study*. Berlin: Springer.

Jeschke, J.M., Kopp, M., and Tollrian, R. (2004). Consumer-food systems: why type I functional responses are exclusive to filter feeders. *Biological Reviews*, **79**, 337-349.

Jones, J.R., and DeByle, N.V. (1985). Fire. In DeByle, N.V. and Winokur, B.P. (Eds.), *Aspen: ecology and management in the western United States*. Fort Collins, CO: US Forest Service General Technical Report.

Judge, S., Lippert, J.S., Misajon, K., Hu, D., and Hess, S.C. (2012). Videographic evidence of endangered species depredation by feral cat. *Pacific Conservation Biology*, **18**, 293-296.

Karanth, K.U., Nichols, J.D., Kumar, N.S., Link, W.A., and Hines, J.E. (2004). Tigers and their prey: predicting carnivore densities from prey abundance. *Proceedings of the National Academy of Sciences of the United States of America*, **101**, 4854-4858.

Karanth, K.U., Gopalaswamy, A.M., Kumar, N.S., et al. (2011). Counting India's wild tigers reliably. *Science*, **332**, 791-791.

Kauffman, M.J., Brodie, J.F., and Jules, E.S. (2010). Are wolves saving Yellowstone's aspen? A landscape-level test of a behaviorally mediated trophic cascade. *Ecology*, **91**, 2742-2755.

Kay, C. (1990). Yellowstone's northern elk heard: a critical evaluation of the 'natural regulation' paradigm. Dissertation, Utah State University.

Knight, D.H. (1994). *Mountains and plains: the ecology of Wyoming landscapes*. New Haven, CT: Yale University Press.

Kopnina, H., Washington, H., Gray, J., and Taylor, B. (2018). The 'future of conservation' debate: defending ecocentrism and the Nature Needs Half movement. *Biological Conservation*, **217**, 140-148.

Kruuk, H. (1972). Surplus killing by carnivores. *Journal of Zoology*, **166**, 233-244.

Kuijper, D.P.J., de Kleine, C., Churski, M., van Hooft, P., Bubnicki, J., and Jędrzejewska, B. (2013). Landscape of fear in Europe: wolves affect spatial patterns of ungulate browsing in Białowieża Primeval Forest, Poland. *Ecography*, **36**, 1263-1275.

Lande, R., Engen, S., and Sæther, B.E. (2003). *Stochastic population dynamics in ecology and conservation*. Oxford: Oxford University Press.

Larsen, E.J., and Ripple, W.J. (2003). Aspen age structure in the northern Yellowstone ecosystem: USA. *Forest Ecology and Management*, **179**, 469-482.

Laundré, J.W., Hernández, L., and Altendorf, K.B. (2001). Wolves, elk, and bison: reestablishing the 'landscape of fear' in Yellowstone National Park, U.S.A. *Canadian Journal of Zoology*, **79**, 1401-1409.

Laundré, J.W., Hernández, L., Medina, P.L., et al. (2014). The landscape of fear: the missing link to understand top-down and bottom-up controls of prey abundance? *Ecology*, **95**, 1141-1152.

Letnic, M., and Ripple, W.J. (2017). Large-scale responses of herbivore prey to canid predators and primary productivity. *Global Ecology and Biogeography*, **26**, 860-866.

Lima, S.L. (1998). Nonlethal effects in the ecology of predator-prey interactions. *BioScience*, **48**, 25-34.

Lima, S.L., and Dill, L.M. (1990). Behavioral decisions made under the risk of predation: a review and prospectus. *Canadian Journal of Zoology*, **68**, 619-640.

Linnell, J.D.C., and Strand, O. (2000). Interference interactions, co-existence and conservation of mammalian carnivores. *Diversity and Distributions*, **6**, 169-176.

Linnell, J.D., Strand, O., and Landa, A. (1999). Use of dens by red *Vulpes vulpes* and arctic

Alopex lagopus foxes in alpine environments: can inter-specific competition explain the non-recovery of Norwegian arctic fox populations? *Wildlife Biology*, **5**, 167-176.

Lodge, D.M., Kershner, M.W., Aloi, J.E., and Covich, A.P. (1994). Effects of an omnivorous crayfish (*Orconectes rusticus*) on a freshwater littoral food web. *Ecology*, **75**, 1265-1281.

Lourenço, R., Penteriani, V., Rabaça, J.E., and Korpimäki, E. (2014). Lethal interactions among vertebrate top predators: a review of concepts, assumptions and terminology. *Biological Reviews*, **89**, 270-283.

Luttbeg, B., and Schmitz, O.J. (2000). Predator and prey models with flexible individual behavior and imperfect information. *The American Naturalist*, **155**, 669-683.

Marco, A., da Graça, J., García-Cerdá, R., Abella, E., and Freitas, R. (2015). Patterns and intensity of ghost crab predation on the nests of an important endangered loggerhead turtle population. *Journal of Experimental Marine Biology and Ecology*, **468**, 74-82.

Marshal, J.P., and Boutin, S. (1999). Power analysis of wolf-moose functional responses. *Journal of Wildlife Management*, **63**, 396-402.

Marshall, K.N., Hobbs, N.T., and Cooper, D.J. (2013). Stream hydrology limits recovery of riparian ecosystems after wolf reintroduction. *Proceedings of the Royal Society of London B: Biological Sciences*, **280**, 20122977.

Mattson, D.J., Blanchard, B.M., and Knight, R.R. (1991). Food habits of Yellowstone grizzly bears, 1977-1987. *Canadian Journal of Zoology*, **69**, 1619-1629.

May, R.M. (1976). Models for two interacting populations. In May, R.M. (Ed.), *Theoretical ecology: principles and applications* (pp. 49-70). Philadelphia, PA: W.B. Saunders.

McGregor, H.W., Legge, S., Jones, M.E., and Johnson, C.N. (2015). Feral cats are better killers in open habitats, revealed by animal-borne video. *PLoS ONE*, **10**, e0133915.

Mech, L.D., and Peterson, R.O. (2003). Wolf-prey relations. In Mech, L.D. and Boitani, L. (Eds.), *Wolves: behavior, ecology, and conservation* (pp. 131-160). Chicago, IL: University of Chicago Press.

Melis, C., Jedrzejewska, B., Apollonio, M., et al. (2009). Predation has a greater impact in less productive environments: variation in roe deer, *Capreolus capreolus*, population density across Europe. *Global Ecology and Biogeography*, **18**, 724-734.

Melis, C., Basille, M., Herfindal, I., et al. (2010). Roe deer population growth and lynx predation along a gradient of environmental productivity and climate in Norway. *EcoScience*, **17**, 166-174.

Melis, C., Nilsen, E.B., Panzacchi, M., Linnell, J.D.C., and Odden, J. (2013). Roe deer face competing risks between predators along a gradient in abundance. *Ecosphere*, **4**, art 111.

Menge, B.A., and Sutherland, J.P. (1976). Species diversity gradients: synthesis of the roles of predation, competition, and temporal heterogeneity. *The American Naturalist*, **110**, 351-369.

Messier, F. (1991). The significance of limiting and regulating factors on the demography of moose and white-tailed deer. *Journal of Animal Ecology*, **60**, 377-393.

Messier, F. (1994). Ungulate population models with predation: a case study with the North American moose. *Ecology*, **75**, 478-488.

Messier, F. (1995a). Is there evidence for a cumulative effect of snow on moose and deer populations. *Journal of Animal Ecology*, **64**, 136-140.

Messier, F. (1995b). Trophic interactions in two northern wolf-ungulate systems. *Wildlife Research*, **22**, 131-145.

Middleton, A.D., Kauffman, M.J., McWhirter, D.E., et al. (2013a). Linking anti-predator behaviour to prey demography reveals limited risk effects of an actively hunting large carnivore. *Ecology Letters*, **16**, 1023-1030.

Middleton, A.D., Morrison, T.A., Fortin, J.K., et al. (2013b). Grizzly bear predation links the loss of native trout to the demography of migratory elk in Yellowstone. *Proceedings of the Royal Society of London B: Biological Sciences*, **280**, 20130870.

Morgan, H.R., Hunter, J.T., Ballard, G., Reid, N.C.H., and Fleming, P.J.S. (2017). Trophic cascades and dingoes in Australia: does the Yellowstone wolf-elk-willow model apply? *Food Webs*, **12**, 76-87.

Morton, S.R. (2017). On pessimism in Australian ecology. *Austral Ecology*, **42**, 122-131.

Moseby, K.E., Neilly, H., Read, J.L., and Crisp, H. (2012). Interactions between a top order predator and exotic mesopredators in the Australian rangelands. *International Journal of Ecology*, **2012**, 250352.

Murdoch, W.W. (1966). Community structure, population control, and competition – a critique. *The American Naturalist*, **100**, 219-226.

National Research Council. (2002). *Ecological dynamics on Yellowstone's northern range*. Washington, DC: National Academies Press.

Newsome, T., Ballard, G.-A., Crowther, M., et al. (2015). Resolving the value of the dingo in ecological restoration: could a reintroduction experiment help? *Restoration Ecology*, **23**, 201-208.

Newsome, T.M., Greenville, A.C., Ćirović, D., et al. (2017). Top predators constrain mesopredator distributions. *Nature Communications*, **8**, 15469.

Ng'weno, C.C., Maiyo, N.J., Ali, A.H., Kibungei, A.K., and Goheen, J.R. (2017). Lions influence the decline and habitat shift of hartebeest in a semiarid savanna. *Journal of Mammalogy*, **98**, 1078-1087.

Nilsen, E.B., Gaillard, J.M., Andersen, R., et al. (2009a). A slow life in hell or a fast life in heaven: demographic analyses of contrasting roe deer populations. *Journal of Animal Ecology*, **78**, 585-594.

Nilsen, E.B., Linnell, J.D.C., Odden, J., and Andersen, R. (2009b). Climate, season, and social status modulate the functional response of an efficient stalking predator: the Eurasian lynx. *Journal of Animal Ecology*, **78**, 741-751.

O'Kane, C.A.J., and Macdonald, D.W. (2016). An experimental demonstration that predation influences antelope sex ratios and resource-associated mortality. *Basic and Applied Ecology*, **17**, 370-376.

Oksanen, L., and Oksanen, T. (2000). The logic and realism of the hypothesis of exploitation ecosystems. *The American Naturalist*, **155**, 703-723.

Orians, G., Cochran, P.A., Duffield, J.W., et al. (1997). Wolves, bears and their prey in Alaska. *Biological and Social Changes in Wildlife Mangaement* (pp. 99-114). New York, NY: FWS.

Owen-Smith, N. (2011). Accommodating environmental variation in population models: metaphysiological biomass loss accounting. *Journal of Animal Ecology*, **80**, 731-741.

Owen-Smith, N., and Mason, D.R. (2005). Comparative changes in adult vs. juvenile survival affecting population trends of African ungulates. *Journal of Animal Ecology*, **74**, 762-773.

Palomares, F., and Caro, T.M. (1999). Interspecific killing among mammalian carnivores. *The American Naturalist*, **153**, 492-508.

Pasanen-Mortensen, M., Elmhagen, B., Lindén, H., et al. (2017). The changing contribution of top-down and bottom-up limitation of mesopredators during 220 years of land use and climate change. *Journal of Animal Ecology*, **86**, 566-576.

Périquet, S., Valeix, M., Loveridge, A.J., Madzikanda, H., Macdonald, D.W., and Fritz, H. (2010). Individual vigilance of African herbivores while drinking: the role of immediate predation risk and context. *Animal Behaviour*, **79**, 665-671.

Phillips, M.K., and Smith, D.W. (1997). *Yellowstone wolf project: biennial report 1995* and 1996. Yellowstone National Park, WY: Yellowstone Center for Resources.

Polis, G.A., and Strong, D.R. (1996). Food web complexity and community dynamics. *The American Naturalist*, **147**, 813-846.

Polis, G.A., Myers, C.A., and Holt, R.D. (1989). The ecology and evolution of intraguild predation: potential competitors that eat each other. *Annual Review of Ecology and Systematics*, **20**, 297-330.

Pollock, K.H., Nichols, J.D., Simons, T.R., Farnsworth, G.L., Bailey, L.L., and Sauer, J.R. (2002). Large scale wildlife monitoring studies: statistical methods for design and analysis. *Environmetrics*, **13**, 105-119.

Pounds, J.A., Bustamante, M.R., Coloma, L.A., et al. (2006). Widespread amphibian extinctions from epidemic disease driven by global warming. *Nature*, **439**, 161-167.

Preisser, E.L., Bolnick, D.I., and Benard, M.F. (2005). Scared to death? The effects of intimidation and consumption in predator-prey interactions. *Ecology*, **86**, 501-509.

Radloff, F.G.T., and du Toit, J.T. (2004). Large predators and their prey in a southern African savanna: a predator's size determines its prey size range. *Journal of Animal Ecology*, **73**, 410-423.

Read, R. (1958). Silvical characteristics of Plains Cottonwood *Populus sargentii*. Station Paper. Rocky Mountain Forest and Range Experiment Station.

Ripple, W.J., and Beschta, R.L. (2003). Wolf reintroduction, predation risk, and cottonwood recovery in Yellowstone National Park. *Forest Ecology and Management*, **184**, 299-313.

Ripple, W.J., and Beschta, R.L. (2012). Trophic cascades in Yellowstone: the first 15 years after wolf reintroduction. *Biological Conservation*, **145**, 205-213.

Ripple, W.J., and Larsen, E.J. (2000). Historic aspen recruitment, elk, and wolves in northern Yellowstone National Park, USA. *Biological Conservation*, **95**, 361-370.

Ripple, W.J., Larsen, E.J., Renkin, R.A., and Smith, D.W. (2001). Trophic cascades among wolves, elk and aspen on Yellowstone National Park's northern range. *Biological Conservation*, **102**, 227-234.

Ritchie, E.G., and Johnson, C.N. (2009). Predator interactions, mesopredator release and biodiversity conservation. *Ecology Letters*, **12**, 982-998.

Romme, W.H., Turner, M.G., Wallace, L.L., and Walker, J.S. (1995). Aspen, elk, and fire in northern Yellowstone Park. *Ecology*, **76**, 2097-2106.

Schullery, P., and Whittlesey, L.H. (1992). The documentary record of wolves and related wildlife species in the Yellowstone National Park area prior to 1882. National Park Service, Division of Research.

Short, J., Kinnear, J.E., and Robley, A. (2002). Surplus killing by introduced predators in Australia – evidence for ineffective anti-predator adaptations in native prey species? *Biological Conservation*, **103**, 283-301.

Shurin, J.B., Borer, E.T., Seabloom, E.W., et al. (2002). A cross-ecosystem comparison of the strength of trophic cascades. *Ecology Letters*, **5**, 785-791.

Sih, A., Crowley, P., McPeek, M., Petranka, J., and Strohmeier, K. (1985). Predation, competition and prey communities: a review of field experiments. *Annual Review of Ecology and Systematics*, **16**, 269-311.

Sinclair, A.R.E., Mduma, S., and Brashares, J.S. (2003). Patterns of predation in a diverse predator-prey system. *Nature*, **425**, 288-290.

Skalski, G.T., and Gilliam, J.F. (2001). Functional responses with predator interference: viable alternatives to the Holling Type II model. *Ecology*, **82**, 3083-3092.

Slobodkin, L.B., Smith, F.E., and Hairston, N.G. (1967). Regulation in terrestrial ecosystems, and the implied balance of nature. *The American Naturalist*, **101**, 109-124.

Solomon, M. (1949). The natural control of animal populations. *Journal of Animal Ecology*, **18**, 1-35.

Soulé, M.E. (2010). Conservation relevance of ecological cascades. In Terborgh, J. and Estes, J.A. (Eds.), *Trophic cascades: predators, prey and the changing dynamics of nature* (pp. 337-353). Washington, DC: Island Press.

Stephens, P.A., Buskirk, S.W., Hayward, G.D., and Del Rio, C.M. (2005). Information theory and hypothesis testing: a call for pluralism. *Journal of Applied Ecology*, **42**, 4-12.

Terborgh, J., Estes, J., Paquet, P., et al. (1999). The role of top carnivores in regulating terrestrial ecosystems. In Soulé, M.E. and Terborgh, J. (Eds.), *Continental conservation: scientific foundations of regional reserve networks* (pp. 42-56). Washington, DC: Island Press.

Tomas, J., Aznar, F.J., and Raga, J.A. (2001). Feeding ecology of the loggerhead turtle *Caretta caretta* in the western Mediterranean. *Journal of Zoology*, **255**, 525-532.

van Dyk, G., and Slotow, R. (2003). The effects of fences and lions on the ecology of African wild dogs reintroduced to Pilanesberg National Park, South Africa. *African Zoology*, **38**, 79-94.

Van Valkenburgh, B., Hayward, M.W., Ripple, W.J., Meloro, C., and Roth, V.L. (2016). The impact of large terrestrial carnivores on Pleistocene ecosystems. *Proceedings of the National Academy of Sciences of the United States of America*, **113**, 862-867.

Vucetich, J.A., Peterson, R.O., and Schaefer, C.L. (2002). The effect of prey and predator densities on wolf predation. *Ecology*, **83**, 3003-3013.

Wilmers, C.C., Post, E., Peterson, R.O., and Vucetich, J.A. (2006). Predator disease out-break modulates top-down, bottom-up and climatic effects on herbivore population dynamics. *Ecology Letters*, **9**, 383-389.

Wilmers, C.C., Post, E., and Hastings, A. (2007). The anatomy of predator-prey dynamics in a changing climate. *Journal of Animal Ecology*, **76**, 1037-1044.

Winnie, J., and Creel, S. (2017). The many effects of carnivores on their prey and their implications for trophic cascades, and ecosystem structure and function. *Food Webs*, **12**, 88-94.

Woodward, F.I., Fogg, G.E., and Heber, U. (1990). The impact of low temperatures in controlling the geographical distribution of plants. *Philosophical Transactions of the Royal Society B: Biological Sciences*, **326**, 585-593.

CHAPTER SEVENTEEN

Rewilding and the risk of creating new, unwanted ecological interactions

MIGUEL DELIBES-MATEOS
Consejo Superior de Investigaciones Científicas
ISABEL C. BARRIO
University of Iceland
A. MÁRCIA BARBOSA
Universidade de Évora
ÍÑIGO MARTÍNEZ-SOLANO
Museo Nacional de Ciencias Naturales
JOHN E. FA
Manchester Metropolitan University
CATARINA C. FERREIRA
Helmholtz Centre for Environmental Research, Trent University

We are currently experiencing unprecedented environmental changes driven by anthropogenic activities with consequences that include soil erosion, nutrient enrichment, population and species extinctions, and species invasions (Corlett, 2016). These rapid changes generate uncertainties that may compromise the goals and priorities of conservation and management efforts (Wiens and Hobbs, 2015), including rewilding attempts. Some conservationists, including rewilding advocates, subscribe to the ideal that natural processes should be allowed to take their course without human intervention. Others believe that such an approach is too risky so it is more appropriate to actively manage nature (Corlett, 2016). However, rewilding outcomes may become more unpredictable because of uncertainties in future conditions (e.g. climate change, land conversion) and increased frequency of extreme events. In this chapter, we focus on how trophic and passive rewilding initiatives may intensify the risk of unwanted ecological effects. We do not address potential economic and societal implications of rewilding initiatives because these are covered in other chapters (see Chapters 8, 9, and 19). In addition, we show that biological communities can be understood only by considering their evolutionary history, and we warn that ignoring this point in rewilding projects could ultimately risk failure.

Trophic rewilding

Rewilding (Soulé and Noss, 1998; Chapter 5) is aimed at restoring and protecting natural processes in specific wild areas, providing connectivity between such areas, and protecting or reintroducing keystone species ('trophic rewilding'). Trophic rewilding aims at restoring top-down interactions and associated trophic cascades through the reintroduction of species lost to the environment, with the ultimate goal of promoting a self-regulating ecosystem (Svenning et al., 2016; Chapter 5). Unwanted effects of trophic rewilding can be broadly classified into three main categories: ecological, human, and economic (but see Nogués-Bravo et al., 2016 for a more detailed discussion of the far-reaching consequences of rewilding). Here, we will briefly review potential unwanted ecological effects caused by the reintroduction of both top predators and herbivores within rewilding initiatives.

Top predators

Rewilding initiatives are usually based on the reintroduction of large predators because their relationships with species at lower trophic levels maintain stability of their ecosystems (Corlett, 2016). This approach is especially useful when a species is known to have widespread effects over an area and changes its ecology, as was the case with grey wolves (*Canis lupus*) in Yellowstone National Park, a case study that has become known globally and acts as a flagship in favour of trophic rewilding using top predators. However, the unprecedented impact that wolves have had on the park's ecology and geography highlights the need to understand better the uncertainty surrounding rewilding initiatives and the importance to reflect upon potential undesirable outcomes thereof (Paine et al., 1998). In their review paper on 'ecological surprises', Doak and collaborators (2008) showcase some unintended consequences of trophic rewilding. For instance, the reintroduction of rock lobsters (*Jasus lalandii*) to a seamount off the western coast of South Africa provides one of the most astonishing examples of predator–prey role reversals (Barkai and McQuaid, 1988). For reasons that remain uncertain, lobsters disappeared from Marcus Island in the early 1970s. As a result of this, predatory whelk populations apparently increased substantially following the lobsters' disappearance because lobsters preyed on the whelks. To re-establish the species, 1000 lobsters were reintroduced but were immediately attacked and consumed by the now over-abundant whelks, their previous prey; a week later, no live lobsters could be found at Marcus Island (Barkai and McQuaid, 1988). Ecological surprises are inescapable given the panoply of ways species interact with one another (Berger et al., 2001; Laundré et al., 2001; Sterner and Elser, 2002; Hansen et al., 2007). Despite this, virtually none of these potential interactions are typically incorporated into broad community predictions in the trophic rewilding of predators (Doak et al., 2008).

Although evidence has been collected showing the negative consequences of large-bodied species defaunation, the reverse (i.e. the restoration of ecosystem functions after these species return) has been assessed less often (Fernández et al., 2017). Potential repercussions include changes in local diversity and ecosystem functioning (defined as the collective life activities of plants, animals, and microbes and the effects these activities have on the physical and chemical conditions of the environment), and the possibility of catastrophic disease transmission (e.g. Daszak et al., 2000). For example, large carnivores typically depress mesopredator abundance, thus potentially favouring their rodent prey and, under some conditions, potentially increasing the incidence of various zoonotic diseases (e.g. Ostfeld and Holt, 2004). Moreover, trophic rewilding experiments do not normally consider potential interactions with undiscovered species, although it is possible that some small, undiscovered prey (e.g. insects) might support many species in an ecosystem.

Even though the expectation with the reintroduction of predators is that they will trigger top-down cascading effects, under certain ecological conditions heterogeneity at any trophic level can affect levels above or below. For instance, in northern Utah, USA, Bridgeland and collaborators (2010) showed experimentally that an arthropod community structure on a foundation riparian tree mediated the ability of insectivorous birds (top predators) to influence tree growth. These authors found that abiotic growing conditions affected tree growth and herbivore populations, which in turn affected bird foraging patterns that cascaded back to the trees. When the main factor limiting tree growth switched from water availability to herbivory, the avian predators gained the potential to reduce herbivory. Such conditionality is consistent with numerous studies showing how fundamental relationships might switch over time, space, or with addition of another interacting community member (Bailey and Whitham, 2007). This dynamic complexity might preclude the predictability of ecosystem response to the addition or loss of top predators at a given place or time (Bridgeland et al., 2010; Mäntylä et al., 2011). This also poses challenges in terms of understanding potential triggers of species invasions in the context of trophic rewilding. Rewilding might present increased opportunities for non-native species to become established, outcompete native species, and reduce species diversity. The reintroduction of dingoes (*Canis dingo*) has been proposed to help restore degraded rangelands in Australia (Newsome et al., 2015). This proposal is based on results of studies suggesting that dingoes can suppress prey populations (especially medium- and large-sized herbivores) and invasive predators such as red foxes (*Vulpes vulpes*) and feral cats (*Felis catus*) that prey on threatened native species. However, dingoes are themselves mesopredators and there is a high risk of increased predation on threatened native predators (Allen and Fleming, 2012).

On the other hand, eliminating feral cats could release other mammalian predator invaders, such as rats (*Rattus* spp.), from predation pressure, with resulting cascading effects on the ecosystem.

Even very well-documented rewilding experiences, such as that of the wolf in Yellowstone, may not have been able to flag unforeseen outcomes with the same species in other systems. Exemplarily, in the Adirondack ecosystem in New York State, USA, coyotes (*Canis latrans*) are thought to be causing a trophic cascade by limiting populations of herbivorous small mammals in recently burned areas, and this in turn could benefit deer mice (*Peromyscus maniculatus*), while indirectly influencing vegetative composition (Ricketts, 2016). Predation by coyotes has been identified as the greatest cause of mortality for red and swift foxes (*V. velox*) in Kansas and Colorado (Sovada et al., 1998; Kitchen et al., 1999) where they tend to persist when coyote numbers are low. Therefore, coyotes might be filling the wolf's ecological niche today; this means that the reintroduction of wolves in this system could have unknown effects such as increasing populations of foxes, further affecting the trophic system in the Adirondack ecosystem (Ricketts, 2016).

Other unanticipated outcomes of trophic rewilding might be driven by predator–prey interactions, and there are many examples illustrating these. For example, in the Addo Elephant National Park (South Africa), ungulate prey species at risk of predation are more likely to be active diurnally when coexisting with nocturnally active predators, thereby reducing the activity overlap with these predators (Tambling et al., 2015). In the absence of predators, such as following their extirpation, the responses related to predator avoidance can be lost or diluted, which suggests that if predators are reintroduced, prey will likely lack the full spectrum of adaptive behaviours to predation, potentially resulting in dramatic effects for prey communities (Tambling et al., 2015).

It is widely known that the fear large carnivores inspire in mesocarnivores can have powerful cascading effects affecting ecosystem structure and function (Prugh et al., 2009; Ritchie and Johnson, 2009; Ripple et al., 2014; Suraci et al., 2016). However, Clinchy and collaborators (2016) have suggested that mesocarnivores are much more fearful of humans than of large carnivores. Indeed, the numerical suppression of mesocarnivores by humans far exceeds that by large carnivores (Darimont et al., 2015), which suggests that fear of humans could affect mesocarnivore demography and behaviour (Dorresteijn et al., 2015; Oriol-Cotterill et al., 2015; Smith et al., 2015), with implications for rewilding initiatives. For example, in human-dominated landscapes, such as in Europe, the recovery (Chapron et al., 2014) or reintroduction (Manning et al., 2009; Svenning et al., 2016) of large carnivores is unlikely to 'restore' fear to mesocarnivores 'released' from behavioural suppression (Prugh et al., 2009; Ritchie and Johnson, 2009), but will instead add to the elevated fear that mesocarnivores are evidently experiencing of humans (Clinchy et al., 2016).

Large herbivores

Large herbivores play key roles in ecosystems, either through direct impacts on vegetation and/or indirect effects on food web structure and ecosystem functioning. Therefore, the decline of large herbivores can lead to loss of ecological interactions and key ecosystem services (Ripple et al., 2015; Bakker et al., 2016). Modern exclosure experiments and palaeoecological records provide evidence of this (Bakker et al., 2016). The megafaunal extinction at the end of the Pleistocene can be viewed as a natural experiment that highlights the ecological roles played by large herbivores at a global scale (Ripple et al., 2015; Bakker et al., 2016). However, the ecological state shifts caused by herbivore depletion were not the same everywhere (Barnosky et al., 2016). The extent of ecological change after megafaunal loss largely depended on the removal of a number of different effective ecosystem engineers among the lost megafauna, and on soil properties or other abiotic constraints that influence vegetation changes (Barnosky et al., 2016). Thus, given the fact that a number of species and processes are involved, it is important to thoroughly understand the ecological role of each before making predictions on the cascade effects expected in an ecosystem (Barnosky et al., 2016).

Given the known impacts of the introduction and reintroduction of large herbivores on the functioning of an ecosystem, herbivores have been at the centre of many trophic rewilding initiatives. Restoring a diverse and abundant wild large herbivore guild is presumed to help maintain a mosaic of vegetation that will effectively promote landscapes of higher biodiversity (Sandom et al., 2014). A noteworthy example of rewilding with large herbivores is Pleistocene Park in Siberia (Zimov, 2005), where bison and other large herbivores were introduced to restore the grazing-dependent mammoth steppe vegetation. Palatable high-productivity grasses, herbs, and willow shrubs originally dominated these steppes and grazing by high densities of large herbivores is believed to suppress woody growth and accelerate nutrient cycling in these cold ecosystems (Zimov et al., 2012). Thus, as a result of the megafaunal collapse during the Holocene, the mammoth steppe was replaced by a water-logged landscape dominated by moss and shrub tundra (Zimov et al., 1995). Results from experimental enclosures in Pleistocene Park demonstrate that a shift occurs from shrub-dominated to grass-dominated vegetation when high densities of large herbivores are included (Zimov et al., 2012), showing that this process can be used to maintain and recreate lost ecosystems (Zimov et al., 1995). However, predators and a strong hunting pressure are needed to keep the overall number of herbivores relatively low, so that their impact on vegetation and soils is not excessive (Zimov, 2005).

Another rewilding initiative, Oostvaardersplassen in the Netherlands, is the oldest large-scale rewilding area in Europe. The area was designated for industry and agricultural use but converted to a nature reserve in the 1970s (Vera, 2009). To keep the area more open and prevent the area becoming a woodland, park managers introduced primitive cattle and horse breeds in the 1980s, as a replacement for their extinct wild ancestors. In Oostvaardersplassen herbivore populations are limited only by resource availability, as there is no human management, nor any effective wild predator control. Given the relatively high productivity of the area, herbivores attain high densities, which can have negative impacts on biodiversity and ecosystem function (Ims et al., 2007). For instance, the high densities of herbivores in Oostvaardersplassen limit seedling establishment and prevent the regeneration of wood-pastures (Smit et al., 2015; Figure 17.1). In these cases, the existence of grazing refuges, in the form of areas inaccessible to herbivores or as herbivore numbers

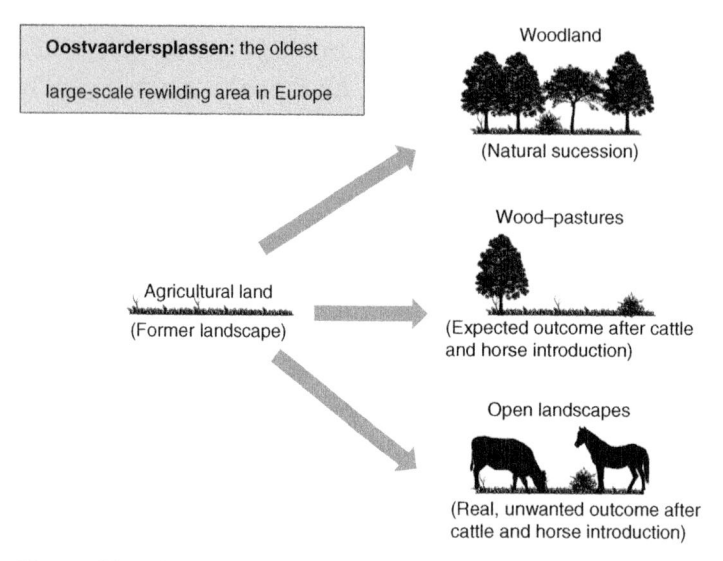

Figure 17.1. Oostvaardersplassen is the oldest large-scale rewilding area in Europe. Primitive breeds of cattle and horses were introduced as a replacement for their extinct wild ancestors to this area formerly designated for agricultural use, to prevent formation of closed woodlands. However, the lack of effective wild predator control or human management and the high productivity of the area allowed herbivores to reach high densities, which prevent seedling establishment and the regeneration of woodland-pastures. The existence of grazing refuges is essential to create windows of opportunity for woody species to establish, and rewilding initiatives aimed at restoring wood-pasture landscapes in productive areas need to manage herbivore densities to create grazing refuges that allow the regeneration of woody species.

temporarily decline, are essential to create windows of opportunity for woody species to establish themselves (Cornelissen et al., 2014). Thus, rewilding initiatives with large herbivores aimed at restoring wood–pasture landscapes in productive areas need to create grazing refuges that allow the regeneration of woody species (Smit et al., 2015).

As shown in the examples above, proposals to conserve grazed ecosystems often focus on introducing herbivores as surrogates of locally extinct herbivores that were deemed important for the maintenance of these ecosystems. However, where a species has gone globally extinct, the restoration of its ecological functions might be achieved only through ecological replacement, that is, the introduction of an exotic, functionally similar species (Seddon et al., 2014). An example of these ecological replacements is the introduction of non-native giant tortoises as replacements for extinct tortoise species in oceanic islands (Hansen et al., 2010). In the case of Aldabran giant (*Aldabrachelys gigantea*) and Madagascan radiated (*Astrochelys radiata*) tortoises, taxonomically and functionally similar to the extinct Mauritian giant tortoises (*Cylindraspis* spp.), their successful establishment improved dispersal and recruitment of endemic tree species in Round Island, Mauritius (Griffiths et al., 2011), and supressed invasive plants (Griffiths et al., 2013). Yet, in some cases plant communities are so severely degraded that the introduction of these ecological replacements alone is insufficient to restore the ecosystem (Griffiths et al., 2013), and large-scale habitat restoration might be additionally required (Gibbs et al., 2014). Taxonomic relatedness and functional equivalence to the native herbivore are important criteria when selecting potential ecological replacements, yet the difficulties in predicting their effects on recipient ecosystems is the main barrier for their widespread use in conservation. However, the introduction of livestock as a surrogate of extinct wild herbivores circumvents the problem of taxonomic relatedness because domestic breeds are derived from wild herbivore ancestors and are therefore taxonomically and, theoretically, functionally similar to wild herbivores. Hence, it has been proposed that grassland conservation could be achieved through grazing of domestic herbivores or native species such as bison (Towne et al., 2005). However, when livestock species are introduced into a co-evolved assemblage of native wild herbivores, they might compete with and even exclude native wild herbivores (Mishra et al., 2002; Madhusudan, 2004).

In any case, how communities respond to the introduction or reintroduction of large herbivores will be determined by the extent to which the recipient ecosystem has been modified. Environmental changes and human activities that have taken place since the extirpation of the herbivore might have produced new communities and novel ecological equilibria (Smith, 2005). Reintroductions of extirpated species have complex effects on plant communities, and can give rise to mixed management outcomes. For example, although the successful reintroduction of the recently extirpated Tule elk (*Cervus elaphus nannodes*) in California effectively reduced the abundance of a highly invasive exotic grass, at the same time the abundance and richness of other non-native taxa increased in the community (Johnson and Cushman, 2007). Similarly, the management of some introduced species is complicated if their impacts threaten native communities. For example, reindeer (*Rangifer tarandus*) introduced to South Georgia by Norwegian whalers in the early 1900s (Leader-Williams et al., 1989) have caused major changes to the vegetation, including favouring the expansion of various exotic plants (Leader-Williams et al., 1987). Part of the explanation for this result is that the South Georgia species-poor vascular flora is not adapted to grazing by vertebrates. However, by feeding on native tussock grassland, reindeer control the expansion of non-native brown rats that use tussock grassland as shelter (Leader-Williams et al., 1989).

Finally, it is important to keep in mind that an underlying assumption of trophic rewilding with large herbivores is that species that share a recent evolutionary history will interact in the same way today and in the future (Caro, 2007). This is less likely under rapid, ongoing environmental changes. For example, the effects of megafauna on vegetation during the Pleistocene may have been exacerbated by the lower CO_2 atmospheric concentrations, which may have further inhibited woody vegetation growth and made it more susceptible to browsing pressure (Malhi et al., 2016). In contrast, with increased levels of atmospheric CO_2, vegetation today may be more able to withstand browsing pressure.

Passive rewilding

The absence of sustained human intervention is central to passive rewilding (Chapter 6). In other words, passive rewilding is based on a 'leave it to nature' philosophy, although any justification for this approach is more philosophical than scientific (Schnitzler, 2014). However, what happens if, for example, large areas of former agricultural land are simply left alone? Over the past decades, land abandonment has occurred in developed countries in Europe and North America (Shengfa and Xiubin, 2017), and its effects on biodiversity have been widely studied. In general, impacts of land abandonment on ecosystem composition and functioning are heterogeneous and depend on a variety of factors (Plieninger et al., 2014). This means that both benefits and detrimental impacts of land abandonment on ecosystems have been documented (Queiroz et al., 2014; Lasanta et al., 2015). The highest proportion of studies reporting negative impacts of land abandonment are found in Europe and Asia (Queiroz et al., 2014). Detrimental impacts are particularly evident in semi-natural habitats that have been traditionally maintained by anthropogenic activities, such as grazing or mowing, and that harbour a remarkably rich biodiversity in terms of both animal and plant species (Carboni et al., 2015). Such ecosystems could be threatened by passive rewilding attempts if the risks of getting unwanted interactions because of land abandonment are ignored.

Detrimental effects of land abandonment on biodiversity have been documented at multiple levels (Figure 17.2). At the species level, these include the decline in species abundance and the modification of species distribution. Multiple studies on different taxa clearly illustrate this point. For instance, the abandonment of traditional activities such as extensive grazing or farming threatens plant species typical of semi-natural habitats. A paradigmatic case is set in Sweden and Norway, where the future distribution of the endemic *Primula scandinavica* is projected to decrease with continued relinquishment of grazing (Wehn and Johansen, 2015; Speed and Austrheim, 2017). The abandonment of traditional human activities has also had negative impacts on many animal species. For example, land abandonment and pine reforestation have led to landscape homogeneity in the Collserola Natural Park (north-east Spain) that might have caused the extinction of six open-habitat gastropod species in the area (Torre et al., 2014). Similarly, the abandonment of low-intensity grazing is associated with the decline in abundance of several ground spider species in Greece (Zakkak et al., 2014), and the threat

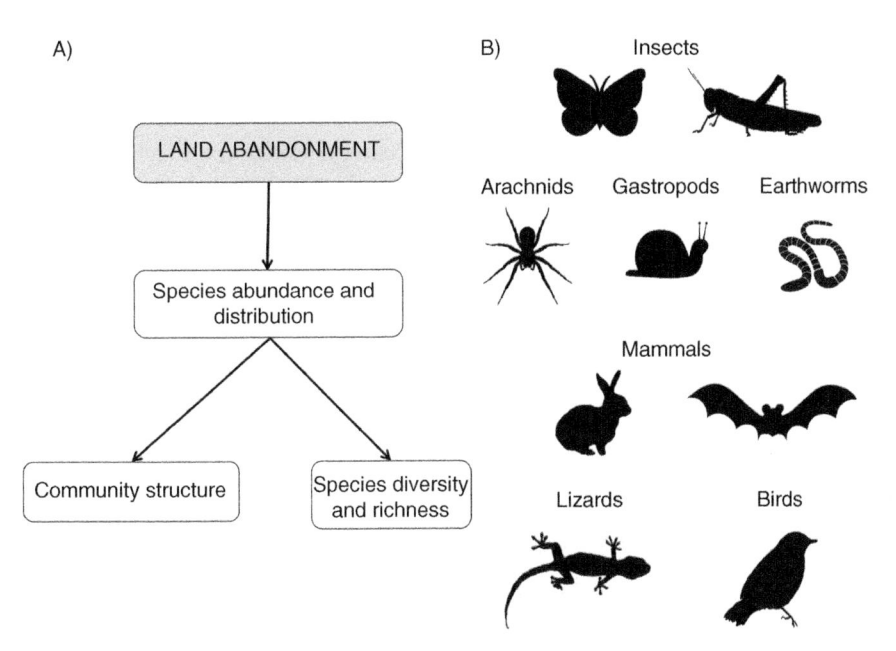

Figure 17.2. (A) Levels at which negative effects of land abandonment have been documented. (B) Examples of groups of animals that have been negatively impacted by land abandonment on some occasions.

to the conservation of endangered, endemic butterfly species in Spain (Munguira et al., 2017). Moreover, land abandonment has caused clear detrimental impacts to vertebrates that primarily use open habitats. For example, studies on avifauna mainly report negative abandonment-related impacts (Queiroz et al., 2014), revealing the decline of many farmland bird species across several European regions (e.g. Zakkak et al., 2015a; Mischenko and Sukhanova, 2016; Regos et al., 2016) and Asia (e.g. Katayama et al., 2015). Also, land abandonment leads to the reduction in abundance of several mammal species. For example, the loss of farmland landscape diversity as a consequence of agricultural intensification and crop abandonment is thought to be the prime factor responsible for the long-term decline of European hare (*Lepus europaeus*) populations across most of its range (Edwards et al., 2000). Similarly, in Greece, the abandonment of agricultural fields has contributed to the decline of lizard species that typically inhabit open agricultural landscapes or prefer open grassy habitats (Zakkak et al., 2015b).

A large body of knowledge suggests that the abandonment of traditional human activities, as expected in passive rewilding attempts, can result in the appearance of negative, unwanted ecological interactions (Figure 17.2). The encroachment of forests as a consequence of agricultural abandonment has resulted in a remarkable increase in ungulate numbers in Europe and North America (e.g. Acevedo et al., 2011), which has negatively affected other species of herbivores through competition. For example, the increasing number of wild boar (*Sus scrofa*) in Spain could have a negative effect on European rabbit (*Oryctolagus cuniculus*) populations (Cabezas-Díaz et al., 2011; Carpio et al., 2014), and as a consequence negatively affects the numerous Iberian rabbit predators (Lozano et al., 2007). Likewise, grazing abandonment has favoured the invasion of the tall grass *Brachypodium genuense* in the central Apennines (Italy), reducing by competitive exclusion the availability of palatable plants for the Apennine chamois (*Rupicapra pyrenaica ornata*), whose numbers have dramatically declined in the area (Corazza et al., 2016). In addition, allowing ecosystems to evolve away from human control, as proposed by passive rewilding advocates (Corlett, 2016), can compromise the constraining of harmful invasive species. In this sense, the lack of management in abandoned lands in Nepal facilitates the spread of invasive plant species, hindering the growth of native vegetation (Jaquet et al., 2015). Furthermore, the abandonment of traditional practices can foster the establishment and spread of invasive species. Abandoned farmsteads support the persistence and spread of formerly cultivated alien plants (Pándi et al., 2014).

As environmental conditions change with time after abandonment, new communities establish, and a shift in species composition occurs (Figure 17.2). In Japan, the succession of grasslands to secondary forests after land abandonment leads to the dominance of tall grasses and woody species that suppress the growth of many threatened grassland plants, which in addition decreases grassland herbivorous insects (Uchida and Ushimaru, 2014). Similarly, in European mountains the abandonment of productive pastures or the decrease in herbage use typically encourages the invasion of coarse tall grasses mostly with competitive stress-tolerant strategies, and leading to the competitive exclusion of subordinate and accidental plant species (Corazza et al., 2016). Also, the loss of grasslands and semi-open formations due to land abandonment changes the composition of animal communities (Figure 17.2). In the southern Balkans, land abandonment caused a shift in the butterfly community from Mediterranean endemics towards species with European or Eurosiberian distribution (Slancarova et al., 2016). Analogous shifts in the community structure of belowground invertebrate species after abandonment have also been documented (e.g. in alpine soils; Steinwandter et al., 2017). Similar patterns have been demonstrated for vertebrate communities. In many European areas where land has been abandoned, forest-dwelling bird species increase at the expense of farmland birds (Zakkak et al., 2015a).

Overall, the abandonment of traditional human practices considered in passive rewilding projects can reduce species diversity and richness (Figure 17.2). In this sense, land abandonment usually leads to vegetation homogenisation and a reduction in landscape heterogeneity (Rey-Benayas et al., 2007). Vegetation homogenisation triggered by secondary succession after abandonment increases fire frequency (Moreira and Russo, 2007). Fire on abandoned land often leads to a further decline in biodiversity, as it enhances the growth of fire-adapted species (Rey-Benayas et al., 2007). Examples of vegetation homogenisation and plant diversity loss as a result of abandonment have been often reported across many regions in Europe (e.g. Persson, 1984; Campagnaro et al., 2017) and Asia (Suzuki et al., 2016; Uchida et al., 2016). However, land abandonment can also threaten animal diversity and species richness (Figure 17.2). In semi-grasslands of Japan, the diversity of both threatened and common butterflies is significantly higher in traditional land-use sites than in those where land has been abandoned (Uchida et al., 2016). Similar findings have been reported in Europe (Loos et al., 2014; Buvobá et al., 2015). The diversity of other invertebrates could be also threatened by the land abandonment supported by advocates of passive rewilding. In the Italian Alps, the number of orthopteran species decreases with increasing time since abandonment (Marini et al., 2009), and ground spider species diversity also declines after abandonment in Greek ecosystems (Zakkak et al., 2014). Passive rewilding might also cause negative consequences for vertebrate diversity. Moreira and Russo (2007) modelled the global impact of abandonment on 554 species of terrestrial vertebrates occurring in Europe and found that, for all groups except amphibians, open habitats or farmland sustained higher species richness. This is consistent with the findings obtained at local or regional scales; e.g. avian species richness and diversity decreased with the secondary succession after land abandonment in south-eastern Europe (Zakkak et al., 2015a). The loss of human-made structures (walls, ponds, farmland buildings) associated with abandonment can also have detrimental impacts on animal species richness. The abandonment of mountainous zones in the Iberian Peninsula has led to the loss of many ponds in Mediterranean dry forests, and such ponds harbour higher bat species richness than nearby areas, including some species of conservation concern like horseshoe bats (*Rhinolophus* spp.) or *Myotis* spp. (Lisón and Calvo, 2014).

Although many studies have demonstrated positive outcomes of land abandonment (Chapter 6), the truth is that negative impacts on biodiversity are also frequent, which are well illustrated in the numerous examples provided in this chapter. Land abandonment can also cause unwanted abiotic consequences, such as soil erosion and desertification, and a reduction in water availability (Rey-Benayas et al., 2007). Overall, this indicates that passive rewilding attempts should not ignore the social and ecological complexity of the areas that are to be restored. Otherwise, the conservation of semi-natural habitats of high nature value will be compromised.

Implications of evolutionary pathways when designing rewilding schemes

The structure and functioning of biological communities can be understood only in the light of their evolutionary history, which documents the past processes that have led to their current configuration. Therefore, in order to be successful, rewilding initiatives should explicitly adopt an evolutionary perspective and take into consideration the time frame associated with specific biotic interactions. The general expectation is that newly established interactions (on an evolutionary timescale) will lead to unpredictable (and often undesired) outcomes (Saul and Jeschke, 2015).

For instance, an important aspect to take into account in rewilding programmes is the maintenance or the possible disruption of existing biotic interactions, such as predator–prey or host–parasite relationships – or even both simultaneously, such as in predator–prey–parasite triangles (e.g. Barbosa et al., 2012). Host–parasite interactions can have profound consequences on a variety of aspects, including population structure, social traits, physiology, macroecology, and evolution (Guilhaumon et al., 2012; Quigley et al., 2012; Greenwood et al., 2016). Predator–prey relationships also play a major role in the functioning of ecosystems. All these interactions can be highly species-specific – i.e. several parasites and predators depend on a single species of host or prey, respectively. The maintenance of biotic interactions can even be affected by the spatial genetic structure (i.e. the geographic distribution of different genetic lineages) of the species involved (Real et al., 2009), which should be carefully taken into account when planning rewilding initiatives. On the other hand, Late Quaternary extinctions, namely of megafauna, radically transformed the habitat structure of many landscapes and the functioning of ecosystems through trophic cascade effects (Malhi et al., 2016). Many plant communities have thereon evolved in the absence of large herbivores, and now lack particular adaptations to persist under their grazing pressure (Johnson, 2009). Thus, the conservation of several plant species can be compromised by the reintroduction of large herbivores within rewilding projects, which can negatively affect vegetation structure and composition (as described above).

All these complex interactions need to be taken into account when designing rewilding schemes. The co-evolutionary history of the taxa involved is a critical piece of information when aiming at predicting possible outcomes of new interactions at the community level. The incorporation of an evolutionary perspective on rewilding approaches is limited by our incomplete knowledge of the Tree of Life (especially in terminal branches) and, more importantly, of fully resolved networks of species interactions in natural communities, which are the two major pillars on which any attempt to infer the evolutionary history of biotic interactions should be based. The study of the fossil record can provide key insights on the history of species interactions, but it is inherently incomplete. All these levels of uncertainty are finally translated to the practical stage (implementing rewilding programmes), leaving much room for speculation and discussion.

In Europe, where people and nature have interacted for millennia, it is estimated that about 50 per cent of wildlife species now depend on agricultural habitats (Kristensen, 2003), and thus the attempts to conserve most of these species contrast with the enhanced dichotomy between nature and human culture implied by rewilding (Linnell et al., 2015). This debate about the need to conserve ecological interactions that have evolved over long periods of time is not new. Some authors have argued that species that were introduced out of their native ranges in the distant past have effectively become part of the ecosystem they invaded, taking up the ecological roles of species that have either become totally or at least functionally extinct. Consequently, they question the need to remove these non-native species from invaded ecosystems, on the basis that long-term key ecological interactions need to be preserved (e.g. Lees and Bell, 2008). A better integration of phylogenetic and ecological studies of species interaction networks will be needed to make progress on this exciting but as yet open debate.

Conclusions

This chapter shows a number of examples of potential unwanted outcomes of rewilding initiatives, largely explained by the complexity of natural systems, by the extent to which the ecosystem target of rewilding has been modified, or because our understanding of the relevant ecosystem dynamics is limited (Baker et al., 2016). However, it is important to account for these uncertainties, as they will likely increase with future environmental changes. In this context, several modelling tools (such as ensemble ecosystem modelling: Baker et al., 2016) and natural experimental settings (Mech et al., 2017) can provide additional insights on how to resolve some of the unknowns surrounding ecosystem responses to trophic rewilding in a structured and quantitative way. Overall, it is essential that rewilding initiatives do not ignore that species composition and ecological interactions of any given ecosystem reflect the history of how it was assembled.

Acknowledgements

Special thanks go to Drs Phil Seddom and Nathalie Pettorelli for their helpful comments. M. Delibes-Mateos was supported by V Plan Propio de Investigación of the University of Seville, I.C. Barrio by a postdoctoral fellowship funded by the Icelandic Research Fund (Rannsóknasjóður, grant no. 152468-051) and AXA Research Fund (15-AXA-PDOC-307), A.M. Barbosa by FCT (Portugal) and FEDER/COMPETE 2020 through contract and exploratory project IF/00266/2013/P1168/CT0001, and by funds POCI-01-0145-FEDER-006821 to research unit UID/BIA/50027, and C.C. Ferreira by a Marie Curie Outgoing International Fellowship for Career Development (PIOF-GA-2013-621571) within the 7th Framework Programme of the European Union.

References

Acevedo, P., Farfán, M.A., Márquez, A.L., Delibes-Mateos, M., Real, R., and Vargas, J.M. (2011). Past, present and future of wild ungulates in relation to changes in land use. *Landscape Ecology*, **26**, 19-31.

Allen, B.L., and Fleming, P.J.S. (2012). Reintroducing the dingo: the risk of dingo predation to threatened vertebrates of western New South Wales. *Wildlife Research*, **39**, 35-50.

Bailey, J.K., and Whitham, T.G. (2007). Biodiversity is related to indirect interactions among species of large effect. In Ohgushi, T., Craig, T.P., and Price, P.W. (Eds.), *Ecological communities: plant mediation in indirect interaction webs* (pp. 306-328). Cambridge: Cambridge University Press.

Baker, C.M., Gordon, A., and Bode, M. (2016). Ensemble ecosystem modeling for predicting ecosystem response to predator reintroduction. *Conservation Biology*, **2**, 376-384.

Bakker, E.S., Gill, J.L., Johnson, C.N., et al. (2016). Combining paleo-data and modern exclosure experiments to assess the impact of megafauna extinctions on woody vegetation. *Proceedings of the National Academy of Sciences of the United States of America*, **113**, 847-855.

Barbosa, A.M., Thode, G., Real, R., Feliu, C., and Vargas, J.M. (2012). Phylogeographic triangulation: using predator-prey-parasite interactions to infer population history from partial genetic information. *PLoS ONE*, **7**, e50877.

Barkai, A., and McQuaid, C. (1988). Predator-prey role reversal in a marine benthic ecosystem. *Science*, **242**, 62-64.

Barnosky, A.D., Lindsey, E.L., Villavicencio, N.A., et al. (2016). Variable impact of late-Quaternary megafaunal extinction in causing ecological state shifts in North and South America. *Proceedings of the National Academy of Sciences of the United States of America*, **113**, 856-861.

Berger, J., Swenson, J.E.,, and Persson, I.L. (2001). Recolonizing carnivores and naive prey: conservation lessons from Pleistocene extinctions. *Science*, **291**, 1036-1039.

Bridgeland, W.T., Beier, P., Kolb, T., and Whitham, T. (2010). A conditional trophic cascade: birds benefit faster growing trees with strong links between predators and plants. *Ecology*, **91**, 73-84.

Buvobá, T., Vrabec, T., Kulma, M., and Nowicki, P. (2015). Land management impacts on European butterflies of conservation concern: a review. *Journal of Insect Conservation*, **19**, 805-821.

Cabezas-Díaz, S., Virgós, E., Mangas, J.G., and Lozano, J. (2011). The presence of a 'competitor pit effect' compromises wild rabbit (*Oryctolagus cuniculus*) conservation. *Animal Biology*, **61**, 319-334.

Campagnaro, T., Frate, L., Carranza, M.L., and Sitzia, T. (2017). Multi-scale analysis of alpine landscapes with different intensities of abandonment reveals similar pattern changes: implications for habitat conservation. *Ecological Indicators*, **74**, 147-159.

Carboni, M., Dengler, J., Mantilla-Contreras, J., Venn, S., and Török, P. (2015). Conservation value, management and restoration of Europe's semi-natural open landscapes. *Hacquetia*, **14**, 5-17.

Caro, T. (2007). The Pleistocene re-wilding gambit. *Trends in Ecology & Evolution*, **22**, 281-283.

Carpio, A.J., Guerrero-Casado, J., Ruiz-Aizpurua, L., Vicente, J., and Tortosa, F.S. (2014). The high abundance of wild ungulates in a Mediterranean region: is this compatible with the European rabbit? *Wildlife Biology*, **20**, 161-166.

Chapron, G., Kaczensky, P., Linnell, J.D.C., et al. (2014). Recovery of large carnivores in Europe's modern human-dominated landscapes. *Science*, **346**, 1517-1519.

Clinchy, M., Zanette, L.Y., Roberts, D., et al. (2016). Fear of the human 'super predator' far exceeds the fear of large carnivores in a model mesocarnivore. *Behavioral Ecology*, **27**, 1826-1832.

Corazza, M., Tardella, F.M., Ferrari, C., and Catorci, A. (2016). Tall grass invasion after grassland abandonment influences the availability of palatable plants for wild herbivores: insights into the conservation of the Apennine chamois *Rupicapra pyrenaica ornata*. *Environmental Management*, **57**, 1247-1261.

Corlett, R.T. (2016). Restoration, reintroduction and rewilding in a changing world. *Trends in Ecology & Evolution*, **31**, 453-462.

Cornelissen, P., Bokdam, J., Sykora, K., and Berendse, F. (2014). Effects of large herbivores on wood pasture dynamics in a European wetland system. *Basic and Applied Ecology*, **15**, 396-406.

Darimont, C.T., Fox, C.H., Bryan, H.M., and Reimchen, T.E. (2015). The unique ecology of human predators. *Science*, **349**, 858-860.

Daszak, P., Cunningham, A.A., and Hyatt, A.D. (2000). Emerging infectious diseases of wildlife: threats to biodiversity and human health. *Science*, **287**, 443-449.

Doak, D.F., Estes, J.A., Halpern, B.S., et al. (2008). Understanding and predicting ecological dynamics: are major surprises inevitable? *Ecology*, **89**, 952-961.

Dorresteijn, I., Schultner, J., Nimmo, D.G., et al. (2015). Incorporating anthropogenic effects into trophic ecology: predator–prey interactions in a human-dominated landscape. *Proceedings of the Royal Society of London B: Biological Sciences*, **282**, 20151602.

Edwards, P.J., Fletcher, M.R., and Berny, P. (2000). Review of the factors affecting the decline of the European brown hare, *Lepus europaeus* (Pallas, 1778), and the use of wildlife incident data to evaluate the significance of paraquat. *Agriculture, Ecosystems and Environment*, **79**, 95-103.

Fernández, N., Navarro, L.M., and Pereira, H.M. (2017). Rewilding: a call for boosting ecological complexity in conservation. *Conservation Letters*, **10**, 276-278.

Gibbs, J.P., Hunter, E.A., Shoemaker, K.T., Tapia, W.H., and Cayot, L.J. (2014). Demographic outcomes and ecosystem implications of giant tortoise reintroduction to Española Island, Galápagos. *PLoS ONE*, **9**, e110742.

Greenwood, J.M., López Ezquerra, A., Behrens, S., Branca, A., and Mallet, L. (2016). Current analysis of host-parasite interactions with a focus on next generation sequencing data. *Zoology*, **119**, 298-306.

Griffiths, C.J., Hansen, D.M., Jones, C.G., Zuël, N., and Harris, S. (2011). Resurrecting extinct interactions with extant substitutes. *Current Biology*, **21**, 762-765.

Griffiths, C.J., Zuël, N., Jones, C.G., Ahamud, Z., and Harris, S. (2013). Assessing the potential torestore historic grazing ecosystems with tortoise ecological replacements. *Conservation Biology*, **27**, 690-700.

Guilhaumon, F., Krasnov, B.R., Poulin, R., Shenbrot, G.I., and Mouillot, D. (2012). Latitudinal mismatches between the components of mammal-flea interaction networks. *Global Ecology and Biogeography*, **21**, 725-731.

Hansen, D.M., Kiesbuy, H.C., Jones, C.G., and Muller, C.B. (2007). Positive indirect interactions between neighboring plant species via a lizard pollinator. *The American Naturalist*, **169**, 534-542.

Hansen, D.M., Donlan, C.J., Griffiths, C.J., and Campbell, K.J. (2010). Ecological history and latent conservation potential: large and giant tortoises as a model for taxon substitutions. *Ecography*, **33**, 272-284.

Ims, R.A., Yoccoz, N.G., Brathen, K.A., Fauchald, P., Tveraa, T., and Hausner, V. (2007). Can reindeer overabundance cause a trophic cascade? *Ecosystems*, **10**, 607-622.

Jaquet, S., Schwilch, G., Hartung-Hofmann, F., et al. (2015). Does outmigration lead to land degradation? Labour shortage and land management in a western Nepal watershed. *Applied Geography*, **62**, 157-170.

Johnson, B.E., and Cushman, J.H. (2007). Influence of a large herbivore reintroduction on plant invasions and community composition in a California grassland. *Conservation Biology*, **21**, 515-526.

Johnson, C.N. (2009). Ecological consequences of Late Quaternary extinctions of megafauna. *Proceedings of the Royal Society of London B: Biological Sciences Biological Sciences*, **276**, 2509-2519.

Katayama, N., Osawa, T., Amano, T., and Kusumoto, Y. (2015). Are both agricultural intensification and farmland abandonment threats to biodiversity? A test with bird communities in paddy-dominated landscapes. *Agriculture, Ecosystems and Environment*, **214**, 21-30.

Kitchen, A.M., Gese, E.M., and Schauster, E.R. (1999). Resource partitioning between coyotes and swift foxes: space, time, and diet. *Canadian Journal of Zoology*, **77**, 1645-1656.

Kristensen, P. (2003). EEA core set of indicators. Revised version April 2003. Adopted version for ECCAA countries May 2003. Copenhagen, Denmark.

Lasanta, T., Nadal-Romero, E., and Arnáez, J. (2015). Managing abandoned farmland to control the impact of re-vegetation on the environment. The state of the art in Europe. *Environmental Science and Policy*, **52**, 99-109.

Laundré, J.W., Hernandez, L., and Altendorf, K.B. (2001). Wolves, elk, and bison: reestablishing the 'landscape of fear' in Yellowstone National Park, USA. *Canadian Journal of Zoology*, **79**, 1401-1409.

Leader-Williams, N., Smith, R.I.L., and Rothery, P. (1987). Influence of introduced reindeer on the vegetation of South Georgia: results from a long-term exclusion experiment. *Journal of Applied Ecology*, **24**, 801-822.

Leader-Williams, N., Walton, D.W.H., and Prince, P.A. (1989). Introduced reindeer on South Georgia – a management dilemma. *Biological Conservation*, **47**, 1–11.

Lees, A.C., and Bell, D.J. (2008). A conservation paradox of the 21st century: the European wild rabbit *Oryctolagus cuniculus*, and invasive alien and an endangered natives species. *Mammal Review*, **38**, 304–320.

Linnell, J.D.C., Kaczensky, P., Wotschikowsky, U., Lescureux, L., and Boitani, L. (2015). Framing the relationship between people and nature in the context of European conservation. *Conservation Biology*, **29**, 978–985.

Lisón, F., and Calvo, J.F. (2014). Bat activity over small ponds in dry Mediterranean forests: implications for conservation. *Acta Chiropterologica*, **16**, 95–101.

Loos, J., Dorresteijn, I., Hanspach, J., Fust, P., Rakosy, L., and Fischer, J. (2014). Low-intensity agricultural landscapes in Transylvania support high butterfly diversity: implications for conservation. *PLoS ONE*, **9**, e103256.

Lozano, J., Virgós, E., Cabezas-Díaz, S., and Mangas, J.G. (2007). Increase of large game species in Mediterranean areas: is the European wildcat (*Felis silvestris*) facing a new threat? *Biological Conservation*, **138**, 321–329.

Madhusudan, M.D. (2004). Recovery of wild large herbivores following livestock decline in a tropical Indian wildlife reserve. *Journal of Applied Ecology*, **41**, 858–869.

Malhi, Y., Doughty, C.E., Galleti, M., Smith, F.A., Svenning, J.C., and Terborgh, J.W. (2016). Megafauna and ecosystem function from the Pleistocene to the Anthropocene. *Proceedings of the National Academy of Sciences of the United States of America*, **113**, 838–846.

Manning, A.D., Gordon, I.J., and Ripple, W.J. (2009). Restoring landscapes of fear with wolves in the Scottish highlands. *Biological Conservation*, **142**, 2314–2321.

Mäntylä, E., Klemola, T., and Laaksonen, T. (2011). Birds help plants: a meta-analysis of top-down trophic cascades caused by avian predators. *Oecologia*, **165**, 143–151.

Marini, L., Fontana, P., Battisti, A., and Gaston, K.J. (2009). Response of orthopteran diversity to abandonment of semi-natural meadows. *Agriculture, Ecosystems and Environment*, **132**, 232–236.

Mech, L.D., Barber-Meyer, S., Blanco, J.C., et al. (2017). An unparalleled opportunity for an important ecological study. *BioScience*, **67**, 875–876.

Mischenko, A.L., and Sukhanova, O. (2016). Response of wader populations in the Vinogradovo foodplain (Moscow region, Russia) to changes in agricultural change and spring flooding. *Wader Study*, **123**, 136–142.

Mishra, C., Van Wieren, S., Heitköning, I.M.A., and Prins, H.H.T. (2002). A theoretical analysis of competitive exclusion in a Trans-Himalayan large-herbivore assemblage. *Animal Conservation*, **5**, 251–258.

Moreira, F., and Russo, D. (2007). Modelling the impact of agricultural abandonment and wildfires on vertebrate diversity in Mediterranean Europe. *Landscape Ecology*, **22**, 1461–1467.

Munguira, M.L., Barea-Azcón, J.M., Castro-Cobo, S., et al. (2017). Ecology and recovery plans for the four Spanish endangered endemic butterfly species. *Journal of Insect Conservation*, **21**, 423–437.

Newsome, T.M., Ballard, G., Crowther, M.S., et al. (2015). Resolving the value of the dingo in ecological restoration. *Restoration Ecology*, **23**, 201–208.

Nogués-Bravo, D., Simberloff, D., Rahbek, C., and Sanders, N.J. (2016). Rewilding is the new Pandora's box in conservation. *Current Biology*, **26**, R83–R101.

Oriol-Cotterill, A., Valeix, M., Frank, L.G., Riginos, C., and Macdonald, D.W. (2015). Landscapes of coexistence for terrestrial carnivores: the ecological consequences of being downgraded from ultimate to penultimate predator by humans. *Oikos*, **124**, 1263–1273.

Ostfeld, R.S., and Holt, R.D. (2004). Are predators good for your health? Evaluating evidence for top-down regulation of zoonotic disease reservoirs. *Frontiers in Ecology and Environment*, **2**, 13-20.

Paine, R.T., Tegner, M.J., and Johnson, A.E. (1998). Compounded perturbations yield ecological surprises. *Ecosystems*, **1**, 535-545.

Pándi, I., Penksza, K., Botta-Dukát, Z., and Kröel-Dulay, G. (2014). People move but cultivated plants stay: abandoned farmsteads support the persistence and spread of alien plants. *Biodiversity and Conservation*, **23**, 1289-1302.

Perrson, S. (1984). Vegetation development after the exclusion of grazing cattle in a meadow area in the south of Sweden. *Vegetatio*, **55**, 65-92.

Plieninger, T., Hui, C., Gaertner, M., and Huntsinger, L. (2014). The impact of land abandonment on species richness and abundance in the Mediterranean Basin: a meta-analysis. *PLoS ONE*, **9**, e98355.

Prugh, L.R., Stoner, C.J., Epps, C.W., et al. (2009). The rise of the mesopredator. *BioScience*, **59**, 779-791.

Queiroz, C., Beilin, R., Folke, C., and Lindborg, R. (2014). Farmland abandonment: threat or opportunity for biodiversity conservation? A global review. *Frontiers in Ecology and the Environment*, **12**, 288-296.

Quigley, B.J.Z., García López, D., Buckling, A., McKane, A.J., and Brown, S.P. (2012). The mode of host-parasite interaction shapes coevolutionary dynamics and the fate of host cooperation. *Proceedings of the Royal Society of London B: Biological Sciences*, **279**, 3742-3748.

Real, R., Barbosa, A.M., Rodríguez, A., et al. (2009). Conservation biogeography of ecologically interacting species: the case of the Iberian lynx and the European rabbit. *Diversity and Distributions*, **15**, 390-400.

Regos, A., Domínguez, J., Gil-Tena, A., Brotons, L., Ninyerola, M., and Pons, X. (2016). Rural abandoned landscapes and bird assemblages: winners and losers in the rewilding of a marginal mountain area (NW Spain). *Regional Environmental Change*, **16**, 199-211.

Rey-Benayas, J.M., Martinis, A., Nicolau, J.M., and Schulz, J.J. (2007). Abandonment of agricultural land: an overview of drivers and consequences. *CAB Reviews: Perspectives in Agriculture, Veterinary Science, Nutrition and Natural Resources*, **2**, 1-14.

Ricketts, A.M. (2016). Of mice and coyotes: mammalian responses to rangeland management practices in tallgrass prairie. PhD Dissertation, Kansas State University, USA.

Ripple, W.J., Estes, J.A., Beschta, R.L., et al. (2014). Status and ecological effects of the world's largest carnivores. *Science*, **343**, 151.

Ripple, W.J., Newsome, T.M., Wolf, C., et al. (2015). Collapse of the world's largest herbivores. *Science Advances*, **1**, e140010.

Ritchie, E.G., and Johnson, C.N. (2009). Predator interactions, mesopredator release and biodiversity conservation. *Ecology Letters*, **12**, 982-998.

Sandom, C.J., Ejrnæs, R., Hansen, M.D.D., and Svenning, J.C. (2014). High herbivore density associated with vegetation diversity in interglacial ecosystems. *Proceedings of the National Academy of Sciences of the United States of America*, **111**, 1-6.

Saul, W.-C., and Jeschke, J.M. (2015). Eco-evolutionary experience in novel species interactions. *Ecology Letters*, **18**, 236-245.

Schnitzler, A. (2014). Towards a new European wilderness: embracing unmanaged forest growth and the decolonisation of nature. *Landscape and Urban Planning*, **126**, 74-80.

Seddon, P.J., Griffiths, C.J., Soorae, P.S., and Armstrong, D.P. (2014). Reversing defaunation: restoring species in a changing world. *Science*, **345**, 406-412.

Shengfa, L., and Xiubin, L. (2017). A global understanding of farmland abandonment: a review and prospects. *Journal of Geographical Sciences*, **27**, 1123-1150.

Slancarova, J., Bartonova, A., Zapletal, M., et al. (2016). Life history traits reflect changes in Mediterranean butterfly communities due to forest encroachment. *PLoS ONE*, **11**, e0152026.

Smit, C., Kruifot, J.L., van Klink, R., and Olff, H. (2015). Rewilding with large herbivores: the importance of grazing refuges for sapling establishment and wood–pasture formation. *Biological Conservation*, **182**, 134–142.

Smith, C.I. (2005). Re-wilding: introductions could reduce biodiversity. *Nature*, **437**, 318.

Smith, J.A., Wang, Y., and Wilmers, C.C. (2015). Top carnivores increase their kill rates on prey as a response to human-induced fear. *Proceedings of the Royal Society of London B: Biological Sciences*, **282**, 2014–2711.

Soulé, M., and Noss, R. (1998). Rewilding and biodiversity: complementary goals for continental conservation. *Wild Earth*, **8**, 19–28.

Sovada, M.A., Roy, C.C., Bright, J.B., and Gillis, J.R. (1998). Causes and rates of mortality of swift foxes in western Kansas. *Journal of Wildlife Management*, **62**, 1300–1306.

Speed, J.D., and Austrheim, G. (2017). The importance of herbivore density and management as determinants of the distribution of rare plant species. *Biological Conservation*, **205**, 77–84.

Steinwandter, M., Schlick-Steiner, B., Seeber, G.U.H., Steiner, F.M., and Seeber, J. (2017). Effects of Alpine land-use changes: soil macrofauna community revisited. *Ecology and Evolution*, **7**, 5389–5399.

Sterner, R.W., and Elser, J.J. (2002). *Ecological stoichiometry: the biology of elements from molecules to the biosphere*. Princeton, NJ: Princeton University Press.

Suraci, J.P., Clinchy, M., Dill, L.M., Roberts, D., and Zanette, L.Y. (2016). Fear of large carnivores causes a trophic cascade. *Nature Communications*, **7**, 10698.

Suzuki, R.O., Kenta, T., Sato, M., Masaki, D., and Kanai, R. (2016). Continuous harvesting of a dominant bracken alters a cool-temperate montane grassland community and increases plant diversity in Nagano. *Ecological Restoration*, **31**, 639–644.

Svenning, J.C., Pedersen, P.B.M., Donlan, C.J., et al. (2016). Science for a wilder Anthropocene: synthesis and future directions for trophic rewilding research. *Proceedings of the National Academy of Sciences of the United States of America*, **113**, 898–906.

Tambling, C., Minnie, L., Meyer, J., et al. (2015). Temporal shifts in activity of prey following large predator reintroductions. *Behavioural Ecology and Sociobiology*, **69**, 1153–1161.

Torre, I., Bros, V., and Santos, X. (2014). Assessing the impact of restoration on the diversity of Mediterranean terrestrial Gastropoda. *Biodiversity and Conservation*, **23**, 2579–2589.

Towne, E.G., Hartnett, D.C., and Cochran, R.C. (2005). Vegetation trends in tallgrass prairie from bison and cattle grazing. *Ecological Applications*, **15**, 1550–1559.

Uchida, K., and Ushimaru, A. (2014). Biodiversity declines due to land abandonment and intensification of agricultural lands: patterns and mechanisms. *Ecological Monographs*, **84**, 637–658.

Uchida, K., Takahashi, S., Shinohara, T., and Ushimaru, A. (2016). Threatened herbivorous insects maintained by long-term traditional management practices in semi-natural grasslands. *Agriculture, Ecosystems and Environment*, **221**, 156–162.

Vera, F.W.M. (2009). Large-scale nature development – the Oostvaardersplassen. *British Wildlife*, **2009**, 28–36.

Wehn, S., and Johansen, L. (2015). The distribution of the endemic plant Primula scandinavica, at local and national scales, in changing mountains environments. *Biodiversity*, **16**, 278–288.

Wiens, J.A., and Hobbs, R.J. (2015). Integrating conservation and restoration in a changing world. *BioScience*, **65**, 302–312.

Zakkak, S., Chatzaki, M., Karamalis, N., and Kati, V. (2014). Spiders in the context of agricultural land abandonment in Greek mountains: species responses, community structure and the need to preserve traditional agricultural practices. *Journal of Insect Conservation*, **18**, 599–611.

Zakkak, S., Radovic, A., Nikolov, S.C., Shumka, S., Kakalis, L., and Kati, V. (2015a). Assessing the effect of agricultural land abandonment on bird communities in south-eastern Europe. *Journal of Environmental Management*, **164**, 171–179.

Zakkak, S., Halley, J.M., Akriotis, T., and Kati, V. (2015b). Lizards along an agricultural land abandonment gradient in Pindos Mountains, Greece. *Amphibia-Reptilia*, **36**, 253–264.

Zimov, S.A. (2005). Pleistocene Park: return of the mammoth's ecosystem. *Science*, **308**, 796–798.

Zimov, S.A., Chupryin, V.I., Oreshko, A.P., Chapin III, F.S., Reynolds, J.F., and Chapin, M.C. (1995). Steppe-tundra transition: a herbivore-driven biome shift at the end of the Pleistocene. *The American Naturalist*, **146**, 765–794.

Zimov, S.A., Zimov, N.S., Tikhonov, A.N., and Chapin, F.S. (2012). Mammoth steppe: a high-productivity phenomenon. *Quaternary Science Reviews*, **57**, 26–45.

CHAPTER EIGHTEEN

Auditing the wild: how do we assess if rewilding objectives are achieved?

RICHARD T. CORLETT
Xishuangbanna Tropical Botanical Gardens, Chinese Academy of Sciences

Rewilding is part of conservation, if not yet part of the conservation mainstream (Murray, 2017). Like other conservation projects, rewilding needs funding and other support from local or national governments, international agencies, bilateral donors, local and international NGOs, corporations, foundations, or wealthy individuals. Funders will want to know that the money and other resources are being used wisely, so they expect targets with timelines and clearly defined deliverables, i.e. the goods or services that will be provided upon the completion of a project. The project will usually end with an audit – formal or informal – during which project outcomes are checked against pre-agreed success criteria. Even if funders do not require that measurable, desired outcomes are specified at the beginning of a project, this is widely considered the best practice in conservation (Conservation Measures Partnership, 2013). The Society for Ecological Restoration, for example, states that restoration objectives need to be 'specific, measurable, achievable, reasonable, and time-bound', and that this should be achieved by the use of 'specific, quantifiable indicators' (McDonald et al., 2016), although, in practice, unpredictability of restoration outcomes continues to be a major problem (Brudvig, 2017).

Rewilding, in contrast, has been described as having targets and timelines that are 'fluid and unscripted' (Law et al., 2017) and outcomes that are indeterminate (Kirby, 2017). The main deliverable is an increase in 'wildness', implying autonomy, spontaneity, self-organisation, and lack of human control. The aim is for nature to take care of itself. Humans can initiate rewilding, but continued intervention is not compatible with the concept. How can the question 'Were the objectives achieved?' be answered if the objectives themselves are unclear? What if the outcomes are unexpected? What if they clash with other conservation objectives? What if they are damaging or dangerous?

Rewilding is still a fairly new concept and both funders and decision-makers mostly lack experience with rewilding projects, so there is still time to think about how the success of these projects should be assessed. If we don't, there is a risk that either audits will impose their own values on rewilding projects, thus potentially undermining their original intentions, or that large amounts of resources will be wasted on projects that achieve nothing or do harm (however this may be defined). This chapter therefore aims to answer three basic questions about auditing rewilding.

1. Are rewilding projects more difficult to audit than other conservation projects?
2. Should rewilding projects be audited?
3. If so, how can this best be done?

What is rewilding?

Although the word 'rewilding' is new, the process by which it was formed, by modification of the word 'wild', fixes its meaning as 'making something wild again'. Despite this, the multiple usages of rewilding in the literature give the impression of a very fluid concept (Jørgensen, 2015; Nogués-Bravo et al., 2016) and current rewilding practices also vary widely. However, the idea of an autonomous, 'self-willed' nature is at the core of most definitions (Prior and Ward, 2016) and also of most non-specialists' understanding of the concept (Deary and Warren, 2017). Interventions – often the reintroduction of large herbivores or apex predators – may be necessary to start the process, but the aim is always to reduce and, ideally, eliminate interventions with time.

In practice, rewilding projects usually aim for an increase in relative wildness – i.e. relative freedom from human interventions – rather than 100 per cent wildness. Indeed, almost any landscape could be rewilded to some extent (Sandom and Macdonald, 2015). The conservation NGO, Rewilding Europe, envisages a scale from 0 (city centres) to 10 (remote wilderness), and suggests that all of Europe could be moved up this scale, including existing 'wilderness' areas with poor connectivity or missing keystone species (www.rewildingeurope.com). Rewilding is thus part of a continuum, with any increase in wildness implying increased autonomy and reduced human control (Müller et al., 2018). Rewilding is also inevitably limited spatially, either by fencing or by gradual transitions into managed areas. Although fencing and wildness appear to be inherently incompatible, there will often be no alternative on a planet shared with 7.5 billion people.

Environmental auditing

An audit is a systematic and independent review of the records of an organisation or project in order to assess whether they are accurate. Audits started in the financial sector, where they have a long history, and in the 1970s and 1980s spread from finance to many other fields (Cook et al., 2016). The term audit is now used for any systematic evaluation of the performance of an institution or project. Audits are retrospective and usually done at the end of a project – or of a phase of a project – and thus differ from monitoring, which is typically carried out more frequently or continuously, although the aims are similar and regular audits can be considered as a type of monitoring.

Environmental auditing is the term used for various types of evaluations of environmental organisations or projects in order to assess their compliance with appropriate laws and regulations, other standards, or project-specific targets. Evidence of compliance is collected by reviewing documents, interviewing personnel, and making direct observations (Jain et al., 2016). Environmental audits may be done voluntarily, in order to improve management, but may also be legally required. These audits usually follow strict protocols, such as those in ISO 19011, and use audit criteria derived from legal or project-specific requirements (Jain et al., 2016). However, although generic standards and criteria allow easy comparisons between projects and tracking of improvements over time, they can be difficult to apply in the diverse and heterogeneous real world in which environmental impacts occur (Cook et al., 2016). It can also be difficult to attribute any observed changes to specific interventions. Auditing complex biological systems is particularly challenging because of the difficulties of collecting adequate data, our incomplete understanding of key processes, and the problems of defining relevant audit criteria.

Can rewilding be audited?

The problem with auditing rewilding is not that it holds people accountable for their actions – it is hard to envisage a legitimate objection to accountability when public funds and land, as well as potential risks to lives and livelihoods, are at stake – but that worthwhile projects may have to be modified in ways that make them more auditable but, as a consequence, less wild. Not everything is auditable: think of love, art, or religion. Is rewilding in this category? The question asked here, therefore, is not 'Can rewilding be made auditable?', but 'Can we audit rewilding?'

If the deliverable is wildness, how is success to be assessed? Human-perceived wildness can be measured, mapped, and predicted from physical variables (Müller et al., 2015; Pheasant and Watts, 2015), but human perceptions do not necessarily correlate with the self-willed ecological wildness that rewilding projects target (Müller et al., 2018). The aim of rewilding is to let nature take over, either from the beginning (passive rewilding) or after initial human interventions (e.g. species introductions or removals) intended to make an autonomous, self-sustaining, nature possible. Ecological uncertainty is a problem in conventional restoration projects (Brudvig, 2017), but there it is a problem to be minimised, while with rewilding the restoration of uncontrolled ecological processes is the major point of the exercise. Despite its nostalgic origins, rewilding has a strong emphasis on the future; the target is a future wildness, informed by the past, rather than the historical 'reference ecosystem' that provides the aspirational target of most restoration projects (McDonald et al., 2016). Moreover, most restoration projects have a focus on vegetation recovery, which is relatively easily measured, while many rewilding projects focus on animals, which are inherently more difficult to assess (Pires, 2017). Interventions can be audited, but this goes against the generally accepted principle that it is better to audit outcomes rather than actions (Conservation Measures Partnership, 2013).

In practice, with the exception of completely passive approaches, the rewilding literature mentions many desired outcomes. Some of these are rather general and diffuse (e.g. 'promoting self-regulation', 'maintaining habitat heterogeneity', 'a sense of place'), so not easily audited, but rewilding projects involving reintroductions (e.g. trophic and Pleistocene rewilding) explicitly target the restoration of ecosystem processes – predation, seed dispersal, or grazing are most often mentioned – and some of these processes can potentially be measured and audited. Where rewilding projects are within fenced areas, ecosystem processes can be compared with adjacent ecosystems outside the fence. Where this has been done, however, it can reveal unexpected outcomes (Mills et al., 2018), reflecting both the lack of empirical research and the complexity of trophic and other interaction networks (Nogués-Bravo et al., 2016; Svenning et al., 2016). Indeed, it is probably impossible, currently, to make predictions on the impacts of reintroducing large carnivores or herbivores that are specific and detailed enough to use as auditable criteria for success (Pires, 2017). The further back in time that the reintroduced species was lost from the area, the greater the uncertainties of the impacts, and these are increased still more if the introductions are non-native ecological replacements for extinct species (N. Fernández et al., 2017; Pires, 2017) or even 'de-extinct' previously native species (Seddon et al., 2014).

The viability of the introduced population over time is a possible intermediate target for audit – downstream of the initial action but upstream of the desired outcome (F.A.S. Fernandez et al., 2017; Zamboni et al., 2017). However, this target conflicts with the frequent assertion in the rewilding literature that reintroductions should be seen as a means to restore ecological processes and not as ends in themselves (Sandom and Macdonald, 2015; Pires, 2017). Moreover, the long-term viability of populations of long-lived, slow-reproducing, wide-ranging large mammals is not easily assessed (F.A.S. Fernandez et al., 2017).

Most rewilding projects also envisage an improved provision of ecosystem services: expanded recreational opportunities and associated employment in tourism support services, cleaner freshwater, reduced soil erosion, a reduction in flooding, increased carbon sequestration, and the removal of air pollutants

(Cerqueira et al., 2015; Navarro and Pereira, 2015; Law et al., 2017). In theory, these services could provide quantifiable targets for auditing, but the complexities of the links between ecosystem attributes and the services provided make detailed predictions difficult. This is even truer of the possible social, psychological, and health benefits.

A more general problem for auditors is that rewilding is slow and nothing substantial may have changed by the end of the funding period. Even in successful projects, the desired positive impacts on biodiversity and processes may not be visible for years. For example, the impact of beaver reintroductions on degraded wetlands can take a decade or more to develop (Law et al., 2017). Ten years after the start of an ambitious rewilding programme in Iberá Nature Reserve, Argentina, self-sustaining populations of two locally extirpated species had been established, of the five species targeted, and there were early signs of restored ecological roles, such as the growth of tree seedlings in tapir faeces (Zamboni et al., 2017). Where multiple reintroductions are needed to restore ecosystem processes, they may need to be spread over many years in an appropriate sequence, so that reintroduced prey populations, for example, are sufficient to support reintroduced predators (Sandom and Macdonald, 2015; F.A.S. Fernandez et al., 2017). Unless auditing actions rather than outcomes is acceptable, there may simply be nothing to report at the end of the funding period. Conventional ecological restoration is also slow and often involves sequencing of interventions, but the interventions typically include planting native plant species, which is not usually done in rewilding projects, so the initial changes are generally larger, more rapid, and more easily assessed.

The values that many rewilding advocates espouse may also be a problem for auditors. 'Scientific conservation', as exemplified by the IUCN 'Categories and Criteria' (IUCN, 2012), has worked to replace subjective judgements by objective criteria that can be quantified and assessed, i.e. to make conservation auditable. Rewilding advocates, in contrast, often invoke aesthetic, cultural, or spiritual values, and our moral responsibilities towards nature, none of which are easily quantified and audited (Cooper et al., 2016). 'A 30 per cent increase in spiritual value' simply does not make sense, and using 'psychological benefits' as a potentially assessable proxy is not a sufficient alternative. Advocates are also not always clear about their own values. See, for example, George Monbiot's inspiring, but sometimes confusing, advocacy of a wildness created by people for people (Monbiot, 2013).

It also needs to be mentioned that a major practical motivation for choosing rewilding, particularly passive rewilding (i.e. without initial interventions), is hard economics – the low costs of a do-nothing approach (Navarro and Pereira, 2015; Zefferman et al., 2018) – and costs are easily audited. However, rewilding advocates are unlikely to be satisfied by low cost as the sole criterion for success!

Should rewilding be audited?

Rewilding is conservation on the offensive. Rewilding projects are therefore likely to attract more attention than those that fit within the widely accepted, defensive, model of conservation that has dominated since the nineteenth century. Rewilding is a provocation – often deliberately so – so disagreements are inevitable, but unless it is carried out behind fences on private land, the support of all major stakeholders is essential for success. Rewilding is also always an experiment, because both the science and the accumulated experience are currently insufficient to foresee the full range of possible outcomes.

Rewilding efforts are likely to increase, so it is important to know what 'works' and what does not, for which at least some form of auditing is essential. Moreover, land, funds, and other resources dedicated to rewilding are land, funds, and other resources that are not available for other uses, including more interventionist – and probably more predictable – approaches to conservation. Is rewilding demonstrably better than these alternatives? In practice, because rewilding is often promoted as a low-cost option for abandoned rural (Navarro and Pereira, 2015) or industrial land (Zefferman et al., 2018), benefits need to be compared with costs. If this is done in dollar terms, aesthetic, cultural, or spiritual benefits may be ignored or undervalued, but this is not a good argument for not assessing the economic cost/benefit ratio. If the economic costs are covered by the economic benefits, then all other benefits are free: if not, then a subjective judgement has to be made about whether the non-material benefits make up the economic deficit. The measurable claims of rewilding project proponents should therefore be scientifically assessed and at least an attempt made to assess the achievement of those desired outcomes that resist quantification.

Moreover, rewilding's romantic basis does not excuse environmental damage or human suffering. It is therefore important to also assess undesired, but more or less predictable, outcomes, such as increased conflicts between people and self-willed large animals. Apex predators may kill people (e.g. lions in Africa) and they often kill domesticated livestock (e.g. wolves wherever alternative prey is rare). Elephants also kill people (e.g. in Sri Lanka and south-west China) and both they and other large herbivores damage crops. Other potential negative impacts from rewilding projects include the spread of fire-prone vegetation and the accumulation of ignitable fuels, potentially threatening lives and property, the loss of valued biocultural landscapes (e.g. most European grassland types) and the species they support, and the spread of invasive species when control interventions are halted (Corlett, 2016).

Given our current poor understanding of the complexities of trophic and other interaction webs, there is also a large potential for ecological surprises: large changes that could not have been predicted from current knowledge and the available information (Pires, 2017). This unpredictability is inherent in most rewilding projects and, indeed, could be considered as a positive indicator of the self-willed nature that rewilding targets. However, while unpredictability would be an interesting topic for research on an isolated island, it can be a cause for concern in real landscapes that also contain people.

Dealing with unexpected outcomes

Rewilding projects must necessarily take a 'trial and error' approach, as the field currently lacks the large corpus of case studies that underpins conventional ecological restoration practices. Predictions of outcomes are based on theory, on information from unplanned and unintentional rewilding, and, to varying degrees, on romantic 'nature knows best' optimism. When things go badly wrong, from a human perspective, additional, unplanned interventions may be unavoidable. Adaptive management, where management actions are treated as experiments and adjustments made in response to the results, is now routinely advocated in traditional conservation projects (e.g. McDonald et al., 2016), but the limits this approach places on natural autonomy conflict with rewilding's emphasis on a self-willed nature. In practice, adaptation in response to unpredicted results is almost inevitable in the early stages of vertebrate reintroductions, and rewilding projects will differ from other approaches largely in a willingness to tolerate a wider deviation from the initial plan and in their intention of foregoing interventions altogether as soon as possible. Moreover, conservation management in human-dominated parts of the world is often mostly about managing people, where the issue of infringing natural autonomy does not arise.

Most unforeseen problems are likely to involve an unexpected magnitude of predictable border problems, such as the human–wildlife conflicts mentioned above. These conflicts are not a problem only for rewilding: millions of people living next to protected areas that support carnivores, primates, or large herbivores already suffer from significant economic losses and injuries (Karanth and Kudalkar, 2017; Seoray-Pillai and Pillay, 2017). However, tolerance of existing populations of large animals may be higher than tolerance of reintroduced ones, so rewilding projects may suffer greater scrutiny. Various mitigation measures have been tried and financial compensation for damage is common, but well-maintained fencing seems to be most widely effective against dangerous and destructive animals (Durant et al., 2015). Other potential disservices from rewilding projects, such as an increase in fire hazard, will need a case-by-case approach, although compensation and insurance schemes may help with a wide range of non-lethal problems.

Conflicts with other conservation targets

Conservationists want to achieve multiple objectives, including the protection of endangered species and ecosystems, the control of invasive species, the provision of ecosystem services, such as clean water and carbon sequestration, and, as this book shows, an increase in wildness. They have succeeded in getting some of these wants written into national or supranational legislation. Conservation actors may thus be legally required to protect endangered species, maintain rare, semi-natural, vegetation types, control specific alien invasive species, and ensure that water supplies are not affected. Even where there is no legal obligation, all conservationists feel a strong moral obligation to prevent species extinctions, although their attitudes to semi-natural habitats, invasive species, and ecosystem services are more varied.

All species native to a region must have persisted there in natural habitats before humans arrived, but the areas available for rewilding today are not large or representative enough to include all facets of the landscape, and thus may not provide the specific habitat conditions that particular species need (Corlett, 2016). Many species in northern Europe have benefitted from traditional management practices that produced anthropogenic habitats, such as coppiced woodlands and hay meadows (Colebourn and Kite, 2017). Some of these practices may have imitated natural processes, such as grazing by large herbivores, that no longer occur, but in other cases it is likely that different processes have produced similar outcomes. In Europe, the species most threatened by rewilding are those that depend on disturbed sites without a closed canopy (Navarro et al., 2015). Hopes that reintroduced large herbivores can maintain analogues of valued semi-natural habitats in abandoned cultural landscapes (Vera, 2009) are almost certainly unrealistic, and planned interventions that mimic previous management practices are probably necessary for this (Fuller et al., 2017). Planned interventions are the opposite of rewilding, but if the aim is 'an increase in relative wildness' rather than 100 per cent wildness, then this aim can surely accommodate continued limited, local, interventions with proven benefits for wild species (Colebourn and Kite, 2017). Where reintroduced herbivores threaten grazing-sensitive plants, then the use of small exclosures can protect these species without disrupting large-scale ecosystem processes (Murray, 2017).

Laws and public opinion on animal cruelty may also constrain rewilding initiatives. At Oostvaardersplassen, a rewilded polder north of Amsterdam, rewilded horses and cattle that look as if they would not survive the winter are killed by rangers, so that they do not die slowly of starvation (Lorimer and Driessen, 2013). Invasive species management can also lead to a clash of values. A study of the 688-ha Knoxville Urban Wilderness in Tennessee identified widespread infestations of invasive species and provided management recommendations for their control, but also noted that 94 per cent of the visitors surveyed did not want more vegetation management (Zefferman et al., 2018). There has always been a degree of tension between 'scientific conservation' and public opinion, but this is likely to be greater with rewilding projects, simply because they are new and not well understood.

Recommendations and research needs

We need to know if rewilding works in the way that its proponents predict it will, so auditing of rewilding projects is inevitable. Auditability will vary between different types of rewilding projects, from very low in passive rewilding projects with no human interventions to relatively high in projects which start with interventions that are explicitly targeted at restoring ecosystem functions and processes. In these cases it may sometimes be possible to define at least some 'specific, measurable, achievable, reasonable, and time-bound' objectives and to use 'specific, quantifiable indicators' (McDonald et al., 2016), but these indicators are likely to miss some key aims of the projects. A more flexible approach would be to create a management committee that includes scientists, managers, and a wide range of other stakeholders, to monitor the project and to ensure that benefits, including non-material ones, continue to outweigh costs. Rewilding projects are all experiments, given our current state of knowledge. There should be a presumption against unplanned interventions, but there will also be red lines – legal, moral, or practical – that should not be crossed. The role of non-specialist stakeholders is crucial here, as they may well see things very differently from scientists and wildlife managers.

As many authors have pointed out, we need large-scale, long-term rewilding trials, ideally replicated, in which various options along the scale from zero intervention to ongoing management are compared (Corlett, 2016). A partial substitute for designed experiments is detailed monitoring of planned rewilding projects from before they start. Indeed, all projects should be viewed as experiments. Enclosures and exclosures can be used for experiments with species that do not require large areas, and fenced areas within rewilding projects can allow alternative interventions to be tried out on a smaller scale.

Finally, it is clear that we cannot have it all, at least not at the same time in the same place. We cannot combine a self-willed nature with other conservation targets in the areas left over after 7.5 billion people have taken their non-random share of the planet. We will need spatial separation between conflicting conservation goals and between these goals and most people, and, except in the largest areas, we will often have to accept the paradox of fences. 'Hybrid landscapes' in which people and wild species coexist are an alternative solution (Prior and Ward, 2016), but they are not always an option. Modern humans may learn to live with the uncomfortable wildness of wolves (Arts et al., 2016), but probably not with the more frequently lethal wildness of lions or elephants, even though these would be legitimate components of Pleistocene rewilding in Europe and North America. A rewilding that excludes dangerous and nuisance species and/or is confined behind fences may offend our ideas of 'wildness', but this is a human perspective, unlikely to be shared by any wild species. Nature runs wild whenever and wherever it can, and the aim of rewilding is to expand these opportunities as much as possible.

References

Arts, K., Fischer, A., and van der Wal, R. (2016). Boundaries of the wolf and the wild: a conceptual examination of the relationship between rewilding and animal reintroduction. *Restoration Ecology*, **24**, 27-34.

Brudvig, L.A. (2017). Toward prediction in the restoration of biodiversity. *Journal of Applied Ecology*, **54**, 1013-1017.

Cerqueira, Y., Navarro, L.M., Maes, J., Marta-Pedroso, C., Pradinho Honrado, J., and Pereira, H.M. (2015). Ecosystem services: the opportunities of rewilding in Europe. In Pereira, H.M. and Navarro, L.M. (Eds.), *Rewilding European landscapes* (pp. 47-64). Cham: Springer International.

Colebourn, P., and Kite, B. (2017). Anthropogenic rewilding: an oxymoron? *In Practice – Bulletin of the Chartered Institute of Ecology and Environmental Management*, **95**, 16-20.

Conservation Measures Partnership. (2013). *Open standards for the practice of conservation. Version 3.0.* Conservation Measures Partnership.

Cook, W., van Bommel, S., and Turnhout, E. (2016). Inside environmental auditing: effectiveness, objectivity, and transparency. *Current Opinion in Environmental Sustainability*, **18**, 33-39.

Cooper, N., Brady, E., Steen, H., and Bryce, R. (2016). Aesthetic and spiritual values of ecosystems: recognising the ontological and axiological plurality of cultural ecosystem 'services'. *Ecosystem Services*, **21**, 218-229.

Corlett, R.T. (2016). The role of rewilding in landscape design for conservation. *Current Landscape Ecology Reports*, **1**, 127-133.

Deary, H., and Warren, C.R. (2017). Divergent visions of wildness and naturalness in a storied landscape: practices and discourses of rewilding in Scotland's wild places. *Journal of Rural Studies*, **54**, 211-222.

Durant, S.M., Becker, M.S., Creel, S., et al. (2015). Developing fencing policies for dryland ecosystems. *Journal of Applied Ecology*, **52**, 544-551.

Fernandez, F.A.S., Rheingantz, M.L., Genes, L., et al. (2017). Rewilding the Atlantic Forest: restoring the fauna and ecological interactions of a protected area. *Perspectives in Ecology and Conservation*, **15**, 308-314.

Fernández, N., Navarro, L.M., and Pereira, H.M. (2017). Rewilding: a call for boosting ecological complexity in conservation. *Conservation Letters*, **10**, 276-278.

Fuller, R.J., Williamson, T., Barnes, G., and Dolman, P.M. (2017). Human activities and biodiversity opportunities in pre-industrial cultural landscapes: relevance to conservation. *Journal of Applied Ecology*, **54**, 459-469.

IUCN. (2012). *IUCN Red List categories and criteria version 3.1*, 2nd edition. Gland, Switzerland: IUCN.

Jain, R.K., Cui, Z.C., and Domen, J.K. (2016). Environmental auditing. In Jain, R.K., Cui, Z.C., and Domen, J.K. (Eds.), *Environmental impact of mining and mineral processing* (pp. 201-227). Boston, MA: Butterworth-Heinemann.

Jørgensen, D. (2015). Rethinking rewilding. *Geoforum*, **65**, 482-488.

Karanth, K.K., and Kudalkar, S. (2017). History, location, and species matter: insights for human-wildlife conflict mitigation from India. *Human Dimensions of Wildlife*, **22**, 331-346.

Kirby, K. (2017). Viewpoint: Are there too many or too few large herbivores in our woods? *In Practice – Bulletin of the Chartered Institute of Ecology and Environmental Management*, **95**, 11-14.

Law, A., Gaywood, M.J., Jones, K.C., Ramsay, P., and Willby, N.J. (2017). Using ecosystem engineers as tools in habitat restoration and rewilding: beaver and wetlands. *Science of the Total Environment*, **605**, 1021-1030.

Lorimer, J., and Driessen, C. (2013). Bovine biopolitics and the promise of monsters in the rewilding of Heck cattle. *Geoforum*, **48**, 249-259.

McDonald, T., Gann, G.D., Jonson, J., and Dixon, K.W. (2016). *International standards for the practice of ecological restoration- including principles and key concepts*. Washington, DC: Society for Ecological Restoration.

Mills, C.H., Gordon, C.E., and Letnic, M. (2018). Rewilded mammal assemblages reveal the missing ecological functions of granivores. *Functional Ecology*, **32**, 475-485.

Monbiot, G. (2013). *Feral: searching for enchantment in the frontiers of rewilding*. London: Allen Lane.

Müller, A., Bøcher, P.K., and Svenning, J.-C. (2015). Where are the wilder parts of anthropogenic landscapes? A mapping case study for Denmark. *Landscape and Urban Planning*, **144**, 90-102.

Müller, A., Bøcher, P.K., Fischer, C., and Svenning, J.-C. (2018). 'Wild' in the city context: do relative wild areas offer opportunities for urban biodiversity? *Landscape and Urban Planning*, **170**, 256-265.

Murray, M. (2017). Wild pathways of inclusive conservation. *Biological Conservation*, **214**, 206-212.

Navarro, L.M., and Pereira, H.M. (2015). Rewilding abandoned landscapes in Europe. In Pereira, H.M. and Navarro, L.M. (Eds.), *Rewilding European landscapes* (pp. 3-23). Cham: Springer International.

Navarro, L.M., Proença, V., Kaplan, J.O., and Pereira, H.M. (2015). Maintaining disturbance-dependent habitats. In Pereira, H.M. and Navarro, L.M. (Eds.), *Rewilding European landscapes* (pp. 143-167). Cham: Springer International.

Nogués-Bravo, D., Simberloff, D., Rahbek, C., and Sanders, N.J. (2016). Rewilding is the new Pandora's box in conservation. *Current Biology*, **26**, R87-R91.

Pheasant, R.J., and Watts, G.R. (2015). Towards predicting wildness in the United Kingdom. *Landscape and Urban Planning*, **133**, 87-97.

Pires, M.M. (2017). Rewilding ecological communities and rewiring ecological networks. *Perspectives in Ecology and Conservation*, **15**, 257-265.

Prior, J., and Ward, K.J. (2016). Rethinking rewilding: a response to Jørgensen. *Geoforum*, **69**, 132-135.

Sandom, C.J., and Macdonald, D.W. (2015). What next? Rewilding as a radical future for the British countryside. In Macdonald, D.W. and Feber, R.E. (Eds.), *Wildlife conservation on farmland* (pp. 291-316) Oxford: Oxford University Press.

Seddon, P.J., Moehrenschlager, A., and Ewen, J. (2014). Reintroducing resurrected species: selecting DeExtinction candidates. *Trends in Ecology & Evolution*, **29**, 140-147.

Seoray-Pillai, N., and Pillay, N. (2017). A meta-analysis of human–wildlife conflict: South African and global perspectives. *Sustainability*, **9**, 34.

Svenning, J.C., Pedersen, P.B.M., Donlan, C.J., et al. (2016). Science for a wilder Anthropocene: synthesis and future directions for trophic rewilding research. *Proceedings of the National Academy of Sciences of the United States of America*, **113**, 898-906.

Vera, F.W.M. (2009). Large-scale nature development – the Oostvaardersplassen. *Bristish Wildlife*, **2009**(June), 28-36.

Zamboni, T., Di Martino, S., and Jiménez-Pérez, I. (2017). A review of a multispecies reintroduction to restore a large ecosystem: the Iberá Rewilding Program (Argentina). *Perspectives in Ecology and Conservation*, **15**, 248-256.

Zefferman, E.P., McKinney, M.L., Cianciolo, T., and Fritz, B.I. (2018). Knoxville's urban wilderness: moving toward sustainable multifunctional management. *Urban Forestry & Urban Greening*, **29**, 357-366.

CHAPTER NINETEEN

Adaptive co-management and conflict resolution for rewilding across development contexts

JAMES R.A. BUTLER
CSIRO
JULIETTE C. YOUNG
NERC Centre for Ecology and Hydrology
MARIELLA MARZANO
Forest Research

Conservation translocation, or 'rewilding', is the deliberate movement of organisms from one site for release into another in order to yield a measurable conservation benefit at the levels of a population, species or ecosystem (IUCN/SSC, 2013). The intention of rewilding, including taxonomic substitution where direct species reintroduction is not feasible, is to restore ecological system processes (Soulé and Noss, 1998). As such, the primary justification for translocations may be national or globally determined conservation objectives, rather than local communities' needs and priorities, which can potentially generate conflict, jeopardising the initiative (Nogués-Bravo et al., 2016).

Because rewilding is a recent concept, popularised originally by Soulé and Noss in 1998, only preliminary guiding principles have been formulated. Foremost are the IUCN/SSC's (2013) guidelines, which focus on identifying and managing the ecological risks from a conservation and restoration biology perspective. Subsequent refinements of these guidelines have also been undertaken with an exclusively ecological bent (e.g. Batson et al., 2015; Robert et al., 2015). The IUCN/SSC guidelines also recommend consideration of the social feasibility of reintroductions: 'planning should accommodate the socio-economic circumstances, community attitudes and values, motivations and expectations, behaviours and behavioural change, and the anticipated costs and benefits of the translocation' (IUCN/SSC, 2013, p. 11). These guidelines suggest mitigating measures such as communication, engagement, and problem-solving, and the establishment of special teams working outside formal bureaucratic hierarchies to respond to management issues. However, considering the critical role of stakeholder relationships in the success of rewilding initiatives, we suggest that the governance of rewilding requires deeper consideration, where 'governance' is defined as 'the norms, institutions and processes that determine how power and responsibilities over natural resources are exercised, how decisions are taken, and how citizens ... participate in and benefit from the management of natural resources' (Campese et al., 2016, p. 1).

There appear to be three areas requiring attention. First is the tendency to regard stakeholder involvement as a subsidiary consideration in rewilding initiatives. For example, the IUCN/SSC (2013) guidelines list six biological feasibility criteria that should be met before social and regulatory issues are considered. Similarly, Nogués-Bravo and colleagues (2016) list economic and societal conflicts as potential consequences of rewilding, but only after biological diversity and biocontrol concerns. In Gray et al.'s (2017) framework for assessing the feasibility of large carnivore reintroductions, management capacity and local community support are the third and fourth steps in their decision tree. While perhaps unintentional, the priority given to ecological issues implies the predominance of external experts, their knowledge and goals above social considerations and local stakeholders' roles. This overlooks the recognition that collaborative approaches to natural resource management can pre-empt conflict (Davies and White, 2012; O'Brien et al., 2013), and that governance is strengthened by the involvement of all stakeholders as equal partners (Phillipson et al., 2009; Redpath et al., 2017).

Second, perhaps due to the heritage of rewilding in conservation biology, is the distinction between ecological and social considerations. This is despite the fact that natural resources and efforts to manage them are increasingly being conceptualised as social–ecological systems (Walker et al., 2006), defined as societal and ecological subsystems in mutual interaction (Gallopin, 1991). Social–ecological systems are characterised by non-linear dynamics due to thresholds in variables, and cross-scale reinforcing feedback loops that amplify interactions within the system, which can potentially cause sudden and unexpected outcomes and shifts to alternative system states (Walker et al., 2004). Human components of social–ecological systems are as inherently important as ecological factors, and understanding the influence of stakeholders' behaviour, values, power, politics, and rights on system dynamics is critical for any systems-based thinking (Brown and Westaway, 2011; Armitage et al., 2012; Stone-Jovicich, 2015).

Third, there is a continuum of social and ecological contexts within which rewilding may occur (Figure 19.1), and which governance must account for. While the greatest attention is being given to rewilding in developed nations, where most extirpations have occurred (Carey, 2016),

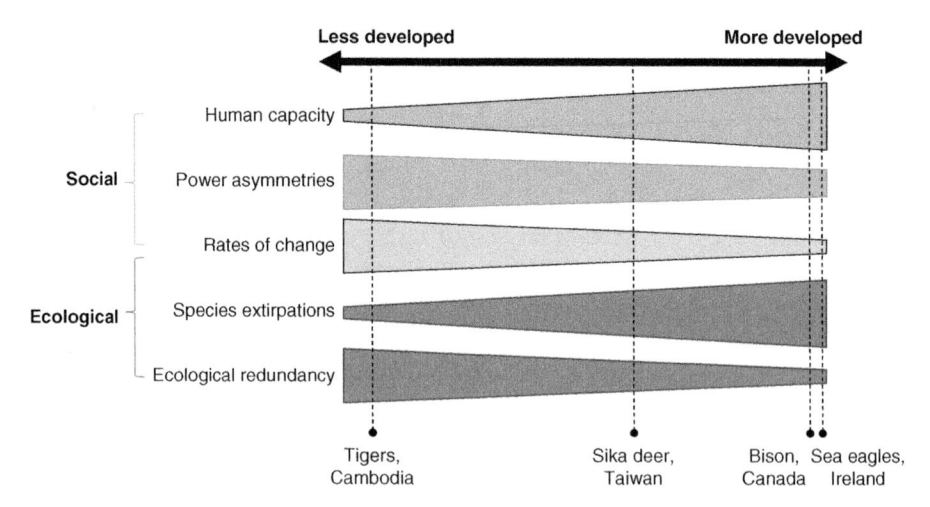

Figure 19.1. A heuristic of the relative differences in the social and ecological contexts for rewilding, on a continuum between less and more developed countries and regions, informed by Walker (1992), Armitage and Johnson (2006), Ensor (2011), and Butler and colleagues (2014). The indicative positions of the four rewilding case studies considered in this chapter are also shown. (A black and white version of this figure will appear in some formats. For the colour version, please refer to the plate section.)

future initiatives may be necessary in developing countries, where socio-economic transitions are threatening many species ranges and existence. In these situations the rewilding process may face greater governance challenges, because human capacity (including information and knowledge) among all stakeholders is lower, and asymmetries in power between interest groups are acute (Ensor, 2011). Relative to more developed nations, population and economic growth are faster, and consequently societal and ecological change is rapid (Armitage and Johnson, 2006; Butler et al., 2014). The justification for rewilding may also be different from a conservation perspective, because many developing countries are located in the tropics or subtropics, where ecological redundancy is higher (Walker, 1992). In addition, conservation values and motivations may differ from Western societies due to more immediate poverty alleviation priorities and cultural norms (Nilsson et al., 2016a, 2016b).

The purpose of this chapter is to address the evident gap in the governance of rewilding. We aim (1) to assess case studies of rewilding across development contexts from a governance perspective (Figure 19.1); (2) to review contemporary principles for the effective governance of social–ecological systems and contested development interventions; and (3) using the common challenges distilled from the case studies, to adapt the principles into a generic approach for planning and implementing rewilding.

Case studies of rewilding and its governance

Because rewilding is a recent phenomenon there is a paucity of published examples, and fewer that explicitly evaluate their governance. Most initiatives have been established over several decades in the more developed world, providing the opportunity for some longitudinal analysis, but those in the developing world are limited to scoping or preparatory assessments. Here we review four case studies of reintroductions from across the continuum of development contexts (Figure 19.1) in order to highlight their differences and commonalities, and to inform a generic governance approach for rewilding.

Sea eagles in Ireland

O'Rourke (2014) describes the reintroduction of the white-tailed sea eagle into Killarney National Park, Ireland, by National Park authorities. The project team consisted of wildlife and ecology experts who largely neglected consultation and information provision with local communities. Although there had been some public meetings and media reporting, detailed consultation and awareness-raising about the project with communities only occurred after it had started. While sea eagles were promoted as a beneficial species for sheep farmers because they could control other predators of sheep such as grey crows and foxes, farmers and community members were opposed to the reintroduction due to the risk of increased predation on lambs by the eagles. In addition, there were fears about future conservation designations for sea eagles which would restrict alternative land uses (e.g. wind farms) and hence future income sources. There were already tensions between local farmers and the National Park authorities over the management of other species (e.g. deer), and the impact of conservation designations on land use. Conversely, the local eco-tourism sector supported the reintroduction, causing strained relations within the community.

O'Rourke (2014) concludes that in areas with a strong agricultural heritage there can often be a friction between wild and domestic animal interests. The author suggests that successful rewilding initiatives in these contexts require a change in farmer attitudes, and the development of trust between stakeholders, although this takes time and effort. Ultimately, O'Rourke considers that social acceptability of rewilding necessitates giving local communities equal stake and ownership in any decision that affects their livelihoods and land use.

Bison in Canada

Clark et al. (2016) and Jung (2017) have documented the Canadian government's reintroduction of bison in south-western and north-western Canada in 1988–1995, and the unexpected problems caused by the species years after their establishment. The reintroductions were considered successful from an ecological perspective, but two decades later a series of concerns were raised by Champagne and Aishihik First Nations' (CAFN) people in the south-western case. The reintroduction had taken place without any prior consultation with CAFN, whose land claims in the region were established a year after the programme began. A participatory socioeconomic impact assessment with the CAFN was subsequently initiated, involving workshops and interviews with community members. Bison were perceived to have impacted the habitats of other valued species, posed risks to human safety, and had resulted in increased disturbance from recreational bison hunters. In the north-western case, movement by adult male bison away from their original range had resulted in aggressive encounters with people in settlements and crop damage, and a negative impact on social acceptance. As a result, wildlife officers had to shoot four of the 23 problem animals.

For the CAFN there was a particular concern about destruction of medicinal plants by bison, which had knock-on effects in the form of communities having to travel further to find these resources, which posed difficulties for less-mobile individuals, as well as fears of being attacked by other wild animals. However, a participatory approach to examining the problems led to wider discussions over the impacts of climate and environmental change on livelihoods and ecology in the area, and a move away from 'blaming the bison' (Clark et al., 2016, p. 4). Discussions also revealed emerging disquiet about other issues, such as the influx of bison hunters from other areas on to CAFN traditional lands. However, since this response, a collaboration has been established between National Park authorities, First Nations' authorities, and communities that encourages knowledge sharing and co-management, and generating practical solutions (e.g. fencing, changes to hunting seasons, and the utilisation of bison), which has restored the social acceptability of the initiative.

Sika deer in Taiwan

Another example involves the reintroduction of the Formosan sika deer in Kenting National Park, Taiwan (Yen et al., 2015). A restoration programme was initiated in 1994, and successfully re-established a wild sika population. The National Park also encompasses human settlements, and there has been a history of discord between residents and the park authorities over use of the protected area (e.g. agricultural activities, hunting, fishing, and settlement). However, eco-tourism related to wildlife viewing has been developed in some areas of the park, yielding economic benefits for resident-led schemes, which have provided important livelihood options in the typically low-income communities within and surrounding the park.

A survey carried out in 2010 among 228 local community residents found strong support for the presence of wild sika, but there were concerns from farmers about deer damage to crops. However, Yen et al. (2015) note that as the sika population grows, movement out of the park and increased interactions with people are likely to occur, and hence attitudes may change if authorities do not engage communities in the programme. This will be exacerbated by existing underlying tensions between residents and authorities over the governance of the park. The authors identified a clear demand from survey participants for the development of community-run eco-tourism initiatives to reduce tensions over park administration, and any potential conflicts over future deer damage. They also recommended the engagement of communities in park co-management, and improved communication by park authorities about the reintroduction.

Tigers in Cambodia

While there is not a similar depth of experience of rewilding in less-developed countries, rapid loss of habitat and overexploitation is forcing the consideration of reintroductions of large mammals in some locations. For example, Asian tigers are functionally extinct in Cambodia, and reintroductions are being planned by the national government's 2016 Cambodia Tiger Action Plan (Gray et al., 2017). To support this policy, Gray et al. (2017) present a framework for undertaking ecological, management, and social feasibility studies of large carnivore reintroductions in less-developed tropical contexts. The framework involves a four-step decision tree of assessments: first, considering whether there is a landscape suitable for a reintroduction; second, the quantification of potential prey; third, evaluating whether there is sufficient management capacity; and fourth, quantifying the levels of local community support for the reintroduction. The authors carried out a feasibility assessment using the Srepok Wildlife Sanctuary as a test case.

The third step applied the Management Effectiveness Tracing Tool (METT; Hockings et al., 2003), which evaluates protected area management planning, legal status and law enforcement, resources, budgeting and staff, education and community relations, and eco-tourism potential. To undertake the METT, a workshop was held with park management and rangers, local community members, and an international NGO's staff. This was augmented by the Conservation Assured Tiger Standards (CA|TS) tool, which is designed to assess management effectiveness specifically for tiger reserves (Pasha et al., 2014). The CA|TS assessment was undertaken by a working group consisting of provincial government and local NGO staff with knowledge of the site. The fourth step was carried out by interviewing households in 12 communities surrounding the reserve. The inhabitants were mostly from the Bunong ethnic minority.

Gray et al.'s (2017) results showed that the human and institutional capacity for managing the tiger reintroduction did not meet their targets, due to inadequate staffing, financial planning and zonation, and insufficient eco-tourism activities to generate positive economic benefits for communities. In terms of community acceptance, less than 75 per cent expressed support for the reintroduction, which did not achieve the assessment's target. Also, the lack of park authorities' conflict mitigation strategies was considered a barrier to successful implementation. Consequently, even though these aspects were the third and fourth steps of the decision tree, the authors concluded that the reintroduction should not go ahead until the management and social limitations had been addressed. Furthermore, the broader issues of rapid urbanisation and linked demand for wildlife products, many harvested illegally from potential tiger prey species, require national government action including strengthened legal enforcement and targeted behaviour change campaigns to reduce consumption by urban populations.

Differences and common themes

Although the information provided by the case studies' literature is patchy, there is some evidence that supports our heuristic of social–ecological differences across the development continuum (Figure 19.1). On the human capacity spectrum, the METT and CA|TS assessments revealed that management effectiveness was poor in the less-developed case of Cambodian tiger reintroduction, while in the more developed case of bison in north-western Canada the wildlife authorities were clearly well resourced and able to control problem animals. In terms of the power asymmetry spectrum, the local communities in the Cambodia tiger case were from the Bunong ethnic minority, and hence possibly marginalised by management. In the middle of the spectrum, the communities in the Taiwan sika deer case were recognised to be poor, but were empowered to some extent, as evidenced by their demands for the development of community-run eco-tourism initiatives. At the more developed extreme, the Canadian First Nations in the bison case had established land claims, which may have enabled them to negotiate co-management arrangements with parks authorities. In terms of rates of change, the Cambodian tiger case illustrated the social and ecological flux typical of developing countries, with rapid urbanisation and growing markets for wildlife products resulting in the illegal harvest of tiger prey species.

Despite these probable differences, there were four common challenges for the governance of rewilding in complex social–ecological systems: the presence of multiple private and public stakeholders; their engagement in decision-making about the reintroduction, and its varied consequences; existing tensions between livelihoods and conservation; and other system drivers.

Table 19.1. Four governance challenges common to the rewilding case studies, plus solutions and innovations.

Case study	Governance challenges				5. Solutions and innovations
	1. Multiple private and public stakeholders	2. Engagement and consequences	3. Existing tensions between livelihoods and conservation	4. Other system drivers	
Sea eagles, Ireland	Farmers, eco-tourism operators, tourists, National Park authorities, scientific experts	Consultation after reintroduction; farmer opposition, tourism support	Farmers and park authorities conflict over other species, conservation designations	Wind farms and alternative land-use options	
Bison, Canada	First Nations' community, bison hunters, farmers, National Park authorities	Consultation after reintroduction; co-management	First Nations' land claim	Community health, environmental and climate change, bison hunting	Co-management, utilising bison, changes to hunting season, fencing
Sika deer, Taiwan	Farmers, fishers, hunters, eco-tourism operators, tourists, National Park authorities	Community survey after reintroduction	Communities and park authorities discord over use and access to park	Low income of communities within and around park	Co-management, community-run eco-tourism
Tigers, Cambodia	Ethnic minority communities, eco-tourism operators, tourists, National Park authorities, provincial government, local and international NGOs	Community survey before reintroduction; insufficient support to proceed	Lack of management and enforcement capacity, illegal wildlife harvesting	Ethnic minority rights, rapid urbanisation, demand for wildlife products, illegal wildlife harvesting	Tiger–human conflict mitigation strategies, eco-tourism development, national enforcement and urban awareness-raising

A fifth theme was the solutions and innovations that had emerged to address the challenges (Table 19.1).

In all case studies there were at least four clear stakeholder groups, both private (e.g. farmers, eco-tourism operators) and public (e.g. National Park authorities, provincial government, NGOs), with differing values and objectives. These stakeholders could be further distinguished as 'on-site' (i.e. those who stand to lose the most from land-use decisions) or 'off-site', and from 'local' to 'international' in scale (Brown et al., 2001; de Groot, 2006; Butler et al., 2013). Their engagement in decision-making differed, with varied outcomes. Only for tigers in Cambodia were local villagers engaged before the reintroduction, and their lack of support contributed to the decision to postpone and redesign the initiative. For sea eagles in Ireland, bison in Canada, and sika in Taiwan, there was no concerted engagement until after the reintroduction had occurred. In Ireland, this amplified existing tensions between farmers and conservation authorities. In Canada and Taiwan, subsequent bison and sika deer problems were addressed through participatory processes initiated by National Park authorities. There were existing tensions between local livelihoods and conservation, such as the conflict over conservation designations for other species in Ireland, and discord about local residents' access and use of the National Park in Taiwan. Wider drivers and changes also contributed to this context, such as First Nations' land claims and rights in Canada, and rapid urbanisation and the related growing demand for wildlife products derived from potential tiger prey in Cambodia.

Finally, with the exception of sea eagles in Ireland, there was clear evidence of solutions and innovations which could ameliorate some of the existing or emerging conflicts, or address wider system drivers (Table 19.1). In the case of bison in Canada, a co-management process had been initiated which triggered novel activities regarding bison utilisation, changes to hunting seasons and fencing. Co-management was also recommended as a solution to sika deer problems and wider park management access issues in Taiwan. The feasibility assessment for tigers in Cambodia highlighted the need for national government action on illegal wildlife harvesting and urban awareness-raising campaigns.

Contemporary principles for the governance of rewilding

As the case studies show, rewilding is essentially an experiment. To be successful, it requires social acceptance by stakeholders about a risk that may yield unexpected benefits, costs, and conflict, and these have to be monitored and managed over the long term (IUCN/SSC, 2013; Nogués-Bravo et al., 2016). In combination with other drivers, the reintroduction may cause the social–ecological system within which it occurs to alter (Wiens and Hobbs, 2015; Corlett, 2016), generating sudden and unprecedented challenges for stakeholders (Armitage et al., 2009; Abel et al., 2016). While not conceived or analysed as social–ecological systems, the case studies could be interpreted in these terms.

For example, the sea eagle reintroduction in Ireland may increase predation of lambs, reducing flock productivity to the point where sheep farming is no longer economically viable. Having passed this threshold, farmers may diversify into other forms of income, such as electricity generation from wind farms, which fundamentally alter the aesthetic attraction of the Killarney National Park, reducing tourist numbers and undermining the economic viability of local eco-tourism businesses. This could result in a shift from a system characterised by diversified livelihoods based on traditional agriculture, conservation, and tourism to one reliant on energy infrastructure and production. Exogenous sociopolitical drivers such as European Union and Irish energy and conservation policies could accelerate or hinder this landscape transformation. If it were to occur, the consequences in terms of the maintenance of biodiversity and cultural values and escalations in stakeholder conflict and community tensions could be far-reaching.

In this section we review two contemporary and overlapping governance approaches which are applicable to such rewilding experiments: adaptive co-management of complex social–ecological systems, and social licence to operate for new and contested developments.

Adaptive co-management

Due to their inherent complexity, the design of governance for social–ecological systems has become an important and growing field of research (e.g. Folke et al., 2005; Armitage et al., 2009; Butler et al., 2016a; Plummer et al., 2017). Adaptive co-management (ACM) has recently evolved as a successful approach. In contrast to conventional, centralised 'command-and-control' governance, it combines the iterative co-learning, knowledge generation, and problem-solving of adaptive management with the stakeholder power-sharing and conflict resolution of co-management (Armitage et al., 2009; Berkes, 2009; Keith et al., 2011). Folke et al. (2002, p. 8) define ACM as 'a process by which institutional arrangements and ecological knowledge are tested and revised in a dynamic, ongoing, self-organized process of trial-and-error'.

ACM is advocated for the management of complex social–ecological systems because it encourages cross-scale social networks, integration of multiple knowledge types to solve complex and unprecedented problems, and reflexivity through continual evaluation and learning, which together enhance capacity to anticipate and respond to uncertainty and shocks (Olsson et al., 2004; Armitage et al., 2009; Fabricius and Cundill, 2014). In addition, due to its focus on collaboration and power-sharing, ACM is conducive to resolving conflict between stakeholders, aided by its co-learning, solution-orientated innovation, and trust-building elements (Gutiérrez et al., 2016).

ACM is underpinned by stakeholder engagement and participation, which is one of the fundamental prerequisites for the achievement of sustainable development (UNCED, 1992), and has long been advocated in decision-making (Renn, 2006), policy implementation (Eden, 1996), policy evaluation (Fischer, 1995), conflict resolution (Gutiérrez et al., 2016), and poverty alleviation (Chambers, 1994). The assumption is that 'greater participation will allow more inclusive inputs into decision-making processes, which in turn will lead to better decisions [and] ... more informed forms of representation' (Gaventa, 2004, p. 9). Such participation also brings new or different knowledge types to the decision-making process, which improves the technical quality of decisions and more holistic conceptualisations of the environment (Berkes, 2009) through the inclusion of local actors' ways of knowing, their values (Beierle and Konisky, 2001) and interests (Primmer and Kyllonen, 2006). Successful knowledge co-production is an 'inclusive, accountable, fair and open process, enabling actors to engage in learning through iterative dialogue' (Hegger et al., 2012, p. 62), which enhances local community empowerment and self-determination (Sillitoe, 1998). In turn, decisions that are agreed upon collectively and acknowledge local concerns and knowledge have a higher chance of being socially and politically accepted (Harrison and Burgess, 2000; McCool et al., 2000; Marzano et al., 2017), potentially pre-empting conflict and enhancing the likelihood of positive social and ecological outcomes (Young et al., 2010).

While the 'what' of ACM is clear, it has been critiqued for the lack of detail on the 'how', and limited published evidence of clear outcomes (Rist et al., 2013; Fabricius and Cundill, 2014; Plummer et al., 2017). To some extent this is defensible, because ACM is an emergent property of a social–ecological system, often occurring in response to an exogenous shock or resource crisis (e.g. Olsson et al., 2004, 2006; Butler et al., 2008; Plummer, 2009; Butler, 2011). Consequently there is no blueprint for the process of ACM, as each instance will be context-specific and self-organising (Plummer et al., 2012). However, some initiatives have attempted to engineer ACM by creating a structure and process founded on its principles of multistakeholder engagement and learning (e.g. Cundill and Fabricius, 2010; Smedstad and Gosnell, 2013; Butler et al., 2016a, 2016b).

There are examples of successful conflict mitigation due to ACM, for example between dugong hunting and conservation (Butler et al., 2012), seal conservation and salmon fishery interests (Butler et al., 2015a), dolphin conservation and illegal netting (Butler et al., 2017b), and in UNESCO Biosphere Reserves (Plummer et al., 2017). These examples have also identified key prerequisites for the maintenance of conflict resolution via ACM, including long-term government support for the process, strong leadership and champions, cross-scale partnerships, and the appearance of 'windows of opportunity' to modify policies and institutions (Young et al., 2012; Butler et al., 2015a). These are now being mainstreamed into conservation conflict efforts (e.g. Redpath et al., 2013, 2017; Young et al., 2016), where the focus is shifting from conflict resolution, which emphasises compromise and jointly agreed outcomes, to conflict transformation, which identifies and addresses the root sociopolitical sources of conflict (Mitchell, 2002). This requires the transformation of institutions and discourses within which actors frame their positions (Ramsbotham et al., 2014), which the co-learning facet of ACM can potentially facilitate.

Social licence to operate

The term 'social licence to operate' (SLO) emerged in the 1990s to describe the informal acceptance, approval, or trust that a local community extends to a corporate entity or industry developing operations within its geography, with a specific application to mining (Lacey and Lamont, 2014). The concept has since been extended to other industries, such as forestry (Moffat et al., 2016). More recently, SLO has been applied to the introduction of biotechnology in biosecurity, such as Sterile Insect Technology in tropical Australia to control mosquito-borne dengue fever (Butler et al., 2017a).

SLO is a useful paradigm because it highlights the need for development proponents to acknowledge and address social concerns about a novel proposal, and is the starting point for dialogue between stakeholders (Moffat et al., 2016). It also emphasises the need for a relationship based on trust and transparency to be cultivated between the proponents and local communities, and hence good governance and social justice (Lacey and Lamont, 2014). SLO implies that an agreement will be reached between a developer and communities which mirrors the 'licence' granted by government to the company to undertake operations, with its necessary safeguards (Moffat et al., 2016).

However, SLO frames relationships between developers and communities simplistically, and could enable business to claim that it is addressing community concern in order to gain reputational capital, while real issues are avoided (Owen and Kemp, 2013; Moffat et al., 2016). Also, while the objective of SLO is clear, the process for achieving it is difficult to define and open to multiple interpretations (Parsons and Moffat, 2014). Furthermore, most communities are heterogeneous (Ojha et al., 2016), making it problematic to identify who 'grants' the SLO (Banks, 2002). Combined with the shifting dynamics of communities, any SLO will have to be continually revisited and renewed (Moffat et al., 2016).

Kendal and Ford (2018) have assessed the relevance of SLO to threatened species programmes. Conservation interventions are likely to be more complicated than a development intervention, because stakeholders tend to range from local to global, and have a greater spectrum of attitudes on environmental issues (Ford and Williams, 2016). Because conservation initiatives are usually government-led and therefore acting in the public rather than the private interest, more complex partnerships are required between the public sector and local stakeholders (Ojha et al., 2016). Regardless, SLO is a useful paradigm in conservation because it emphasises the need for practitioners to develop trusting relationships with local and other participants, to recognise and address the diversity of their views, and to anticipate and address potential conflict through transparent governance (Kendal and Ford, 2018). Clearly these fundamental principles overlap those of ACM.

A generic governance approach for rewilding

In this section we adapt and apply the overlapping governance principles and processes of ACM and SLO to rewilding. The common challenges distilled from the four case studies (Table 19.1) provide generic issues which our approach is designed to account for. To provide a framework, we present our approach as three components: structure, process, and outcomes.

Structure

While there is no blueprint for ACM, it is possible to engineer an ACM process by encouraging multistakeholder engagement and learning (Cundill and Fabricius, 2010; Smedstad and Gosnell, 2013; Butler et al., 2016a, 2016b). It is this approach that we suggest for rewilding initiatives. Our structure is the well-known adaptive management cycle, involving the steps of plan, design, implement, monitor and evaluate, and revise (Williams et al., 2009). We simplify this into three steps: (1) plan and design, (2) implement activities, and (3) monitor and evaluate. To this we add the prior establishment of a facilitation team to act as brokers among the multiple private and public stakeholders from the different scales (Figure 19.2). It should be emphasised that the establishment and maintenance of a facilitation team requires adequate resourcing, something which is often overlooked by funding agencies (Butler et al., 2016b, 2017c).

The role of the facilitation team is to identify and engage stakeholders, organise activities that enable dialogue and consensus-building, broker knowledge and information, and mediate or resolve conflict. Also referred to as 'bridging' or 'brokering' organisations (or individuals), the team should be regarded as independent and trustworthy by all stakeholders (Olsson et al., 2004; Armitage et al., 2009; Cundill and Fabricius, 2010). The team acts as a conduit between stakeholders from different levels, creating the social networks across scales that are critical to harnessing knowledge and generating innovation (Olsson et al., 2004). Hence team members must be skilled in cross-

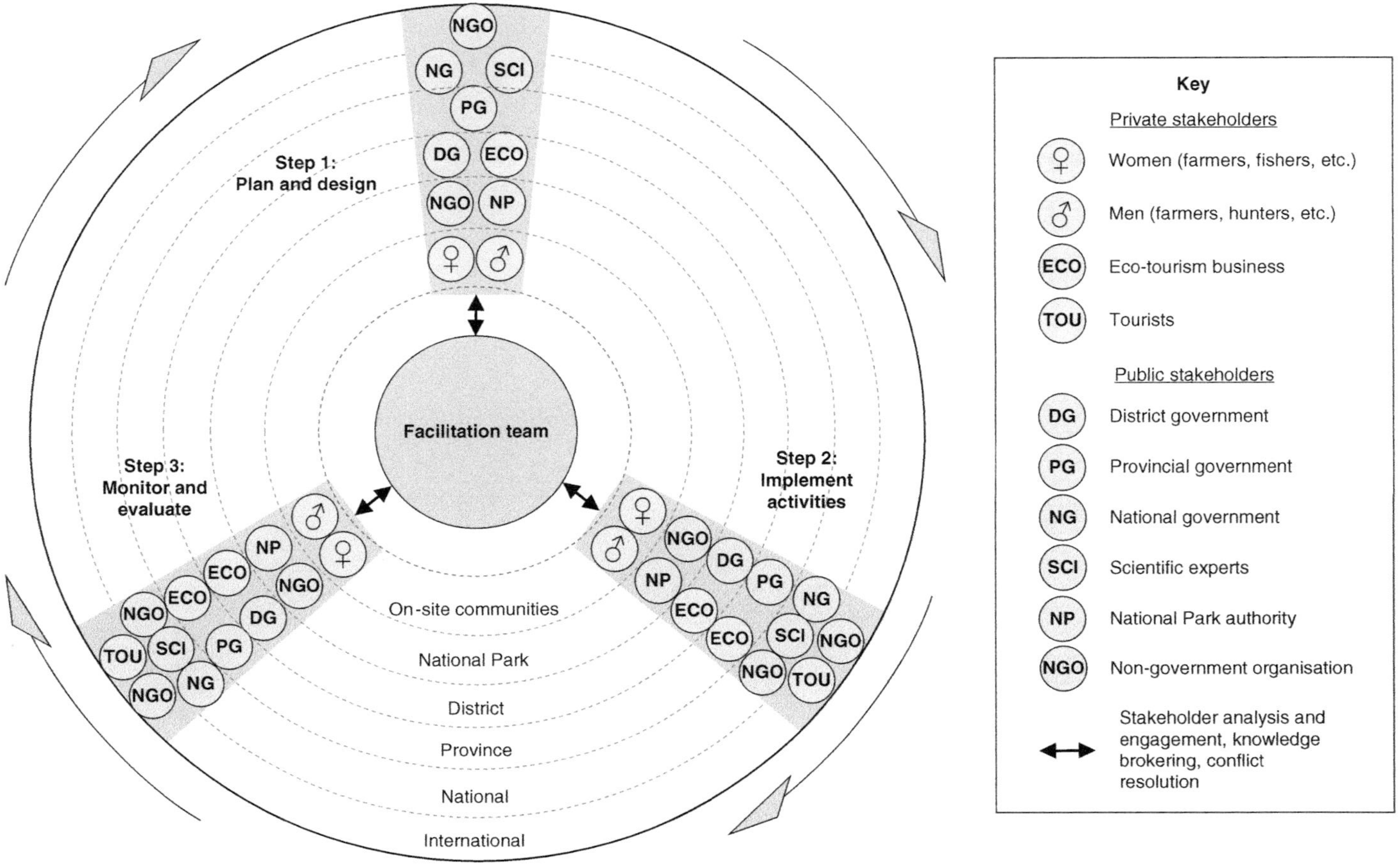

Figure 19.2. The ACM structure and process for the governance of rewilding. Likely stakeholders in a rewilding initiative are differentiated as private or public, and by their scale, from on-site to international. (A black and white version of this figure will appear in some formats. For the colour version, please refer to the plate section.)

sectoral communication, mediation, conflict resolution, event organisation, and facilitation (Butler et al., 2017c).

The facilitation team's first task is to carry out a stakeholder analysis for the rewilding location and its social–ecological system. There are numerous suitable methodologies (e.g. Shultz et al., 2007; Reed et al., 2009; Baird et al., 2014). To address potential power asymmetries between local communities and other actors, this exercise must analyse political dynamics and ensure that weaker or marginalised stakeholders are adequately represented, and that the most powerful are willing to share decision-making (Armitage et al., 2009; Butler et al., 2015b). It may also be necessary to create a steering committee representing the major stakeholder groups to provide the political legitimacy for the governance structure, and to formally link to national policy processes (Butler et al., 2016a, 2016b). Stakeholders likely to be engaged are women and men from on-site communities, eco-tourism businesses and tourists, and the relevant government and National Park authorities, scientific experts and NGOs (Figure 19.2).

Process

Step 1: Plan and design. Step 1 is planning and designing the rewilding initiative. There may be legal requirements which predetermine the format of this activity, particularly where public lands such as National Parks are concerned, or locations including First Nation or Aboriginal land rights (Pratt Miles, 2013). Encouraging stakeholders to participate, and understanding their incentives to do so, can be problematic, and contains its own ethical and political tensions (Cooke and Kothari, 2001; Stringer et al., 2006; Hurlbert and Gupta, 2015) which the facilitation team must have skills to manage (Butler et al., 2017c). However, fundamental to ACM is the creation of a forum that can engage stakeholders in open dialogue, and where different knowledge and information can be considered and respected equally.

To frame the reintroduction within a social–ecological context, we suggest that Step 1 should carry out the co-learning activity illustrated in Figure 19.3. This enables the invited stakeholders and their knowledge to 'move together in an interactive, iterative process in which everyone enhances the understanding of everyone else' (Brown, 2008, p. 48). Referring to the system and challenge concerned, four questions are addressed in succession: 'what is?', 'what should be?', 'what could be?', and 'what can be?', resulting in an agreed set of actions. When applied to rural livelihood adaptation through a series of multistakeholder workshops in Indonesia and Papua New Guinea, this activity generated innovative ideas, empowerment, and new partnerships (Butler et al., 2015b).

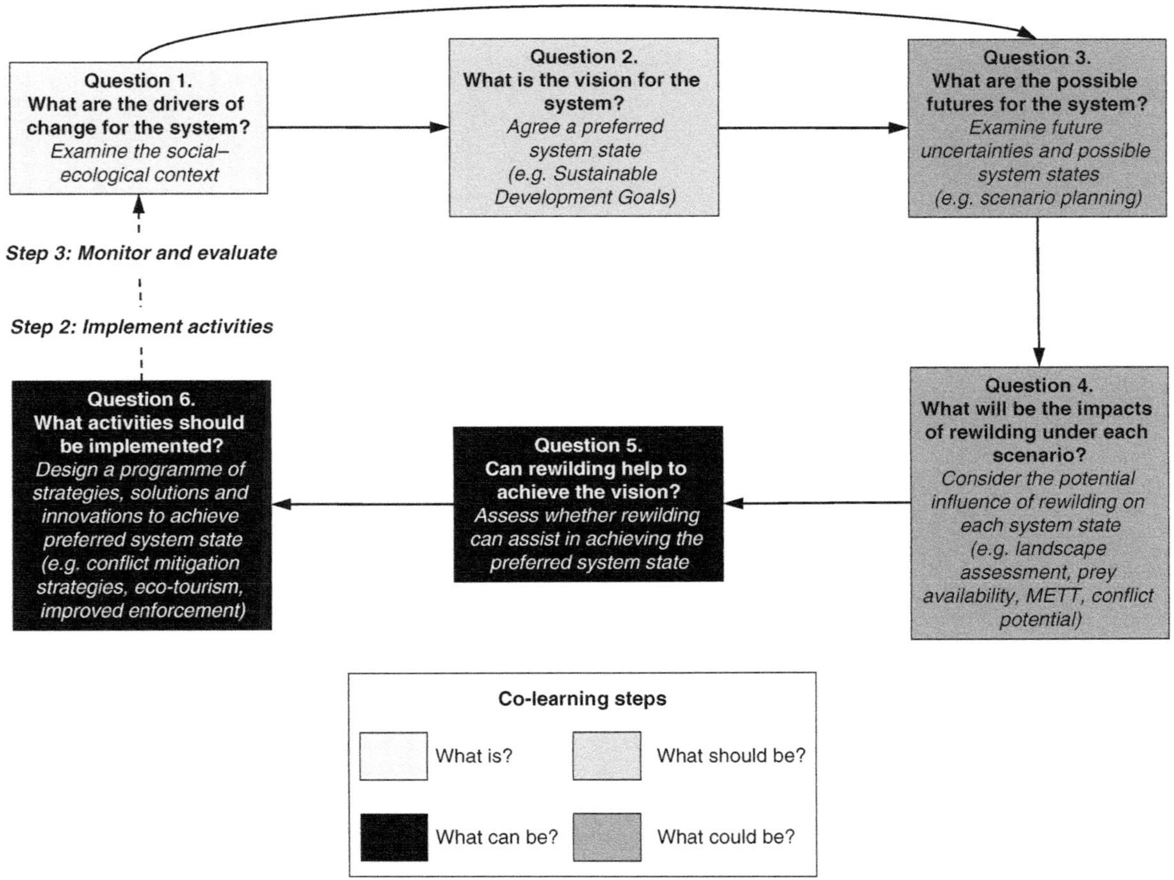

Figure 19.3. The co-learning activity recommended for *Step 1: Plan and design* in a rewilding initiative, adapted from Brown (2008) and Butler et al. (2015b). After the first iteration of Step 1, this is repeated in successive ACM cycles following *Step 2: Implement activities* and *Step 3: Monitor and evaluate*.

The first question addresses the drivers of change influencing the system, thus establishing the social–ecological context and 'what is?' (Figure 19.3). The second question establishes a vision for the system, and hence a consensus on 'what should be?' One prompt for this discussion could be the United Nation's Sustainable Development Goals (United Nations, 2015), which include the achievement of terrestrial biodiversity, human health and well-being outcomes by 2030. The third question examines potential future system states given trends and uncertainties in the primary drivers identified in the first question. Scenario planning is an effective and well-established tool for this exercise (e.g. Bohensky et al., 2011; Oteros-Rozas et al., 2015; Butler et al., 2016c).

The remaining three questions then cast the rewilding initiative into the system context and stakeholders' agreed vision (Figure 19.3). The fourth question considers the potential influence and role of rewilding on each future system state. At this stage, various tools and information already established in rewilding and restoration science could be applied, including landscape suitability assessments, prey availability (for carnivores), and management effectiveness (e.g. METT). Based on these assessments, the fifth question assesses whether the initiative complements or impedes the attainment of the stakeholders' vision, thus asking 'what can be?' If rewilding is compatible with the vision, or requires further preparation or prior experimentation, the sixth and final question seeks to agree a programme of strategies and innovative solutions which can be rolled out in Step 2: Implement activities. These might include the development and testing of conflict mitigation strategies, ecotourism development, or policy interventions to tackle broader system issues such as urban demand for illegal wildlife products.

Step 2: Implement activities. Step 2 implements the activities identified by Step 1. Fundamental to this is multistakeholder engagement in learning-by-doing experiments (Armitage et al., 2009; Plummer, 2009; Plummer et al., 2012). Each may involve a different set of actors, and possibly others additional to those identified in Step 1 (Figure 19.2). In other examples, innovative livelihood strategies developed by the planning activities have been trialled as pilot studies, creating their own 'bridgeheads' for ACM within communities (Butler et al., 2016a, 2016b).

Step 3: Monitor and evaluate. Step 3 is the monitoring and evaluation of the activities. While regarded as separate and stand-alone in adaptive management, for ACM this should be engrained within all activities to create a culture of ongoing reflection and learning (Armitage et al., 2009), enabled by the facilitation team (Figure 19.2). There may be different forms of monitoring and evaluation applied to different aspects of the initiative. For example, an overall Theory of Change (ToC) could be developed, which articulates a vision of change, and systematically describes the sequence of activities, outputs, outcomes, and impacts to achieve it, and the assumptions about the relationships between interventions and change (Vogel, 2012). A ToC is recommended for conceptualising complex systems problems, and guiding evaluation of project impact (Bours et al., 2013). If used in a participatory process among multiple stakeholders it can be a powerful co-learning tool because it openly tests their assumptions and world views (Vogel, 2012). Similarly, a ToC could be developed for individual activities or experiments implemented in Step 2.

Another complementary approach to ToC is the assessment of programme outcomes in terms of ACM and SLO (see below). If carried out in a participatory process which engages stakeholders from multiple scales to reflect and learn (Figure 19.2), this has the added advantage of catalysing action to improve the design of the programme in the subsequent ACM cycle (Butler et al., 2015a, 2016a; Plummer et al., 2017).

Outcomes

A key aspect of Step 3: Monitor and evaluate should be to assess the outcomes of the rewilding initiative, or its preliminary stages if it has not been fully implemented. However, any evaluation should also consider the effectiveness of the governance process and necessary adjustments in terms of ACM and SLO principles.

Various efforts have been made to design evaluations of ACM (Fabricius and Currie, 2015). Plummer and Armitage (2007) devised a framework to measure outcomes under three components: ecosystem condition, livelihoods, and process. Armitage et al. (2009) further identified preconditions for the continuation of effective ACM, such as leaders championing the process, training and capacity-building, and a policy environment supportive of collaborative management. These frameworks, and methods of applying them, have since been trialled in South African community-based natural resource management (Cundill and Fabricius, 2010), UNESCO Biosphere Reserves (Plummer et al., 2017), and livelihood climate adaptation in Indonesia (Butler et al., 2016a). Butler et al. (2015a) further modified them to include indicators to track the resolution of conservation conflict, combined with a participatory evaluation methodology.

The primary outcome sought through SLO is community acceptance of a mining or other development, exhibited as levels of trust, transparency, and conflict resolution. While it is recognised that monitoring these outcomes is important (Roche and Bice, 2013), so far no governance-focused frameworks exist. Only bespoke approaches have been developed, such as Anglo American's 'Connected Communities' project, which allows local on-site stakeholders to use mobile phones to provide real-time feedback on community perceptions of the mining operation (Anglo American, 2017). Using 'reflexivity' technology, a questionnaire asks communities to

Table 19.2. *An indicator framework for evaluating (A) rewilding outcomes and (B) preconditions for ongoing ACM, and alignment with ACM and SLO principles, adapted from Butler et al. (2015a). Additional or modified indicators are italicised.*

Indicator	ACM and SLO principles
A. Outcomes	
1. New institutional arrangements	New institutions
2. New institutions codified in law	New institutions
3. *Reintroduction management plan*	New institutions
4. Legitimisation of policies and actions	Learning
5. Changes in perceptions and actions	Learning
6. Engagement and learning across scales	Learning and networks
7. Questioning of routines, values and governance	Learning and conflict resolution
8. Creative ideas for problem-solving	Learning and conflict resolution
9. Agreed upon sanctions	Conflict resolution
10. No party asserting its interests to the detriment of others	Conflict resolution
11. *Reintroduction outcome acceptable to all parties*	Conflict resolution
12. *Acceptable conservation status of reintroduced species*	Ecosystem condition
13. *Acceptable conservation status of all other species*	Ecosystem condition
B. Preconditions	
1. Presence of a bridging organisation or individual	New institutions
2. Commitment to long-term institution-building	New institutions
3. Adaptable portfolio of management resources	Learning
4. Provision of training and capacity-building	Learning
5. Stakeholders drawing on and sharing diverse knowledge	Learning
6. Formal and regular evaluation with stakeholders	Learning
7. High quality of information and resources	Learning
8. Leaders prepared to champion the process	Politics and power
9. Supportive policy environment	Politics and power
10. Transparency of stakeholders' goals and values	Politics and conflict resolution
11. Trust among stakeholders	Politics and conflict resolution
12. Participation of all impacted stakeholders	Politics and conflict resolution

evaluate dust and noise, the effectiveness of social enterprise projects, employment opportunities, and skills training (CSIRO, 2017). Considering the overlaps between ACM and SLO's principles and intended outcomes, we consider that ACM's more mature evaluation frameworks are appropriate for the measurement of both.

Hence, to accommodate the conservation conflict challenges inherent in rewilding, we recommend applying Butler et al.'s (2015a) indicator framework for ACM outcomes and preconditions (Table 19.2). Furthermore, this framework is useful because it is tailored to a participatory methodology that further catalyses stakeholder learning and action (Butler et al., 2015a, 2016a). We have adapted the framework by adding outcome indicators specific to rewilding: 'reintroduction management plan'; 'reintroduction outcome acceptable to all parties'; 'acceptable conservation status of reintroduced species'; and 'acceptable conservation status of all other species' (Table 19.2). Tools such as community surveys could be applied at this stage to assess social acceptance of the reintroduction. However, if at Step 1 consensus to begin the reintroduction was not reached, then some of these indicators would not be used. Finally, to focus on conflict transformation, which identifies and addresses the root sociopolitical sources of conflict, the outcome indicator 'questioning of routines, values and governance' could examine stakeholders' underlying perceptions of the drivers of conflict, and whether the process has succeeded in altering them.

Conclusions

Rewilding is a novel and controversial concept, described by Nogués-Bravo et al. (2016) as a 'Pandora's box', yet guidelines for its governance are relatively immature. Despite the critical importance of stakeholder engagement, adaptive experimentation, and learning, the landmark IUCN/SSC guidelines (2013) only advise proponents to consider community perceptions, costs, and benefits, and do not incorporate governance approaches that are tailored to complexity, uncertainty, and conflict. In this chapter we have attempted to address this gap by proposing a governance approach based on contemporary models of ACM and SLO, which are evolving to address the management of complex social–ecological systems and contested development, respectively. As such, we hope to have shifted the focus of rewilding from a discussion focused on conservation biology and externally driven conservation objectives to one that recasts reintroductions in terms of social–ecological systems and the sustainable development of livelihoods and landscapes.

Although we only reviewed four case studies, there is some evidence that there are differences in the social and ecological contexts for rewilding between more- and less-developed countries or regions. As illustrated by the case of tiger reintroduction in Cambodia, human and management capacity is likely to be lower in less-developed contexts, and rates of change are rapid, creating highly dynamic systems within which rewilding takes place. On-site communities may also be more marginalised, and hence power asymmetries between them and conservation interests are potentially acute. In more developed situations such as for sea eagles in Ireland and bison in Canada, the capacity of all stakeholders may be higher, but tensions between conservation and existing livelihoods remain entrenched and highly political. Despite these probable differences, we found four common challenges across the case studies which provide a foundation for the design of a generic governance approach: multiple private and public stakeholders, the manner of engagement and its varied consequences, existing tensions between livelihoods and conservation, and other broader system drivers.

Our proposed approach builds primarily on ACM, which has evolved as a governance model for complex social–ecological systems, and overlapping principles of SLO, which focuses on creating partnerships and trust between development proponents and local communities. As both approaches are themselves evolving, and have not yet been applied to rewilding, there is no blueprint for their application. Consequently, our proposed components of structure, process, and outcomes are not prescriptive, and deliberately only form a skeleton to be tested. However, by being generic and plastic our approach is likely to be applicable across the development continuum. Nonetheless, the indicator framework proposed in Table 19.2 for monitoring and evaluating rewilding outcomes and preconditions for ongoing ACM contains the key principles of our approach, which should be adhered to by any initiative.

Interestingly, there was some evidence of these outcomes in the case studies. For example, creative solutions had emerged in response to the problems caused by the bison reintroduction in Canada, made possible by the co-management arrangements initiated between the National Park authorities and the First Nations' communities. This platform enabled discussions to move away from 'blaming the bison' to addressing wider sustainable development issues such as the impacts of climate and environmental change on livelihoods and ecology. The tiger case study in Cambodia and its reintroduction feasibility assessment also demonstrated clear elements of ACM, because it involved multistakeholder engagement, the integration of different knowledge types, and learning and action to address wider system issues such as the impacts of rapid urbanisation on potential tiger prey. Furthermore, Sutton (2015) suggests four preconditions to successful sea eagle reintroductions in Ireland: leadership and champions; autonomy and accountability within implementation teams; goal-setting and evaluation; and public relations and stakeholder inclusivity to enable conflict resolution. Similar recommendations have been made for sea eagle reintroduction programmes in Scotland by Young et al. (2016). While not evident in the Killarney National Park case study, these principles chime with the preconditions for our ACM approach (Table 19.2). This evidence suggests that in some cases ACM is evolving organically, perhaps as a result of the reintroductions acting as a shock or crisis, triggering collaborative responses.

Hence, perhaps through implicit trial-and-error, some rewilding initiatives are adopting ACM akin to our proposed approach. However, we would argue that by applying our approach in advance, significant transaction costs could be avoided in controversial reintroductions, where stakeholder conflict may otherwise escalate to irreversible levels. In less-contentious cases, our approach would still promote transparent governance, and enable stakeholders to attain sustainable development outcomes while accounting for future uncertainties and shocks. Therefore, we hope that the ACM principles and processes presented here, which combine those of SLO, can be established in rewilding guidelines as an important prerequisite to any proposed reintroduction. However, we must emphasise that to be effective any participatory governance approach requires adequate and consistent resourcing, and these costs should not be underestimated.

Acknowledgements

We thank one anonymous reviewer and Samantha Stone-Jovicich, whose suggestions greatly improved earlier drafts of this chapter.

References

Abel, N., Wise, R.M., Colloff, M., et al. (2016). Building resilient pathways to transformation when no-one is in charge: insights from Australia's Murray–Darling Basin. *Ecology and Society*, **21**, 23.

Anglo American. (2017). www.angloamerican.com/about-us/our-stories/connected-communities.

Armitage, D., and Johnson, D. (2006). Can resilience be recognised with globalization and increasingly complex resource degradation in Asian coastal regions? *Ecology and Society*, **11**, 2.

Armitage, D.R., Plummer, R., Berkes, F., et al. (2009). Adaptive co-management for social–ecological complexity. *Frontiers in Ecology and the Environment*, **7**, 95–102.

Armitage, D., Béné, C., Charles, A.T., Johnson, D., and Allison, E.H. (2012). The interplay of well-being and resilience in applying a social-ecological perspective. *Ecology and Society*, **17**, 15.

Baird, J., Plummer, R., and Pickering, K. (2014). Priming the governance system for climate change adaptation: the application of a social-ecological inventory to engage actors in Niagara, Canada. *Ecology and Society*, **19**, 3.

Banks, G. (2002). Mining and the environment in Melanesia: contemporary debates reviewed. *The Contemporary Pacific*, **14**, 39–67.

Batson, W.G., Gordon, I.J., Fletcher, D.B., and Manning, A.D. (2015). Translocation tactics: a framework to support the IUCN guidelines for wildlife translocations and improve the quality of applied methods. *Journal of Applied Ecology*, **52**, 1598–1607.

Beierle, T.C., and Konisky, D.M. (2001). What are we gaining from stakeholder involvement? Observations from environmental planning in the Great Lakes. *Environment and Planning C – Government and Policy*, **19**, 515–527.

Berkes, F. (2009). Community conserved areas: policy issues in historic and contemporary context. *Conservation Letters*, **2**, 19–24.

Bohensky, E., Butler, J.R.A., Costanza, R., et al. (2011). Future takers or future makers? A scenario analysis of climate change and the Great Barrier Reef. *Global Environmental Change*, **21**, 876–893.

Bours, D., McGinn, C., and Pringle, P. (2013). *Monitoring and evaluation for climate change adaptation: a synthesis of tools, frameworks and approaches*. Phnom Penh: SEA Change CoP, and Oxford: UKCIP.

Brown, K., and Westaway, E. (2011). Agency, capacity and resilience to environmental change: lessons from human development, well-being, and disasters. *Annual Review of Environment and Resources*, **36**, 321–342.

Brown, K., Tompkins, E., and Adger, W.N. (2001). *Trade-off analysis for participatory coastal zone decision-making. overseas development group*. Norwich: University of East Anglia.

Brown, V.A. (2008). *Leonardo's vision: a guide to collective thinking and action*. Rotterdam: Sense Publishers.

Butler, J.R.A. (2011). The challenge of knowledge integration in the adaptive co-management of conflicting ecosystem services provided by seals and salmon. *Animal Conservation*, **14**, 599–601.

Butler, J.R.A., Middlemas, S.J., McKelvey, S.A., et al. (2008). The Moray Firth Seal Management Plan: an adaptive framework for balancing the conservation of seals, salmon, fisheries and wildlife tourism in the UK. *Aquatic Conservation: Marine and Freshwater Ecosystems*, **18**, 1025–1038.

Butler, J.R.A., Tawake, L., Tawake, A., Skewes, T., and McGrath, V. (2012). Integrating traditional ecological knowledge and fisheries management in the Torres Strait, Australia: the catalytic role of turtles and dugong as cultural keystone species. *Ecology and Society*, **17**, 34.

Butler, J.R.A., Wong, G., Metcalfe, D., et al. (2013). An analysis of trade-offs between multiple ecosystem services and stakeholders linked to land use and water quality management in the Great Barrier Reef, Australia. *Agriculture, Ecosystems and Environment*, **180**, 176–191.

Butler, J.R.A., Suadnya, W., Puspadi, K., et al. (2014). Framing the application of adaptation pathways for rural livelihoods and global change in Eastern Indonesian islands. *Global Environmental Change*, **28**, 368–382.

Butler, J.R.A., Young, J.C., McMyn, I., et al. (2015a). Evaluating adaptive co-management as conservation conflict resolution: learning from seals and salmon. *Journal of Environmental Management*, **160**, 212–225.

Butler, J.R.A., Wise, R.M., Skewes, T.D., et al. (2015b). Integrating top-down and bottom-up adaptation planning to build adaptive capacity: a structured learning approach. *Coastal Management*, **43**, 346–364.

Butler, J.R.A., Suadnya, I.W., Yanuartati, Y., et al. (2016a). Priming adaptation pathways through adaptive co-management: design and evaluation for developing countries. *Climate Risk Management*, **12**, 1–16.

Butler, J.R.A., Bohensky, E.L., Darbas, T., Kirono, D.G.C, Wise, R.M., and Sutaryono, Y. (2016b). Building capacity for adaptation pathways in eastern Indonesian islands: synthesis and lessons learned. *Climate Risk Management*, **12**, A1–A10.

Butler, J.R.A., Bohensky, E.L., Suadnya, W., et al. (2016c). Scenario planning to leap-frog the Sustainable Development Goals: an adaptation pathways approach. *Climate Risk Management*, **12**, 83–99.

Butler, J.R.A., Bohensky, E., Carter, L., et al. (2017a). *A 'total system health' approach to tuberculosis control in the transboundary Torres Strait*. Brisbane: CSIRO Land and Water.

Butler, J.R.A., McKelvey, S.A., McMyn, I.A.G., Leyshon, B., Reid, R.J., and Thompson, P.M. (2017b). Does community surveillance mitigate by-catch risk to coastal cetaceans? Insights from salmon poaching and bottlenose dolphins in Scotland. *Oceanography and Fisheries*, **3**, 555603.

Butler, J.R.A., Darbas, T., Addison, J., et al. (2017c). A hierarchy of needs for achieving impact in international research for development projects. In Schandl, H. and Walker, I. (Eds.), *Social science and sustainability* (pp. 109–129). Melbourne: CSIRO Publishing.

Campese, J., Nakangu, B., Silverman, A., and Springer, J. (2016). The Natural Resource

Governance Framework Assessment Guide: Learning for Improved Natural Resource Governance. NRGF Paper. IUCN and CEESP, Gland, Switzerland.

Carey, J. (2016). Rewilding. *Proceedings of the National Academy of Sciences of the United States of America*, **113**, 806–808.

Chambers, R. (1994.). Participatory Rural Appraisal (PRA): challenges, potentials and paradigm. *World Development*, **22**, 1437–1454.

Clark, D.A., Workman, L., and Jung, T.S. (2016). Impacts of reintroduced bison on First Nations People in Yukon, Canada: finding common ground through participatory research and social learning. *Conservation and Society*, **14**(1), 1–12.

Cooke, B., and Kothari, U. (2001). *Participation: the new tyranny*. London: Zed Press.

Corlett, R.T. (2016). Restoration, reintroduction, and rewilding in a changing world. *Trends in Ecology & Evolution*, **31**(6), 453–462.

CSIRO. (2017). www.csiro.au/en/Research/MRF/Areas/Community-and-environment/Social-licence-to-operate/Anglo-American

Cundill, G., and Fabricius, C. (2010). Monitoring the governance dimension of natural resource co-management. *Ecology and Society*, **15**(1), 15.

Davies, A., and White, R. (2012). Collaboration in natural resource governance: reconciling stakeholder expectations in deer management in Scotland. *Journal of Environmental Management*, **112**, 160–169.

De Groot, R.S. (2006). Function-analysis and valuation as a tool to assess land use conflicts in planning for sustainable, multi-functional landscapes. *Land Use and Urban Planning*, **75**, 175–186.

Eden, S. (1996). Public participation in environmental policy: considering scientific, counter-scientific and non-scientific contributions. *Public Understanding of Science*, **5**, 183–204.

Ensor, J. (2011). *Uncertain futures: adapting development to a changing climate*. Rugby: Practical Action Publishing.

Fabricius, C., and Cundill, G. (2014). Learning in adaptive management: insights from published practice. *Ecology and Society*, **19**(1), 29.

Fabricius, C., and Currie, B. (2015). Adaptive co-management. In Allen, C.R. and Garmestani, A.S. (Eds.), *Adaptive management of social–ecological systems* (pp. 147–179). Dordrecht: Springer Science Business Media.

Fischer, F. (1995). *Evaluating public policy*. Chicago, IL: Nelson-Hall.

Folke, C., Carpenter, S., Elmqvist, T., et al. (2002). Resilience and sustainable development: building adaptive capacity in a world of transformations. The Environmental Advisory Council to the Swedish Government Scientific Background Paper, Stockholm.

Folke, C., Hahn, T., Olsson, P., and Norberg, J. (2005). Adaptive governance of social–ecological systems. *Annual Review of Environment and Resources*, **30**, 441–473.

Ford, R.M., and Williams, K.J.H. (2016). How can social acceptability research in Australian forests inform social licence to operate? *Forestry*, **89**, 512–524.

Gallopin, G.C. (1991). Human dimensions of global change: linking the global and the local processes. *International Social Science Journal*, **130**, 707–718.

Gaventa, J. (2004). Representation, community leadership and participation: neighbourhood renewal and local governance. Prepared for the Neighbourhood Renewal Unit, Office of the Deputy Prime Minister.

Gray, T.N.E., Crouthers, R., Ramesh, K., et al. (2017). A framework for assessing readiness for tiger *Panthera tigris* reintroduction: a case study from eastern Cambodia. *Biodiversity and Conservation*, **26**, 2383–2399.

Gutiérrez, R.J., Wood, K.A., Redpath, S.M., and Young, J.C. (2016). Conservation conflicts: future research challenges. In Mateo, R., Arroyo, B., and Garcia, J.T. (Eds.), *Current trends in wildlife research* (Vol. 1, pp.

267-282). Wildlife Research Monographs. Cham: Springer.

Harrison, C., and Burgess, J. (2000). Valuing nature in context: the contribution of common-good approaches. *Biodiversity and Conservation*, **9**, 1115-1130.

Hegger, D., Lamers, M., Van Zeijl-Rozema, A., and Dieperink, C. (2012). Conceptualising joint knowledge production in regional climate change adaptation projects: success conditions and levers for action. *Environmental Science and Policy*, **18**, 52-65.

Hockings, M., Dudley, N., MacKinnon, K., Whitten, T., and Leverington, F. (2003). *Reporting progress in protected areas: a site-level management effectiveness tracking tool.* Washington, DC: World Bank/WWF Alliance for Forest Conservation and Sustainable Use.

Hurlbert, M., and Gupta, J. (2015). The split ladder of participation: a diagnostic, strategic, and evaluation tool to assess when participation is necessary. *Environmental Science and Policy*, **50**, 100-113.

IUCN/SSC. (2013). *Guidelines for reintroductions and other conservation translocations. Version 1.0.* Gland, Switzerland: IUCN Species Survival Commission.

Jung, T.S. (2017). Extralimital movements of reintroduced bison (*Bison bison*): implications for potential range expansion and human–wildlife conflict. *European Journal of Wildlife Research*, **62**(3), 35.

Keith, D.A., Martin, T.G., McDonald-Madden, E., and Walters, C. (2011). Uncertainty and adaptive management for biodiversity conservation. *Biological Conservation*, **144**, 1175-1178.

Kendal, D., and Ford, R.M. (2018). The role of social license in conservation. *Conservation Biology*, **32**, 493-495.

Lacey, J., and Lamont, J. (2014). Using social contract to inform social licence to operate: an application in the Australian coal seam gas industry. *Journal of Cleaner Production*, **84**, 831-839.

Marzano, M., Dandy, N., Allen, W., et al. (2017). The role of the social sciences and economics in understanding and informing tree biosecurity policy and planning: a global synthesis. *Biological Invasions*, **19**(11), 3317-3332.

McCool, S.F., Guthrie, K., and Smith, J.K. (2000). Building consensus: legitimate hope or seductive paradox? USDA Forest Service Rocky Mountain Research Station Research Paper.

Mitchell, C. (2002). Beyond resolution: what does conflict transformation actually transform? *Peace and Conflict Studies*, **9**(1), Article 1.

Moffat, K., Lacey, J., Zhang, A., and Leipold, S. (2016). The social licence to operate: a critical review. *Forestry*, **89**, 477-488.

Nilsson, D., Baxter, G., Butler, J.R.A., Wich, S.A., and McAlpine, C.A. (2016a). How do community-based conservation programs in developing countries change human behaviour? A realist synthesis. *Biological Conservation*, **200**, 93-103.

Nilsson, D., Gramotnev, G., Baxter, G., Butler, J.R.A., Wich, S.A., and McAlpine, C.A. (2016b). What motivates communities in developing countries to adopt conservation behaviours? A Sumatran orangutan case study. *Conservation Biology*, **30**(4), 816-826.

Nogués-Bravo, D., Simberloff, D., Rahbek, C., and Sanders, N.J. (2016). Rewilding is the new Pandora's box in conservation. *Current Biology*, **26**, R83-R101.

O'Brien, L., Marzano, M., and White, R.W. (2013). Participatory interdisciplinarity: towards the integration of disciplinary diversity with stakeholder engagement for new models of knowledge production. *Science and Public Policy*, **40**(1), 51-61.

O'Rourke, E. (2014). The reintroduction of the white-tailed sea eagle to Ireland: people and wildlife. *Land Use Policy*, **38**, 129-137.

Ojha, H.R., Ford, R., Keenan, R.J., et al. (2016). Delocalizing communities: changing forms of community engagement in natural resources governance. *World Development*, **87**, 274-290.

Olsson, P., Folke, C., and Berkes, F. (2004). Adaptive co-management for building resilience in social–ecological systems. *Environmental Management*, **34**(1), 75–90.

Olsson, P., Gunderson, L.H., Carpenter, S.R., et al. (2006). Shooting the rapids: navigating transitions to adaptive governance of social–ecological systems. *Ecology and Society*, **11**(1), 18.

Oteros-Rozas, E., Martín-López, B., Daw, T., et al. (2015). Participatory scenario-planning in place-based social–ecological research: insights and experiences from 23 case studies. *Ecology and Society*, **20**(4), 32.

Owen, J.R., and Kemp, D. (2013). Social licence and mining: a critical perspective. *Resources Policy*, **38**, 29–35.

Parsons, R., and Moffat, K. (2014). Constructing the meaning of social licence. *Social Epistemology*, **28**, 340–363.

Pasha, M.K.S., Stolton, S., Baltzer, M., and Belecky, M. (2014). Conservation Assured| Tiger Standards (CA|TS): A Multifunctional Protected Area Management Tool to Aid Implementation of International Conventions, Multilateral Treaties, Global Initiatives and National Action. Conservation Assured, Petaling Jaya.

Phillipson, J., and Lowe, P. (2006). The scoping of an interdisciplinary research agenda. *Journal of Agricultural Economics*, **57**, 163–164.

Plummer, R. (2009). The adaptive co-management process: an initial synthesis of representative models and influential variables. *Ecology and Society*, **14**(2), 24.

Plummer, R., and Armitage, D.R. (2007). A resilience-based framework for evaluating adaptive co-management: linking ecology, economics and society in a complex world. *Ecological Economics*, **61**, 62–74.

Plummer, R., Crona, B., Armitage, D.R., Olsson, P., Tengö, M., and Yudina, O. (2012). Adaptive co-management: a systematic review and analysis. *Ecology and Society*, **17**(3), 11.

Plummer, R., Baird, J., Dzyundzyak, A., Armitage, D., Bodin, O., and Schultz, L. (2017). Is adaptive co-management delivering? Examining relationships between collaboration, learning and outcomes in UNESCO Biosphere Reserves. *Ecological Economics*, **140**, 79–88.

Pratt Miles, J.D. (2013). Designing collaborative processes for adaptive management: four structures for multi-stakeholder collaboration. *Ecology and Society*, **18**(4), 5.

Primmer, E., and Kyllonen, S. (2006). Goals for public participation implied by sustainable development, and the preparatory process of the Finnish National Forest Programme. *Forest Policy and Economics*, **8**, 838–853.

Ramsbotham, O., Woodhouse, T., and Miall, H. (2014). *Contemporary conflict resolution*. Cambridge: Polity Press.

Redpath, S., Young, J., Evely, A., et al. (2013). Understanding and managing conflicts in biodiversity conservation. *Trends in Ecology & Evolution*, **28**(2), 100–109.

Redpath, S., Linnell, J., Festa-Bianchet, M., et al. (2017). Don't forget to look down – collaborative approaches to predator conservation. *Biological Reviews*, **92**(4), 2157–2163.

Reed, M.S., Graves, A., Dandy, N., et al. (2009). Who's in and why? A typology of stakeholder analysis methods for natural resource management. *Journal of Environmental Management*, **90**, 1933–1949.

Renn, O. (2006). Participatory processes for designing environmental policies. *Land Use Policy*, **23**, 34–43.

Rist, L., Felton, A., Samuelsson, L., Sandstrom, C., and Rosvall, O. (2013). A new paradigm for adaptive management. *Ecology and Society*, **18**(4), 63.

Robert, A., Colas, B., Guigon, I., et al. (2015). Defining reintroduction success using IUCN criteria for threatened species: a demographic assessment. *Animal Conservation*, **18**, 397–406.

Roche, C., and Bice, S. (2013). Anticipating social and community impacts of deep sea mining. In Baker, E. and Beaudoin, Y. (Eds.), *Deep sea minerals: deep sea minerals and the green economy* (Vol. 2, pp. 59-80). Fiji: Secretariat of the Pacific Community.

Schultz, L., Folke, C., and Olsson, P. (2007). Enhancing ecosystem management through social–ecological inventories. Lessons learned from Kristianstads Vattenrike Biosphere Reserve. *Environmental Conservation*, **34**, 140-152.

Sillitoe, P. (1998). The development of indigenous knowledge. *Current Anthropology*, **39**, 223-252.

Smedstad, J.A., and Gosnell, H. (2013). Do adaptive co-management processes lead to adaptive co-management outcomes? A multi-case study of long-term outcomes associated with the National Riparian Service Team's place-based riparian assistance. *Ecology and Society*, **18**(4), 8.

Soulé, M., and Noss, R. (1998). Rewilding and biodiversity: complementary goals for continental conservation. *Wild Earth*, **8**, 19-28.

Stone-Jovicich, S. (2015). Probing the interfaces between the social sciences and social–ecological resilience: insights from integrative and hybrid perspectives in the social sciences. *Ecology and Society*, **20**(2), 25.

Stringer, L.C., Dougill, A.J., Fraser, E., Hubacek, K., Prell, C., and Reed, M.S. (2006). Unpacking 'participation' in the adaptive management of social–ecological systems: a critical review. *Ecology and Society*, **11**(2), 39.

Sutton, A.E. (2015). Leadership and management influences the outcome of wildlife reintroduction programs: findings from the Sea Eagle Recovery Project. *PeerJ*, doi: 10.7717/peerj.1012

UNCED. (1992). Agenda 21: Programme for Action for Sustainable Development. United Nations Conference on the Environment and Development, Rio de Janeiro, June 1992.

United Nations. (2015). Transforming our World: the 2030 Agenda for Sustainable Development. https://sustainabledevelopment.un.org/post2015/transformingourworld

Vogel, I. (2012). *Review of the use of 'Theory of Change' in international development*. Report for the UK Department for International Development.

Walker, B.H. (1992). Biodiversity and ecological redundancy. *Conservation Biology*, **6**(1), 18-23.

Walker, B.H., Holling, C.S., Carpenter, S.C., and Kinzig, A.P. (2004). Resilience, adaptability and transformability. *Ecology and Society*, **9**, 5.

Walker, B.H., Anderies, J.M., Kinzig, A.P., and Ryan, P. (2006). *Exploring resilience in social–ecological systems: comparative studies and theory development*. Melbourne: CSIRO Publishing.

Wiens, J.A., and Hobbs, R.J. (2015). Integrating conservation and restoration in a changing world. *BioScience*, **65**, 302-312.

Williams, B.K., Szaro, R.C., and Shapiro, C.D. (2009). *Adaptive management: the US Department of the Interior technical guide*. Washington, DC: US Department of the Interior.

Yen, S.-C., Chen, K.H., Wang, Y., and Wang, C.-P. (2015). Residents' attitudes toward reintroduced sika deer in Kenting National Park, Taiwan. *Wildlife Biology*, **21**(4), 220-226.

Young, J., Marzano, M., White, R.M., et al. (2010). The emergence of biodiversity conflicts from biodiversity impacts: characteristics and management strategies. *Biodiversity and Conservation*, **19**(14), 3973-3990.

Young, J., Butler, J.R.A., Jordan, A., and Watt, A.D. (2012). Less government intervention in biodiversity management: risks and opportunities. *Biodiversity and Conservation*, **21**(4), 1095-1100.

Young, J.C., Thompson, D., Moore, P., MacGugan, A., Watt, A.D., and Redpath, S.M. (2016). A conflict management tool for conservation agencies. *Journal of Applied Ecology*, **53**(3), 705-711.

CHAPTER TWENTY

The future of rewilding: fostering nature and people in a changing world

SARAH M. DURANT and NATHALIE PETTORELLI
Institute of Zoology
JOHAN T. DU TOIT
Utah State University

Rewilding started as a philosophical, rather than a scientific, approach to nature management at the turn of the twenty-first century. However, as we have seen in the preceding chapters, the concept of rewilding has grown substantially since its original formation and is emerging as a scientific discipline. As the science underpinning rewilding has come of age, so too have its human dimensions, including economic, social, cultural, and psychological aspects. In this volume we have sought to review the variety of scientific approaches to rewilding, together with its multiple human dimensions, to understand how rewilding works in practice.

As many authors have noted, rewilding can embody a more flexible approach to nature conservation than more traditional approaches because of its focus on the maintenance of dynamic ecological processes, rather than striving for a predefined ecological end-point. Such a flexible strategy is particularly appropriate under a changing climate. However, this focus on process rather than end-point also indicates another implicit, but no less fundamental, shift. Whereas previous approaches to conservation have been dominated by the establishment and maintenance of wilderness, a Western concept that separates nature from people (Chapter 3), a focus on process requires no such duality of people versus nature. Rewilding thus provides an opportunity to develop new approaches to conservation that are more holistic and see nature and humans as intertwined, and not distinct from each other.

In this concluding chapter we draw from the preceding chapters to review rewilding approaches, their capacity to address complex and multidimensional conservation challenges, and how we might expect rewilding science, and its associated philosophy, to evolve in the future.

Approaches to rewilding

Multiple philosophies underpin multiple visions for rewilding, resulting in a variety of different approaches that vary regionally and culturally. Central to any interpretation is the core meaning of the term 'wild' as 'self-willed', (Chapter 2) with an implicit emphasis on process, rather than end-point. In general, rewilding interventions aim to move the biotas of defined spaces, including their human inhabitants, along trajectories of increasing wildness towards becoming self-organising and sustainable social–ecological systems. Such systems are characterised by continual adaptive change that generates ecological resilience (Gunderson and Holling, 2002). It is therefore helpful to regard wildness as a relational concept, as well as one that is able to include people as part of, not separate to, wildness (Chapter 3). The origins of rewilding lie in North America, where it was bound up with the concept of large open wilderness as designated in the USA's Wilderness Act (Chapter 2). The term was first used in the early 1990s in the context of the recovery of native keystone species to regain functional and wild ecosystems. This concept of wilderness has spread south in the Americas, and large reserves in Latin America have also been established to halt habitat fragmentation and restore fragile ecosystems, although the term 'rewilding' is rarely used there.

Rewilding has taken on different interpretations as it has taken hold in places beyond the Americas. Outside of the USA, Europe is now perhaps the region where the term is most widely used, and where aspects of rewilding have become embedded in regional policy and in the activities of civil society (Chapter 2). This has culminated in the establishment of the Wild Europe Initiative and a resolution on 'Wildness in Europe' in the European Parliament in 2009. An active civil society, represented by various conservation and environmental organisations, has helped push Europe along a trajectory towards increased promotion of native species recovery, including rewilding interventions, with the Habitats and Birds Directives providing an enabling policy environment. In Europe, rewilding often sits alongside other land uses, with large carnivores, such as wolves (*Canis lupus*), lynx (*Lynx lynx*), and bears (*Ursus arctos*), moving between agricultural land, protected areas, and other patches of suitable habitat in the landscape. The European version of the rewilding concept is thus more comfortable with humans becoming part of, and contributing to, rewilded environments than the American version (Chapter 12).

Perspectives differ again in Asia and Africa, where widespread poverty increases the challenges faced by rural communities (Roser and Ortiz-Ospina, 2017), reducing people's resilience when living alongside multiple species of problematical large mammals. In both regions the term rewilding is rarely used, despite efforts to recover ecosystem function and restore keystone species, both of which deliver rewilding objectives. In Asia there is an emphasis on connectivity and protected areas, particularly targeted at Asia's remaining large mammals and carnivores (Chapter 2), whereas Africa, which possesses more large mammal species than any other continent (Ripple et al., 2016), still has to confront much of the degradation of its ecosystems arising from the loss of keystone species and their habitats. Here, protection, to minimise further degradation, has higher priority than restoration or rewilding and also has significant economic drivers, given the large tourist incomes that can be generated by wildlife, particularly within the existing network of protected areas. African Parks, Transfrontier Conservation Areas, and Peace Parks movements have spearheaded private/public partnerships for the management of some reserves, to recover their economic and biological potential and protect charismatic keystone species that restore ecosystem processes as well as attract tourist dollars.

Australia, and oceanic island environments, presents a special case for rewilding (Chapter 5). In Australia most of the indigenous megafaunal species went extinct soon after the arrival of humans, and there are no remaining marsupial analogues that could be used for trophic rewilding. On oceanic islands, large herbivores or predators were never present and current faunas include many species, such as ground-nesting and flightless birds, which are vulnerable to invasive species from the mainland (Chapter 5). In such environments, it is difficult to conceive of a low-management rewilded nirvana without complete eradication of, and then constant protection from, multiple invasive species. However, some measures of success in restoring habitat complexity have been achieved on certain oceanic islands by introducing substitutes for extinct giant tortoises (Chapters 5 and 7). On the Australian continent, where complete eradication of invasive species is an impossibility, there is the option of using feral exotic species in a science-based approach to serve the functions once performed by the indigenous megafauna. Australia is also probably the continent with the strongest achievement in explicitly embedding social goals into its rewilding agenda (Chapter 2).

More recently, rewilding has started to address the most human-modified environment of all – the city (Chapter 9). As more and more people become city-dwellers, there is increasing awareness of the need to improve health and well-being in urban environments. Provision of better access to nature and the ecosystem services it provides have been central to delivering such improvements. The growth of urban rewilding represents a significant step in the development of the rewilding concept, taking it away from the wilderness areas of its conception. The new vision of the 'zoopolis' explicitly advocates for an ethical and caring environment for non-human animals and other elements of nature within the city (Chapter 14). Clearly, if cities can be rewilded as well as wilderness, then rewilding is applicable to all social–ecological systems across the full spectrum of human occupancy, from the most wild to the most urban.

Before rewilding there was restoration (Chapter 7), but a fundamental difference between these two approaches is that restoration tends to focus on prior conditions and rewilding on future conditions (Chapter 1). Rewilding outcomes may result in conditions similar to those that prevailed at some stage in the past, but these outcomes occur because of the recovery of ecological process, and not by design. In Chapter 11, Marcus Hall uses the analogy of art curation to compare restoring with rewilding and identifies restorers as traditionally bound by a supposedly 'objective' past, whereas rewilders face no such constraints, and focus on wild conditions, not past conditions. Interestingly, he argues that Pleistocene rewilding is, essentially, a restoration, not a rewilding process, because of the high levels of intervention required, and can never be truly wild. This opens a discussion about what degree of human intervention is acceptable in rewilding, and what it means to be wild. Under the scientific definition from Pettorelli and colleagues (2018), focused as it is on process, Pleistocene rewilding could, indeed, be rewilding, provided the system could become self-sustaining without continual maintenance. In practice, however, there is overlap between rewilding and restoration, and both are parts of a continuum (Chapter 18), whereby all spaces can become wilder, or more like prior ecosystems, or both.

The multidimensionality of rewilding

The origins of rewilding are firmly rooted in ecology, as the understanding of how each species contributes to an ecosystem draws firmly on precepts from fundamental ecology. However, all the chapters, to a greater or lesser degree, address some level of human dimension to rewilding. This is because rewilding, despite its intrinsic ecological basis, is done by, and for, humans. Kim Ward underlines the importance of embracing these multiple human dimensions of rewilding (Chapter 3). She argues that rewilding should break with past 'anti-humanist' conservation approaches that were focused on pristine wilderness that can only exist without people. Her arguments are particularly relevant to the twenty-first century, where growing urban populations are increasingly unconnected to, and disengaged with, nature. While there is a place for wilderness in providing refuge for many persecuted and problematic megafauna, there is a danger in any approach that frames people as separate from nature, as it provides an illusion that nature is detached from us, whereas in reality the fates of people and nature are inescapably intertwined. If rewilding is to be embraced, or at least accepted, by society at large and by local communities living alongside or in rewilded environments, then the multidimensionality of rewilding needs to be addressed. To do this then rewilding, as a science-based practice, needs to become multidisciplinary and draw on and contribute to other fields, particularly those rooted in the humanities and social sciences.

As this volume demonstrates, rewilding affects our societies, our communities, our culture, our economies, and our well-being. Some of these impacts can be positive and some negative (Chapter 9). However, as we start to consider the human dimensions to rewilding, it becomes clear that the diversity of approaches to rewilding are part of its strength, as different approaches may be suited to different situations (Chapter 7). For example, ecological rewilding may be particularly suited to the urban context, bringing a little bit of the wild into the city (Chapter 9) and allowing urban communities to better connect and engage with nature. On the other hand, passive rewilding could be suited to agricultural communities that are hesitant about deliberate restoration, but more comfortable in accepting natural recolonisation (Chapter 13). This capacity for the inclusion of human values in rewilding is another feature which distinguishes it from restoration, which is seen as a primarily ecological or biological activity (Chapter 7).

A key strength of rewilding is its appeal to the public imagination, as demonstrated by George Monbiot's popular book *Feral* (2013). Although this book has sparked heated debate about the management of nature across the UK, there is no question that rewilding resonates with many members of the public. However, the mixed reception of Monbiot's book also demonstrates that rewilding can polarise debate. For example, in the UK there is public disagreement over rewilding arguments to reforest moorland uplands, in line with Britain's history as a forested isle (Chapter 12). These moorlands, now often intensively burned and managed primarily for grouse shooting, have substantially lower biodiversity than the forests they used to support. However, reforestation would completely change the landscapes to which many people have become strongly attached. The fear of such fundamental changes in local environments can, understandably, induce antagonistic feelings within the communities that live and visit these spaces (Chapter 10). Understanding these feelings, and their underlying deep-rooted attitudes, beliefs, and values, are crucial to securing public acceptance of change in any rewilding initiative.

People's values, beliefs, and attitudes provide the lens through which they perceive nature and conservation, and hence underpin their responses to rewilding (Chapter 10). We are only just beginning to develop the tools to unpack these underlying psychological and social responses to nature and their influence on reactions to rewilding (Chapters 8 and 10). Likewise, despite increasing attention over recent years, we still know very little about the benefits, and costs, of nature to human well-being (Chapter 9). Moreover, the research around these issues is largely limited to Western cultures. How communities from different cultures and environments depend on, and benefit from, contact with nature, and how this supports their well-being, is even less well understood (Keniger et al., 2013;

Chapters 8 and 9). Many values, beliefs, and attitudes will be shaped by upbringing and culture, but as Nicole Bauer and Aline von Atzigen (Chapter 8) show, attitudes can change with new knowledge or new environments, although some more easily than others.

There is often a disconnect between public perceptions of rewilding landscapes and wildlife with what is best ecologically. For example, leaving dead wood in woodlands delivers a variety of important ecological services, including provision of microhabitats, nutrient recycling, and carbon storage, but may not always meet with public approval (Chapter 8). Yet, a simple intervention of providing education materials to explain the importance of dead wood in ecosystems helps improve public acceptance, and demonstrates the importance of knowledge in changing public attitudes to rewilding initiatives. Other rewilding initiatives can result in less tractable problems such as in the Oostvaardersplassen in the Netherlands, where concerns about emaciated cattle and horses dying in public view each winter necessitates euthanasia by park rangers as an ongoing intervention (Chapter 18). This illustrates potential challenges in gaining public acceptance in circumstances where rewilding may result in animal suffering. Tensions between public perceptions and conservation science are exacerbated in rewilding projects as their impacts are difficult to predict and may often be poorly understood (Chapter 18). The need for social acceptance for rewilding underlines the importance of public engagement in decision-making and management.

Rewilding involving large carnivores is especially problematic for gaining social acceptance because attitudes towards large carnivores are often deep-rooted, and may be based more on values about what carnivores represent than what they actually do (Chapter 13). Securing acceptance of such potentially problematic species requires engaging at all levels with local communities – psychologically, socially, culturally, and politically – and even then it is difficult to change deep-seated values and beliefs (Chapter 10). Sometimes it may never be possible to secure public acceptance, but where acceptance is possible, practitioners should assume responsibility for ensuring that the most vulnerable members of communities are protected from impacts on their livelihoods. James Butler and colleagues (Chapter 19) provide a promising mechanism for securing social agreement and developing co-management strategies and systems of governance with active and ongoing local community engagement. An agreed and adequately resourced strategy for problem animal control is likely to be a crucial part of such an agreement.

Building the science to move rewilding forwards
As a relatively new approach to environmental stewardship, rewilding has yet to garner an evidence base from which its performance can be evaluated and

its use prescribed. A recent analysis identified five areas that should be targeted for scientific research: (1) target setting and implementation; (2) risk assessment; (3) economic costs and benefits; (4) identification and characterisation of likely social impacts; and (5) monitoring and evaluation (Pettorelli et al., 2018). This volume makes a substantial contribution to advancing that research agenda but a discipline of 'rewilding science' remains to be formalised. The definition of rewilding, based as it is on ecological process, rather than an end-point, poses a particular challenge for scientists, as we lack sufficient tools to measure such processes (Pettorelli et al., 2018). This challenge needs to be addressed if we are able to improve monitoring and evaluation of rewilding interventions. However, Richard Corlett's (Chapter 18) caution that we should not ask 'Can rewilding be made auditable?', but 'Can we audit rewilding?' is apt. To be meaningful, science must be developed around rewilding, and rewilding should not be adjusted to make scientific research more amenable or tractable.

Across the scientific disciplines relevant to rewilding, ecological science is in the strongest position because the overall success or failure of rewilding has historically been defined in relation to ecosystem processes. Nevertheless, as Miguel Delibes-Mateos and colleagues (Chapter 17) point out, there is substantial scientific evidence on the impacts of defaunation on ecosystem functions and services, but very little on the responses of those functions and services after the return of missing species (or substitutes for them). The length of time a species has been absent will affect the likelihood that a rewilding intervention will restore the prior ecological function of that species within the present ecosystem (Chapter 5). Where there has been long-term and substantive change in an ecosystem, which may be compounded by ongoing change to its environment imposed by large-scale anthropogenic processes such as climate change, the resurrection of any former state of that ecosystem is unlikely. Instead, the only way for such spaces to become wilder may be through the development of novel ecosystems, which poses additional challenges for rewilding science (Pettorelli et al., 2018).

In order for rewilding to deliver benefits to people and communities, it should move away from conservation mistakes of the past (Chapter 3), and operate in a manner that is ethically and socially just. All too often, elite capture can mean that most of the benefits of conservation accrue to a wealthy elite, while marginalised vulnerable people may benefit little, and could even be left to pay many of the costs (Adams, 2004). Philanthropic approaches to conservation whereby wealthy individuals buy up land and set it aside for conservation may look good on paper, but can also generate real or perceived problems for local communities. In the UK, philanthropists buying up large tracts of land in the Scottish Highlands for rewilding have led to fears that this may destroy local businesses and undermine local rural communities (Chapter 12). Understanding the legitimate fears of local communities, as well as patterns of distribution of costs and benefits for people living in, or close to, rewilded environments will be key to developing delivery mechanisms to ensure that benefits accrue to those that are most marginalised, while the costs are paid by those in the best position to pay. The need for a socially just approach to rewilding applies equally well to cities as to rural environments. Those areas in cities closest to green spaces often become the most expensive, leading to green gentrification, and exacerbations of inequities in access to nature which compromises contributions to human well-being (Chapter 9). Improving our understanding of these socio-ecological processes in order to help find new mechanisms that support rewilding, yet maintain environmental justice and access to nature for all, should be a priority.

It has to be accepted that, because of the complexity of social–ecological systems and the continually changing environment, each rewilding intervention is an experiment (Chapter 18); this underlines the importance of careful auditing of each rewilding project. Rewilding audits will allow the growth of an evidence base to improve our understanding as to how different interventions impact the functions and processes of ecosystems as well as the wider socioecological system. However, we are never likely to be able to predict the full range of possible outcomes across different sites, even as the science grows with time. The time span needed for rewilding interventions to take effect may stretch across decades, raising further challenges for research (Chapter 18). Thus there will always be high levels of uncertainty and potential surprises (Chapter 3). An accumulating evidence base from well-designed monitoring and evaluation programmes integral to every rewilding intervention is essential to better understand and manage the uncertainty and risks of each successive rewilding intervention.

Rewilding for resilience in a changing world

We are entering an Anthropocene era of rapid environmental change due to the impacts of humans on the planet (Dirzo et al., 2014). The resultant changes in climate will fundamentally and irreversibly alter ecosystems, making it difficult to continue to maintain current patterns of biodiversity and ecosystem functions. Rewilding, with its focus on underlying processes and changing states, provides an approach to maintaining the resilience of social–ecological systems as they adapt to these continually changing environments (Chapter 5). Managing wildlife for resilience requires an adaptive management philosophy that embraces uncertainty and variability, and encourages novelty (Allen et al., 2011). Proponents of rewilding are more comfortable with the concept of novel assemblages of species that may be better suited to a changed climate than traditional conservationists, who are tied to the conservation of present (or restoring past) assemblages (Chapter 7). Although large-scale Pleistocene rewilding is impractical, the idea inherent in this approach that the 'functional properties of large animals are ecologically more important than their taxonomic identities' (Chapter 4) is appropriate for maintaining ecosystem processes in a changing environment. Thus, rewilding may well be the approach to conservation that is needed for the twenty-first century, providing a dynamic and adaptable pathway to increase wildness in the face of rapidly changing climate and a human population that is predicted to grow to 10 billion (United Nations, 2017).

Finding new ways to improve ecological resilience is particularly important in the urban environment. Cities currently support 54 per cent of the global population, and this proportion is predicted to increase to 66 per cent by 2050 (United Nations, 2014). A twenty-first-century approach to conservation needs to be relevant to city ecosystems and their ability to support their increasing human populations, as well as to rural and wild environments. However, by focusing on a continuum of wildness and autonomy (Chapter 3), rewilding can apply equally well to cities as to wildlands. Nevertheless, in extremely modified environments, rewilding will usually need careful design and substantially more ongoing management than in wilderness if it is to improve delivery of ecosystem services. Better-designed cities will also help improve the delivery of sustainable benefits from nature to city-dwellers that help increase human well-being, while also contributing to biodiversity conservation (Chapter 14).

Over the coming decades increasing awareness about issues of equity in access to nature and environmental justice is likely to result in increasing pressure on governments to provide access to green spaces and nature, as their multiple benefits to human well-being become better known (Chapter 9). Thus, while the percentage of the global population living in cities increases, access to green spaces might also increase, not decrease, in urban environments. Active engagement with ecological projects provides additional increases in subjective and social well-being (Chapter 9), raising the possibility that citizen rewilding curators may reap further benefits from the act of curation itself. Cities may therefore provide valuable opportunities to explore the limits of wildness and its benefits and costs to the inhabitants of these extremely modified environments.

Rewilding Europe provides a useful example of what can be possible outside the urban environment and in more traditional rewilding spaces (Chapter 3). The European approach, in its acknowledgement that abiotic, biotic, and social features together create a 'sense of the place', explicitly includes people within rewilding spaces. Oostvaardersplassen, mentioned above, also provides an interesting example of potential opportunities through rewilding. Here, cattle and horses that are not truly wild but are derived from ancient breeds modified by humans are allowed to graze together with deer and geese across land reclaimed by humans from the sea. The programme demonstrates both a novel ecosystem – with its own characteristics of wildness and autonomous nature – as well as the generation of the ecological functions and services that are required to fit the scientific definition. It proves that anywhere can be made wilder. However, it also shows that, in confined spaces that are much smaller than the original ranges of its megafaunal inhabitants, and in the absence of large predators, active management is required to safeguard ecosystem processes and secure public acceptance through the explicit consideration of animal welfare to prevent starvation (Chapter 18).

As well as changing 'what' we do, rewilding also presents an opportunity to change 'how' we do conservation to better address human and environmental needs on our changing planet. Rewilding has the potential to deliver a progressive and socially just approach to conservation, connecting people with, rather than separating them from, nature (Chapter 12). This would allow a break from some of the damaging impacts of the fortress approach to conservation that dominated much of the twentieth century (Adams, 2004). For this to work, new models of community engagement are required, and the social licence to operate (SLO) provides an example of a useful mechanism to achieve buy-in from communities and agreement on systems of equitable co-management (Chapter 19). The SLO, combined with adaptive co-management of rewilding interventions, requires oversight by an independent facilitation team or management committee to act as brokers among multiple stakeholders, and to help ensure independent and transparent systems of governance (Chapters 18 and 19). Independent oversight can also be used to ensure that benefits from a rewilding intervention continue to outweigh costs, and guard against elite capture of benefits. However, there may also be a need to weigh the importance of independence and honest brokership in this central committee versus the advantages of local stakeholder engagement and ownership across multiple sectors. Such community-led approaches, together with the focus on ecosystem process and sustainable management that underpin rewilding (Pettorelli et al., 2018), provide a useful socioecological framework to help advance a people and nature frame that has become dominant in the current decade (Mace, 2014).

Finally, there is also a need for pragmatism and political leadership. Rewilding can be divisive. The dynamism and the possibility of change that are fundamental to rewilding may also generate problems for local communities. Inhabitants of rural communities are often firmly rooted in their sense of place and see their landscapes as timeless and unchanging (Chapter 8). Rewilding could trigger ecological processes that bring about substantial change as ecosystems reach alternative stable states or, if continual climatic change makes a stable state elusive, then constant adaptation and change in response to changing environmental conditions. Human values will inevitably determine the goal of any rewilding intervention, and local communities have to be at the heart of decision-making to help engage their support for rewilding and, where necessary, bring acceptance and adaptation to any resulting environmental change (Chapter 12). Key to changing strongly held beliefs and attitudes will be the revitalisation of local economies, with new, nature-based enterprises that generate income and livelihoods at rewilded sites, with knock-on effects for local employment and businesses.

Conclusions

We have seen from this volume that the original conception of rewilding as limited to large connected tracts of wilderness devoid of humans is no longer adequate for the multiple approaches to rewilding that exist today. Wildness, to a greater or lesser degree, can be part of all human-modified landscapes. Moreover, ecological and social systems cannot exist in isolation from each other when the global human population has reached a critical tipping point and serious attention must now be paid to safeguarding the resilience of integrated social–ecological systems. Can rewilding be used to develop new models of coexistence and engagement with nature? Large carnivores have a long history of often intense conflict with humans, yet the resurgence of large carnivores in Europe from populations that were once threatened with extinction (Chapron et al., 2014) demonstrates that nature is adaptable and resilient. Once people stop killing them, large carnivores can flourish and adapt to anthropogenic environments (Chapter 13). Indeed, in Ethiopia, spotted hyaenas (*Crocuta crocuta*) and golden wolves (*Canis anthus*) are positively associated with anthropogenic impacts (Yirga et al., 2017), while leopards (*Panthera pardus*) thrive even in one of the world's largest cities, Mumbai (Braczkowski et al., 2018). Conversely, the spatial scale of large carnivore ecology forces us to confront the reality that there is inadequate space left for truly wild and self-sustaining populations free from human impact (Chapter 13). For people and nature to thrive, perhaps we need to rewild ourselves and better accept the wildlife in our midst. As John Linnell and Craig Jackson note, large carnivores provide the ultimate litmus test of humanity's commitment to coexist with nature (Chapter 13).

At a time when natural ecosystems are under pressure as never before, people are becoming increasingly disconnected from nature (Miller, 2005). This separation of people from nature risks reducing public engagement and support for conservation, with potentially catastrophic impacts on our planet's support systems (Miller, 2005). Rewilding has already demonstrated its capacity to inspire people about wild nature and hence has the potential to help re-engage people with conservation. Now, as rewilding develops a solid foundation in multidisciplinary science, its capacity to deliver measurable outputs for policy-makers and practitioners is also improving (Chapter 1). Some proponents of rewilding will continue to see it as a path to pristine wilderness where people are separate from nature, but from this volume it is clear that others see rewilding as a more inclusive conservation strategy that integrates social, ethical, and ecological considerations (Chapter 3). When rewilding is seen as a continuum rather than a final state (Chapter 18), allowing all spaces to become a bit more wild, it gives rewilding a wider relevance, increasing opportunities for everyone to engage and connect with nature, even in the most human-modified environments. The wilder spaces generated by rewilding can then deliver multiple benefits to people while allowing diverse biological communities to thrive.

This volume has reviewed the current state of the rewilding concept. In so doing, it hints at the promise that rewilding could hold for a changing planet if we want it. Although rewilding was conceived as a philosophical approach to damaged nature, its growth and public appeal increasingly call for a science-based foundation that is able to inform policy and management to deliver rewilding outcomes. However, the establishment of a scientific discipline around rewilding raises further questions about its disciplinary scope. Indeed, rewilding can be identified in purely ecological terms, but the concept has the capacity for delivering something bigger and more multidimensional than this ecological prescription. Rewilding could become a pragmatic approach to conservation that is inclusive of people, is forward-looking, and is dynamic and adaptable. Ultimately, rewilding could grow to become a dominant approach to support people and wild nature in a rapidly changing environment for the twenty-first century.

References

Adams, W.M. (2004). *Against extinction: the story of conservation*. London: Earthscan.

Allen, C.R., Cumming, G.S., Garmenstani, A.S., Taylor, P.D., and Walker, B.H. (2011). Managing for resilience. *Wildlife Biology*, **17**, 337–349.

Braczkowski, A.R., O'Bryan, C.J., Stringer, M.J., Watson, J.E., Possingham, H.P., and Beyer, H.L. (2018). Leopards provide public health benefits in Mumbai, India. *Frontiers in Ecology and the Environment*, **16**, 176–182.

Chapron, G., Kaczensky, P., Linnell, J.D.C., et al. (2014). Recovery of large carnivores in Europe's modern human-dominated landscapes. *Science*, **346**, 1517–1519.

Dirzo, R., Young, H.S., Galetti, M., Ceballos, G., Isaac, N.J.B., and Collen, B. (2014). Defaunation in the Anthropocene. *Science*, **345**, 401–406.

Gunderson, L.H., and Holling, C.S. (2002). *Panarchy: understanding transformations in human and natural systems*. Washington, DC: Island Press.

Keniger, E.L., Gaston, J.K., Irvine, N.K., and Fuller, A.R. (2013). What are the benefits of interacting with nature? *International Journal of Environmental Research and Public Health*, **10**, 913–935.

Mace, G.M. (2014). Whose conservation? *Science*, **345**, 1558–1559.

Miller, J.R. (2005). Biodiversity conservation and the extinction of experience. *Trends in Ecology & Evolution*, **20**, 430–434.

Monbiot, G. (2013). *Feral*. London: Allen Lane.

Pettorelli, N., Barlow, J., Stephens, P.A., et al. (2018). Making rewilding fit for policy. *Journal of Applied Ecology*, **55**, 1114–1125.

Ripple, W.J., Chapron, G., Lopez-Bao, J.V., et al. (2016). Saving the world's terrestrial megafauna. *BioScience*, **66**, 807–812.

Roser, M., and Ortiz-Ospina, E. (2017). Global extreme poverty. https://ourworldindata.org/extreme-poverty (accessed 2 May 2018).

United Nations. (2014). World urbanization prospects: the 2014 revision, highlights. Department of Economic and Social Affairs, Population Division (ST/ESA/SER.A/352).

United Nations. (2017). World population prospects: the 2017 revision, key findings and advance tables. Working Paper No. ESA/P/WP/248. Department of Economic and Social Affairs, Population Division.

Yirga, G., Leirs, H., De Iongh H.H., et al. (2017). Densities of spotted hyaena (*Crocuta crocuta*) and African golden wolf (*Canis anthus*) increase with increasing anthropogenic influence. *Mammalian Biology*, **85**, 60–69.

Index

Locators in **bold** refers to tables; those in *italic* to figures.

abandonment **7**, 8, 48, 99–100, 108–113
  and biodiversity 106–108
  case studies 113–116
  classification of approaches 110–113
  cost-benefits 106–108
  definitions 100–101
  drivers 101–103
  future scenarios 116–117
  mapping/modelling 103–106, *104*
  restoration ecology 128–129
  risks of 362–366, *363*
  and soil erosion 108
  spatial ecology 110–112
  temporal/time factors *112*, 113
  three C's model *111*, 111
  see also passive rewilding
abiotic factors, ecosystem restoration 248
abundance, species 331
active rewilding
  Britain 227–228
  carnivores **249-252**, 249–252, 254–257
  urban rewilding 286
  see also reintroductions
adaptive co-management (ACM) 395–397, *399*, 405–407
  co-learning activity *401*
  indicator frameworks **404**
  outcomes 403–405
  process 400–403
  structures 398–400
aesthetic images of rewilding 146–147, 281–282
affective components of rewilding 189–190
Africa
  carnivore introductions 253
  conceptualisations of rewilding 27–29, 414–415
African wild dog (*Lycaon pictus*) **249-252**, 255
age factors, attitudes to rewilding 155
age-specific predation 328–330
agri-environment indicators, EU 103
agriculture
  intensification 99

  mastery of nature orientations 158
Alladale Wilderness Reserve 231
America *see* North America
American elk (*Cervus canadensis*), Yellowstone National Park 82–83, 335–344, *336*, *337*
American National Park project 40–42
analogues for extinct fauna *see* proxies
animal welfare
  auditing 382–383
  future of rewilding 418, 422
  trophic rewilding 88
  wilderness conceptualisations 49–50
Anthropocene
  art of rewilding 215
  carnivore reintroductions 248, 269
  future of rewilding 420–423
  Pleistocene rewilding 58, *64*, 64–65, *66*, 67–68
  trophic rewilding 76, 87–92, *91*
  urban rewilding 175–176, 285
  see also climate change; human impacts
anthropocentric values 189; *see also* value orientations
arcadian images of rewilding 148–149, 155
arctic fox (*Vulpes lagopus*) 331
art of rewilding 201–202
  hubris of rewilding 214–215, 218
  restoration vs. art 206–208
  restoration vs. rewilding 208–210
  role of artificial intelligence 215–217
  wilderness/wildness 210–212
  see also curation metaphor; *Earth art* project
artificial intelligence, role in rewilding 215–217
Asia, conceptualisations of rewilding 24–26, 414–415
Asian elephant (*Elephas maximus*) 317–318
aspen (*Populus tremuloides*) 335–344, *336*, *337*
assessment of outcomes *see* outcome evaluation
Atlas of Forest and Landscape Restoration Opportunities (World Resources Institute) 131
attitudes to rewilding 142, 156, 417
  carnivore introductions 263

definitions **143**, 144
and ecological knowledge 156, 157, 187, 188, 194, 418
empirical evidence 143-144
future research needs 156-159
human-nature relationships **143**, 144, 145, 147-149
individual differences 185-187, *193*
and landscape management practices 187-189, **192**
nature/wildness/rewilding conceptions 146
perceptions of nature 146-147
psychology of 183, 184, 185-187, *193*
public participation 159-160
value orientations **143**, 144-145
visions of nature **143**, 144, 145-146
wilderness/wildness 146-147, 148, 149-156
auditing *see* environmental auditing; outcome evaluation
aurochs (*Bos primigenius*), ecological replacements 316-317
Australia 123-124
attitudes to rewilding 153
conceptualisations of rewilding 26-27, 415
urban rewilding 167
Australian brush turkey (*Alectura lathami*) 167
autonomous vehicles, role in urban rewilding 295-298
autonomy discourse 44-45
attitudes to rewilding 146-147
Britain 226-227
wilderness/wildness 45

back-breeding 316
aurochs 316-317
baseline states 4
moving targets 209-210
psychology of rewilding 195
restoration ecology 209
and trophic rewilding 81
bears *see* black bear; brown bear; grizzly bear
beaver (*Castor fiber*) 77, 80-81, 195, 210-212
Britain 228, 231-232
translocation of species 313-314
Bekoff, Marc (*Rewilding our Hearts*) 184
beliefs 189-190; *see also* attitudes to rewilding
biocentric values 189; *see also* value orientations
biodiversity 1
Britain 227-228
definitions 73
globalisation of conservation efforts 65-66
historical development of concept 19
and land abandonment 106-108, 362-363
psychology of rewilding 197
top-down control of ecosystems *345*
trophic rewilding *76*, 77-78, 80-81, 83
urban rewilding 165, 171-172
and wilderness/wildness 15, 16-18
biophilia 144
Biophillic Cities (Timothy Beatley) 287
biophobia 144

biotic factors, ecosystem restoration 248
birds, risks of rewilding 357
bison (*Bison bison*)
governance of rewilding 389-390, **393**, 406
Pleistocene rewilding 62
Romanian Danube delta 50
Yellowstone National Park 82-83
black bear (*Ursus americanus*) **249-252**
Bolson tortoise (*Gopherus flavomarginatus*) 59-60
borderland, wildness as 44-45, 50
bottom-up control/bottom-up processes
Britain 227
ecosystems 56, *57*
see also vegetation succession
Brazil, Cerrado-Pantanal corridors project 21
Brexit, impact on rewilding projects 223, 240
British case study 222-224
definitions of rewilding 225-228
future prospects 240-242
geography of rewilding 229-232
human dimensions 234-238
outcomes 238-240
practical projects 229-232, *230*
spatial scales 232-233
visions/management strategies 233-234
see also United Kingdom
Broken Circle *Earth art* project 212-214, *213*, 218
brown bear (*Ursus arctos*), introductions **249-252**
attitudes to 154
Britain 231
brush turkey, Australian (*Alectura lathami*) 167
Burke, E. (*A Philosophical Enquiry into the Origins of Our Ideas of the Sublime and the Beautiful*) 37

CAFN *see* Champagne and Aishihik First Nations' people
Callaeidae family, New Zealand 314
Cambodia, tiger introductions 391-392, **393**
camera traps 288, *290*
Canada, bison 389-390, **393**, 406
captive animal sources, carnivores 254-255
capture methodology, carnivores 255
carbon storage and sequestration
Britain 239
land abandonment 108
see also climate change
carnivore introductions 9-10, 248, 270, 418
active rewilding **249-252**, 249-252, 254-257
attitudes to rewilding 154
and biodiversity 18
Britain 228
conflicts with human interests 248, 254, 261-265
cost-benefits 267-268
ecological impacts 77
examples/case studies **249-252**

carnivore introductions (cont.)
  future of rewilding 423
  habitat needs 260-261
  historical development of rewilding concept 19
  mortality and illegal killing 265-266
  passive rewilding **249-252**, 249-252, 257-258
  Pleistocene rewilding 61-62
  risks of rewilding 356-359
  South Africa 253
  spatial scales 254, 259, 268-269
  special features of large carnivores 252-254
  translocation of species 312-313
  Wales 237
  see also specific predators by name
Carrifran Wildwood project, Scotland 114-115
Cartesian dualism *see* dualisms
Cerrado-Pantanal corridors project, Brazil 21
challenges of rewilding 4, 9-10, 174-176
Champagne and Aishihik First Nations' (CAFN) people, Canada 389-390, **393**
change *see* climate change; environmental change; urban transformation
charismatic species 46, 65, 68, 217, 267, 415
cheetah (*Acinonyx jubatus*) **249-252**, 255
Chernobyl Nuclear Power Plant, Ukraine 116
Chihuahuan Desert, Bolson tortoise reintroductions 59-60
cities as biodiversity 'hotspots' 165, 171-172; *see also* urban rewilding
City Beautiful movement, USA 282
climate change
  attitudes to rewilding 188
  psychology of rewilding 195-196
  relevance of historical ecosystems 132
  trophic rewilding 87, 90
  urban rewilding 298-299
cloning, de-extinction methods 316
cognitive components of rewilding 189-190; *see also* attitudes to rewilding
collaborative ethos 175-176; *see also* community engagement
colonialism 34, 38
colonisation, abandoned land *see* vegetation succession
community engagement 50
  Britain 236-237
  environmental change 422
  urban rewilding 175-176
Community Nature Conservancies, Africa 28
Community Nature Conservancies, India 25
Community of Arran Seabed Trust (COAST) 229
competition, top-down control of ecosystems 331
complexity *see* ecosystem complexity
conflict resolution, humans 386-388, **393**, 394
  Cambodian tiger introductions 392
  Canadian bison introductions 389-390
  Irish white-tailed sea eagle introductions 389
  Taiwanese sika deer introductions 390-391
  see also adaptive co-management
conflicts, human-wildlife *see* human-wildlife conflicts
connectedness to nature 186
connectivity, protected areas 8, 12, 29
  African projects 28
  Asian projects 24
  Britain 232-233
  historical development of rewilding concept 19
  wilderness/wildness 17-18
  see also corridors
conservation introductions 303, 310
conservation policy
  future of rewilding 413
  social constructions of nature 35-37
  top-down control of ecosystems 345
  trophic rewilding 84, 87-89
  and wilderness/wildness 39-40, 45-51
  see also governance of rewilding
conservation professionals *see* experts
conservation targets, outcome evaluation 381-383
conservation translocations 303
  definition 304
  projects utilising 305-307
  success rates 307-309
  see also translocation of species
constructivism/social constructions of nature 35-37
continuum, rewilding 376, 416, 424
control, concept of 13; *see also* self-willed land
corridors, habitat
  abandonment *111*, 111
  Pleistocene rewilding 55-56
  trophic rewilding 84
  see also connectivity
cortisol levels, and interactions with nature 170
cost-benefits
  abandonment 106-108
  carnivore introductions 267-268
  see also funding rewilding
cougar (*Puma concolor*) **249-252**
coyote (*Canis latrans*) 284, *290*, 292, 293-294, 358
culling animals *see* lethal control
cultural factors, rewilding 39-40
  attitudes to rewilding 148, 157
  psychology of rewilding 185-187, 197
Cuningar Loop, River Clyde, Scotland case study 113-114
curation metaphor, rewilding as 201-202, 211-212, 215, 217-218, 416; *see also* Earth *art* project

danger *see* risks of rewilding
de-extinction 304, 316

dead wood, value of 153, 155–156, 159, 418
decision-making processes
  carnivore introductions 264
  psychology of rewilding 187
  urban rewilding 175–176
  see also conservation policy; governance; structured decision-making
deer 62; *see also* red deer; roe deer; sika deer; white-tailed deer
definitions 1, 4, **7**, 8, 12–13, 376
  abandonment 100–101
  attitudes to rewilding **143**, 144, 150
  British case study 225–228
  historical development of concept 18–20
  naturalness 195
  restoration ecology 123–125
  translocation of species 304
  trophic rewilding 73–75
  urban rewilding 166–167, 168
  wilderness/wildness 13–18, 42–43, 51, 195
  wildlife management 189
deindustrialisation, and urban rewilding 286
demographics, species 331
designed ecosystems, restoration ecology 135–136
development continuum 387, *388*, 392–394, 405–406
digital images, urban rewilding 287–293, *290*, *292*
dingo (*Canis dingo*) 357–358
disease transmission, risks of rewilding 357
dispersal biology, carnivore introductions 257–258
disturbance, impact of large herbivores 77
divine, transcendental nature of wilderness 38
domesticated wilderness 38
drones, information communications technologies 290–293
dualisms 413
  othering of nature 42–43
  social constructions of nature 36
Dyna-CLUE model, land abandonment 103, *104*

Earth art project
  American installation *see* Spiral Jettty, Great Salt Lake
  European installation 212–214, *213*, 218
Earth First! 46
eco-imperialism 65
ecological baselines *see* baseline states
ecological boredom 165
Ecological Corridor of the Americas 21
ecological health of landscapes 184
ecological knowledge, and attitudes to rewilding 156, 157, 187, 188, 194, 418
ecological memory, trophic rewilding 86–87
ecological replacement
  definition 304
  outcomes *318*
  risks 318–319

translocation of species 303, **306**, 314–318
ecological restoration *see* restoration ecology
ecological rewilding 8
  restoration ecology 128–132, *130*
  translocation of species 303
ecological risks *see* risks of rewilding
ecosystem complexity
  conservation targets 382
  environmental auditing 378
  risks of rewilding 366, 367–368
  top-down control 325–326, 332, 345–346
ecosystem disservices, urban rewilding 174
ecosystem engineers 1, 313–314
  definition 304
  translocation of species 303
  trophic rewilding 75–77
  see also beaver
ecosystem impacts
  evidence base 419
  Pleistocene rewilding 56–58, *57*, 60–65
  restoration 248
  translocation of species 311
  trophic rewilding 81–83
ecosystem productivity, top-down control 330
ecosystem services 1
  auditing 378–379
  Britain 235, 239
  land abandonment 116–117
  restoration ecology 128–129, *130*
  top-down control *345*
  trophic rewilding 88
  urban rewilding 167
eco-tourism 49–50
  attitudes to rewilding 185
  Britain 226, 235–236, 239–240
  carnivore introductions 267–268
educational status, and attitudes to rewilding 155; *see also* ecological knowledge
elephant (*Elephas maximus*), as replacement 317–318
elk *see* American elk
empirical evidence *see* evidence base
engagement, public *see* community engagement
entropy 201, 207
environmental auditing 376–377; *see also* outcome evaluation
environmental change
  future of rewilding 420–423
  risks of rewilding 361–362
  see also climate change
environmental factors, human-nature relationships 148
environmental health of landscapes 184
environmental identity (EID), psychology of 186–187
environmental justice *see* equity
environmental volunteering, urban rewilding 173
equity
  Britain 235–236
  carnivore introductions 263

equity (cont.)
  future of rewilding 419-420, 421
  urban rewilding 174-175
ethical values *see* value orientations
Eurasian lynx *see* lynx
Europe
  attitudes to rewilding 152-153, 159, 222-223
  conceptualisations of rewilding 414
  Earth art project 212-214, *213*, 218
  projects utilising conservation translocations **306**
  restoration ecology 22, 127
  rewilding projects 2, 21-24, 47-51
  see also British case study; Rewilding Europe
European Commission/Parliament
  agri-environment indicators 103
  Brexit 223, 240
  Guidelines/Wilderness Register 16
  Resolution on Wilderness in Europe 21
evaluation of outcomes *see* outcome evaluation
evidence base
  attitudes to rewilding 143-144
  auditing 383-384
  deficits 4, 89-92
  future of rewilding 418-420
  top-down control of ecosystems 333-343
  trophic rewilding 81-83, 89-92, 126
evolutionary perspectives
  Pleistocene rewilding 63-65
  risks of rewilding 366-367
experts, conservation
  attitudes to rewilding 187, 191-195, **192**
  value orientations 190-191
  see also ecological knowledge
extant megafauna, conservation of 65-66
extinct megafauna
  reasons for extinction 57-58, 78-79
  see also ecological replacement; proxies for extinct fauna; Pleistocene rewilding

farmland abandonment *see* abandonment
fear of humans, mesocarnivores 358-359
fear of wildness 50, 185-186, 187-188
fenced conservation areas
  auditing 381, 383-384
  carnivore introductions 253
fenced farmland, carnivore introductions 261-262
Feral (Monbiot) 184, 222, 417
Field of Dreams hypothesis 127
fitness, top-down control of ecosystems 332
flood control 158
Florida panther (*Puma concolor*) **249-252**
focal species 18; *see also* charismatic species; keystone species
food security, and land abandonment 116-117
forests
  attitudes to 158, 159
  planting in Britain 234

fortress conservation strategies 226, 422
fox (*Vulpes vulpes*), Norway 331
functional images, attitudes to rewilding 146-147, 148
functional responses, top-down control 326, 327-328
funding rewilding 10
  Britain 236
  outcome evaluation 375-376
future of rewilding 413, 423-424
  differing approaches 414-416
  environmental change 420-423
  evidence base 418-420
  multidimensionality 416-418

Galapagos Islands, giant tortoises 86-87
Garden City movement, UK 282
gardens, urban rewilding 172
Garo Green Spine project, Asia 24
gendered constructions of nature 38, 39
genetically modified organisms (GMOs) 316, 317-319
gentrification, green 175, 284
geography of rewilding, Britain 229-232; *see also* spatial scales
Germany
  attitudes to rewilding 152
  projects utilising conservation translocations **306**
giant hogweed (*Heracleum mantegazzianum*) 110
giant tortoises
  ecological replacements 125-126, 315-316
  Galapagos Islands 86-87
  risks of rewilding 361
global rewilding projects 2
globalisation, and biodiversity conservation 65-66
GMOs *see* genetically modified organisms
Good Agricultural and Environmental Condition (GAEC) land 106
Gorongosa National Park, Mozambique 28
governance of rewilding 386-388, 405-407
  case studies 389-392, **393**
  common themes 392-394
  environmental change 422-423
  generic approaches 398-405, *399*, *401*, **404**
  principles/models 394-398
  psychology of rewilding 183-184, 187-189, **192**
  see also adaptive co-management; conservation policy
Great Salt Lake *see* Spiral Jettty (Great Salt Lake)
Greater Manas project 24
green gentrification 175, 284
greenhouse gases *see* carbon storage and sequestration; climate change
grey wolf (*Canis lupis*) **249-252**
  attitudes to rewilding 153-154
  Britain 228, 231
  psychology of rewilding 197

restoration ecology 123, 125
risks of rewilding 356
translocation of species 312-313
trophic cascades, Yellowstone 335-344, *336*, 337
trophic rewilding 82-83
Yellowstone case study 266
grizzly bear (*Ursus arctos horribilis*) 123, 332
guided entropy 207

habitat corridors *see* corridors
habitat heterogeneity
　Britain 227-228
　land abandonment 365
　trophic rewilding 86
habitat needs, carnivore introductions 260-261
habitat quality, translocation outcomes 310
habitat use, top-down control of ecosystems 332
health benefits, interactions with nature 167-170, 187
herbivore introductions
　Britain 228, 234
　ecological impacts 77
　European rewilding conceptualisations 47, 49-50
　risks of rewilding 359-362, *360*
　see also specific species by name
heterogeneity *see* habitat heterogeneity
hihi (*Notiomystis cincta*), translocation of species 311
Himalayan balsam (*Impatiens glandulifera*) 110
historical development of rewilding concepts 12-13, 29
　African 27-29
　Asian 24-26
　Australian 26-27
　definitions 13-16
　European 21-24
　Latin American 20-21
　Pleistocene rewilding 55-56
　rewilding 18-20
　trophic rewilding 75-80
　urban transformation 281-284
　wilderness and biodiversity 16-18
　wilderness and rewilding 13
historical ecosystems, relevance in face of change 132-133; *see also* baseline states; Pleistocene rewilding
historical fidelity, restoration ecology 208
Homo sapiens see human impacts
hubris of rewilding 214-215, 218
huia (*Heteralocha acutirostris*) 314
human constructions of nature 35-37
human dimensions, rewilding 9, 234-238, 416-418
human-excluding narratives
　Britain 225-226
　ecological rewilding 131
　indigenous peoples 38, 41-42
　wilderness/wildness 40, 45-46, 47
human impacts 29
　Pleistocene extinctions 57-58
　see also Anthropocene; climate change
human intervention, psychology of rewilding 183-184
human microbiomes, urban rewilding 171-172
human-nature relationships
　attitudes to rewilding **143**, 144, 145, 147-149
　psychology of rewilding 195-196
　urban rewilding 166-167
　see also social-ecological systems
human overkill hypothesis 64, 78-79
human population growth, and land abandonment 116-117
human risk *see* risks of rewilding
human well-being *see* well-being
human-wildlife conflicts
　auditing 380, 381
　carnivore introductions 248, 254, 261-265
　trophic rewilding 88-89
　wilderness conceptualisations 50
hunter-gatherer societies 13
hunting, recreational 262, 267-268
　to replace top predators 67
hybrid landscapes 383-384

iconic species *see* charismatic species
identity and landscape
　Britain 225
　psychology of 186-187
iNaturalist social networking site 294-295
inclusive images of rewilding 146-147
India 25
indigenous peoples, impact of rewilding 38, 41-42
indigenous species ranges
　definition 304
　translocation of species 303, 310
individual differences, rewilding attitudes 185-187, *193*
information communications technologies, role in urban rewilding 284-285, 287-293
interspecies cloning 316
international rewilding projects 2
International Union for the Conservation of Nature (IUCN)
　definitions of wilderness/wildness 15-16
　guidelines for rewilding 386-387
　Large Carnivore Initiative for Europe 252
　reintroduction specialist group 252, 305
　rewilding task force 1
　success rates, conservation translocations 308
introductions, ecological design considerations 84-87; *see also* reintroductions
Ireland, sea eagle study 389, **393**, 395, 406
IUCN *see* International Union for the Conservation of Nature
ivory trade 65

jaguar (*Panthera onca*), Mexico 21
Japanese knotweed (*Fallopia japonica*) 110
John Muir Trust, Britain 235–236, 237
justice, environmental *see* equity

Kavango Zambezi Transfrontier Conservation Area (KAZA) 27
Kenting National Park, Taiwan 390–391, **393**
Kerala, India 25
keystone species
  and biodiversity 18
  Britain 223
  definition 304
  European rewilding conceptualisations 48
  historical development of rewilding concept 19
  Pleistocene rewilding 55–56
  restoration ecology 123
  translocation of species 303
  trophic rewilding 75–77
kill rates, predator–prey theory 327
Killarney National Park, Ireland 389, **393**, 395, 406
killing of carnivores, illegal 265–266
killing, mechanisms of top-down control 330–331
knowledge, ecological *see* ecological knowledge
kokako (*Callaeas* sp.) 314
Komodo dragon (*Varanus komodoensis*) 124

land abandonment *see* abandonment
land-take 99
landscape management, Britain 233–234; *see also* adaptive co-management; conservation policy; governance
Large Carnivore Initiative for Europe (IUCN) 252
large carnivores *see* carnivore introductions
large herbivores *see* herbivore introductions
Latin American projects 20–21
Lawton Report, Britain 224
legal definitions, wilderness/wildness 14–15
legislation
  environmental protection 283
  trophic rewilding 84
  see also conservation policy
leopard (*Panthera pardus*) **249–252**
lethal control/culling 265–266
  Oostvaardersplassen 211–212, 418
  outcome evaluation 382–383
lion (*Panthera leo*) **249–252**, 255
  African projects 28
literature search, rewilding 5
live feeds, webcams 290–293
livestock predation, carnivore introductions 261–262
Living Planet Report (WWF) 29
local attitudes to rewilding 157–158
lynx (*Lynx lynx*) 228, 231, **249–252**

Madeira, abandonment case study 115
mammoth (*Mammuthus primigenius*) 317–318
Management Effectiveness Tracing Tool (METT), Cambodian tiger introductions 391
management strategies, Britain 233–234; *see also* adaptive co-management; conservation policy; governance
manicured nature 183–184
Manifesto for Pleistocene rewilding 46
marine rewilding 229
masculinity, constructions of nature 38, 39
mastery of nature orientations 158
Mauritian Islands 125–126
megafauna
  conservation of 65–66
  definitions 73–75
  trophic rewilding 75–80
  see also carnivore introductions; extinct megafauna; herbivore introductions
Mehrangarh Fort, Rajasthan 24
Mesoamerican Biological Corridor 21
mesocarnivores, fear of humans 358–359
Mexican wolf (*Canis lupus baileyi*) **249–252**
Mexico, jaguar-based projects 21
microbiome, human 171–172
moa-nalo (extinct birds), replacement with tortoises 315–316
Monbiot, George (*Feral*) 184, 222, 417
monitoring programmes
  adaptive co-management 402–403
  carnivore introductions 256–257
  trophic rewilding 89
Mont Blanc: Lines Written in the Vale of Chamouni (Shelly) 37–38
mortality rates, carnivore introductions 265–266
mosaic forest hypothesis 47
mountain lion (*Puma concolor*) 123, **249–252**
moving targets 209–210; *see also* baseline states
Muir, John 38, 39–40
multidimensionality, rewilding 416–418

National Elephant Corridor project 24
National Parks, historical perspectives 283; *see also* specific parks by name
National Rewilding Forum, Australia 26–27
native species, proxies for 117
natural recovery of populations *see* passive rewilding
naturalistic landscapes, psychology of 183–184
naturalness, incompatibility with wildness 195
nature, social constructions 35–37
nature-based tourism *see* eco-tourism
nature deficit disorder 165, 234
nature, perceptions
  attitudes to rewilding 146–147
  psychology of rewilding 182–183
neoliberalism, ecosystem services 345

Nepal, Terai Arc Landscape Programme 24, 25
Netherlands
  Earth art project 212-214, *213*, 218
  perceptions of nature 146-147
  projects utilising conservation translocations **306**
  visions of nature concept 145-146
  see also Oostvaardersplassen
network society 284
New Urban Agenda (UN General Assembly) 165
New York High Line railway, Manhattan 135
New Zealand, translocation of species 311, 314
NIMBY (Not In My Backyard) effect 157-158
non-governmental organisations (NGOs) 19, 391, **393**
  Britain 235-236
  Rewilding Australia 26
  Rewilding Britain 222, 226
  Rewilding Europe 48-50, 226, 421-422
  Wild Europe Initiative 16, 21-24, 414
  see also governance
non-human autonomy 44-45
non-intervention management 109-110
North America 45-47, 123
  attitudes to rewilding 151-152, 159, 222-223
  bison 389-390, **393**, 406
  City Beautiful movement 282
  Europeans settlers 37-39
  rewilding conceptualisations 18-21, 414
  wilderness conceptualisations 39-40
  see also Earth art project; Pleistocene rewilding; Yellowstone National Park
Norway, abandonment case study 115
novel ecosystems 4
  restoration ecology 133-136
  trophic rewilding 81, 89, 90
nutrient transport, impact of large fauna 61

oceanic islands, trophic rewilding 85-86
Oostvaardersplassen (OVP), Netherlands 47-48, 49-50
  designed ecosystems 135
  future of rewilding 421-422
  herbivore introductions *360*, 361
  lethal control/culling 418
  Pleistocene rewilding 58-59
  rewilding as curation 217-218
  trophic rewilding 82-83
  wilderness vs. wildness 211-212
othering of nature 38, 42-43
outcome evaluation 375-376, 419, 420
  conservation targets 381-383
  criteria for auditing 377-379
  definitions 376
  environmental auditing 376-377
  need for auditing 380-381
  recommendations/research needs 383-384
  unexpected outcomes 381
overgrazing, by deer 62, 109-110, 115

parks, urban 167, 171-172, 282
Paseo Pantera (Path of the Panther) 20
passive management of artworks 207
passive rewilding 8, 84, 99, 124-125
  approaches to 108-113
  Britain 227
  carnivores **249-252**, 249-252, 257-258
  definitions 100
  risks of rewilding 362-366, *363*
  spatial ecology 110-112
  temporal/time factors *112*, 113
  and urban rewilding 286
  see also abandonment
pastoral idyll, Britain 225
pastoralism, development of 13
payments for ecosystem services (PES) 235
philanthropic approaches 236, 419-420
A Philosophical Enquiry into the Origins of Our Ideas of the Sublime and the Beautiful (Burke) 37
photography 282-283
  camera traps 288-290
  webcams 290-293
pine marten (*Martes martes*) 228, 232
pioneers, European settlers in America 39
place, attachment to 225
planetary-scale computation, role in urban rewilding 285
Pleistocene Park, Siberia 59, 359-360
Pleistocene rewilding 8, 20, 55, 67-68, 124, 416
  case studies 58-60
  conservation of extant megafauna 65-66
  dealing with grassland encroachment 66, 67
  ecological basis 56-58, *57*
  ecosystem impacts 56-58, *57*, 60-65
  functional outcomes and applications 66-67
  historical development of concept 55-56
  relevance of historical ecosystems 132, 210
  species geographic range/size relationship *64*, 64-65
  translocation of species 303
Polesky State Radioecological Reserve, Belarus 116
policy considerations *see* conservation policy
political values, rewilding 39-40
population density, human 223
population growth of fauna, top-down control 329-330
predation 1
predator-prey interactions, impact of rewilding 356, 358, 366-367
predator-prey theory 326-327, 344
predators *see* carnivore introductions; top-down control
preservation policies, wilderness/wildness 39-40
prey-dependent theory 328
Primula scandinavica 363
pristine environments 34, 35, 38, 424
  historical perspectives 283
  wilderness vs. wildness 212

process art 207
productivity of ecosystems, and top-down control 330
professionals *see* experts
protected areas (PAs)
  IUCN definitions 15-16
  size criteria 23, 28
  see also connectivity
proxies for extinct fauna 56, 85
  aurochs 316-317
  de-extinction 316
  giant tortoises 125-126, 315-316
  novel ecosystems 134
  restoration ecology 210
  risks of rewilding 361
  translocation of species 310
  woolly mammoth 317-318
  see also ecological replacement
proxies for for native species 117
psychology of rewilding 182, 196-198
  climate change 195-196
  ecological health of landscape 184
  individual differences in attitudes 185-187, *193*
  landscape management 183-184, 187-189, **192**
  survey of conservation experts 191-195, **192**
  value orientations 189-190
  wilderness/wildness 182-185
public attitudes to rewilding *see* attitudes to rewilding
public engagement *see* community engagement

quaking aspen (*Populus tremuloides*), Yellowstone National Park 335-344, *336, 337*
quantification of outcomes *see* outcome evaluation

ratio-dependent predator-prey theory 328
rationalism, social constructions of nature 36-37
recovery, historical development of concept 18-20
recreational hunting *see* hunting
red deer (*Cervus elaphus*)
  Oostvaardersplassen 82-83
  Scottish Highlands 109-110, 115
red fox (*Vulpes vulpes*), Norway 331
red kite (*Milvus milvus*), Britain 232
red squirrel (*Sciurus vulgaris*), Britain 232
red wolf (*Canis rufus*) **249-252**
refugia 16
reindeer (*Rangifer tarandus*) 362
reinforcement
  definition 304
  translocation of species **306**, *318*, 319
Reintroduction Specialist Group (RSG), IUCN Species Survival Commission 252, 305

reintroductions 7
  definition 304
  psychology of rewilding 187-188
  risks 318-319
  translocations 303, **306**, 312-314, *318*
  see also active rewilding; carnivore introductions; herbivore introductions
relational conceptualisations of wildness 44, 414
release methodology, carnivore introductions 256
remote sensing
  mapping/modelling land abandonment 105-106
  and urban rewilding 287-293, *290, 292*
renaissance 281-284
research studies *see* evidence base
resilience, ecological 420-423
restoration ecology 4, 8, 123-125, 136-137, 416
  designed ecosystems 135-136
  ecological rewilding 128-132, *130*
  ecosystem services 128-129, *130*
  European projects 22, 127
  historical development of rewilding concept 18-20
  management strategies *134*
  novel ecosystems 133-136
  relevance of historical ecosystems 132-133
  restoration of nature vs. art 206-208
  vs. rewilding 208-210
  trophic rewilding 124, 125-128
Rewilding Australia 26
Rewilding Britain 222, 226
Rewilding Europe 48-50, 226, 421-422
Rewilding Institute 46
Rewilding our Hearts (Bekoff) 184
rhino horn trade 65
right to roam legislation, Britain 236
risks of rewilding 355, 367-368
  and attitudes to rewilding 157-158
  carnivore introductions 262
  evolutionary perspectives 366-367
  IUCN/SSC guidelines 386-387
  outcome evaluation 380, 383-384
  passive rewilding 362-366, *363*
  translocation of species 318-319
  trophic rewilding 88-89, 356-362, *360*
roadkill, urban rewilding 296-297
rock lobster (*Jasus lalandii*) 356
roe deer (*Capreolus capreolus*) 329-330
Romania
  attitudes to rewilding 152
  human-nature relationships 147-148
  rewilding project 49-50
Romanticism 37-40, 281-282

saddleback (*Philesurmus* sp.) 314
safety *see* risks of rewilding
satellite data *see* remote sensing
Scar Close nature reserve, Yorkshire Dales 114

science of rewilding 418-420; *see also* evidence base
Scotland/Scottish Highlands 419-420
　beaver reintroduction 313-314
　British case study 225
　ecological rewilding 129-131
　geography of rewilding 229-232
　non-intervention management 109-110, 115
sea eagles (*Haliaeetus* sp.) case study 389, 393, 395, 406
sea lions/seals, webcam data 292-293
seed dispersal, impact of large herbivores 77
self-regulating ecosystems, definition 73
self-willed land 13, 14-15, 16, 414
semi-natural landscapes, European 49
Sheffield urban parks, UK 171
Shelly, Percy Bysshe 37-38
Sierra Club 39-40
sika deer, Kenting National Park, Taiwan 390-391, **393**
Sistine Chapel debate, restoration of artworks 206-208
site-specific attitudes to rewilding 157-158
size criteria *see* spatial scales
size-specific predation 328-330
Smithson, Robert *see* Spiral Jettty *Earth art* project
snapshots of the past 209; *see also* baseline states
social benefits, interactions with nature 167-170
social constructions of nature 35-37
social-ecological systems 7, 414
　concept of control 13
　future of rewilding 423-424
　governance of rewilding 387, *388*
social justice *see* equity
social licence to operate (SLO) 397-398, 405-407, 422
social networking sites, urban rewilding *292*, 293-295
social perspectives/contexts
　governance of rewilding 387, *388*, 392-394
　human-nature relationships 148-149
　IUCN/SSC guidelines for rewilding 386-387
　psychology of rewilding 197-198
　translocation of species 311-312
　trophic rewilding 87-89
socioeconomic status, and attitudes to rewilding 155
soil erosion, and land abandonment 108
South Africa, carnivore introductions 253
spatial scales, protected areas 23, 28
　Britain 232-233
　carnivore introductions 254, 259, 268-269
　passive rewilding 110-112
　Pleistocene rewilding *64*, 64-65
　restoration ecology 136-137
　trophic rewilding 86
species reintroductions *see* reintroductions
species selection, trophic rewilding 92

Species Survival Commission (SSC), IUCN 252, 305
Spiral Hill, Netherlands *Earth art* project 212-214, *213*, 218
Spiral Jettty, Great Salt Lake, *Earth art* project 201-202
　historic levels of Great Salt Lake *205*
　interpretations of the artwork 202-206
　photographs *202*, *205*
　restoration of artworks 206-208
sport *see* hunting, recreational
spotted hyaena (*Crocuta crocuta*) **249-252**
staffing/staff training, carnivore introductions 257
stakeholder involvement 387, 394
statistical indices, abandonment 103-106, *104*
stewardship of nature orientation, forests 158
stress levels, interactions with nature 170
structured decision-making
　definition 304
　translocation of species 309
sublime 37-38
succession of vegetation *see* vegetation succession
survey of attitudes, conservation experts 191-195, **192**
Switzerland
　attitudes to rewilding 153
　human-nature relationships 147
　projects utilising conservation translocations **306**
symbolic meanings
　carnivore introductions 263-264
　rewilding 182
　see also psychology of rewilding

Tauros cattle project 50
taxonomic substitution 8, 386
　ecological replacement of birds with tortoises 315-316
　Pleistocene rewilding 66
　trophic rewilding 85
technology, role in rewilding *see* urban transformation
temporal/time factors, abandonment *112*, 113
Terai Arc Landscape Programme, Nepal 24, 25
Thoreau, Henry David 39-40, 43
three C's argument (cores, carnivores and corridors)
　abandonment *111*, 111
　Pleistocene rewilding 46, 55-56
tiger (*Panthera tigris*) **249-252**
　Cambodia 391-392, **393**
top-down control of ecosystems 325-327, 344-346
　age/size-specific predation 328
　Britain 227
　carnivore introductions 248
　ecosystem productivity 330
　evidence base 333-343
　functional responses/trophic interactions 327-328

top-down control of ecosystems (cont.)
  mechanisms 330-331
  outcomes 331-332
  Pleistocene rewilding 56, 57, 61-62
  predator-prey theory 326-327, 344
  rationale for conservation policy 345
  trophic rewilding 73, 75-80
  Yellowstone National Park 335-344, *336*, *337*
top predators *see* carnivore introductions
tortoises, Chihuahuan Desert 59-60; *see also* giant tortoises
tourism *see* eco-tourism
transcendental nature of wilderness 38
transformation, urban *see* urban transformation
translocation of species 303, 320
  conservation translocation projects 305-307, **306**
  conservation translocation success rates 307-309
  definitions 304
  ecological replacements 303, **306**, 314-318
  outcomes 309-312, *318*
  reinforcement **306**, *318*
  reintroductions 303, **306**, 312-314
  risks 318-319
  taxonomy of species translocated *305*
transport methodology, carnivore introductions 255-256
Tree for Life organisation 22
tree planting, Britain 234
trophic cascades 73, 335-344, *336*, *337*, 345-346
trophic interactions, top-down control 327-328
trophic rewilding 8, 66, 92-93
  and biodiversity 76, 77-78, 80-81, 83
  case studies 74-75
  definitions 73-75
  ecological basis 75-80
  ecological design considerations 84-87
  ecosystem impacts 81-83
  empirical evidence 81-83, 126
  functional outcomes 83
  goals 80-81
  implementation considerations 84
  monitoring programmes 89
  research needs 89-92, *91*
  restoration ecology 124, 125-128
  risks and human-wildlife conflicts 88-89
  risks of rewilding 356-362, *360*
  social perspectives 87-89
  translocation of species 303
  see also carnivore introductions; herbivore introductions

Uncommon Ground (conference and book) 286
unintentional rewilding 82
United Kingdom 417
  carnivore introductions 237
  Garden City movement 282
  projects utilising conservation translocations **306**
  see also British case study; Scotland/Scottish Highlands
United States, Wilderness Act (1964) 14-15; *see also* North America; Yellowstone National Park
urban rewilding 165-166, 176-177, 280-281, 415
  attitudes to 136, 148-149, 155
  categorisation of benefits model 168-174, *169*
  challenges of 174-176
  definitions 166-167, *168*
  environmental change 421
  health and social benefits 167-170
  human-nature relationships 148-149, 166-167
  incidental interactions 170, 171-172
  intentional interactions 172, 173-174
  less wild nature 170, 172
  wilder nature 171-172, 173-174
urban transformation 280-281, 298-299
  autonomous vehicles 295-298
  digital images/remote sensing 287-293, *290*, *292*
  historical perspectives 281-284
  information communication 284-285, 287-293
  social networking sites *292*, 293-295
  wilderness/wildness 281-284, 286-287
utilitarian approaches to nature 148-149

value orientations **143**, 144-145, 417
  carnivore introductions 263-264
  outcome evaluation 379, 382-383
  psychology of rewilding 189-190
  restoration ecology 136
  rewilding 39-40
  see also attitudes to rewilding
vegetation succession 1
  abandoned land 100, 107, 128-129
  restoration ecology 124-125, 128-129
  see also passive rewilding
visions of nature **143**, 144, 145-146
volunteering, environmental 173

Wales, carnivore introductions 237
webcams 290-293
well-being, and interactions with nature 167-170, 187
welfare *see* animal welfare
white-tailed deer (*Odocoileus virginianus*) 167
white-tailed eagle (*Haliaeetus albicilla*)
  Britain 228, 232
  governance of rewilding 389, **393**, 395, 406
wild boar (*Sus scrofa*), Britain 228, 231
Wild Country project, Australia 26
Wild Europe Initiative (WEI) 16, 21-24, 414
wild population sources, carnivore introductions 254-255
Wilderness Act (US, 1964) 14-15, 414

Wilderness Movement, America 39–40
Wilderness Society, Australia 26
wilderness/wildness 7, 34–35, 50, 413
  American National Park project 40–42
  attitudes to rewilding 146–147, 148, 149–156
  big wilderness 55
  and biodiversity 15, 16–18
  and conservation/preservation policies 39–40, 45–51
  definitions 13, 15–16, 195, 225–228
  human excluding narratives 40
  imaginaries of 35
  and naturalness 195
  psychology of rewilding 182–185
  vs. restoration 210–212
  Romanticism 37–40
  social constructions of nature 35–37
  and urban rewilding 281–284, 286–287
  wildness as autonomy 45
  wildness as borderland 44–45, 50
  wildness as relational concept 44
  wilderness vs. wildness 42–43, 51, 210–212
wildfires, megafaunal extinction 79
Wildlands Network 12, 17, 21, 46, 123
wildlife comebacks/immigration 82
wildlife gardening, urban rewilding 173–174
wildlife–human conflicts *see* human–wildlife conflicts
wildlife management, definition 189; *see also* adaptive co-management; conservation policy; governance
wildlife value orientations (WVOs) 154
willows (*Salix* sp.), Yellowstone National Park 63
wolf *see* grey wolf; red wolf
wolverine (*Gulo gulo*) **249–252**
woody encroachment of grasslands 66, 67
woolly mammoth (*Mammuthus primigenius*) 317–318
Worldwide Fund for Nature (*Living Planet Report*) 29

Yakutia Republic, Siberia, Pleistocene Park 59
Yellowstone National Park, USA 40–42
  grey wolf case study 266
  trophic cascades 335–344, *336*, *337*
  willows 63

zoöpolis 287